SYSTOLIC ALGORITHMS

TOPICS IN COMPUTER MATHEMATICS

A series edited by David J. Evans, Loughborough University of Technology, UK

Volume 1 PRECONDITIONING METHODS: THEORY AND APPLICATIONS
Edited by David J. Evans

Volume 2 DISCRETE MATHEMATICS FOR ENGINEERS
By O.P. Kuznetsov and G.M. Adel'son Vel'skii

Volume 3 SYSTOLIC ALGORITHMS
Edited by David J. Evans

This book is part of a series. The publisher will accept continuation orders which may be cancelled at any time and which provide for automatic billing and shipping of each title in the series upon publication. Please write for details.

SYSTOLIC ALGORITHMS

Edited by
DAVID J. EVANS
Loughborough University of Technology, UK

GORDON AND BREACH SCIENCE PUBLISHERS
Philadelphia Reading Paris Montreux Tokyo Melbourne

Gordon and Breach Science Publishers

5301 Tacony Street, Drawer 330
Philadelphia, Pennsylavania 19137
United States of America

Post Office Box 161
1820 Montreux 2
Switzerland

Post Office Box 90
Reading, Berkshire RG1 8JL
United Kingdom

3-14-9, Okubo
Shinjuku-ku, Tokyo 169
Japan

58, rue Lhomond
75005 Paris
France

Private Bag 8
Camberwell, Victoria 3124
Australia

The articles published in this book first appeared in the *International Journal of Computer Mathematics*, Volume 21, Numbers 3 to 4; Volume 22, Number 1; Volume 25, Numbers 3 to 4; Volume 30, Numbers 1 to 2; Volume 33, Numbers 1 to 2; Volume 37, Numbers 1 to 2; *Parallel Computing*, Volume 3, Number 4; Volume 4, Number 1; Volume 10; Volume 14, Number 1; Volume 15; Volume 16; and in *Integration*, Volume 8.

We thank Elsevier Science Publishers B.V., PO Box 103, 1000 AC Amsterdam, The Netherlands, for granting us permission to reproduce articles from the journals *Parallel Computing* and *Integration*.

Library of Congress Cataloging-in-Publication Data

Systolic algorithms / edited by David J. Evans.
p. cm. -- (Topics in computer mathematics ; v. 3)
Includes index.
ISBN 2-88124-804-7
1. Computer algorithms. I. Evans, David J. II. Series.
QA76.9.A43S9 1991
005.1--dc20 91-23899
CIP

CONTENTS

PREFACE

The Flynn classification of parallel architectures (SIMD, MIMD, etc.) is well known but not exhaustive. More recently advances in semi-conductor technology, systolic architectures have emerged as efficient computational structures encompassing both multiprocessing and pipelining concepts. Systolic arrays are special purpose synchronous architectures consisting of simple processors or cells locally and regularly interconnected. Data streams flow through the cells in such a way that they interact at each connection. An analogy with biology exists where the word 'systole' describes the heart contraction rhythm to pump blood around the body. Similarly systolic arrays pump data to give a regular data flow in the network.

Algorithms suitable for implementation in systolic arrays can be found in many applications such as digital signal and image processing, linear algebra, pattern recognition, linear and dynamic programming and graph problems, etc. Such algorithms are computationally intensive and preferably require systolic architectures for their implementation in real time environments.

The common factors in these algorithms are the requirement of large throughput while processing wide data bandwidths on systolic arrays requiring only a few types of simple processing elements. The processing is characterized by repeated computations involving only a few types of relatively simple operations that are common to many input data items.

Thus, a systolic algorithm schedules computations in such a way that a data item is not only used when it is input but is also re-used as it moves through the pipelines in the systolic array. This results in balancing the processing and input/output bandwidths, especially in compute-bound problems in a way not portrayed in other forms of parallel processing.

This volume in the Topics in Computer Mathematics series attempts to provide insights into the systolic implementation process and to illustrate the different techniques and theories that contribute to the design of systolic algorithms.

Finally I should like to express my gratitude to the contributors for their prompt and cooperative support and Mrs. Judith Poulton for her excellent typing and skillful organisation in compiling this volume.

David J. Evans
Editor

INTRODUCTION

Systolic Algorithms

D. J. EVANS

Department of Computer Studies, Loughborough University of Technology, Loughborough, Leicestershire, UK

In this survey paper, a brief review of the background and environment of the development of the Systolic approach in parallel processing is given. Then, the major areas of current systolic systems research are outlined together with their cross fertilisation with other related areas of research.

KEY WORDS: Systolic approach, parallel processing, soft systolic algorithms, abstract math. model, cellular automata, systolic networks.

C.R. CATEGORIES: F1.1, B7.1, F1.2.

1. INTRODUCTION

The systolic approach in parallel processing came as a product of a certain environment, that encompassed the needs, (the possible applications); the means, (the appropriate technology); and the background knowledge for its realisation. The needs can be outlined as the ever-increasing tendency for faster and more reliable computations, especially in areas like real-time signal processing and large-scale scientific computation. The means were provided by the remarkable advances in VLSI technology and automated design tools. Finally, the background includes the applications of parallel processing in the form of parallel algorithms and the design of parallel computers; as well as the theory of cellular automata. These aspects are now briefly discussed, especially in their special relation to systolic architectures.

2. APPLICATIONS OF SYSTOLIC APPROACH

Systolic systems were introduced by H. T. Kung, C. E. Leiserson [1–3] as high-performance, special-purpose VLSI computer systems that are suitable for specific application requirements or to off-load computations that are especially taxing to general-purpose computers.

The rationale behind the use of special-purpose systems, as opposed to general-purpose, is very carefully explained in [4, 5]. In areas such as real-time signal processing and large-scale scientific computation the trade-off balance between generality and performance comes down on the side of special-purpose devices, because of the stringent time requirements. Thus, a systolic engine can function as a peripheral device attached to a host system, as shown in Figure 1.

The host system may not be a computer: in the case of real-time signal processing, systolic systems are suitable for sensor devices accepting a sampled signal and then passing it on, after some processing, to other systems for further processing in radar, sonar applications, [6, 7]. In the case of large-scale scientific computation, systolic systems can be used as a "hardware library", for certain numerical algorithms, equivalent to the software libraries currently available (see Figure 2) [7, 8]. Alternatively, they can be utilized to "matricialize" the internal arithmetic units of more general-purpose supercomputers.

However, apart from these traditional application areas, an increasing number of computations seem to benefit from the systolic approach. The common characteristics of all these processes is that they are **compute-bound problems**, i.e., with large amount of com-

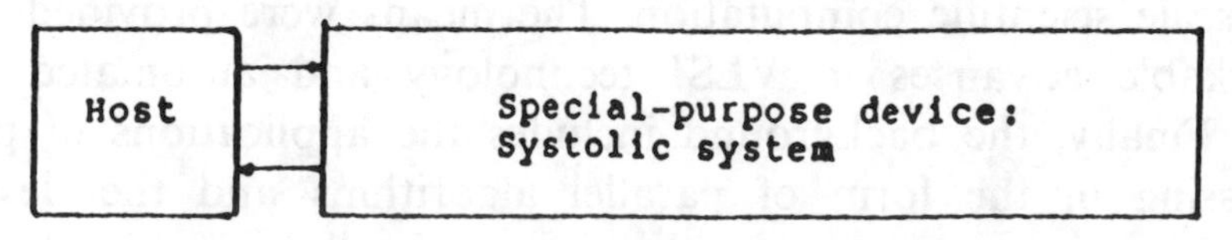

Figure 1 Systolic system as special-purpose device.

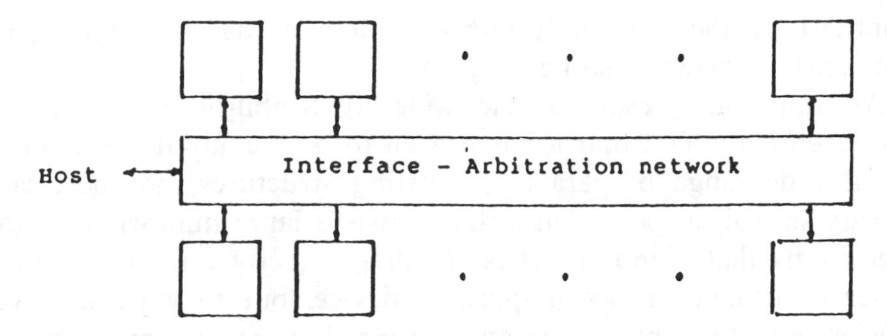

Figure 2 Hardware library design.

putation versus input/output (i/o) communication. Usually in compute-bound problems, multiple operations are performed on each data-item in a repetitive manner. In contrast, problems with a large amount of (i/o) communication versus computation are called **i/o-bound**. A survey of the applications of systolic systems can be found in [1, 4, 9]. Table 1 gives a representative selection of systolic

Table 1 Selection of systolic applications

Signal processing	Signal processor for recursive filtering, Implementation of Kalman filters, Discrete Fourier Transform (DFT), Convolution (multi-dimensional), Linear algebra machines in digital processing.
Numerical problems	Finite element analysis, Singular value decomposition, Linear time solution of Toeplitz systems, Orthogonal equivalence transformations, Least-squares (adaptive beam forming), Eigenvalues and generalized inverses, (symmetric matrices), Iterative algorithms.
Shapes and patterns	Pattern matching, Feature extraction and pattern classification, Stereo matching, Algorithms for recti-linear polygons.
Words and relations	Largest common subsequence problem, Dictionary machines, Relational Database operations, Connected word recognition.
Automata	Tree acceptors, Trellis automata, Binary tree automata, Design rule checker.
General	Shortest path problem, Algebraic path problem (including matrix inverse), Fundamental sorting problems, Linear-time Greatest Common Divisor (GCD) computation, Priority queues.

applications; the table is further extended during the discussion of the areas of current systolic research.

An important result of the wide applicability of the systolic approach is the fact that it has proven to be a computational model for a wide range of parallel processing structures, not necessarily strictly special-purpose. Thus, there exists a large number of systolic algorithms that is not practical to map directly onto hardware in order to produce a special-purpose device, but they perform very efficiently when implemented on appropriate parallel computers.

3. SYSTOLIC ALGORITHMS AND PARALLEL PROCESSING

Now we briefly turn our attention to the contribution of parallel processing in the development of the systolic concept. Initially we examine the application of traditional parallel computing techniques on systolic systems and then the relation of systolic systems with other models of parallel computers is briefly outlined. For a more general introduction to parallel processing see [10–12].

Systolic systems combine pipelining, array-processing and multi-processing to produce a high-performance parallel computer system. This combination is exemplified with the help of Figure 3, which is a typical arrangement of a systolic system. A linear array (pipeline) of *n* processors (cells, in the systolic terminology) is connected with the host system, via the boundary cells. The number of cells in the array is determined by the maximum attainable i/o bandwidth of the host. All processors perform their computation simultaneously and each cell exchanges information (data, control) with its neighbouring cells, for further processing.*

In the simplest case, all processors perform the same computation, on a different set of data (array processing), and then they pass data to the right-hand-side cell, while they accept data from the left-hand-

*The name "systolic" is taken from the Greek word systole (*σνστολη*). The physiology terms systole (*σνστὸλη*) and diastole (*διαστολη*) indicate the successive contraction and expansion of the heart, by means of which blood is pumped to the different organs of the human body. The function of the memory in Figure 3 is analogous to that of the heart: it pulses information (instead of blood) through the pipeline.

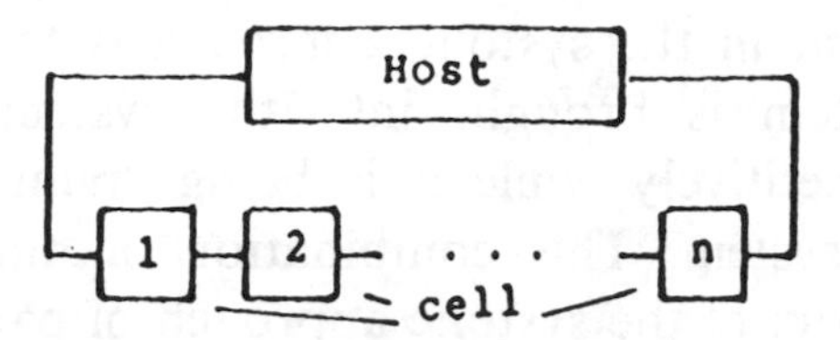

Figure 3 Linear systolic array.

side cell (pipelining). The left boundary cell accepts input from the host and the right boundary cell sends output to the host. In more complicated systolic systems, the data flow can be multidirectional and at different speeds.

The array can expand in two or more dimensions, or take the form of a systolic tree (see Figure 4). Furthermore, the processors need to perform identical computations (multiprocessing). It is common to classify a systolic system according to its communication geometry [13]; thus, "linear systolic array" can be used instead of "systolic system with linear array communication geometry". Further the terms "systolic array" and "systolic system" may interchange since the array interconnection is the most common.

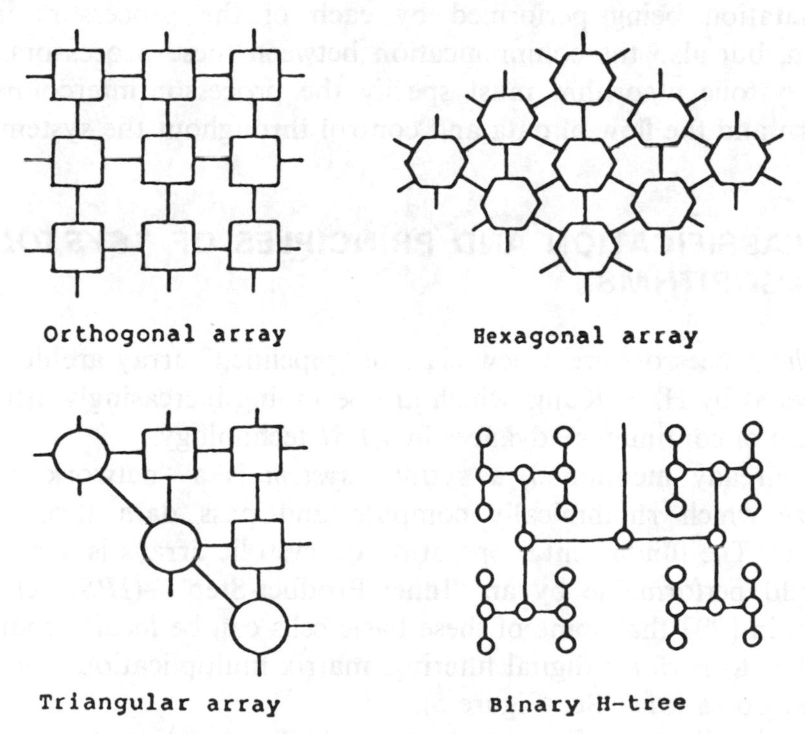

Figure 4 Systolic system communication geometries.

The central point in the systolic approach is to ensure that once an information item is brought into the system it can be used effectively and repetitively while it is being "pumped" from cell to cell through the system. This combination of multiprocessing and pipelining is the crux of the systolic approach of parallel processing.

The relation of systolic systems with the more traditional models of parallel processing is addressed in [13]. Some important differences, include the number and the complexity of the processing elements involved, as well as the generality of the architecture. Whereas most parallel computer concepts which have been pursued so far involve a relatively small number of high-level processors, the systolic systems suggest the design of parallel processing systems with very large numbers of relatively simple processing elements. Furthermore the systolic systems are algorithmically-specialised, and therefore can achieve a better balance between computation and communication, since the communication geometry and the computation performed by each processor are unique for the specific problem to be solved.

Thus, a systolic algorithm must explicitly define not only the computation being performed by each of the processors in the system, but also the communication between these processors. That is, a systolic algorithm must specify the processor interconnection pattern and the flow of data and control throughout the system.

4. CLASSIFICATION AND PRINCIPLES OF *"SYSTOLIC"* ALGORITHMS

Systolic processors are a new class of "pipelined" array architectures, pioneered by H. T. Kung, which are becoming increasingly attractive because of continuous advances in *VLSI* technology.

As already mentioned, a *systolic system* is a "network of processors which rhythmically compute and pass data through the system". The fundamental operation on systolic arrays is a multiply-and-add performable by an "Inner-Product-Step"—(*IPS*) cell. It is shown in [22] that some of these basic cells can be *locally* connected together to perform digital filtering, matrix multiplication, and other related operations. (See Figure 5).

The systolic array features the important properties of modularity,

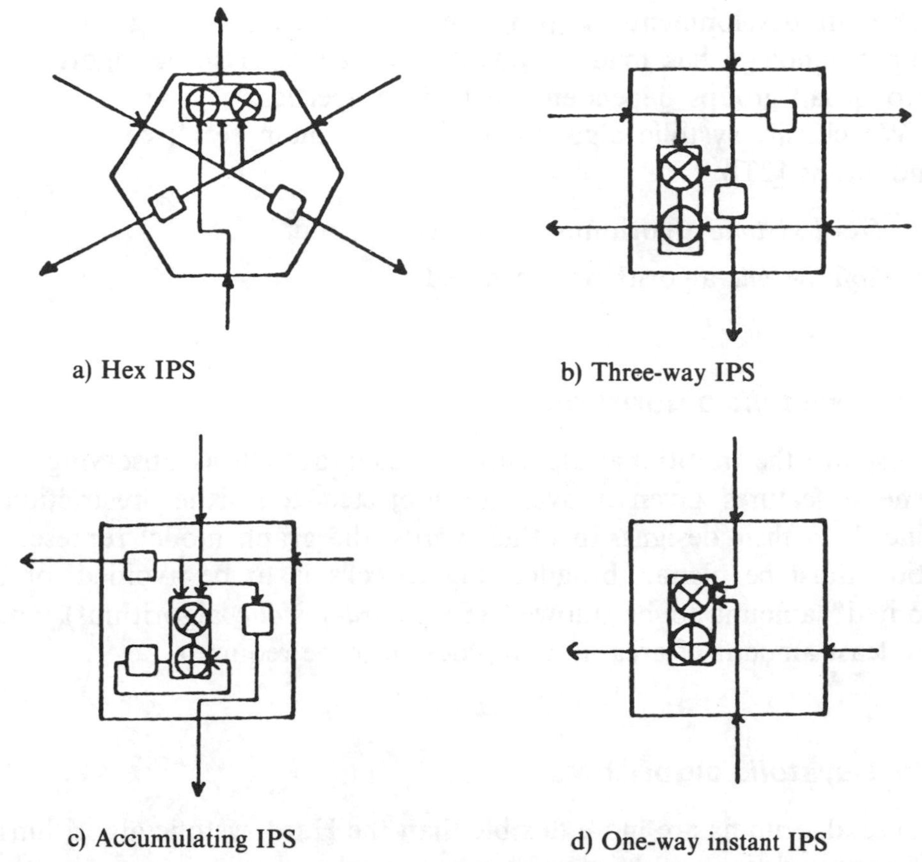

Figure 5 IPS geometries.

regularity, local interconnection, a high degree of pipelining, and highly synchronized multiprocessing. The data movements in a systolic array are often described in terms of the "snapshots" of the activities as the *unit* of time is considered the time necessary to achieve a multiply-and-add operation.

One of the major challenging research items, therefore, has become the development of algorithms that can be mapped into and executed efficiently by a special-purpose computer system. Algorithms that match with systolic systems, utilizing extensive pipelining and multiprocessing, are called *systolic algorithms*. As we have previously mentioned, systolic algorithms, in a general comparison with *SIMD* and *MIMD* algorithms, are the most structured and *MIMD* algorithms are the least structured. More specifically, systolic algorithms deal with simple and frequently interacting task modules, while the situation is reversed for *MIMD* algorithms.

Recent developments in programming languages along with the chip technology has made it possible to classify systolic algorithms into broad groups dependent on their properties.

We classify systolic algorithms into two main sets (see Bekakos and Evans [23]):

i) *Hard-systolic* algorithms:—denoted S_H, and
ii) *Soft-systolic* algorithms:—denoted S_S.

Hard-systolic algorithms

These are the traditional algorithms which, as well as observing the general features given above, are subjected to further restrictions placed on their designs. In other words, the graph model representation must be planar, broadcasting to cells is to be avoided, or a limited* amount to be allowed (*Semi-hard-systolic* algorithms), and the least amount of area in a chip design to be required.

Soft-systolic algorithms

These algorithms are more flexible than the Hard-systolic algorithms, since non-planar graphs may be represented and area is not, directly, a major consideration (this translates to storage used in a program). In addition, they do not have to be fabricable (but must be simulatable in some appropriate programming language, e.g., *OCCAM*, *CONCURRENT PROLOG*), and broadcasting is not to be avoided. Intuitively such an algorithm may not be suitable for chip implementation, but it can be performed on a suitable parallel computing structure.

It is evident that all Hard-systolic algorithms are special cases of Soft-systolic ones and so can also be simulated in the same programming languages. Further it is also apparent that some Soft-systolic algorithms will be very close to being Hard-systolic, but under the strict definitions would not be purely classed as such, but as *Hybrid-systolic* algorithms:—denoted S_{Hy}.

---

*If over long distances clock skew occurs and data can become unsynchronized.

Hybrid-systolic algorithms

They will represent a grey area of algorithms being in a state of "migration" between Soft and Hard and altering technological conditions over time. In specific, algorithms which allow local broadcasting (not necessarily between nearest-neighbour cells), limited non-planarity or large amounts of non-planarity (but in a control sense, with regular connection structures), could be considered as candidates for this category of algorithms.

All the above definitions will become increasingly important as *FGCS** evolve. The relations between these classes of algorithms, in a set theory manner, are:

i) $S_H \cup S_S = S$: $S \equiv \{\text{All systolic algorithms}\}$

ii) $S_H \subseteq S_{Hy} \subseteq S_S$;

the important question arising is whether $S_H = S_S$, because if this is the case then, all Soft-systolic algorithms can be in essence fabricable.

In the following paragraph a mathematical model for the verification of systolic networks will be introduced (see Melhem and Rheinboldt [24], along with the data sequences to represent the data appearing on the communication links at successive time intervals, and the causal operators which model the computations performed by a cell of the network. The latter concept was primarily inspired by corresponding approaches in systems theory (see Faurre and Depeyrot [25]).

5. AN ABSTRACT MATHEMATICAL MODEL FOR THE VERIFICATION OF *"SYSTOLIC"* NETWORKS

In all theoretical models of *VLSI* circuits two parameters are of vital importance, *size* and *speed.* Since *VLSI* is essentially two-dimensional, the *size* of a circuit is best expressed in terms of its area. Sufficient area must be provided in a circuit layout for each gate and each wire. Gates are not allowed to overlap each other at all, and only two (or perhaps three) wires can pass over the same point.

The *speed* of a synchronous *VLSI* circuit can be measured by the

**FGCS*—Fifth Generation Computer Systems.

number of clock pulses it takes to complete its computation. The actual size of this time unit, however, is a technological* variable.

The speed of the *VLSI* circuit may be adversely affected by the presence of very long wires, unless special measures are taken. In many *VLSI* processes, a minimum-sized transistor cannot send a signal from one end of the chip to the other in one clock period. Today, to accomplish such unit-delay cross-chip communication, and to achieve large fan-outs, special "driver" (*amplifier*) circuits are employed.

In general, an efficient systolic array should exhibit a *linear-rate pipelineability*, i.e., it should achieve $O(m)$ speed-up, in terms of processing rates, where m is the number of PE's. The term *Efficiency* (E) will denote the fraction of processor cycles during which a typical processor is actively employed in the array.

Let us now proceed with the definition of the main elements of the mathematical model for the verification of systolic networks.

Abstract model of data, causal relations

We define a data sequence to be an infinite sequence whose elements are members of the set $R_\delta = R \cup \{\delta\}$.

Notation

$R \equiv \{\text{Real numbers}\}$
$\delta =$ "don't care element" (or, "dummy element") $\notin R$.

We extend any operator defined on R to R_δ either:

i) By adding the rule that the result of any operator involving δ is δ (class of δ-regular operators), or

ii) by treating δ as a special symbol that affects the result of the operation (class of non-δ-regular operators).

DEFINITION Operations

i) δ-regular operators, e.g., $\delta\text{"op"}x = x\text{"op"}\delta = \delta \; \forall \, x \in R_\delta$,

*For the superconducting technology of Josephson junctions, a clock period of 1–3 *ns* is achievable today, using a process for which the area unit is $25\,\mu\text{m}^2$ (see Ketchen [26]).

ii) non-regular operators, e.g., binary operator $\oplus$ such that for any $x, y \in R_\delta$, $x \oplus y = x + y$ if $x, y \neq \delta$, $x \oplus \delta = \delta \oplus x = x$.

DEFINITION Let N be the set of positive integers. Then any data sequence η is defined as a mapping from N to R_δ; that is, the image element $\eta(i)$, $i \in N$, is the ith element in the sequence. The set of all data sequences, that is the set of all such mappings, will be denoted by $R_\delta^* \equiv \{\eta | \eta: N \to R_\delta\}$.

Remark Any arithmetic operation on R_δ is extended to R_δ^* by applying the operation elementwise to the elements of the sequences, with δ being the result of any undefined operation, e.g., if "op" is a binary operation defined on R_δ, then $\forall \eta_1, \eta_2 \in R_\delta^*$, η_1"op"$\eta_2 = \eta_3$, where $\forall i \in N$

$$\eta_3(i) = \begin{cases} \eta_1(i)\text{"op"}\eta_2(i), & \text{if } \eta_3(i) \text{ is defined} \\ \delta, & \text{otherwise.} \end{cases} \quad \square$$

DEFINITIONS

d_1: We can also use scalar operations on sequences, e.g., *scalar product*: for $\eta \in R_\delta \wedge \omega \in R$, $\zeta = \omega . \eta \in R_\delta^*$ for which $\zeta(i) = \omega\eta(i)$, $i \in N$.

d_2: *Bounded Data Sequence* set: $R_\delta^* \supset \bar{R}_\delta \equiv \{$All sequences having only a finite number of non-δ-elements$\}$.

d_3: *Termination Function*: $T_f: \bar{R}_\delta \to N$ such that, for $\eta \in \bar{R}_\delta$, $T_f(\eta)$ is the position of the last non-δ-element in η; in other words: for any

$$\eta \in \bar{R}_\delta,\ T_f(\eta) = i \underset{\text{def}}{\Leftrightarrow} \eta(i) \neq \delta \text{ and } \eta(j) = \delta \text{ for } j > i.$$

In addition to the operators extended from R_δ to $\bar{R}_\delta$, we may also define operators directly on $\bar{R}_\delta$.

d_4: The *n-ary Sequence Operator* (Γ): is a transformation Γ: $[\bar{R}_\delta]^n \to \bar{R}_\delta$, where $[\bar{R}_\delta]^n = \bar{R}_\delta \times \bar{R}_\delta \times \cdots \times \bar{R}_\delta$ is the cartesian product space of n copies of $\bar{R}_\delta$.

d_5: *Shift* and *Spread Operators* (Ω^k, θ'):

$$\Omega^k \xi = \eta \text{ and } \theta' \xi = \zeta,$$

where

$$\eta(i)=\begin{cases}\delta & \text{if } i\leqq k\\ \xi(i-k) & \text{if } i>k,\end{cases}$$

$$\zeta(i)=\begin{cases}\xi((i+r)/(r+1)), & i=1, r+2, 2r+3,\ldots,(n-1)r+n,\ldots,\\ \delta & \text{otherwise.}\end{cases}$$

More descriptively, Ω^k inserts k δ-elements at the beginning of a sequence, while θ^r inserts r δ-elements between every two elements of a sequence. For example, if $\xi=a_1,a_2,a_3,a_4,\delta,\delta,\ldots$, then $T_f(\xi)=4$ and

$$\xi(i)=a_i,\ 1\leqq i\leqq T_f(\xi),$$

$$\Omega^3\xi=\delta,\delta,\delta,a_1,a_2,a_3,a_4,\delta,\delta,\delta,\ldots,$$

$$\theta^2\xi=a_1,\delta,\delta,a_2,\delta,\delta,a_3,\delta,\delta,a_4,\delta,\delta,\ldots.$$

It is clear that we can define a sequence operator by combining previously defined sequence operators. For example, we might define an operator Γ: $\bar{R}_\delta\times\bar{R}_\delta\times\bar{R}_\delta\rightarrow\bar{R}_\delta$ as follows:

$$\Gamma(\xi,\eta,\zeta)=\Omega[\xi+\eta^*\zeta],$$

where square brackets are used for grouping and parentheses for enclosing the arguments of the operator.

d_6: *Causal Operators*: Any n-ary sequence operator Γ: $[\bar{R}_\delta]^n\rightarrow\bar{R}_\delta$, which satisfies the causality property in the sense that the ith element of any of its operands can only affect the jth element of its image for $j>i$. More formally, assume that for any $\eta_r\in\bar{R}_\delta$, $r=1,2,\ldots,n$, the image under Γ is $\xi=\Gamma(\eta_1,\ldots,\eta_r,\ldots,\eta_n)$. Then Γ is a causal operator if by replacing any operands η_r by another sequence η_r' satisfying

$$\eta_r'(t)=\eta_r(t),\ 1\leqq t<i,$$

the resulting image $\xi'=\Gamma(\eta_1,\ldots,\eta_r',\ldots,\eta_n)$ satisfies

$$\xi'(t)=\xi(t),\ 1\leqq t\leqq i.$$

Namely, the value of $\xi(i)$ depends only on the first $i-1$ elements of η_r, $1 \leqq r \leqq n$. In the case that the ith element of the image sequence $\xi(i)$ depends only on the first i elements of the operands η_r, $1 \leqq r \leqq n$, then we are talking about *weakly causal operators.*

Abstract systolic network model

The systolic model is defined to be composed of the following components:

i) A loopless multigraph $G(V, E, \phi_-, \phi_+)$, which in turn is composed of
   a) $V \equiv \{\text{Nodes or cells}\}$
   b) $E \equiv \{\text{Directed edges}\}$
   c) two functions ϕ_-, ϕ_+: $E \rightarrow V$ satisfying the condition that for any (edge) $e \in E$, $\phi_-(e) \neq \phi_+(e)$ (i.e., prevents direct loops). The nodes $\phi_-(e)$ and $\phi_+(e)$ are the "source" and "destination" node, respectively, of edge $e \in E$.

   The notation V_S, V_T, V_I will be used for the subsets of V defined as:
   1) $V_S \equiv \{\text{Source nodes (no edges directed IN)}\}$
   2) $V_T \equiv \{\text{Sink nodes (no edges directed OUT)}\}$
   3) $V_I \equiv \{\text{Interior nodes (not a source or sink)}\}$;

   certainly, the condition $V_S \cup V_T \cup V_I = V$ is always satisfied.
ii) A colouring function col: $E \rightarrow C_E$, where C_E is a given finite set of colours.
   Essentially, input edges to the same node receive different colours, as do output edges, e.g., $y = \text{col}(e)$ denotes edge e has colour y.
iii) For each edge $e \in E$, a sequence $\xi_e \in \bar{R}_\delta$ is specified.
iv) For each interior node $v \in V$ with IN-degree m and OUT-degree n, n causal m-ary operators Γ_v^i: $[\bar{R}_\delta]^m \rightarrow \bar{R}_\delta$ are given, which specify the "node I/O description". More specifically, if η^j, $j = 1, 2, \ldots, m$, and ξ^i, $i = 1, 2, \ldots, n$, are the sequences associated with the IN and OUT edges of v, respectively, then the n

relations

$$\xi^i = \Gamma^i_v(\eta^1, \eta^2, \ldots, \eta^m),\ i = 1, 2, \ldots, n,$$

are the I/O description of v. The different IN and OUT edges of v are distinguished in the I/O description by their colours.

Remark Each interior node represents a computational cell*/processor and each source/sink node corresponds to an input/output cell for the overall network. Each edge $e \in E$ is a unidirectional communication link or channel. □

Finally, given a systolic network based on the graph $G \equiv \{V, E, \phi_-, \phi_+\}$, a subset $V'_I \subset V_I$ of interior nodes is said to be a homogeneous set if:

i) All the nodes in V'_I have identical IN- and OUT-degrees, say m and n, respectively.
ii) The m colours of the IN edges of any $v \in V'_I$ are identical and so are the n colours of the OUT edges of v.
iii) The node I/O descriptions of any $v \in V'_I$ are generic [23].

To conclude, a network is said to be *homogeneous* if the set of interior nodes V_I, in its graph G, is a homogeneous set. More generally, a network is said to be k-partially homogeneous if there exists a partition $\bigcup_{i=1}^{k} V^i_I$ of V_I into k non-empty homogeneous subsets V^i_I, $i = 1, 2, \ldots, k$.

6. SYSTOLIC SYSTEMS AND CELLULAR AUTOMATA

The concept of a large number of primitive processors leads to another factor that has contributed to the development of the systolic concept. From a theoretical point of view, systolic systems can be traced back to cellular automata of Von-Neumann, the Mealy Machine and Moore Machine [2, 14, 15].

*The computations performed by the cells are modelled by a system of difference equations involving operations on the various data sequences. The input/output descriptions, which describe the global effect of the computations performed by the network, are obtained by solving this system of difference equations [24]).

Automata theory is basically a mathematical model about machines and what they can accomplish at a low level of computation. It has mainly been applied to the design of electrical circuits with digital hardware, the logic of nervous systems in man and animals, and the underlying logic of protein synthesis in cells. This mathematical model seems to gain increasing scientific interest in the investigation of physical systems, showing complex, self-organising and "chaotic" behaviour. Cellular automata, as well as systolic systems, seem to have better capabilities to map physical systems and their parallel space and time evolution into computer architecture, [11].

Automata theory is very important for the comprehension of systolic systems because it lends a ready-made theory about what such machines can achieve. Automata themselves can be represented by labelled directed graphs, with machine states represented as nodes, and arcs defining state transitions. Consequently the function of a simple systolic cell can be represented by such a graph. Inputs and outputs can also be encoded on arcs and from here it is a small step to connect inputs and outputs of a number of machines and operate them in parallel to create a systolic system.

However, the specification of systolic systems by means of automata theory definitions leads to an overly complicated structure. Consequently systolic arrays have their own simpler and relatively abstract graph specification, which collapses whole machines to nodes and i/o histories to sequences on arcs [16, 17].

7. SYSTOLIC ARCHITECTURES AND *VLSI*

Until the advent of *VLSI*, the development of parallel computers with large numbers of processors had been limited by the prohibitively high costs of production. With the use of *VLSI* in circuits, size and cost of processing logic, memory and communication hardware was dramatically reduced, and it became feasible to combine the principles of automata theory with the traditional parallel processing techniques to produce highly parallel *VLSI* architectures, such as the systolic systems. This enterprise can take two forms: either produce a special-purpose *VLSI* chip implementing a specific systolic algorithm; or combine programmable *VLSI* processors to produce a systolic architecture capable of performing one or more algorithms.

For a general introduction to *VLSI*, see [3, 19, 20]; while reviews of parallel *VLSI* architectures are given in [21, 22 and 27].

Now if we attempt to map a complex systolic system directly on a silicon wafer, it is immediately confined to a two-dimensional plane. *VLSI* is achieved by a combination of circuit design with high resolution photographic techniques, where it is convenient to place wires on rectangular grids, and limit the number of parallel layers of semi-conductor material containing wires and circuit elements. Hence, a two-dimensional graph is termed planar if it can be drawn in the plane with no arcs intersecting at places other than nodes (cells). *VLSI* presents additional problems, as the size of wires and transistors approach the limits of photographic resolution, for it becomes impossible to achieve further miniaturization and the actual circuit area becomes a key issue. Furthermore, the chip area is limited in order to maintain high yield, and the number of connections to the outside world (pins) is limited by the finite size of the chip perimeter.

Some of these limitations are alleviated when systolic algorithms are implemented on processor arrays. For example, the actual chip design is not an issue any more, since it is a programmable processor. Further, the interconnections need not be strictly planar. However, in both cases, simplicity and regularity remain factors of utmost importance for an efficient systolic design. In the first case because they ensure the design of cost-effective, special-purpose *VLSI* chips. In the second case because of the promising proposal to harness the programming complexity of parallel computers with a large number of cooperating processors. Simplicity and regularity in systolic architectures are ensured by means of the following techniques: there are only a few types of relatively simple cells, where only local and regular communication is allowed, mainly of nearest-neighbour type [1].

The replication of a processor in large numbers makes the design cost-effective and easy to produce; however, exactly how simple a cell might be is a question that can only be answered on a case by case basis. For example, if a systolic system is to be implemented on a single chip, each cell should probably contain only simple logic circuits plus a few words of memory. On the other hand for board array implementations each cell could reasonably contain a high-performance arithmetic unit, plus a few thousand words of memory and a local control unit. Further, for processor array implementations, each cell can be a simple microcomputer. There is, of course, always a trade-off between simplicity and flexibility, in terms of control and programming overheads as well as system performance.

In principle, systolic systems totally avoid long-distance or irregular interconnections; typical examples are given in Figure 4. The only global communication (besides power and ground) is the system clock. Alternatively the need for global clock distribution can be avoided if self-timed, asynchronous schemes are implemented, based on data-driven protocols. A consequence of that characteristic is that systolic systems are completely modular and expandable, and can be easily adjusted to the problem size or other external factors.

References

[1] H. T. Kung, Why systolic architectures?, *IEEE Computer* **15** (1982), 37–46.

[2] C. E. Leiserson, Area efficient VLSI computation, Ph.D. Thesis, Department of Computer Sciences, CMU, 1981.

[3] C. A. Mead and L. Conway, *Introduction to VLSI Systems*, Addison-Wesley, 1980.

[4] A. L. Fisher and H. T. Kung, Special purpose VLSI architectures: General discussions and a case study, In: *VLSI and Modern Signal Processing*, T. Kailath *et al* (eds.), Prentice-Hall, 1985, 153–169.

[5] M. J. Foster and H. T. Kung, The design of special purpose VLSI chips, *IEEE Computer* **13** (1980), 26–40.

[6] P. M. Dew, L. J. Manning and K. McEvoy, *A Tutorial on Systolic Array Architectures for High Performance Processors*, University of Leeds, Technical Report 205, 1986.

[7] Y. Robert, *Systolic Algorthms and Architectures*, Technical Report 397, IMAG, France, 1986.

[8] D. Heller, Partitioning big matrices for small systolic arrays, In: *VLSI and Modern Signal Processing*, T. Kailath *et al.* (eds.), Prentice-Hall, 1985, pp. 185–199.

[9] H. T. Kung, *A Listing of Systolic Papers*, Department of Computer Science, CMU, 1985.

[10] D. J. Evans, *Parallel Processing Systems*, Cambridge University Press, 1982.

[11] F. Hossfeld, *Strategies for Parallelism in Algorithms*, Lecture notes in IBM summer school in Parallel Computing, 1986.

[12] K. Hwang and F. A. Briggs, *Computer Architecture and Parallel Processing*, McGraw-Hill, 1984.

[13] H. T. Kung, The structure of parallel algorithms, *Advances in Computers* **19** (1980), 65–112.

[14] C. Moraga, On a case of symbiosis between systolic arrays, *Integration, the VLSI Journal* **2** (1984), 243–253.

[15] C. Moraga, *Systolic Algorithms*, Technical Report, Department of Computer Science, University of Dortmund, FRG, 1984.

[16] H. T. Kung and W.T. Lin, An algebra for VLSI algorithim design, *Proc. of Conf. on Elliptic Problem Solvers*, 1983, 141–160.

[17] U. Weiser and A. Davis, A wavefront notation tool for VLSI array design, In: *VSLI Systems and Computations*, H. T. Kung, *et al.* (eds.), Computer Science Press, 1981, pp. 226–234.

[18] E. E. Swartzlander, VLSI architecture, In: *VLSI Fundamentals and Applications*, D. F. Barbe (ed.), Springer-Verlag, 1982, 178–221.

[19] J. D. Ullman, *Computational Aspects of VLSI*, Computer Science Press, 1984.

[20] L. S. Haynes, *et al.*, A survey of highly parallel computing, *IEEE Computer* **15** (1982), 9–24.

[21] C. L. Seitz, Concurrent VLSI architectures, *IEEE Trans. on Computers* **C-33** (1984), 1247–1265.

[22] H. T. Kung and C. E. Leiserson, Systolic arrays (for VLSI), In: *Proc. Sparse Matrix Symp.*, I. A. Duff and G. W. Stewart (eds.), SIAM, 1978, pp. 256–282.

[23] M. P. Bekakos and D. J. Evans, The exposure and exploitation of parallelism on fifth generation computer systems, In: *Parallel Computing*, M. Feilmeier *et al.* (eds.), 1980, North-Holland, pp. 425–442.

[24] R. G. Melhem and W. C. Rheinboldt, A mathematical model for the verification of systolic networks, *SIAM J. Comput.* **13** (1984), 541–565.

[25] P. Faurre and M. Depeyrot, *Elements of System Theory*, North-Holland, Amsterdam, 1977.

[26] M. B. Ketchen, AC powered Josephson miniature system, In: *Proc. Int. Conf. Circuits Comput.*, IEEE Comput. Soc., 1980, pp. 874–877.

[27] K. G. Margaritis, *A Study of Systolic Algorithms for VLSI Processor Arrays and Optical Computing*, Ph.D. Thesis, Loughborough University, 1987.

The Cut Theorem—A Tool for Design of Systolic Algorithms

KEVIN McEVOY and PETER M. DEW

Department of Computer Studies, The University, Leeds LS2 9JT, UK

This paper describes a thorough mathematical treatment of retiming transformations, and discusses the advantages which they provide for the design of systolic algorithms. This approach to synchronous design originated with Leiserson and Saxe [14], but more complete and accurate statements of the retiming results are presented here. This is achieved through a precise formulation of the model of computation. The importance of the Cut Theorem in systolic design has been emphasized on many occasions (e.g. S. Y. Kung and others at the First International Workshop on Systolic Arrays [23]), but the statements and proofs of this theorem have so far been somewhat imprecise. An exact and general formulation of the Cut Theorem is given, and there is discussion of its use as a design tool (including the design of two-level pipelined arrays). This approach provides some insight into the definition of the term systolic, and into the classification of systolic array algorithms on their communication pattern and dataflow properties.

KEY WORDS: Systolic, retiming, cutset, VLSI, synchronous.

C.R. CATEGORIES: C.1.2 Computer Systems Organisation—Multiple Data Stream Architectures; F.1.1 Theory of Computation—Models of Computation.

1. INTRODUCTION

An important contribution to the formalism and understanding of the principles of systolic algorithm design has been made by Leiserson and Saxe [14]. They outline an approach to the design of systolic algorithms which allows an algorithm to be expressed in a convenient form and then transformed to a systolic algorithm by the addition of delays. The importance of this approach is that the transformation is context independent, and therefore can be applied to any synchronous algorithm. The purpose of this paper is to explore the extent of these results through a precise mathematical treatment.

Signal flow graphs are often used to describe synchronous networks in signal processing literature [25]. Although they are adequate for straightforward computations like correlation, they are not precise enough to describe an arbitrary synchronous algorithm. The following are two more mathematical approaches which are among those more commonly used. The first is based on a *functional notation* where the computation is expressed as a set of simultaneous recursive equations. Each step in the recursion is performed by a separate processor operating in a synchronous pipeline mode. Important contributions in the use of functional languages for the design of synchronous VLSI circuits have been made by Sheeran [26], Brookes [2], and Thompson and Tucker [27]. The work by Thompson and Tucker provides a formal framework for the specification of synchronous algorithm using a *functional language* based on *primitive recursive functions*.

The second approach is to use a model based on the directed or the undirected graph. This is the type of model which is presented in this paper, and it is in essence a mathematically precise statement of a signal flow graph. The model is best described as *clocked* communication of processors and is constructed from three types of primitive: *cells* (function elements to perform computations); *channels* (for cummunication between cells); and *delays* (which in a hardware implementation would model registers). Each cell is given a functional definition that fixes the *level of abstraction* of the design. This level of abstraction can be systematically refined by replacing a cell by a full computation graph which computes the function defining the cell (but on a more primitive data structure). The retiming transformations can be expressed precisely on this model, and the model is defined in Section 2.1.

The Retiming Lemma (Leiserson and Saxe [14]) defines a method for moving around the delays in a synchronous algorithm without changing the computational behaviour of the algorithm, and its most common use is in transforming synchronous algorithms into *temporally local* algorithms. (A temporally local algorithm is a synchronous algorithm in which there is at least one delay on every channel). In Section 2.2 the Retiming Lemma is stated in the terms of the model, and it is followed by a discussion of the use of the lemma in inducing temporal locality in Section 2.3. The Cut Theorem is a retiming result which was introduced by Kung and Lam in [12] and improved by S. Y. Kung in [13]. The statement of the Cut Theorem in Section 3.1 is more general than that given by Kung and Lam; a proof is given. It is shown how as retiming transformations the Retiming Lemma and the Cut Theorem are in fact equivalent in power; however the result of several applications of the Cut Theorem may be achieved in one step by the Retiming Lemma. A hierarchical approach is necessary in any design technique, and this is discussed in terms of two-level pipelining in systolic design in Section 3.3. In Section 4 the model is used to shed some light on the meaning of the word systolic itself, together with further comment of the classification of systolic arrays and performance issues. The conclusion contains some indication as to further uses of these techniques.

2. RETIMING

Before defining the retiming transformations, we first introduce a model of computation appropriate for the expression of systolic or synchronous algorithms, and the retiming transformations of such algorithms.

2.1 The model of computation

Figure 1(a) shows a typical representation of a synchronous algorithm to compute the following FIR filter or one-dimensional convolution

$$y_i = \sum_{j=0}^{n-1} w_j x_{i+j}.$$

The rectangular boxes in the figure denote registers used to delay the x and y streams, and the circles denote the functional units. The

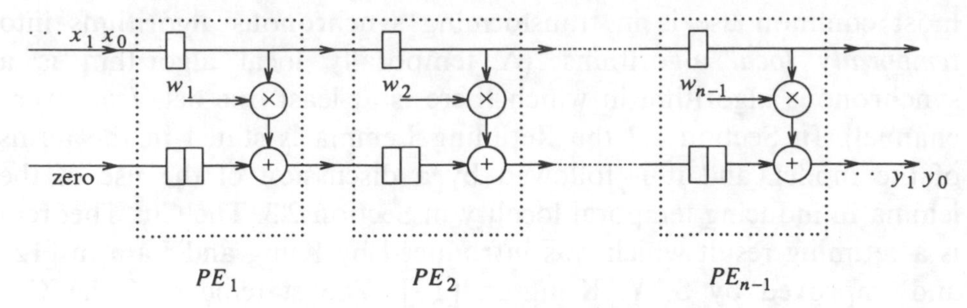

Figure 1(a) Systolic array algorithm for 1-d convolution.

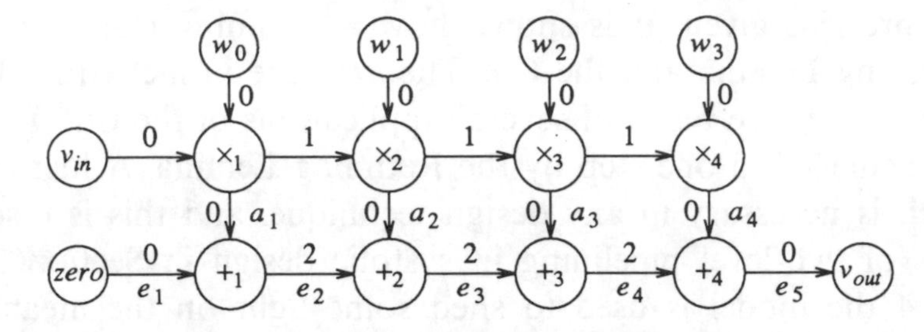

Figure 1(b) The convolution algorithm as a computation graph.

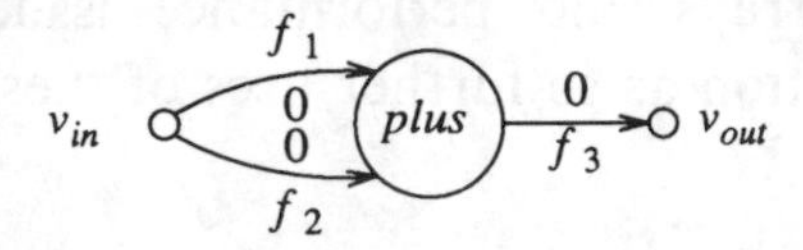

Figure 1(c) A primitive computation graph representing a plus cell.

algorithm is synchronous; that is, during each clock cycle data is passed from one register to the next. In this section of the paper a more precisely defined representation of such a diagram is presented; a representation with a formal semantics which will support mathematical proof about properties of synchronous algorithms.

The communication topology of the algorithm is modelled by a directed multigraph which is represented as a four-tuple $D = \langle V, E, tl, hd \rangle$, where V and E are the sets of cells and channels respectively, and the functions $tl{:}E \rightarrow V$ and $hd{:}E \rightarrow V$ define the adjacency relationship of the graph by associating a head and tail with each edge of the graph. Two distinguished cells v_{in} and v_{out} must exist in every graph; these correspond to the host node used by Leiserson and Saxe—they perform no computation and exist solely to identify the input and output channels.

A synchronous algorithm is constructed from a set of primitive processors. These processors form the data structure of the algorithm and in this model the syntax of the data structure will be presented as a set of *primitive computation graphs.* A primitive computation graphs is defined to be a computation graph with only one node other than v_{in} and v_{out}. To define the functionality (or semantics) of these primitive computation graphs standard techniques from universal algebra are used—under a given interpretation (or algebra) A, each primitive computation graph σ has an associated semantic function σ_A (e.g., addition or multiplication). The data structure may contain different data sets (e.g. boolean, integer), and this is expressed by allowing the algebra to be many-sorted. A function st defines a sort on each channel in the graph, so that the channels are strongly typed. An algorithm is defined by linking up primitive computation graphs from the data structure in some acceptable manner. The syntax for the representation of such acceptable linking is as follows. A function gf associates a primitive computation with each cell in a given computation graph. To check that the association is acceptable, and to complete the definition of the functionality of the cell v, a bijection cn_v pairs off both the channels into v with the channels into $gf(v)$, and the channels out of v with the channels out of $gf(v)$.

Having defined the communication topology and the data structure the third major component is the modelling of delay to introduce sequential computation into the model. A function $d{:}E\to\mathbb{N}$ defines a natural number on each channel (and might model a number of registers on a wire in a hardware implementation of the algorithm). Channels with zero delay (ripple carries) are permitted.

So the syntax of a computation graph is defined to be a five-tuple $G=\langle D, gf, cn\, st, d\rangle$. Figure 1(b) shows the digraph, D, of the computation graph representation of the convolution algorithm shown in Figure 1(a). In this case st takes the same value on every channel (presumably reals or integers). The values of the delay function d are labelled on the channels. For each of the cells labelled $+_i$ the value of $gf(+_i)$ is the primitive computation graph shown in Figure 1(c). The value of gf on the $\times_i$ cells will be a primitive computation graph with a similar syntax to the *plus* graph, but different semantics. The other four primitive computation graphs in the data structure will be constants outputting the values w_i. As an example of the definition of cn consider the $+_i$ cells. For $i=1,2,3,4$, $cn_{+_i}(a_i)=f_1$, $cn_{+_i}(e_i)=f_2$, and $cn_{+_i}(e_{i+1})=f_3$.

A computation on a computation graph, G, is described as follows. During each clock cycle one set of inputs is read onto the channels incident on v_{in}, and there is a flow of one data element along every channel in the graph. There is no delay within the cells, but the interpretation of the delay function d is that data reaching the tail of a channel, e, will leave the head of that channel exactly $d(e)$ clock cycles later. During every clock cycle each cell outputs to its output channels by evaluating its associated semantic function on the values on its input channels. This computational behaviour is defined precisely using a *state transition semantics* to produce a functional semantics for each computation graph of the form

$$\Omega_G : \text{IN}_G^+ \rightarrow \text{OUT}_G$$

where $I \in \text{IN}_G$ (respectively $O \in \text{OUT}_G$) is an assignment of values to the channels incident on v_{in} (respectively v_{out}). Note that as sequential computation is part of the synchronous model, it is possible that the value of OUT_G depends on the whole string of previous inputs (this is the case in, for example, a counter) and so the domain of the function Ω_G is not IN_G, but rather IN_G^+ (the set of non-empty finite strings of members of IN_G). The definition of Ω_G is recursive, and if the computation graph contains a cycle in which the delay of every edge in that cycle is zero, then the recursion is not well-founded. This behaviour models the race conditions in a synchronous circuit containing a cycle with no registers, and so it is an integral part of a model of synchronous circuits. It follows that there exist computation graphs, G, for which Ω_G is a *partial* function (and clearly the design of such computation graphs is to be avoided at all costs).

Our purpose in presenting this model here is to allow a mathematical presentation of the retiming results. Further details on this model, together with extensions into hierarchical definition of graphs can be found in McEvoy [19]. An alternative method of representing delays is to use a *delay operator* Z^d or Z^{-d}; this method has the attraction that the delays can be manipulated algebraically to derive and retime signal flow graphs, and is discussed in Kung and Lin [10]. Other important contributions to the use of mathematical techniques in the design of synchronous and systolic algorithms can be found in the work of Moldovan [22], Fortes and Moldovan [7], Quinton [24], Delosme and Ipsen [5], Harman and Tucker [8],

Martin and Tucker [18], Chen [3], Melhem and Rheinboldt [21], Evans [6], Megson [20] and Li and Wah [15].

2.2 The Retiming Lemma

The complex pipelining schemes used in systolic arrays complicate algorithm design, and in particular validation that an algorithm correctly performs the desired computation can become very difficult. The standard "snapshots" methods are clearly inadequate for all but the most simple of algorithms. An alternative design method has been developed. This is to design the algorithm using as little delay as possible (making liberal use of ripple carries and broadcasts); the algorithm can be more easily validated or verified at this stage. Then the "proven correct" retiming results presented below are used (possibly through automated tools) to introduce the delay necessary to improve the performance of a systolic array. For example, it is relatively easy to validate that if the computation graph shown in Figure 2 is supplied with the input stream $\langle x_i : i \in \mathbb{N} \rangle$, and its output stream is $\langle y_i : i \in \mathbb{N} \rangle$ then (after initialisation) the one-dimensional convolution

$$y_{i+3} = w_0 x_i + w_1 x_{i+1} + w_2 x_{i+2} + w_3 x_{i+3}$$

is computed (given that each *ip* node *w* computes an inner-product step function, $s_{\text{out}} = s_{\text{in}} + w \times x_{\text{in}}$). Pipelining of the algorithm can then be introduced using the retiming results presented below (producing in the case of Figure 2 an algorithm equivalent to that in Figure 1). The essential effect of pipelining is to increase the throughput of the circuit at the expense of increased latency.

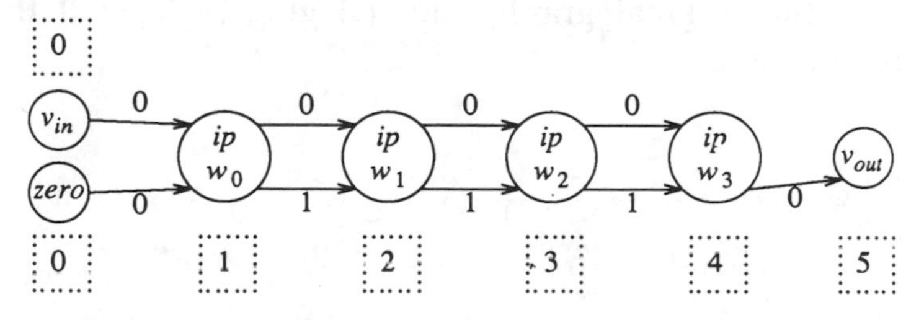

Figure 2 A simple computation graph for 1-d convolution with values of the delay function shown as edge labels and values of the lag function shown in the dotted boxes. The top rail carries the input stream and the bottom rail the results.

There are two basic transformations for altering the delays in a computation graph. The first is to retime the graph according to the Leiserson and Sax Retiming Lemma. However, if the graph does not satisfy the conditions of this lemma, then it is possible to increase all the delays by a constant factor, k say, and then apply the Retiming Lemma. This latter transformation has been called *k-slowing* or *k-interleaving* the graph.

RETIMING LEMMA (Leiserson and Saxe) *Let $G=\langle D, gf, cn, st, d\rangle$ be a computation graph with $D=\langle V, E, tl, hd\rangle$. Given any function*

$$\mathrm{lag}: V \rightarrow \mathbf{Z}$$

(where $\mathbf{Z}$ is the set of integers) define a new delay function by

$$wt(e)=d(e)+\mathrm{lag}(hd(e))-\mathrm{lag}(tl(e)).$$

If $wt(e) \geqq 0$ for all $e \in E$, then the computation graph $H=\langle D, gf, cn, st, wt\rangle$ is called a retiming *of G, by which it is meant that the graph H computes the same function as G modulo certain initialisation conditions.* □

Essentially, the lemma just states that it is possible to add any delay to the input edges of a node providing that we subtract the same delay from its output edges (a formal proof is given in McEvoy [19]). A simple application of the Retiming Lemma is shown in Figure 3 with the values of the lag function shown in the dotted boxes. In this case both G and H compute the same function $\Omega_H=\Omega_G$. In general, if $\mathrm{lag}(v_{\mathrm{out}})>\mathrm{lag}(v_{\mathrm{in}})$ then the latency (a measure of the overall delay in the graph) of the Retimed Graph will be greater than the latency of the original graph, and if $\mathrm{lag}(v_{\mathrm{out}})<\mathrm{lag}(v_{\mathrm{in}})$ then the latency will be decreased.

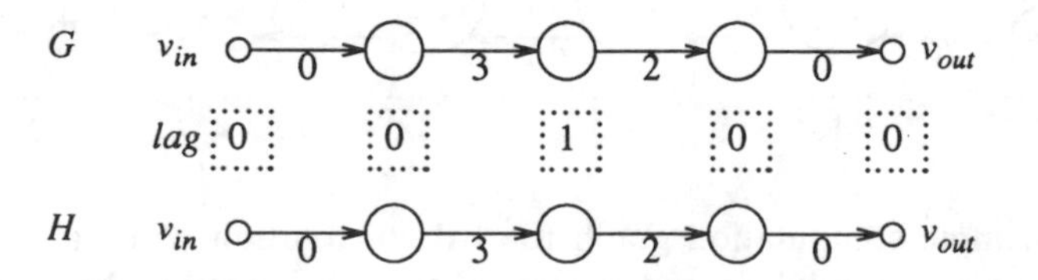

Figure 3 A simple retiming of the graph G.

The second transformation is that of k-slowing or k-interleaving—multiplying every delay in the graph by k. The effect of this transformation is that the computation which was originally carried out on the input stream $\langle x_i : i \in \mathbb{N} \rangle$ is now carried out independently on the k streams $\langle x_{ki} : i \in \mathbb{N} \rangle, \langle x_{ki+1} : i \in \mathbb{N} \rangle, \ldots, \langle x_{ki+k-1} : i \in \mathbb{N} \rangle$. This transformation is not as satisfactory as a simple retiming, as the problem to be solved may not present k independent streams to be interleaved before processing, and in this case many of the processors may be effectively idle (or performing pointless computations) during certain clock cycles. If no interleaving of the problem is possible than the input and output streams are said to have a *spacing* of k clock cycles. However if the problem can be successfully interleaved, then after use of the Retiming Lemma the computation time can be reduced when averaged over k clock cycles.

2.3 Temporal locality

Temporal locality is one of the most important properties of a systolic array, as the existence of a delay between every pair of nodes ensures that the clock cycle can be reduced to the maximum time needed to perform the computation in a node and latch the result. The retiming results can be used to transform any reasonable computation graph into one which is temporally local. This is achieved through moving around the delay within cycles, and the class of computation graphs for which this transformation is not possible is those which contain a cycle in which every edge has delay zero. In a synchronous environment such cycles introduce race conditions and non-deterministic behaviour, and for such a computation graph, G, Ω_G is a partial function. Such computational behaviour cannot exist in a temporally local algorithm, and so as retiming transformations do not change computational behaviour it follows that temporal locality cannot be induced into such an algorithm.

DEFINITION A computation graph is said to be *temporally local* if the only edges in the graph which have delay zero are those incident on either v_{in} or v_{out}. □

Leiserson and Saxe define a systolic array algorithm as one that can be written as a temporally local computation graph, but this simple definition ignores some of the other important properties of systolic algorithms as originally stated by Kung [9]. An attempt is made to characterise some of these properties later in this paper. An important result is that any deterministic computation graph can be retimed into another computation graph which has temporal locality—the spacing of the computations may be different, but both graphs are essentially computationally equivalent.

THEOREM *If every cycle in a computation graph contains at least one channel with non-zero delay, then the computation graph can be retimed to be temporally local.*

Proof There are two cases depending on the delay in each cycle C of the graph.

Case I $d^*(C) \geqq |C|$ for every cycle C in the graph, where $d^*(C)$ is the sum of the delays of the channels in the cycle, and $|C|$ is the number of channels in the cycle. In this case the lag function is defined by

$$\text{lag}(v) = \max\{|P| - d^*(P) : P \text{ is a path in } D \text{ with head at } v\}$$

or *zero* if no such path exists. The condition $d^*(C) \geqq |C|$ means that any cycles in the path contribute a negative quantity to the total $|P| - d^*(P)$ and so the maximum value is attained for a *simple path.* Any graph has only a finite number of simple paths and so the maximum is attained. The values of the lag function for the one-dimensional convolution example shown in Figure 2 are given in the dotted boxes. For this example the lag function is equal to the longest path reaching node v with zero weight. Following the Retiming Lemma let

$$wt(e) = d(e) + \text{lag}(hd(e)) - \text{lag}(tl(e))$$

then either $\text{lag}(tl(e)) = 0$ in which case $\text{lag}(hd(e)) \geqq 1 - d(e)$ and $wt(e) \geqq 1$, or

$$\text{lag}(tl(e)) = |P| - d^*(P)$$

for some path P with head at $tl(e)$. From the definition of the lag function it follows that

$$\text{lag}(hd(e)) \geqq |P| + 1 - d^*(P) - d(e)$$

and hence $wt(e) \geqq 1$.

Case II There is a cycle C in the graph for which $d^*(C) < |C|$. In this case a simple retiming cannot produce a graph which is temporally local. (In fact, the condition of Case I can be shown to be a necessary and sufficient condition for the existence of a simple retiming which is temporally local.) For example, consider the graph (taken from Kung [13]) shown in Figure 4. The lag function assigned using the rule from Case I (considering only simple paths) is shown in the dotted boxes; however this produces retimed delays, $wt(e)$, which are negative for those edges that feed back the results of the inner product step to the corresponding cells on the top rail. To remove zero delays from such a graph, we must first increase each delay in the graph by some factor k chosen so that the condition of Case I is satisfied. The total delay in each cycle will increase by a factor of k, but the length of the cycle remain unchanged, and so it is always possible to find a large enough k to cover all the finitely many simple cycles (given that the hypothesis stated that there is no cycle in the graph which has a total delay of zero). □

Although it is always possible to use these methods to make an algorithm temporally local, if k-slowing is necessary then the performance of the algorithm may not be improved. For example to retime

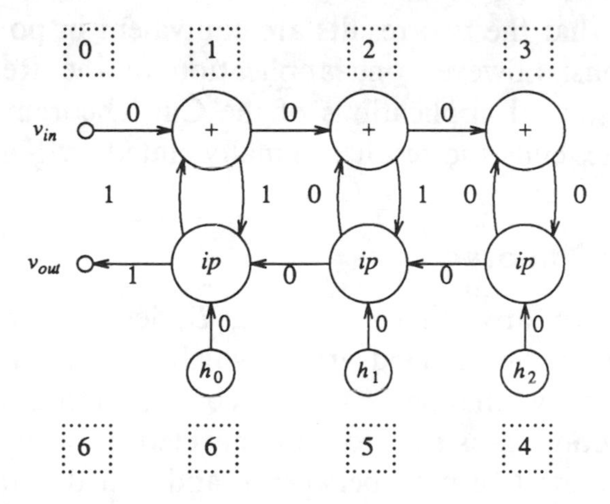

Figure 4 A computation graph containing a large number of cycles.

the algorithm of Figure 4 each delay must first be increased by a factor of six. Therefore each input stream must have a spacing of six clock cycles, and so if it is not possible to interleave several input streams, then on average only one processor of the six will be active in any one clock cycle, which is clearly unacceptable. This example illustrates one of the difficulties in having cycles in a computation graph, and so it is good design practice to minimise the number of cycles in an algorithm, and, for example, consider using ring algorithms where there is just one major cycle (see Kung and Lam [12]).

3. THE CUT THEOREM AND ITS APPLICATIONS

Despite its power and usefulness, the Retiming Lemma is not a convenient tool for algorithm design. An application of the Retiming Lemma requires the assignment of a lag to every node in the graph, and in large graphs this process may introduce considerable overheads. An alternative retiming transformation was introduced by Kung and Lam [12], and later extended by Kung [13]. This transformation, known as the Cut Theorem, only changes the delays of those channels in a cutset for the underlying graph. For this reason the result of applying the theorem is easier to predict, and so a designer can more easily evaluate the applicability of the theorem to a given situation. (It is also true that the regularity of systolic arrays allows cutsets to be very easily identified.) Nevertheless it is shown below that the two results are equivalent in power as retiming transformations; however one application of the Retiming Lemma may replace several applications of the Cut Theorem. The following presentation extends the result originally stated by Kung and Lam.

3.1 The Cut Theorem

DEFINITION For any directed graph, D, let the undirected graph underlying D be the directed graph which is constructed from D by adding, for every channel $e \in E$, a corresponding channel in the opposite direction. D is said to be connected if for every pair of cells u, v in D there exists a path between u and v in the undirected graph underlying D; otherwise D is *disconnected*. If $D = \langle V, E, tl, hd \rangle$ is connected, then a *cutset* for D is a set of channels $C \subseteq E$ such that the graph (called $D-C$) which is formed by removing C from D is disconnected. A cutset C for D is *minimal* if no strict subset of C is a cutset for D. □

A cutset C partitions the nodes of D into two disjoints sets, V_{in}^C and V_{in}^C, where V_{in}^C is the node set of the *connected component* of $D-C$ containing v_{in}; that is, $v \in V_{\text{in}}^C$ if and only if there is a path from v_{in} to v in the undirected graph underlying $D-C$. Kung and Lam used a more restrictive definition of a cutset, where the graph is divided into a *source set*, V_s and a *destination set*, V_d. The source set is the same as V_{in}^C except that v_{out} is not a member of V_s (i.e. all cuts which do not disconnect v_{in} from v_{out} are excluded). $V_d = V - V_s$. They also introduced the restriction that edges in the cutset must be directed from the source set to the destination set, which essentially restricts the application of the lemma to those algorithms with unidirectional flow. The following result is more general.

CUT THEOREM *Let $G = \langle D, gf, cn, st, d\rangle$ be a computation graph, and C a minimal cutset for D. Given $\beta \in \mathbf{Z}$ (known as the* retiming constant), *define $wt: E \to \mathbf{Z}$ by*

$$wt(e) = \begin{cases} d(e) + \beta & \textit{if} \quad e \in C \quad \textit{and} \quad tl(e) \in V_{\text{in}}^C \\ d(e) - \beta & \textit{if} \quad e \in C \quad \textit{and} \quad tl(e) \notin V_{\text{in}}^C \\ d(e) & \textit{if} \quad e \notin C \end{cases}$$

If for all $e \in E$, $wt(e) \geqq 0$ then $\langle D, gf, cn, st, wt\rangle$ is a retiming of G.

Proof Given any channel e, it is easy to show that $e \notin C$ if and only if either both its head and tail are in V_{in}^C or both its head and tail are in $V - V_{\text{in}}^C$ (but minimality of the cutset is necessary to prove this). Define a lag function lag: $V \to \mathbf{Z}$ according to the rule

$$\text{lag}(v) = \begin{cases} 0 & \text{if} \quad v \in V_{\text{in}}^C \\ \beta & \text{otherwise} \end{cases}$$

Then for channels not in the cutset, the lag function has the same value at the head and tail nodes. Edges $e \in C$ with $tl(e) \in V_{\text{in}}^C$ have

$\text{lag}(hd(e)) - \text{lag}(tl(e)) = \beta$ and those with $hd(e) \in V_{\text{in}}^{C}$ have $\text{lag}(tl(e)) - \text{lag}(hd(e)) = \beta$ (again using the minimality of the cut). Therefore for all $e \in E$,

$$wt(e) = d(e) + \text{lag}(hd(e)) - \text{lag}(tl(e))$$

and so the result follows from the Retiming Lemma. □

It is important that any cutset used in the Cut Theorem must be minimal; that is, it must not be possible to partition the graph using any strict subset of C. Consider, for example, Figure 1(b). It can be seen that $\{e_2, a_2, e_3\}$ and $\{e_3, a_3, e_4\}$ are minimal cutsets, and the theorem could be applied consecutively to these cutsets with a retiming constant of one in each case. In particular the delay on the channel e_3 would be changed to 1 and then back to 2 again. The union of these sets is also a cutset, but not minimal; if $\{e_2, a_2, e_3, a_3, e_4\}$ were to be used as a cutset in the theorem with a retiming constant of one, then the result would be the same as the two independent consecutive applications, except that the delay on e_3 would be changed to 1. Clearly both transformations cannot be correct, and it is not difficult to see that the second transformation does not preserve computational behaviour.

The proof of the theorem shows that any application of the Cut Theorem can also be carried out using the Retiming Lemma, and so the Retiming Lemma is at least as powerful as the Cut Theorem. In fact, the converse is also true, as any retiming carried out using the Retiming Lemma with a lag function *lag* can be simulated by a sequence of applications of the Cut Theorem with one cutset, C_v, for every node v in the graph; C_v is the set of edges adjacent to v, and the retiming constant used with the cutset C_v is $lag(v)$. It follows that the Cut Theorem and the Retiming Lemma are equivalent as retiming transformations.

Given the regularity of systolic arrays it is often the case that the Cut Theorem is applied to a number of parallel cutsets. Provided that the cutsets do not intersect, it is possible to include all the cutsets in one application of the theorem.

COROLLARY *Let* $G = \langle D, gf, cn, st, d \rangle$ *be a computation graph, and let* $\{C_1, C_2, \ldots, C_m\}$ *be a set of m pairwise disjoint* sets of channels, such*

---

*That is, every pair i, j with $1 \leq i \leq j \leq m$ satisfies $C_i \cap C_j = \emptyset$.

that each individual C_n *is a minimal cutset for* D. *Given* $\beta_1, \beta_2, \ldots, \beta_m \in \mathbf{Z}$ *define* $wt: E \to \mathbf{Z}$ *by*

$$wt(e) = \begin{cases} d(e) + \beta_n & \text{if} \quad e \in C_n \quad \text{and} \quad tl(e) \in V_{\text{in}}^{C} \\ d(e) - \beta_n & \text{if} \quad e \in C_n \quad \text{and} \quad tl(e) \notin V_{\text{in}}^{C} \\ d(e) & \text{otherwise} \end{cases}$$

If for all $e \in E$, $wt(e) \geqq 0$ *then* $\langle D, gf, cn, st, wt \rangle$ *is a retiming of* G.

Proof As the cutsets are disjoint, the result is just the same as a sequence of m applications of the theorem, treating each cutset individually. □

3.2 A design example

The importance of the Cut Theorem is that it allows us to be more systematic in the design of systolic array algorithms. To illustrate the process of designing a systolic array algorithm using the Cut Theorem, consider the infinite impulse response filter

$$y_i = w_0 x_i + w_1 x_{i-1} + w_2 x_{i-2} + h_0 y_{i-1} + h_1 y_{i-2}.$$

A computation graph for this filter is shown in Figure 5. The x data stream is input along the top rail and the results are fed back along

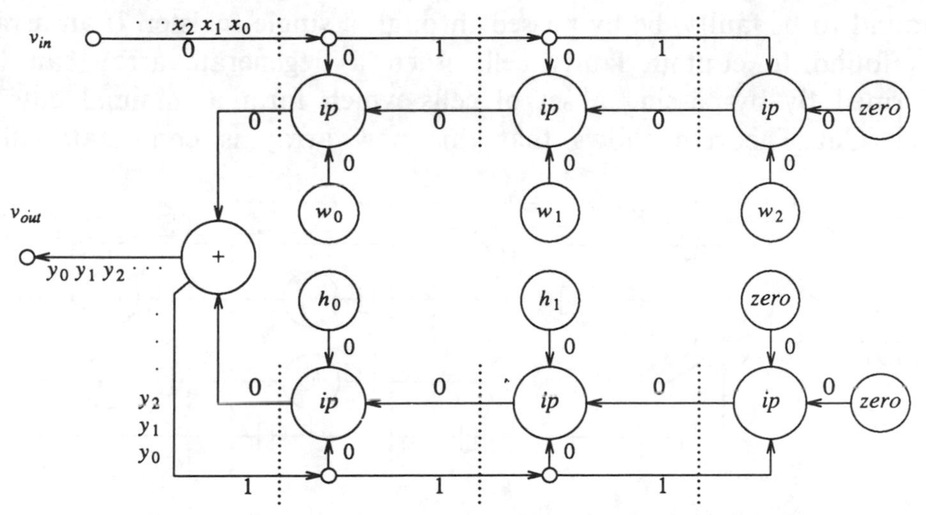

Figure 5 Computation graph for IIR filter (showing cutsets).

the bottom rail. Each cell performs an inner product step with the partial results flowing from left to right along a ripple carry. The results of the rows are then summed using an adder cell. It is easy at this stage to convince oneself that the array performs the correct computation, but because of the channels with zero delay the graph is not systolic.

Since the graph has a cycle of length 6 with total delay 3, it is not possible to retime the graph directly. The first stage is to *two-slow* the graph by multiplying all the delays by two, and so doubling the spacing of the input and output streams. The two-slowed graph can now be retimed by applying the Cut Theorem to the seven disjoint minimal cutsets shown by the vertical dashed lines; a retiming constant of one is used in each case. The delay on all of the channels which are adjacent to neither v_{in} nor v_{out} is now one. The systolic architecture shown in Figure 6 can be derived from the retimed graph. The results appear in every other cycle, and since only 50% of the PEs are active at any one time we need only one inner product step unit per PE. This is an interesting example, as although k-slowing an algorithm is usually considered detrimental to performance, in this particular case two-slowing has decreased by one half the number of inner product processors necessary to implement the algorithm, and so the area needed to lay out the algorithm has been considerably reduced. This is an example of an area/time trade-off.

Kung and Lam introduced the Cut Theorem as a tool for fault tolerant design. Each cell in an array is designed so that it can, if found to be faulty, be by-passed through a single register. If an array is found to contain faulty cells, then a degenerate array can be formed by by-passing a set of cells which form a minimal cutset; the Cut Theorem shows that this new array is computationally

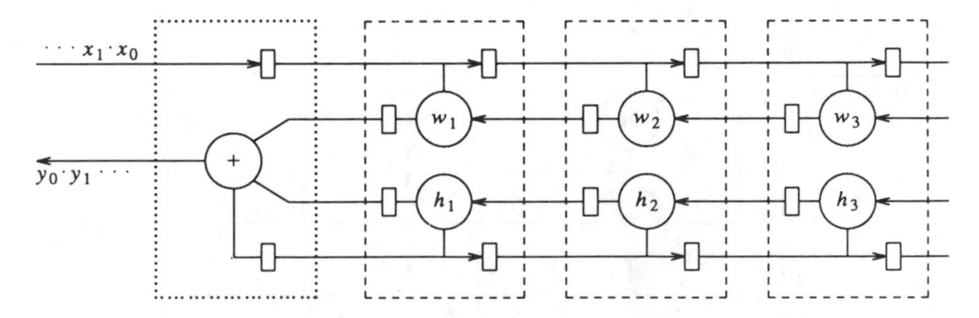

Figure 6 Systolic architecture for IIIR filter.

equivalent to an array similar to the original one, but of a smaller size. However, this technique is only applicable to arrays with unidirectional flow.

3.3 Two-level pipelining systolic algorithms

Many problems from applications such as signal processing require a very high throughput, so that the period of the computation is more important than the computation time itself. Recently the period of such computations has been reduced by the use of processors which are themselves pipelined (e.g., the Weitek 32-bit floating-point chips used in the CMU Warp machine). In the design of algorithms to run on such machines care must be taken to allow for the fact that the elements of the data structure now contain delay, so that their outputs are not just a function of the inputs during that clock cycle. Figure 7 shows a high-level representation of a pipelined adder (the representation is high-level in that the individual stages in the addition algorithm are not represented).

The Cut Theorem is a useful tool in the design of algorithms for such pipelined machines. It is possible to design an algorithm for delay-free processors (as in the previous sections) and then use the Cut Theorem to introduce the necessary delay. The problem is very similar to that of introducing temporal locality, and indeed the following result is essentially a generalisation from $m=1$ of the theorem in the section on Temporal Locality.

LEMMA *Let G be a computation graph in which the cells are delay-free. A necessary and sufficient condition for the existence of a simple retiming of G in which every cell has m pipeline stages is that every cycle C in the graph satisfies* $d^*(C) \geqq m \times |C|$. □

If this condition is not satisfied then the total delay in the graph must first be increased by a suitable factor.

As an example, consider the implementation of the convolution algorithm of Figure 2 on a machine such as the CMU Warp which

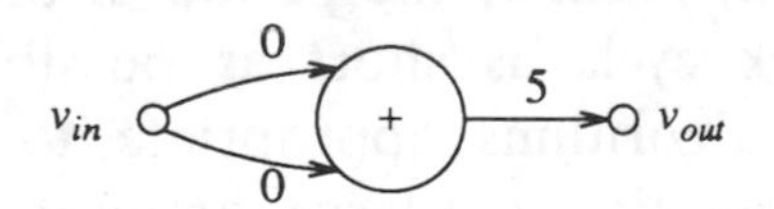

Figure 7 A model of a pipelined adder with five pipelined stages.

has processor elements with five pipeline stages. The first step is to decompose the inner product cells into "plus" and "multiplication" cells. That is, the inner-product element of the data structure is substituted for by a computation graph which computes inner-product (shown in Figure 8), and is expressed over a data structure with more basic functional units. The result of this graph substitution is shown in Figure 9. If the Cut Theorem is now applied to the three vertical disjoint cutsets shown in Figure 9 with a retiming constant of one, then the result is a computation graph for the systolic array shown in Figure 1. To express the algorithm on the pipelined machine a retiming constant of five on the horizontal cut, and four on the vertical cuts must be used. The resulting graph has a delay of five between the plus cells, and this can be 'subsumed' into the plus cells to model a pipelined adder (assuming that the semantics of the model allow delay within the cells). Similarly the delay on the vertical channels can be "subsumed". The result is the computation graph of Figure 10 which models the array shown in Figure 11.

4. SYSTOLIC ARRAYS

Having defined a precise mathematical model of synchronous computation, a natural use for it would be to lay down a list of properties of computation graphs which characterise exactly the systolic algorithms. However, this would appear difficult to do, as five years after Kung advocated the systolic approach to design in *Why Systolic Architectures*, [9], it appears that there is still no concise, definitive and universally accepted characterisation of the phrase *systolic array* (as noted during the panel sessions at the *First International Workshop on Systolic Arrays*, [23]). The original aims put forward by Kung appear to have been the following: (i) to solve the I/O bandwidth problem by ensuring that data items entering the array are "pumped" to as many processors as is possible and appropriate; (ii) to solve the design problem by keeping algorithms *simple* and *regular*; (iii) to make the period of the computation short by keeping the clock cycle as short as possible. These properties have made systolic algorithms appropriate to problems requiring simple regular computation on large amounts of data at a high throughput (e.g. signal processing applications).

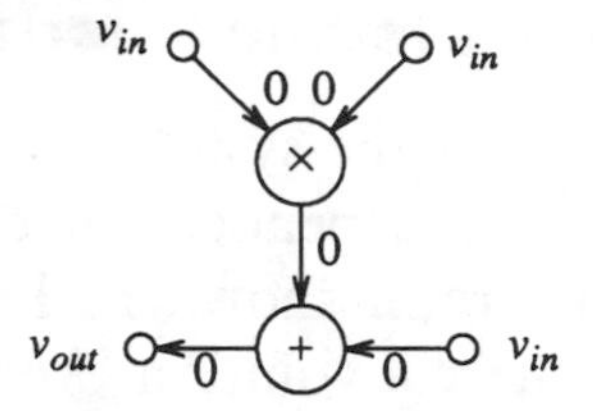

Figure 8 Computation graph for inner product.

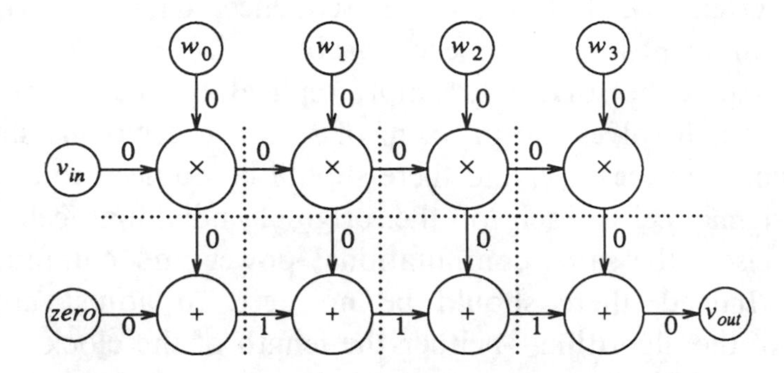

Figure 9. Decomposition of the inner product nodes (with cutsets shown as dotted lines).

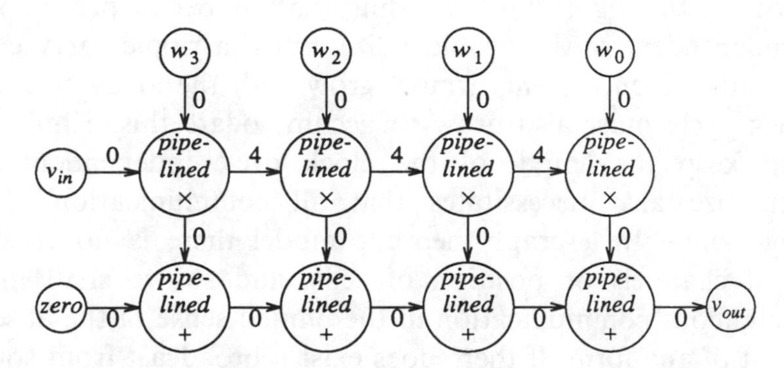

Figure 10. Convolution with pipelined arithmetic cells.

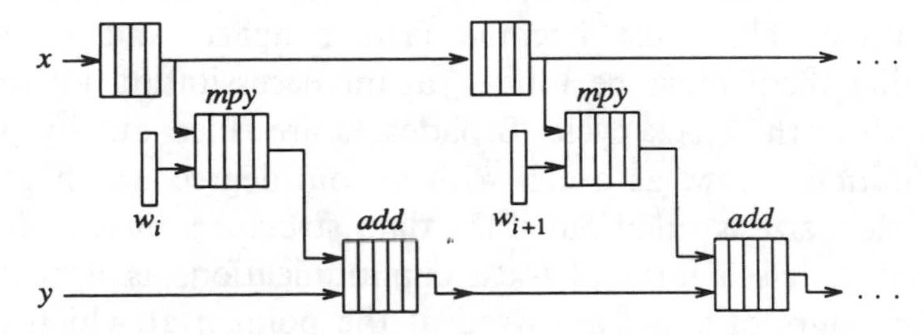

Figure 11. A two-level systolic array algorithm for 1-d convolution (taken from Kung [11]).

4.1 Characteristics of a systolic algorithm

An important characteristic of the systolic solution to a problem is a uniform approach which is independent of the problem size n. For example the computation graph shown in Figure 2 is not intended to be a solution only to the convolution problem with four weights, but rather a typical example from a sequence of arrays for any number of weights. So a systolic algorithm is not a single computation graph G, but rather a sequence $G_1, G_2, \ldots, G_n, \ldots$ of computation graphs—one for each problem size n. The problem of size $n+1$ should be solved by simply replicating one of the cells in the graph which solves the problem of size n. Indeed this should be the only change necessary, and there should be no need to adjust any of the original cells—each of the original individual cells should retain precisely the same computational power and communication topology. Indeed, there should be no need to adjust any other property of the algorithm—neither the length of the clock cycle, nor the I/O bandwidth.

It is this property which necessitates temporal locality. All the properties of the algorithm (excluding the number of processors) are to be independent of the problem size; but if a ripple carry exists in the algorithm then its length must grow with the array size, and so the clock cycle must also grow to accommodate this. Similarly, the need to keep the length of the clock cycle independent of the problem size also necessitates that all communication is *local-neighbour* only. In a graph-theoretic model there is no concept of length of channels or position of cells, and so we shall interpret local-neighbour communication in the limited sense of the absence of broadcast of any form. If there does exist a broadcast from some cell v, then (by definition of a broadcast) as the array size grows the number of cells adjacent to v must increase; as the area of the array grows these cells must become farther apart, and so the wires connecting them must be longer, again necessitating an increase in the length of the clock cycle. Broadcasts are ruled out by part (ii) of the definition below, as a cell with an out-degree which grows with the problem size is ruled out if the data structure must be finite. Also part (iv) models a form of local communication, as it requires that the neighbours of a cell are fixed at the point n at which the cell is added to the array. However, if the algorithm is temporally local, the same processors are used for every problem size, and there is only local-neighbour communication then it is clear that the length of the clock-cycle will be independent of the problem size.

Note that the question of whether an algorithm is systolic cannot be discussed independently of the data structure upon which it is implemented. Consider the computation graph shown in Figure 1(b). This algorithm is not temporally local as the vertical channels have delays of zero; however, if the algorithm was described at a higher level of abstraction by replacing the three cells $\times$, $+$ and w_i in the data structure by a single *inner product* cell then the algorithm would indeed by systolic. We now put the following forward as a definition of systolic.

DEFINITION A systolic solution for the problem P is a sequence $\mathbf{G}=\langle G_{a_n}:n\in\mathbb{N}\rangle$ of computation graphs such that there is one computation graph for each problem size a_n, and the following conditions are satisfied. ($\{a_n:n\in\mathbb{N}\}$ is the set of problem sizes.)

i) For every $n\in\mathbb{N}$, G_{a_n} is a solution to the problem of size a_n.
ii) Each computation graph in the sequence is temporally local.
iii) Each computation graph in the sequence is defined over the same data structure Σ, and should use every element of Σ. Σ can be partitioned into boundary cells and internal cells, and there will typically be only a small number of each. Boundary cells are either adjacent to v_{in} or v_{out} (and perform I/O operations), or they form the end cells in a row (either performing feedback, introducing constants, or initiating or completing carry chains).
iv) For each n, $G_{a_{n+1}}$ can be formed from G_{a_n} by replicating either one or more cells in G_{a_n}, depending on the dimensionality of the array. If $G_{a_{n+1}}$ is formed from G_{a_n} by replicating a constant number (often one) of internal cells then the systolic solution is linear or one-dimensional. If it is necessary to replicate $O(\sqrt{m})$ internal cells (where m is the total number of cells in G_{a_n}) and a constant number of boundary cells then the systolic solution is said to be two-dimensional. In general, if G_{a_n} has $O(m^k)$ cells (with $k\geqq 2$), and $O(m^{k-1})$ internal cells and $O(m^{k-2})$ boundary cells must be replicated to form $G_{a_{n+1}}$ from G_{a_n}, then the solution is said to be k-dimensional. Having formed $G_{a_{n+1}}$ from G_n in this manner, repeated replication of the same cell or cells forms $G_{a_{n+2}}, G_{a_{n+3}}, \ldots$ □

Systolic algorithms have been defined according to the dimensionality of the array upon which the algorithm is defined. This is the standard method of classification, and a further classification called *degenerate two-dimensional array* has also been suggested. A degenerate two-dimensional array is a two-dimensional array in which the size of one of the two dimensions is fixed, and independent of the size of the array (e.g. Figure 9); it is best considered as a hierarchical refinement of a pure one-dimensional array, as can be seen by comparing Figure 9 with Figure 2.

Further classification can be defined through the speed and direction of the data streams which are flowing through the array. This is referred to as the *dataflow* of the algorithm. The dataflow can be described as either *stationary* or *moving*, depending upon whether the result of a computation is accumulated in a cell, or whether a cell evaluates a partial result and passes this on to an adjacent cell. A linear algorithm can be further classified into either *unidirectional* or *bidirectional*, depending on whether the flow of data and results is all in the same direction.

It is well known that many problems which have systolic solutions, have many different systolic solutions. The classic example of this is the original example of convolution as presented by Kung [9]. It is now shown how the Cut Theorem can be of use in showing the equivalence of some of these algorithms.

Consider the algorithm which was originally shown in Figure 2 and is now repeated in Figure 12(a). If the Cut Theorem is applied on the cutsets with a retiming constant of one, then a systolic unidirectional algorithm for convolution is obtained [Figure 12(b)]. Although it is not clear how to immediately transform this systolic algorithm to a bidirectional one, consider the ripple carry in the upper rail of Figure 12(a). This carry simply transmits each data element in the x-stream along the array within a single clock cycle. There is no reason why this ripple-carry should not be transmitted in the opposite direction [Figure 12(c)]; given the synchronous nature of the model, it will make no difference at all to the function computed. The algorithm is now bidirectional, but not temporally local. To make it temporally local (and so systolic), the algorithm is first 2-interleaved [Figure 12(d)], and then the Cut Theorem applied to the same cuts as before, again with a retiming constant of one. So a second, and quite different, systolic algorithm [Figure 12(e)] has been derived from the same starting point.

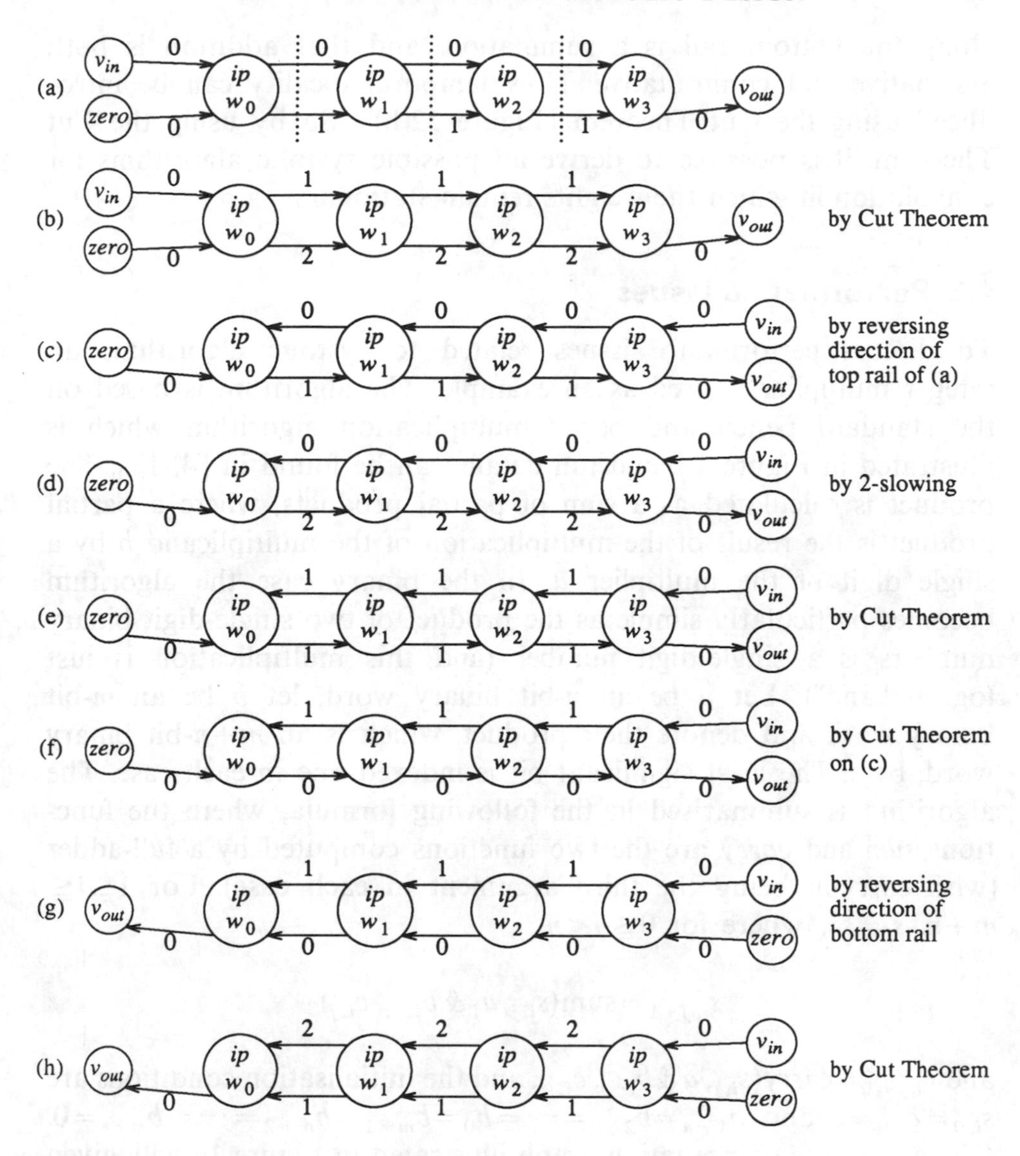

Figure 12. The development of algorithms for 1-d convolution. The top rail carries the input stream and the bottom rail the results.

The algorithm in Figure 12(e) is not a good algorithm for convolution in that it has been 2-slowed and it is bidirectional. It is possible to use these techniques to return to a unidirectional flow which is different to the original algorithm. By using the Cut Theorem an algorithm with a ripple carry along the bottom rail can be produced [Figure 12(f)]. It is now possible to reverse the direction of the data flow along the bottom rail [Figure 12(g)]. That this can be done depends crucially on the fact that the computation along the bottom rail is a summation, and that addition is both associative and commutative. Now temporal locality can be introduced using the Cut Theorem [Figure 12(h)]. So by using the Cut Theorem, it is possible to derive all possible systolic algorithms for convolution in which the weights remain stationary.

4.2 Performance issues

To discuss performance issues related to systolic algorithms an integer multiplier is used as an example. The algorithm is based on the standard pencil and paper multiplication algorithm which is illustrated in Figure 13, and full details can be found in [4, 17]. The product is calculated as a sum of partial products, where a partial product is the result of the multiplication of the multiplicand b by a single digit of the multiplier a. In the binary case the algorithm becomes particularly simple as the product of two single-digit binary numbers is a single-digit number (and this multiplication is just logical "and"). Let a be an n-bit binary word, let b be an m-bit binary word and denote their product, which is an $m+n$-bit binary word, by s. The least significant bit is indexed one in each case. The algorithm is summarised in the following formula, where the functions *sum* and *carry* are the two functions computed by a full-adder (with carry-in being the third argument in each case). For $1 \leqq i \leqq m+n$, $s_i = s_{i,n}$ where for $0 \leqq j < n$,

$$s_{i,\,j+1} = \text{sum}(s_{i,\,j}, a_j \,\&\, b_{i-j}, c_{i,\,j})$$

and $c_{i,j+1} = \text{carry}(s_{i,j}, a \,\&\, b_{i-j}, c_{i,j})$, and the initialisation conditions are $s_{i,0} = c_{i,0} = 0$ and $b_{1-n} = b_{2-n} = \cdots = b_0 = b_{m+1} = b_{m+2} = \cdots = b_{m+n} = 0$. Given $n=2$ the computation graph illustrated in Figure 14 will, given the initialisation conditions, perform this algorithm.

b	4756
a	223
	14268
	9512
	9512
s	1060588

Figure 13. A standard pencil and paper multiplication in denary (carries not shown).

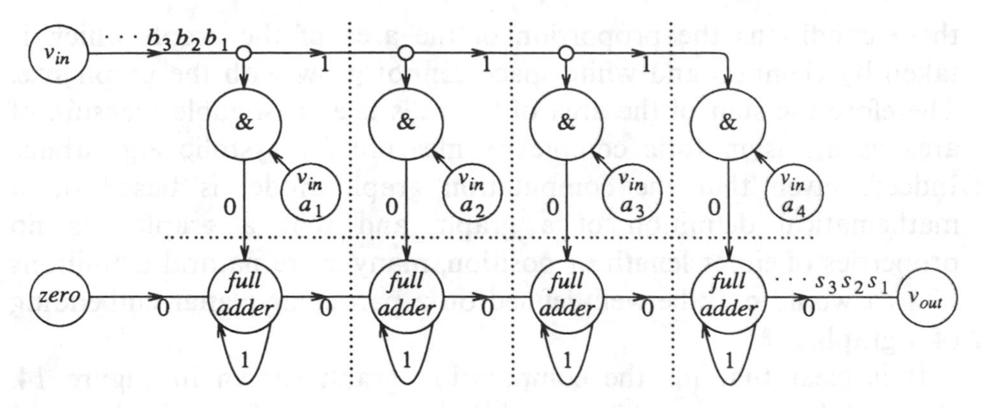

Figure 14. A computation graph for a serial-parallel integer multiplier (showing cutsets).

The standard performance measures for parallel algorithms are *computation time*, *T*, *area*, *A* and *period*, *P*. If comparisons are to be made between different computation graphs, then the performance measures must be absolute, rather than, for example, clock cycles for time; however, in a synchronous model such as the one used here, the computation time will be the product of the number of clock cycles and the length of the clock cycle. Each cell in the data structure is assigned a value for computation time, area and (if two-level pipelining is to be considered) period. A complexity model can be developed so that from these values performance measures for any computation graph on that data structure can be calculated. In constructing this complexity model decisions have to be made on such problems as how to charge for time of transmission along a channel, and for the area occupied by a channel. These are serious problems which have been discussed in the complexity theory of VLSI (e.g. Brent and Kung [1]) but no solutions are put forward here.

The computation time is the time between the first bit, a_1 or b_1, being input at v_{in} and the last bit, s_{m+n}, being output at v_{out}. The period of the computation graph is the time between two successive b_1's being input at v_{in} in a stream of multiplicands. A simple measure of area would be the sum of the areas of the cells used in the graph. As a general model of VLSI complexity this would not be satisfactory, as it does not take account of any charge for area for channels or for white space. However, the definition of systolic bans broadcasts and non-local-neighbour communication, and under these conditions the proportion of the area of the graph which is taken by channels and white space cannot grow with the graph size. Therefore the sum of the area of the cells is a reasonable measure of area as an asymptotic complexity measure for systolic algorithms. Indeed, given that the computation graph model is based on a mathematical definition of a graph, and that a graph has no properties of either length or position, many more natural definitions of area would only be well-defined on a particular planar embedding of a graph.

It is clear that for the computation graph shown in Figure 14, $A=\mathrm{O}(n)$, because as n is increased the number of columns in the array must be increased accordingly. Also, given that s is output serially, it follows that both the computation time and the period are at least $m+n$ clock cycles. Now the ripple-carry on the lower row of the array requires that the clock cycle must increase linearly with n, and so it follows that $T=\mathrm{O}((m+n)n)$ and $P=\mathrm{O}((m+n)n)$.

We can use the Cut Theorem to improve the performance of this algorithm by applying the theorem on the four cutsets shown with a retiming constant of one in each case. The length of the clock cycle is now independent of n, and so $P=\mathrm{O}(m+n)$. However the retiming has increased the latency in the system in that it is no longer true that the input of b_1 and the output of s_1 occur in the same clock cycle; there is a delay of n clock cycles between b_1 being input and s_1 being output. Therefore the computation time is $m+2n-1$ clock cycles and $T=\mathrm{O}(m+n)$.

As presented the computation graph is not systolic as the n bits of the word a are broadcast to the n & cells. An alternative would be to read a in serially, and preload the bits so that effectively the a_i cells become constant cells rather than input cells. The preload requires at most n clock cycles, and so will not affect the asymptotic performance measures. The final computation graph is shown in Figure 15.

It is natural to enquire whether the performance of this computation graph is in any sense optimal. In the literature on VLSI complexity the AT^2 performance measure has been accepted as a reasonable performance measure for VLSI algorithms. If $m=n$ then the algorithm under consideration satisfies $AT^2=\mathrm{O}(n^3)$. Brent and Kung [1], have shown that in a comparable model of computation multiplication of two n-bit binary words requires $AT^2=\Omega(n^2)$.

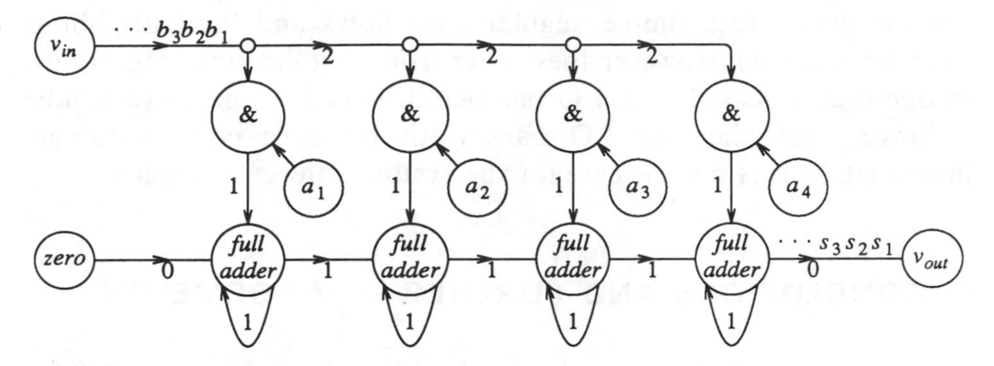

Figure 15. A systolic serial-parallel integer multiplier.

However, if we restrict the problem of multiplication to those solutions where I/O is carried out serially (so that the I/O bandwidth does not increase with n) then it is clear that $T=\Omega(n)$. Also, if we accept that to solve the problem of multiplication $a_i \,\&\, b_j$ must be calculated for each bit a_i of a and each bit b_j of b at some time during the computation, then it follows that either each bit of a or each bit of b must be stored on the graph (i.e. no replication of inputs), and so $A=\Omega(n)$. Therefore under these conditions multiplication requires $AT^2=\Omega(n^3)$, and so the algorithm under consideration is optimal under this performance measure.

The above discussion is of importance because it illustrates the crucial relationship between the I/O bandwidth and the performance of an algorithm. The basic requirement that the systolic array algorithm is to have limited I/O bandwidth to and from the array affects the performance. By relaxing the I/O restriction we can obtain faster multiplier circuits using a *hierarchical algorithm* [16, 28]. These algorithms are based on the observation that integer multiplication can be rewritten as:

$$s=2^n ce+2^{n/2}(cf+de)+df$$

where $a=2^{n/2}c+d$, and $b=2^{n/2}e+f$. Then n-bit multiplication has been replaced by four $n/2$-bit multiplications. A complete layout of the multiplier occupies $O(n^2 \log^2 n)$ area and has a computational time of $O(\log^2 n)$, [28], giving $AT^2=O(n^2 \log 6n)$. A 2-multiplication scheme (recursing on the a only) achieves a time $O(\log n)$, [16]. These hierarchical algorithms enjoy many of the advantages of systolic arrays (e.g. simple, regular data flows and local neighbour interconnections). However they differ from systolic array algorithms in one crucial aspect—the I/O bandwidth is O($2n$), and so to handle a longer multiplier the I/O bandwidth must be correspondingly increased. This is not the case for the systolic integer multiplier.

5. CONCLUSIONS AND FURTHER DEVELOPMENTS

We have described a formal model of synchronous computations within which it is possible to define systolic algorithms, and certain retiming transformations of synchronous algorithms. These transformations form a design tool which can be applied to any synchronous algorithm, and is intended to help the designer by abstracting away some of the many burdensome timing details involved in synchronous design. The designer should express the algorithm in some simple format where it is easy to check that the right data is in the right place at the right time; then through use of these formally verified tools, it is possible to transform this algorithm into one with the desired timing properties, without any further verification being necessary. Using these techniques any reasonable synchronous algorithm can be transformed to be temporally local. The Cut Theorem in particular is a useful tool, as it is easy to understand and straightforward to apply. The theorem can be used as a tool in the transformation of a bidirectional systolic algorithm into a unidirectional one, and vice versa. This is because the direction of a ripple carry can in some circumstances be reversed, and the Cut Theorem can be used to introduce ripple carries so that the direction can be reversed, and then used again to remove the ripple carry.

The Cut Theorem is also an invaluable tool in the design of algorithms for pipelined processors. The design of large systolic systems involving two-level pipelining (or indeed, many levels of pipelining) requires a top-down approach to design. An algorithm is initially defined at a high level of abstraction, and then systematically refined step by step through implementing further details of the data structure. For top-down design to be carried out within a formal model, the model will need to support hierarchical definition. The current model can be so extended by using a recursive definition of a computation graph, $G=\langle D, gf, cn, st, d\rangle$, in which gf can associate with each cell of D a computation graph from any level lower in the recursion. The data structure itself forms the bottom level in the recursion. Such an extension to the definition is discussed in McEvoy [19].

As a computation graph has already been defined over an arbitrary data structure, it is possible to consider the refinement of data structures (either the refinement of data elements or implementation of function elements) in similar terms. However, this generalisation of the model will require one small but important change in the semantics of the data structure. Function refinement is to be modelled by the substitution of primitive computation graphs by computation graphs, and the behaviour of a computation graph, G, has been defined as a function $\Omega_G\colon \mathrm{IN}_G^+ \to \mathrm{OUT}_G$. Therefore if a computation graph is to be substituted for a primitive computation, then in general a primitive computation graph, σ, must be allowed the full history-sensitive functionality of $\sigma_A\colon \mathrm{IN}_\sigma^+ \to \mathrm{OUT}_\sigma$, rather than the purely functional $\sigma_A\colon \mathrm{IN}_\sigma \to \mathrm{OUT}_\sigma$ (where σ_A is the particular semantic interpretation of σ). These ideas extend naturally to describe the computational behaviour of computation graphs as operators on strings or streams (infinite strings) of data.

References

[1] R. P. Brent and H. T. Kung, Area-time complexity of binary multiplication, *J. of Association for Computing Machinery* **28**(3) (July 1981), 521–534.

[2] S. D. Brookes, *Reasoning About Synchronous Systems*, Department of Computer Science Report CMU-CS-84-145, Carnegie-Mellon University, Pittsburgh, USA, 1984.

[3] M. C. Chen, A design methodology for synthesizing parallel algorithms and architectures, *Jrnl of Parallel and Distributed Computing* 3 (1986), 461–491.

[4] P. D. Danielsson, The serial-parallel convolver, *IEEE Trans. on Computers* **C-33** (1984), 652–667.

[5] J. Delosme and I. Ipsen, Efficient systolic arrays for the solution of Toeplitz systems: An illustration of a methodology for the construction of systolic architectures in VLSI. In: *Systolic Arrays, Papers presented at the First International Conference on Systolic Arrays*, Oxford, July 1986 (W. Moore, A. McCabe and R. Urquhart, eds.), Adam Hilger, 1987, pp. 37–46.

[6] D. J. Evans, Designing efficient systolic algorithms for VLSI parallel processor arrays. In: *Parallel Computers and Computer Vision* (M. Brady and I. Page, eds.), Oxford University Press (to appear).

[7] J. A. B. Fortes and D. I. Moldovan, Parallelism detection and transformation techniques useful for VLSI algorithms, *Jrnl of Parallel and Distributed Computing* **2** (1985), 277–301.

[8] N. A. Harman and J. V. Tucker, *The Formal Specification of a Digital Correlator*, Centre for Theoretical Computer Science Report, Leeds University, Leeds, UK, 1987.

[9] H. T. Kung, Why systolic architectures, *IEEE Computer* **15**(1) (January 1982), 37–46.

[10] H. T. Kung and W. T. Lin, An algebra for systolic computation. In: *Elliptic Problem Solvers 2* (*Conf. Proc.*) (G. Birkhoff and A. Schoenstadt, eds.), Academic Press, 1984, pp. 141–160.

[11] H. T. Kung, Systolic algorithms for the CMU Warp processor. In: *Proceedings of the Seventh International Conference on Pattern Recognition*, International Association for Pattern Recognition, July 1984, pp. 576–577.

[12] H. T. Kung and M. S. Lam, Wafer-scale integration and two-level pipelined implementations of systolic arrays, *Journal of Parallel and Distributed Computing* **1**(1) (1984), 32–63.

[13] S. Y. Kung, On supercomputing with systolic/wavefront array processors, *Proceedings of IEEE* **72**(7) (July 1984), 867–884.

[14] C. E. Leiserson and JB. Saxe, Optimizing synchronous systems, *J. VLSI and Computer Systems* **1**(1) (1983), Computer Science Press, pp. 41–67.

[15] G.-J. Li and B. W. Wah, The design of optimal systolic arrays, *IEEE Trans. on Computers* **c-34**(1) (January 1985).

[16] W. K. Luk and J. E. Vuillemin, Recursive implementation of fast multipliers. In: *Prov. of VLSI 83* (F. Anceau and E. J. Aas, eds.), North-Holland, 1983.

[17] R. F. Lyon, Two complement pipeline multipliers, *IEEE Trans. on Comm.* **COM-24**(4) (April 1976), 418–425.

[18] A. R. Martin and J. V. Tucker, The concurrent assignment representation of synchronous systems. In: *Parallel Architectures and Languages Europe: Parallel Languages*, Vol. II (A. J. Nijman and P. C. Treleaven, eds.), Springer-Verlag Lecture Notes in Computer Science 259 (1987), pp. 369–386.

[19] K. McEvoy, *A Formal Model for the Hierarchical Design of Synchronous and Systolic Algorithms*, Report 7.86, Centre for Theoretical Computer Science, The University, Leeds, June 1986.

[20] G. Megson, *Novel Algorithms for the Soft-Systolic Paradigm*, Ph.D. Thesis, Loughborough University of Technology, UK (1987).

[21] R. G. Melhem and W. C. Rheinboldt, A mathematical model for the verification of systolic networks, *SIAM Journal of Computing* **13**(3) (August 1984).

[22] D. I. Moldovan, ADVIS: A software package for the design of systolic arrays, *IEEE Trans. on CAD* **CAD-6** (1987), 33–40.

[23] W. Moore, A. McCabe and R. Urquhart (eds.), *Systolic Arrays*, Papers presented at the First International Conference on Systolic Arrays, Oxford, July 1986. Adam Hilger, 1987.

[24] P. Quinton, An introduction to systolic architectures. In: *Future Parallel Computers* (P. Treleaven and M. Vanneschi, eds.), Springer-Verlag Lecture Notes in Computer Science 272, 1987, pp. 387–400.

[25] L. R. Rabiner and B. Gold, *Theory and Application of Digital Signal Processing*, Prentice-Hall, 1975.

[26] M. Sheeran, *The Design and Verification of Regular Synchronous Circuits*, Programming Research Group Report PRG-39, Oxford, University Computing Laboratory, Oxford, 1985.

[27] B. C. Thompson and J. V. Tucker, Theoretical considerations in algorithm design. In: *Fundamental Algorithms for Computer Graphics* (R. A. Earnshaw, ed.), NATO ASI Series, Vol. F17. Springer-Verlag, 1985, pp. 855–878.

[28] H. C. Yung, C. R. Allen, H. K. E. Liesenberg and D. J. Kinniment, A recursive design methodology for VLSI: Theory and example, *Integration* **2** (1984), 213–225.

POLYNOMIAL AND ROOT FINDING METHODS

A Systolic Ring Architecture for Solving Polynomial Equations

K. MARGARITIS and D. J. EVANS

Department of Computer Studies, Loughborough University of Technology, Loughborough, Leicestershire, UK

In this paper some systolic designs are presented for the implementation of Bernoulli's method for polynomial root solving. From a linear systolic array with feedback a systolic ring is derived that calculates the coefficients of Newton's theorem. The ring is incorporated in a systolic system for calculating the dominant zeros of polynomial equations. The design has been simulated soft-systolically in an OCCAM program.

KEY WORDS: Systolic ring, Bernoulli's method, zeros of polynomial equations.

C.R. CATEGORIES: F1.1, F2.1, B7.1, G1.5.

1. INTRODUCTION

The method of Bernoulli for the calculation of dominant zeros of polynomial equations of the form,

$$f(x)=x^n+a_1x^{n-1}+a_2x^{n-2}+\cdots+a_n=0. \tag{1}$$

is summarised in [3] as follows:

Consider the Eq. (1) for a_i, $i=1,2,\ldots,n$ all real, and $r_1,r_2,\ldots,r_n$ the roots of the equation, i.e.,

$$f(x)=(x-r_1)(x-r_2)\cdots(x-r_n) \tag{2}$$

then from Newton's theorem we have if,

$$s_p = r_1^p + r_2^p + \cdots + r_n^p, \tag{3}$$

then,

$$s_k = -a_1 s_{k-1} - a_2 s_{k-2} - \cdots - a_{k-1} s_1 - a_k k, \quad k = 1, 2, \ldots, n \tag{4}$$

and,

$$s_{n+j} = -a_1 s_{n+j-1} - a_2 s_{n+j-2} - \cdots - a_n s_j, \qquad j = 1, 2, \ldots . \tag{5}$$

Initially, it is assumed that,

$$|r_1| > |r_2| \geqq |r_3| \geqq \cdots \geqq |r_n|. \tag{6}$$

Then, from (3), s_p can be written as,

$$s_p = r_1^p \left\{ 1 + \left(\frac{r_2}{r_1} \right)^p + \cdots + \left(\frac{r_n}{r_1} \right)^p \right\}; \tag{7}$$

hence,

$$s_p / s_{p-1} \xrightarrow[p \to \infty]{} r_1 . \tag{8}$$

Therefore in this case, the method produces the dominant zero r_1 with successive divisions of the coefficients $s_1, s_2, \ldots, s_p$, as in (8).

Now for the case

$$r_2 = \bar{r}_1 \quad \text{and} \quad |r_2| > |r_3| \geqq |r_4| \geqq \cdots \geqq |r_n|, \tag{9}$$

then the method produces the two dominant roots r_1 and r_2, as follows:

Let

$$\begin{aligned} T_p &= s_p s_{p-2} - S_{p-1}^2, \\ U_p &= s_p s_{p-3} - s_{p-1} s_{p-2}; \end{aligned} \tag{10}$$

then,

$$
\begin{aligned}
T_p/T_{p-1} &\xrightarrow[p\to\infty]{} v^2,\\
U_p/T_{p-1} &\xrightarrow[p\to\infty]{} 2v\cos\theta,
\end{aligned}
\tag{11}
$$

where,

$$
\begin{aligned}
r_1 &= v(\cos\theta + i\sin\theta),\\
r_2 &= v(\cos\theta - i\sin\theta).
\end{aligned}
\tag{12}
$$

The cases where,

$$r_1 = r_2 = \cdots = r_k, \quad |r_k| > |r_{k+1}|, \tag{13}$$

and

$$
\begin{aligned}
r_1 = r_2 = \cdots = r_k &= v(\cos\theta + i\sin\theta),\\
r_{k+1} = r_{k+2} = \cdots = r_{2k} &= v(\cos\theta - i\sin\theta),\\
|r_{2k}| &> |r_{2k+1}|,
\end{aligned}
\tag{14}
$$

are treated in a similar manner to (6) and (9) respectively.

If

$$r_1 = r_2 = \cdots = r_k,\ r_{k+1} = r_{k+2} = \cdots = r_{k+l},\ r_1 = -r_{k+1},\ |r_{k+l}| > |r_{k+l+1}|, \tag{15}$$

then

$$s_p/s_{p-2} \xrightarrow[p\to\infty]{} r_1, \text{ for } p = 2m. \tag{16}$$

If (1) has three or more distinct zeros having the same absolute value Bernoulli's method fails. However, the substitution $x = y - b$, $b \neq 0$ and real, will usually result in a polynomial in y, with no three roots having the same absolute value. A zero of the transformed polynomial equation is a zero of the original equation increased by b.

As a rule, the method converges slowly, and, as p grows, an overflow in the numerical calculations may occur. On the other hand Bernoulli's method produces with relatively simple and regular computation the dominant root of a polynomial equation, especially when it is significantly larger than any of the others [2].

The method can be repeated if the original polynomial is divided by $(x-r_1)$ or $(x-r_1)(x-r_2)$ using Horner's scheme giving a polynomial of degree $n-1$ or $n-2$, to which the method is then applied.

In the following sections systolic designs for the calculation of the coefficients $s_1, s_2, \ldots, s_p$ from (4) and (5) are discussed and an appropriate systolic ring is described in detail. Finally, the systolic ring is integrated into a systolic polynomial equation solver.

2. SYSTOLIC DESIGN DERIVATION

From the description of the Bernoulli method it is obvious that the main computational effort is to produce the coefficient $s_1, s_2, \ldots, s_p$ using the recurrences (4) and (5). Then some post-processing may be necessary and finally a series of divisions to produce the dominant zero(s).

Now recurrences (4) and (5) can be unified in a form more suitable for systolic implementation as follows:

$$y_i^{(1)}=0$$

$$y_i^{(k+1)}=y_i^{(k)}-a_k s_{i-k}, \quad k=1,2,\ldots,n \quad \text{and} \quad s_0=s_{-1}=\cdots=0 \tag{17}$$

$$s_i=y_i^{(n+1)}-a_i i, \quad i=1,2,\ldots \quad \text{and} \quad a_{n+1}=a_{n+2}=\cdots=0.$$

For example, we have for $n=5$,

$$s_1=-a_1 1$$

$$s_2=-a_1 s_1-a_2 2$$

$$s_3 = -a_1 s_2 - a_2 s_1 - a_3 3$$

$$s_4 = -a_1 s_3 - a_2 s_2 - a_3 s_1 - a_4 4 \quad (18)$$

$$s_5 = -a_1 s_4 - a_2 s_3 - a_3 s_2 - a_4 s_1 - a_5 5$$

$$s_6 = -a_1 a_5 - a_2 s_4 - a_3 s_3 - a_4 s_2 - a_5 s_1$$

$$s_7 = -a_1 s_6 - a_2 s_5 - a_3 s_4 - a_4 s_3 - a_5 s_2, \quad \text{etc.}$$

Clearly, (17) is a recurrence with feedback cycles, i.e. the value of s_i depends on $s_{i-1}, s_{i-2}, \ldots, s_1$. This type of recurrence is readily realised as a linear systolic network with bidirectional dataflow, as in Figure 1 (see also [4, 5]). The partial sums are accumulated by y_i while it travels leftwards along the array, until it reaches the feedback cell; there s_i is formed and it travels in the reverse direction to produce the inner products for the following y's and finally to form the output of the array.

s_0 can be interpreted as a control flag that enables the storage of the coefficients $a_1, a_2, \ldots, a_n$ in the appropriate cells, and the accumulation of the values $a_1 1, a_2 2, \ldots, a_n n$ as the partial sums for the corresponding y_i. Thus a data sequence of the form:

$$s_0, a_1, a_1 1, a_2, a_2 2, \ldots, a_n, a_n n,$$

pumped into the array through the y input stream for the first $2n+1$ cycles can set up the array in a systolic manner.

The linear array requires n IPS cells and produces its first valid output (s_1) after $2n+2$ time units, and from then on one valid output every two cycles, i.e. it has a throughput rate of $\frac{1}{2}$. Similarly the processor utilisation is $\frac{1}{2}$ since only half of the cells are active at any one time.

In addition to the low throughput rate and low processor utilisation this systolic network has the additional disadvantage of the bidirectional data flow (see also [1, 4, 6]). This fact makes the implementation of the array using WSI techniques or programmable systolic components difficult, because the performance of the array degrades rapidly with respect to the number of consecutive failed cells that need to be tolerated.

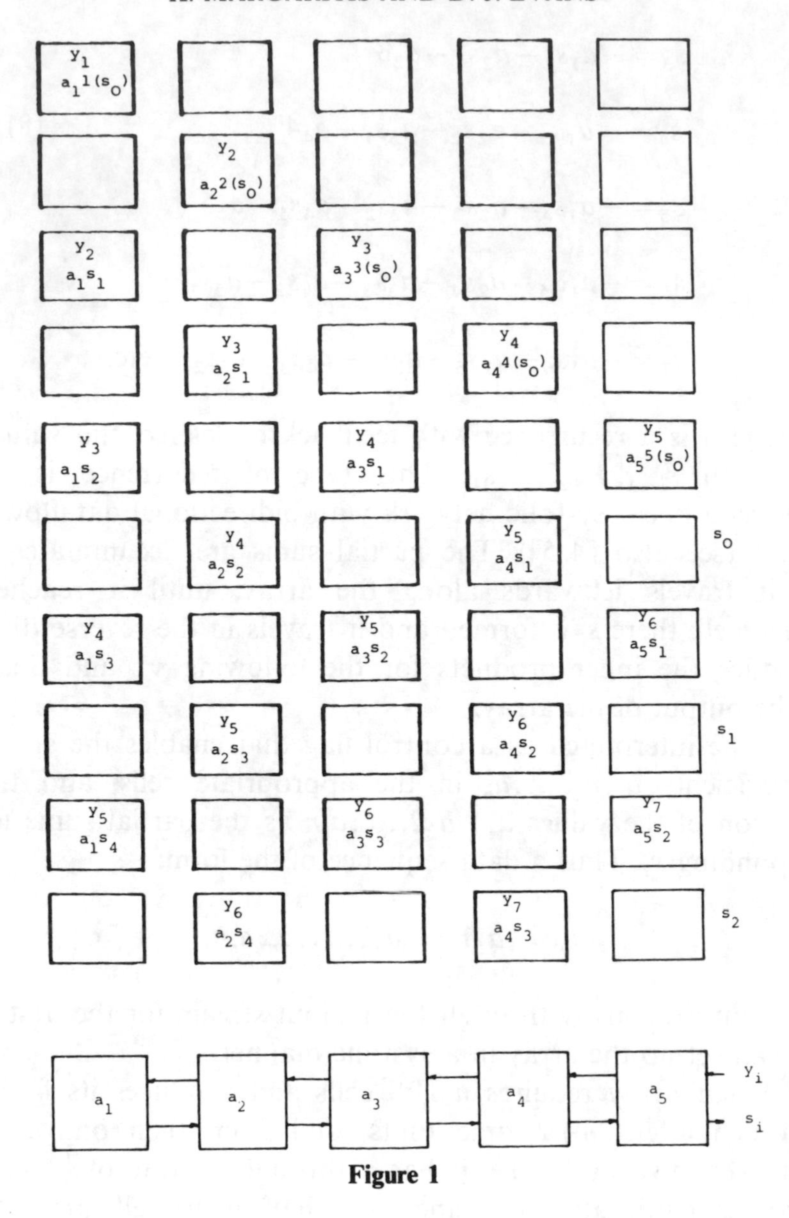

Figure 1

The recurrence relation (18) of size 6 computed by a 5-cell bidirectional linear array in Figure 1 can also be implemented on a 3-cell ring with unidirectional data flow, as in Figure 2 [4]. This result can be derived more systematically if (17) is reformulated as follows:

$$y_i^{(1)} = -a_i i \quad i = 1, 2, \ldots \quad \text{and} \quad a_{n+1} = a_{n+2} = \cdots = 0$$

$$y_i^{(k+1)} = y_i^{(k)} - a_k s_{i-k}, \quad k=1,2,\ldots,n \quad \text{and} \quad s_0 = s_{-1} = \cdots = 0 \tag{19}$$

$$s_i = y_i^{(n+1)}$$

The example of (18) is now written as:

$$\begin{aligned}
s_1 &= -a_1 1 \\
s_2 &= -a_2 2 - a_1 s_1 \\
s_3 &= -a_3 3 - a_2 s_1 - a_2 3 \\
s_4 &= -a_4 4 - a_3 s_1 - a_2 s_2 - a_1 s_3 \\
s_5 &= -a_5 5 - a_4 s_1 - a_3 s_2 - a_2 s_3 - a_1 s_4 \\
s_6 &= \qquad -a_5 s_1 - a_4 s_2 - a_3 s_3 - a_2 s_4 - a_1 s_5 \\
s_7 &= \qquad\qquad -a_5 s_2 - a_4 s_3 - a_3 s_4 - a_2 s_5 - a_1 s_6, \quad \text{etc.}
\end{aligned} \tag{20}$$

It is clear that (19) represents an unrolled ring-like computation; this is more obvious if in s_6 the partial sum $a_1 s_5$ is placed in the leftmost end; and the same happens for $-a_2 s_5$ and $-a_1 s_6$ of s_7: then each column will contain a cyclic sequence of $a_1, a_2, \ldots, a_5$ where all coefficients of a cycle are multiplied with the same s_i; for the next cycle s_i is then replaced with s_{i+n} ($n=5$ here). Again, it is assumed that the computation $-a_1 i$ is replaced by $-a_i s_0$.

The systolic ring of Figure 2 works as follows. The 3 most recently computed results are stored in each of the 3 cells, while the next 3 partial sums travel around the ring to meet these stored values; together with them the coefficients travel around the ring with half the speed of the partial sums. Every two cycles a sum is completed and a new computation begins; the completed sum takes the place of the "oldest" stored result that is produced as an output (denoted by "*" in Figure 2).

The dummy coefficient a_0 controls the output and storage operations of the ring while the initial values $-a_1 1$, $-a_2 2, \ldots, -a_n n$ can

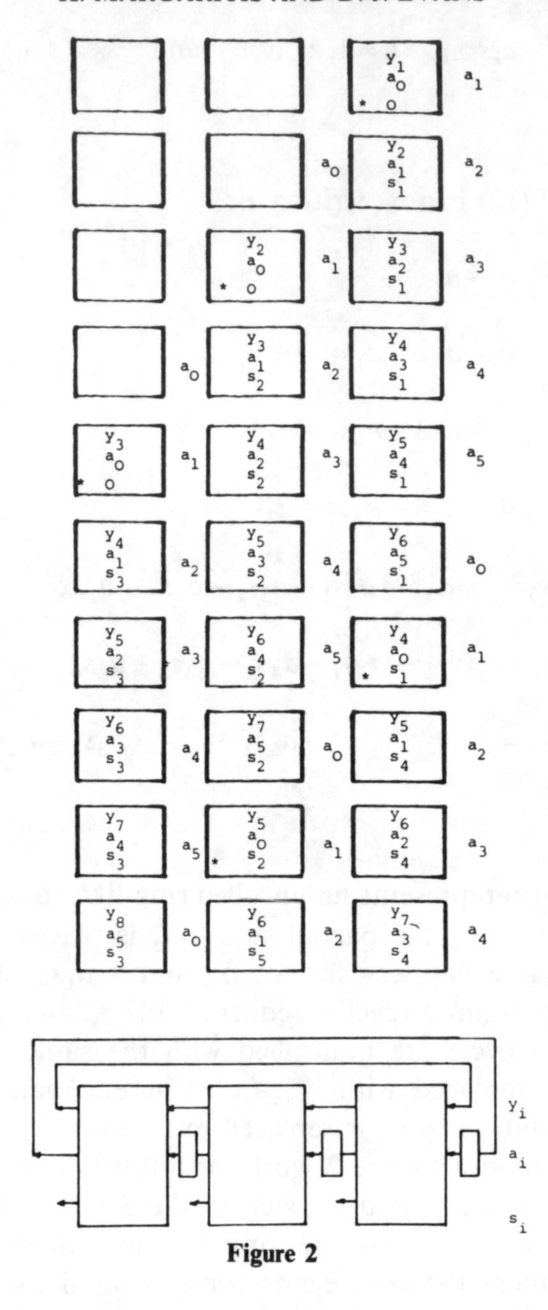

Figure 2

be input onto the ring simultaneously with coefficients $a_1, a_2, \ldots, a_n$ as is explained in the next section.

The systolic ring requires only half of the cells of the linear array and all the cells are active at all times, i.e. the processor utilisation is 1, the throughput rate remains $\frac{1}{2}$ and now the output of the network is collected from all the cells. However the basic advantage of the ring is that it degrades gracefully as the number of detective cells increases: due to the fact that the data flow is now unidirectional the fault tolerance techniques described in [4] can now be applied, for the hardware implementation of the network using WSI techniques or programmable systolic components.

3. IMPLEMENTATION DETAILS

The systolic ring operation for a recurrence of size 5 (odd) is shown in Figure 3. A dummy (zero) coefficient enters the ring to synchronise the calculations in a manner similar to that of Figure 2 (for even recurrence size). In general for a polynomial equation (1) of degree n, the recurrence (19) has size $n+1$ and the systolic ring that implements it requires $(n+1)/2$ IPS cells for $n=$odd and $(n+2)/2$ IPS cells for $n=$even; in addition an equal number of delays is required to accommodate all the coefficients of the polynomial.

During the first $n+1$ cycles, for $n=$odd, or $n+2$ cycles, for $n=$even, the initial values enter the ring as illustrated in Figure 3, from then the normal ring operation is resumed and the first valid result is produced when the polynomial coefficients complete a full circle on the ring. The dummy coefficient a_0 serves as a controlling signal that enables the output of s_1 and the storage of s_{i+n-1} in its place.

Although the output is collected from all the cells of the ring, the valid results can be collected systolically as shown in Figure 4, with the addition of a series of linearly interconnected 2-input and 1-output registers. The inputs of these registers are again controlled by a_0: thus when a valid output from cell i occurs the corresponding register accepts this output; otherwise it propagates the data item received from the preceding register. Thus, after an initial delay, the results are produced in groups of $(n+1)/2$. for $n=$odd, or $(n+2)/2$, for $n=$even, with intermediate intervals of dummy output of the same length.

In Figure 5 an overall system for the implementation of Bernoulli's method is illustrated schematically.

The multiplexer (MUX) "closes" the ring after the initial input cycles; it can be either data-driven, i.e. controlled by the a_0 flag or

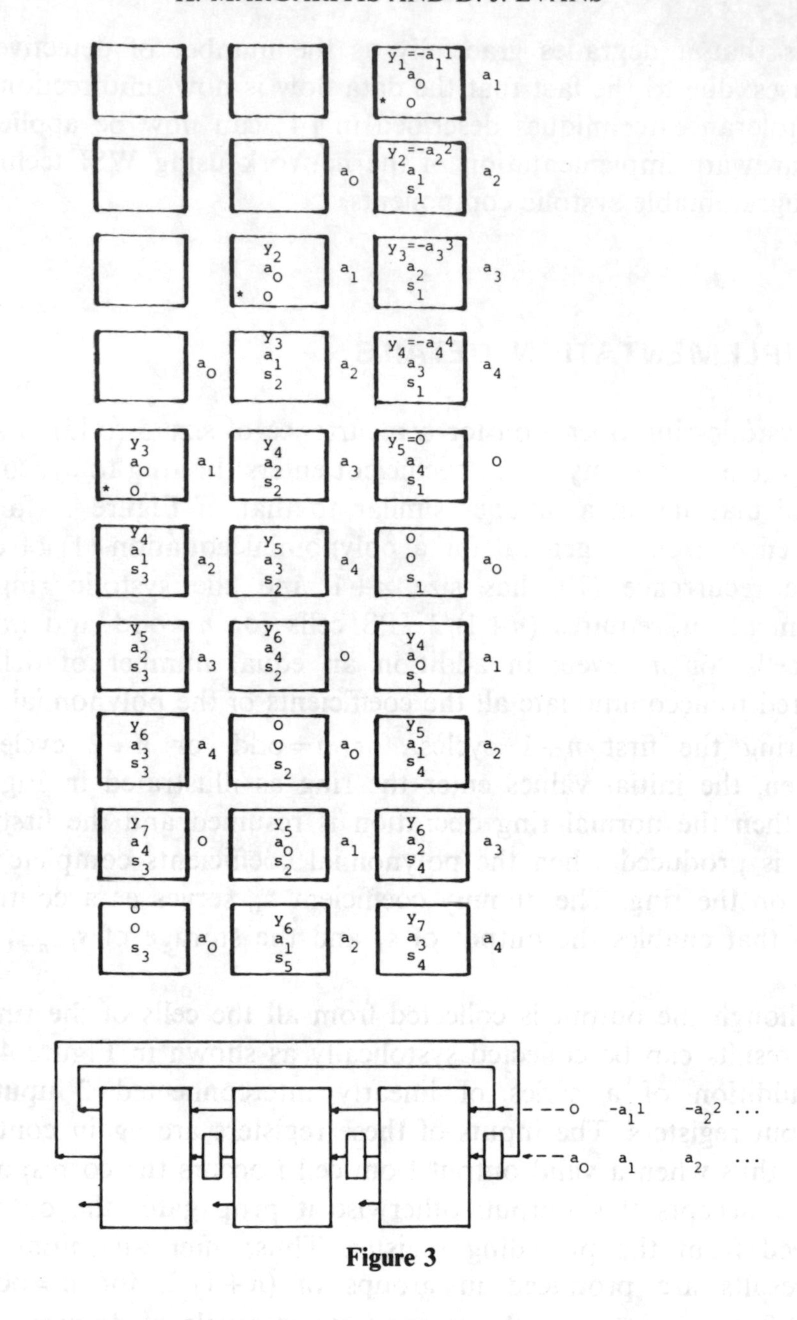

Figure 3

programmed from the host machine. The post-processing element (PP) implements one of the relations (8), (11)–(12), (16), i.e. formulates the successive approximations of the dominant zero(s) of the polynomial equation. In the case of relation (8), PP is a simple divider with registers to store s_p, s_{p-1} and data-driven or host-

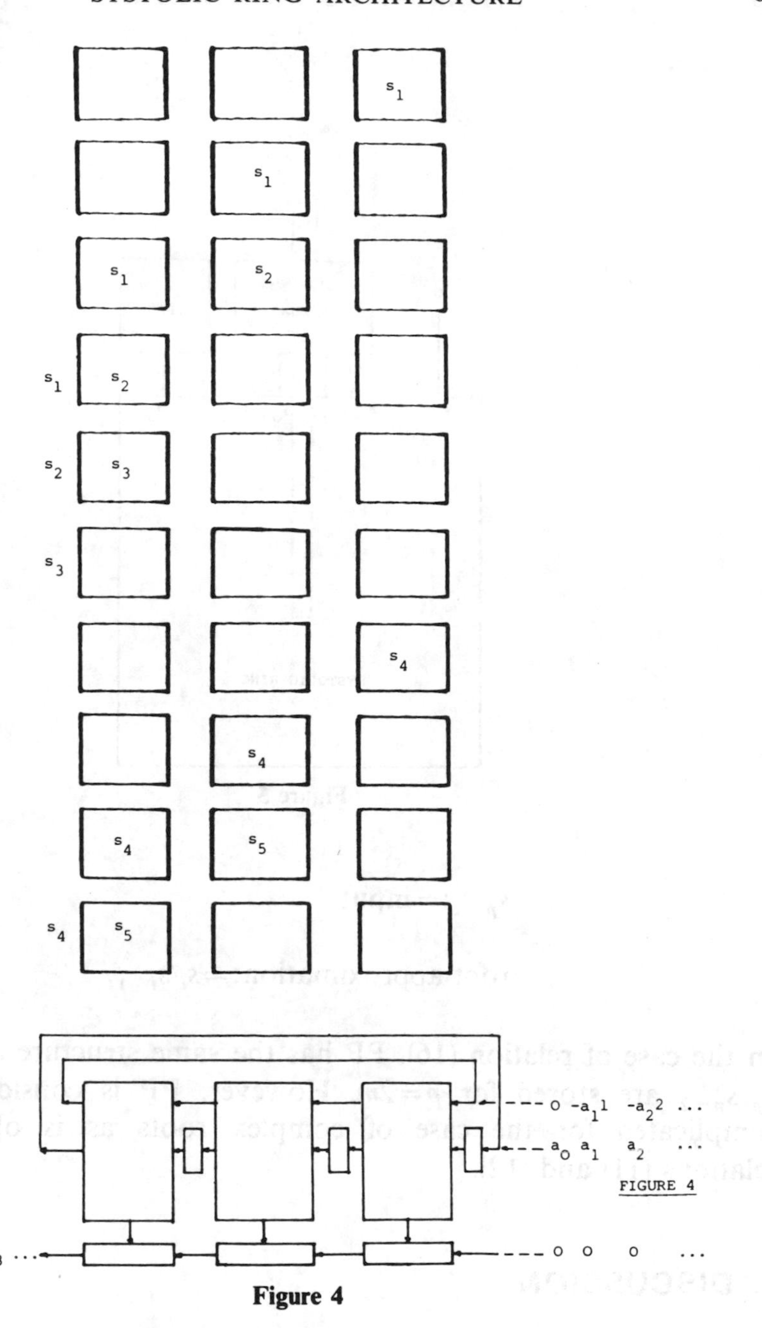

Figure 4

controlled input port from the ring. Its operation can be outlined as follows:

if valid input from ring

$$s_p := s_{p-1}$$

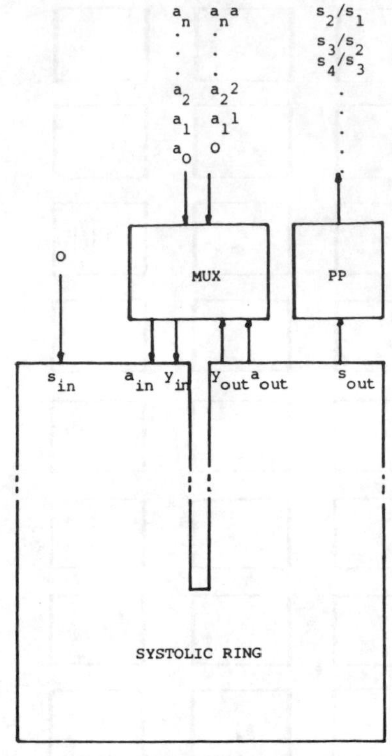

Figure 5

s_{p-1}:=input

root approximation:$=s_i/s_{i-1}$.

In the case of relation (16), PP has the same structure as before but s_p, s_{p-2} are stored for $p=2m$. However, PP is considerably more complicated for the case of complex roots as is obvious from relations (11) and (12).

4. DISCUSSION

This paper describes a systolic implementation of Bernoulli's method for the calculation of dominant zeros of polynomial equations.

The sequence $s_1, s_2, \ldots, s_p$ defined by Newton's theorem is computed by means of a linear array of n IPS cells with bidirectional data flow, and with throughput and processor utilisation equal to $\frac{1}{2}$.

The array is transformed to a systolic ring requiring half of the cells and achieving full processor utilisation.

The systolic ring is integrated into a systolic system which postprocesses the coefficients $s_1, s_2, \ldots, s_p$ for the final calculation of the dominant zero(s) of the polynomial.

The method can be combined with the Synthetic Division to produce a polynomial equation solver. Although Bernoulli's method converges slowly and is used mainly when the dominant zero is significantly larger than any other zero of the polynomial, it provides an area-efficient easily realisable and expandable equation solver due to the simple and regular computation required in the systolic design.

References

[1] D. J. Evans and K. G. Margaritis, Systolic designs for the root-squaring method, *Int. J. Computer Math.* **22** (1987), 43–62.

[2] C. E. Froberg, *Introduction to Numerical Analysis*, Addison-Wesley, 1965.

[3] W. Jennings, *First Course in Numerical Methods*, Macmillan, 1965.

[4] H. T. Kung and M. S. Lam, Wafer-scale integration and two-level pipelined implementations of systolic arrays, *J. of Parallel and Distributed Computing* **1** (1984), 32–63.

[5] C. E. Leiserson, *Area-Efficient VLSI Computation*, Ph.D. Thesis, Dept. of Computer Science, Carnegie Mellon University, CS-82-108, Oct. 1981.

[6] K. G. Margaritis and D. J. Evans, Improved systolic matrix-vector multiplication, Internal Report, Dept. of Computer Studies, Loughborough Univ. of Technology, CS-320, Sept. 1986; presented at the Workshop on VLSI Computation, Univ. of Leeds, Feb. 1987.

Systolic Designs for the Root-Squaring Method

D. J. EVANS and K. MARGARITIS

Department of Computer Studies, Loughborough University of Technology, Loughborough, Leicestershire, U.K.

In this paper some systolic designs are presented for the implementation of the Graeffe root-squaring method for polynomial root solving. From a semi-systolic array, "retiming" transformations are applied to yield a purely systolic array that performs the squaring of the coefficients of an equation. The systolic array is then simulated soft-systolically in an OCCAM program listed in the Appendix. The overall design of a systolic system for the solution of equations based on the Graeffe method is also discussed.

KEY WORDS: Systolic systems, root finding methods, parallel computing.

C.R. CATEGORIES: 5.15, 5.22.

1. INTRODUCTION

The Graeffe root-squaring method, for the solution of polynomial equations of the form,

$$a_0x^n + a_1x^{n-1} + \cdots + a_n = 0, \tag{1}$$

is summarised in [2] as follows:

Consider the Eq. (1) and for the sake of simplicity suppose that all roots are real and distinct. Collecting all even terms on one side and all odd terms on the other, we get on squaring:

$$(a_0x^n + a_2x^{n-2} + a_4x^{n-4} + \cdots)^2 = (a_1x^{n-1} + a_3x^{n-3} + a_5x^{n-5} + \cdots)^2. \tag{2}$$

Now by putting $x^2=y$, a new equation is obtained:

$$b_0y^n+b_1y^{n-1}+\cdots+b_n=0, \tag{3}$$

with,

$$b_0=a_0^2$$

$$b_1=-a_1^2+2a_0a_2$$

$$b_2=a_2^2-2a_1a_3+2a_0a_4$$

$$\vdots$$

$$b_n=(-1)^na_n^2$$

or

$$(-1)^kb_k=a_k^2-2a_{k-1}a_{k+1}+2a_{k-2}a_{k+2}-\cdots. \tag{4}$$

The procedure can then be repeated and is finally terminated when the double products can be neglected, compared with the quadratic terms in the formation of the new coefficients. Suppose that, after m squarings, we have obtained the equation,

$$A_0x^n+A_1x^{n-1}+\cdots+A_n=0, \tag{5}$$

with the roots $q_1,q_2,\ldots,q_n$, while the original equation has the roots $p_1,p_2,\ldots,p_n$. Then the following relation holds between the roots of the old and new polynomials, i.e.,

$$q_i=p_i^{2m}, \quad i=1,2,\ldots,n. \tag{6}$$

Further suppose that,

$$|p_1|>|p_2|>\cdots>|p_n| \quad \text{and} \quad |q_1|\gg|q_2|\gg\cdots\gg|q_n|. \tag{7}$$

Hence,

$$q_1 \cong -A_1/A_0$$

$$q_2 \cong -A_2/A_1$$

$$q_3 \cong -A_3/A_2$$

$$\vdots$$

$$q_n \cong -A_n/A_{n-1}$$

or

$$q_i \cong -A_i/A_{i-1}, \quad i=1,\dots,n. \tag{8}$$

Finally, p_i is obtained by m successive square-root extraction of q_i, i.e., $p_i = \sqrt[2m]{q_i}$, while the sign has to be determined by insertion of the root into the equation.

The Graeffe method can be easily extended to accommodate double and complex roots, as is shown in [1].

The main disadvantage of the method is the fact that the successive squaring may cause an overflow in the numerical calculations; however, the method is still useful for the separation of very close roots. An advantage is that the root-squaring method can provide all n roots of an equation simultaneously with a relatively simple and regular computation.

In the following sections the systolic design for an iteration of the Graeffe method based on (4) is discussed and an appropriate systolic array is described in detail.

Finally, the systolic array is integrated in a systolic equation solver.

2. SYSTOLIC DESIGN DERIVATION

As is obvious from the description of the root-squaring method, the main computational effort is to produce the coefficients, b_i, $i=0,1,\dots,n$, of the new polynomial (3) by using the recurrence (4). This calculation is repeated m times, i.e. we have m steps, until the Eq. (5) is produced. Finally, a series of divisions and square-root extractions produces the roots of the original equation.

Now recurrence (4) can be re-written in a form suitable for systolic implementation as follows:

$$b_k=(-1)^k a_k a_k+(-2)^{k-1} a_{k-1} a_{k+1}+\cdots \tag{9}$$

where,

$$k=0,1,\ldots,n,$$

and

$$a_{-1}=a_{-2}=\cdots=a_{n+1}=a_{n+2}=\cdots=0.$$

For example, for $n=5$ (n odd) we have:

$$\begin{aligned}
b_0&=(-1)^0 a_0 a_0\\
b_1&=(-1)^1 a_1 a_1+(-2)^0 a_0 a_2\\
b_2&=(-1)^2 a_2 a_2+(-2)^1 a_1 a_3+(-2)^0 a_0 a_4\\
b_3&=(-1)^3 a_3 a_3+(-2)^2 a_2 a_4+(-2)^1 a_1 a_5\\
b_4&=(-1)^4 a_4 a_4+(-2)^3 a_3 a_5\\
b_5&=(-1)^5 a_5 a_5.
\end{aligned} \tag{10}$$

Similarly, for $n=6$ (n even)

$$\begin{aligned}
b_0&=(-1)^0 a_0 a_0\\
b_1&=(-1)^1 a_1 a_1+(-2)^0 a_0 a_2\\
b_2&=(-1)^2 a_2 a_2+(-2)^1 a_1 a_3+(-2)^0 a_0 a_4\\
b_3&=(-1)^3 a_3 a_3+(-2)^2 a_2 a_4+(-2)^1 a_1 a_5+(-2)^0 a_0 a_6\\
b_4&=(-1)^4 a_4 a_4+(-2)^3 a_3 a_5+(-2)^2 a_2 a_6\\
b_5&=(-1)^5 a_5 a_5+(-2)^4 a_4 a_6\\
b_6&=(-1)^6 a_6 a_6.
\end{aligned} \tag{11}$$

It can be easily observed that the basic calculation required for the computation of a coefficient b_i is the Inner Product (IP), i.e. "multiplication plus addition", which makes the method suitable for systolic implementation. Another important feature is the pipelineability of the calculations for successive b's: as it is obvious from (10) and (11) the coefficients of the original polynomial are regularly arranged throughout the calculations. Furthermore, the number of IP operations required is binomially distributed over the central coefficients, with a minimum of 1 IP operation for b_0 and b_n and a maximum of $\lceil (n+1)/2 \rceil$ operations for $b_{n/2}$ (for n even) or $b_{(n-1)/2}, b_{(n+2)/2}$ (for n odd). A final point is that the exponent that determines the sign of each IP is arranged to be the same with the subscript of the first component of the IP; thus it can be said that this component "carries" its sign throughout the computation as it moved in different positions for the calculation of successive b's.

The above observations will be used in the derivation of the systolic design implementing the calculations of (10) and (11).

The first requirement for that design is the occurrence of the appropriate data items, in the correct processor (cell) and during the desired computation step (time unit or cycle). A possible data-flow specification and the resulting semi-systolic design are shown in Figure 1. The case of $n=6$ is used as an example for the remainder of the paper since the differences for n odd or n even are insignificant.

The data flow illustrated in Figure 1 is a direct mapping of the recurrence relation (11) onto a linear array of processors; the unused variables are multiplied by zero and thus in every step one coefficient b_i is produced. The data form a reflected wavefront, i.e. enter the array from the right end, travel through the array and then they are reflected back to their input side: a semi-systolic design implementing this data-movement is also shown in Figure 1. Notice that only $\lceil (n+1)/2 \rceil$ cells are required for the formulation of a polynomial of order n. The main operation for the cells is a multiplicative process while the left-most cell reflects the data item and multiplies it by itself. During a time unit all cells perform a multiplication and the results are fanned-in and summed using an adder, to form a b_i. When the number of cells is large the adder can be implemented on a pipelined adder tree to avoid large delays in each cycle (see [3]).

The systolic design of Figure 1 requires $\lceil (n+1)/2 \rceil$ preloading steps so that the array is ready to produce b_0; from then on one result

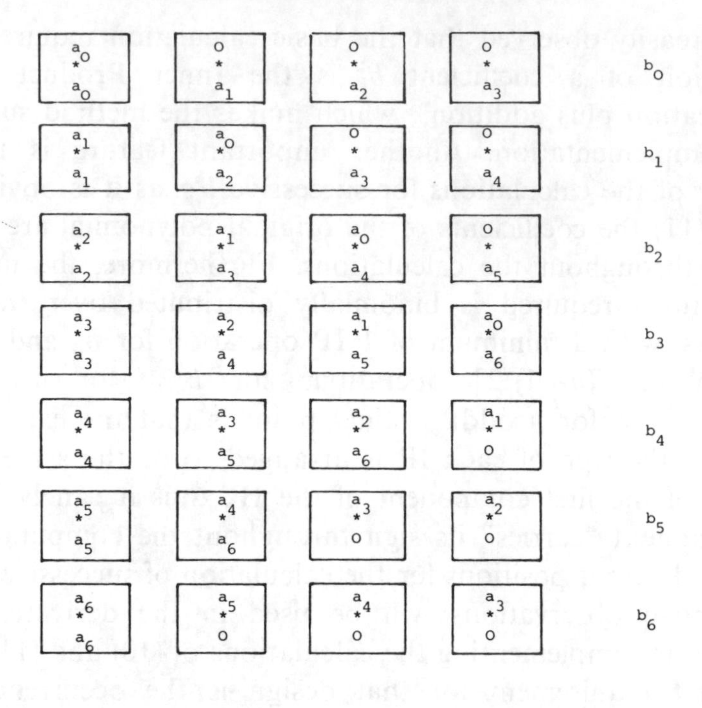

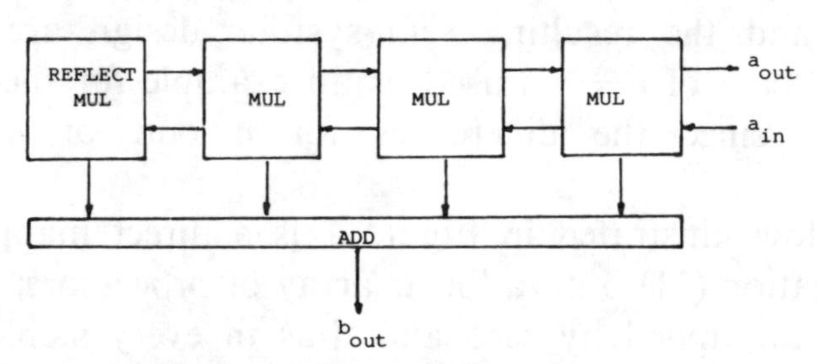

Figure 1 A simple systolic design for the calculation of the coefficients, b_i, $i=0,1,2,\ldots,n$.

may be produced in each cycle, if we assume that there is no delay in the adder: otherwise an additional delay of lgn time units is required before b_0 is produced (see [4]). In general, therefore the semi-systolic design needs $(\lceil(n+1)/2\rceil)+(lgn)+n$ computational steps for the formulation of a polynomial of order n; the initial delay is $(\lceil(n+1)/2\rceil)+(lgn)$ and the throughput is 1 coefficient per cycle.

The main disadvantages of the design of Figure 1 are the fan-in mechanism with the tree-structured adder, together with the three-directional data-flow of the main array. Some of these disadvantages

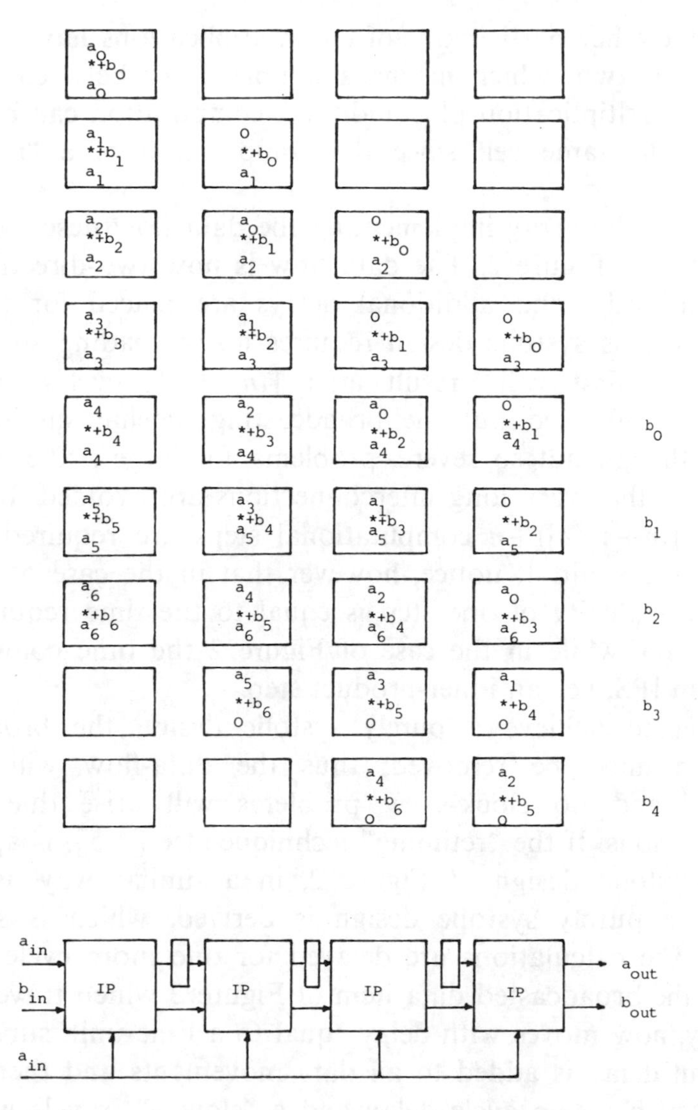

Figure 2 Data flow for semi-systolic array design.

can be removed as shown in Figure 2. If we simply 'shift' the calculations in the time domain, in proportion to their spatial distance from the left-most processor, we can derive the data-movement of Figure 2: the calculations of the second cell (where the first cell is considered the left-most cell) are delayed by one cycle, for the third by two cycles, etc. After this rearrangement the following occurrences are observed: the same "*a*" participates in the calculations of all cells, a fact that implies the suitability of broadcasting that "a"; the other participants of the multiplications have subscripts that differ by two, which means a double delay between the cells; finally the "multiplication plus addition" computation can be accomplished in the same cell since the computation of a "b" is now pipelined.

A semi-systolic array implementing the data flow described above is illustrated in Figure 2. The data flow is now two-directional, the cells are IP cells and additional delays are placed for the slow-moving a's. This systolic design requires no preloading steps and it produces the first valid result after $\lceil (n+1)/2 \rceil$ cycles; no fan-in operation is involved but the broadcasting mechanism introduces similar, although not so severe, problems; for large n the array can be folded so that very long interconnections are avoided. In general therefore $(\lceil (n+1)/2 \rceil)+n$ computational steps are required and the throughput is again 1; notice, however that in the case of Figure 1 the time complexity of one step is equal to the time required by a multiplication, while in the case of Figure 2 the time complexity is equal to an IPS, i.e., an inner-product step.

In order to achieve a purely systolic design the broadcasting mechanism must be removed; thus the data-flow will be unidirectional and no clock-skew problems will arise due to long interconnections. If the "retiming" technique (see [4, 5]) is applied on the semi-systolic design of Figure 2, in a similar way as that of Figure 1, a purely systolic design is derived, which is shown in Figure 3. The calculations are delayed for one more cycle for each cell: thus the broadcasted data item of Figure 2 which travelled with zero delay, now moves with delay equal to a time unit; subsequently a time unit delay is added to all data movements and therefore the b's travel with a two cycle delay and a "slow a" travels with three cycles delay. The data flow is now unidirectional but an additional delay is introduced, the first valid result being produced after $2^*\lceil (n+1)/2 \rceil - 1$ cycles. In general, therefore $(2^*\lceil (n+1)/2 \rceil - 1)+n$ computational steps are required and the throughput remains at 1 with the time complexity of a time unit again 1 IPS.

3. SYSTOLIC ARRAY DESCRIPTION

The systolic array of Figure 3 needs some further refinements as

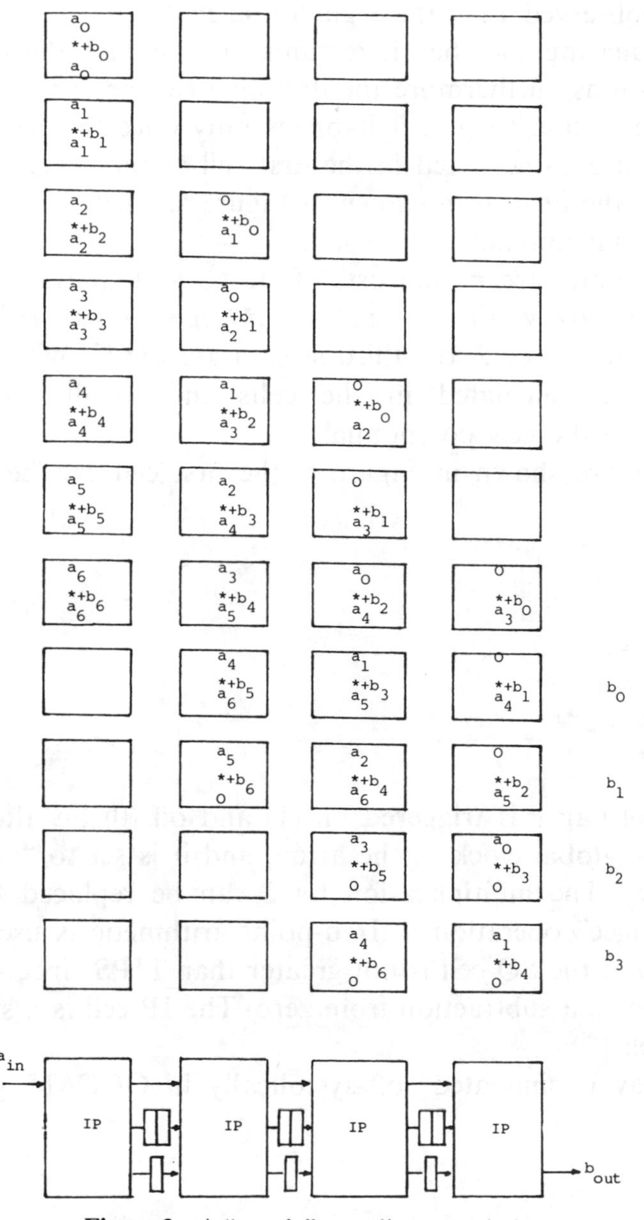

Figure 3 A "purely" systolic array design.

regards the operations to be performed in each cell. As is obvious in Sections 1 and 2, there are some more complications in the computation of the coefficients of the new polynomial, namely, the determination of the sign for each partial IP and the multiplication of the factor 2 so that the double products are produced. In Section

2, it was observed that the sign for each "a" is determined once, upon its squaring and then it remains the same for the remainder of the calculations; furthermore the doubling can be combined with the sign determination so that it happens only once for each "a". These operations are all collected in the first cell of the array and thus the array takes the final form shown in Figure 4, where the detailed data flow is also illustrated.

The first data stream, consists of "fast a's" that travel unchanged through the array; the second stream has "slow a's" which are multiplied by -2 or 2; the third stream are the b's which collect the IPs that are calculated in the cells, and finally produces the coefficients of the new polynomial.

The cells are shown in Figure 5; the first cell can be specified as follows:

if $c=0$
then $t:=-a$
else $t:=a$
a-fast: $=a$
a-slow: $=2*t$
$b:=a*t$

The control flag c is triggered on (1) and off (0) on alternate clock ticks of the global clock of the array, and it is set to "off" upon the input of a_0. The multiplication by 2 can be replaced by a simple "shift-left-once" operation if fixed-point arithmetic is used. The time complexity of the SQ cell is not greater than 1 IPS since the negation can be seen as a subtraction from zero. The IP cell is a simple inner-product cell.

The array is simulated soft-systolically in OCCAM given in the Appendix.

4. OVERALL SYSTEM DESCRIPTION

A systolic system based on the Graeffe root-squaring method should be capable of performing a number of successive polynomial formulations, using the systolic array described in Sections 2 and 3. The number of iterations can be predefined or can be controlled by the system on inspection of the difference in the magnitude of the

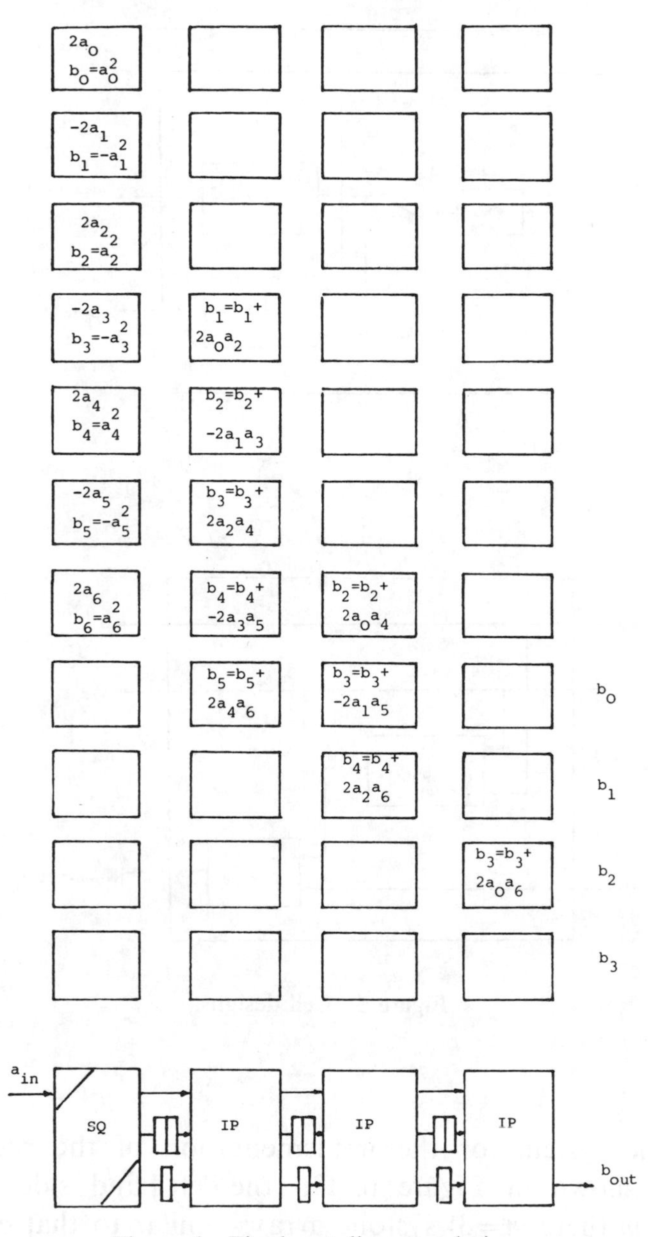

Figure 4 Final systolic array design.

coefficients being produced. Usually three iterations are enough (see [1,2]).

After the final polynomial is formed a series of divisions follows and then successive square-root extractions; the number of square-root operations is the same as that of the polynomial formulations.

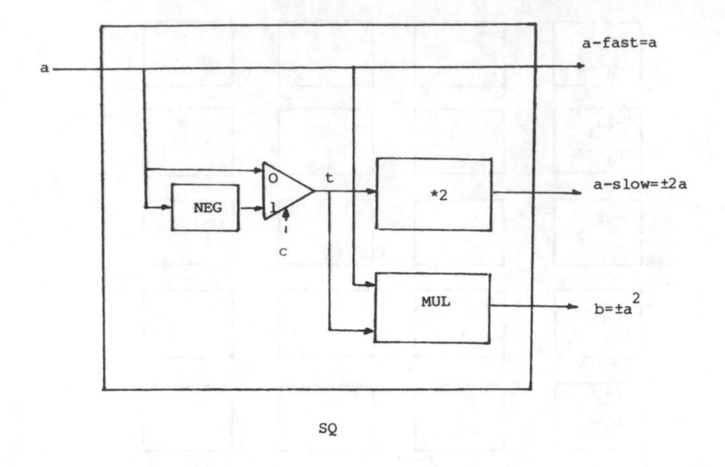

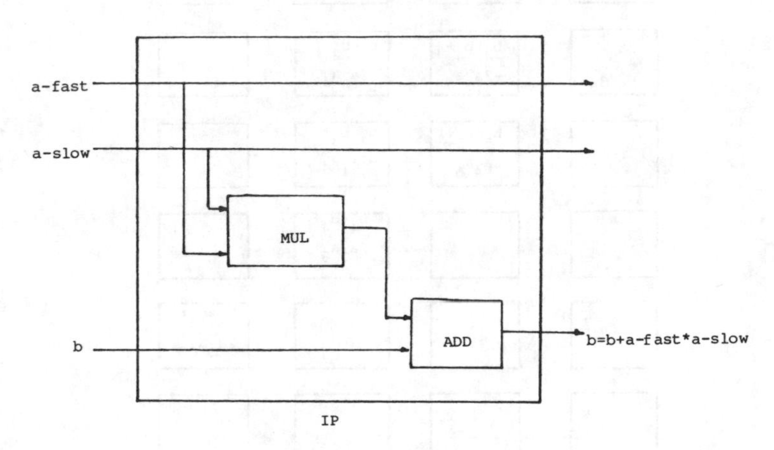

Figure 5 Cell designs.

A systolic system for the implementation of the root-squaring method is shown in Figure 6. On the left-hand side, there is a pipeline of m (here $m=3$) systolic arrays similar to that of Figure 4. The first array accepts as input the coefficients of $f(x)$ and calculates the coefficients of the new polynomial, and thus each array represents a root-squaring iteration. The coefficients of the final polynomial pass through a divider/negator where the roots of the final equation are calculated according to (8): the device is quite simple as illustrated in Figure 7. Before the division operation is the denominator can be checked against zero so that a division by zero is

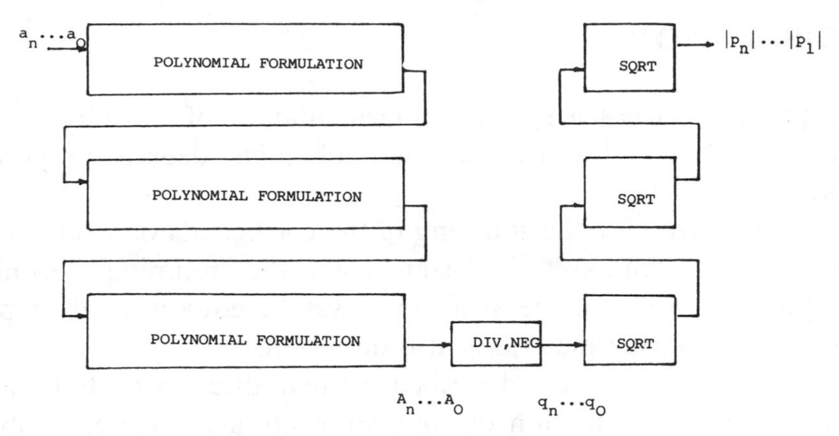

Figure 6 A systolic system for the Graeffe root squaring method.

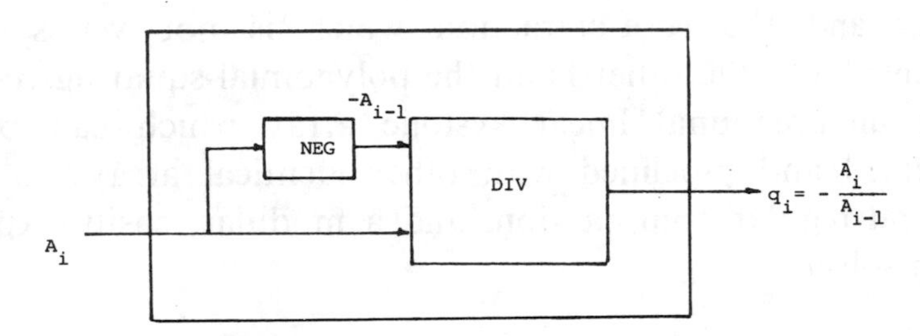

Figure 7 The divider/negator design.

avoided; in that case a division over a suitably small quantity suffices.

Notice that since the coefficients $A_0, A_1, \ldots, A_n$ are produced one every cycle they can form the input to the divider/negator with no delay. Similarly for the roots $q_1, q_2, \ldots, q_n$. A possible pipelined implementation for a square-root device (SQRT) would be suitable so that a smooth dataflow is secured through the successive square-root extractions, which are illustrated in the right-hand side of Figure 6. Thus, the absolute value of the roots of the original polynomial can be produced in the same pipelined fashion.

An alternative configuration for the systolic system would be the use of only one systolic array with a feedback mechanism so that the same array produces the successive polynomials; similarly only one square-root device can be used; however since the pipelining timings cannot be secured a more complicated control of the system is required.

5. DISCUSSION

This paper describes a systolic implementation of the Graeffe root squaring method, for the solution of high degree polynomial equations.

A systolic array for the squaring of the coefficients of a polynomial is derived from semi-systolic designs using the "retiming" technique. Also the integration of the array in a systolic equation solver based on the root-squaring method is also discussed.

Although the method is described for real, discrete roots it can be extended to the calculation of complex roots also. Major problems in the systolic calculation of the roots of a polynomial by means of Graeffe's method are the possible overflow because of the successive squarings and the root-extraction which is not yet systolically implemented. On the other hand the polynomial-squaring recurrence yields a unidirectional linear systolic array which can be easily implemented and pipelined with other identical arrays to provide simple and regular computation, and a modular, easily expandable equation solver.

References

[1] S. Brodetsky and G. Smeal, On Graeffe's method for complex roots of algebraic equations, *Proc. Camb. Phil. Soc.* **22,** 2 (1924).
[2] C.-E. Froberg, *Introduction to Numerical Analysis*, Addison-Wesley, (1964).
[3] H. T. Kung, Why systolic architectures, *IEEE Computer Magazine* **15,** 1 (1982).
[4] C. E. Leiserson, *Area-Efficient VLSI Computation*, Ph.D. Thesis, CS-CMU, 1981.
[5] J. D. Ullman, *Computational Aspects of VLSI*, Computer Science Press, 1983.

Appendix

Occam program

```
— Systolic array for the Graeffe (Root Squaring) Method.
— It performs one iteration of the algorithm, i.e. for a given
  polynomial
—          f(x)=a[0]*x**n+a[1]*x**(n-1)+..+a[n]
— it produces the polynomial
—          g(x)=b[0]*y**n+b[1]*y**(n-1)+..+b[n]
— which has as roots the squares of the roots of f(x).
```

```
external proc get (var v, value s[ ]):
external proc fp.get.n(var float v[ ], value n, s[ ]):
external proc fp.put.n(value float v[ ], value n, s[ ]):

— Max degree of polynomials and max number of ips cells.
def no = 10, so = 5:
— Vectors of coefficients for polynomials f(x), g(x).
var float a[no + 1], b[no + 1]:
— Actual degree of polynomials and actual number of ips cells.
var n, s,
— Overall operation time, and initial delay.
  time, del:
— Channels for fast-moving a's, slow-moving a's and b's.
chan af.c[so + 2], as.c[(3*so) + 1], b.c[(2*so) + 1]:

— The first cell of the array: produces the squared coefficient and
— defines the sign of it and of a-slow, which is also multiplied by
  two.
proc square (chan ain, afout, asout, bout,
               value time) =

  var float a[4], b:
  var neg:

  seq
    — initialisation
    par
      par i = [0 for 4]
        a[i] := 0.0
      b := 0.0
      neg := false
    — main operation
    seq i = [0 for time]
      seq
        — i/o
        par
          ain? a[0]
          afout ! a[1]
          asout ! a[2]
          bout ! b
        — calculation
```

```
        if
          neg
            —negate a-slow on alternate cycles.
          par
            a[3]:= -a[0]
            neg:=false
          true
            par
              a[3]:=a[0]
              neg:=true
        par
          a[1]:=a[0]
          a[2]:=(2.0 * a[3])
          b:=(a[0] * a[3]):
—Inner Product Step Cell: accumulates the inner product
—coefficients in b and propagates a-fast, a-slow and b.
proc ips (chan afin, asin, bin, afout, asout, bout,
         value time)=

  var float a[4], b[2]:

  seq
    —initialisation
    par
      par i=[0 for 4]
        a[i]:=0.0
      par i=[0 for 2]
        b[i]:=0.0
    —main operation
    seq i=[0 for time]
      seq
        —i/o
        par
          afin ? a[0]
          asin ? a[1]
          bin ? b[0]
          afout ! a[2]
          asout ! a[3]
          bout ! b[1]
```

```
        —calculation
        par
          a[2]:=a[0]
          a[3]:=a[1]
          b[1]:=b[0]+(a[0] * a[1]):

—Delay Cell: propagates its input with one cycle delay.
proc delay (chan xin, xout,
            value time)=

  var float x[2]:

  seq
    —initialisation
    par i=[0 for 2]
      x[i]:=0.0
    —main operation
    seq i=[0 for time]
      seq
        —i/o
        par
          xin ? x[0]
          xout ! x[1]
        —calculation
        x[1]:=x[0]:

—Source of the array: pumps the coefficients of the original
—polynomial
proc source (chan aout,
             value float a[ ],
             valye n, time)=
  seq i=[0 for time]
    if
      i<=n
        aout ! a[i]
      true
        aout ! 0.0:
—Sink of the array: collects the coefficients of the new polynomial
—after an initial delay.
```

```
proc sink (chan afin, asin, bin,
           var float b[ ],
           value delay, time) =
  seq i = [0 for time]
    par
      afin ? any
      asin ? any
      if
        i < delay
          bin ? any
        true
          bin ? b[i-delay]:

— Array configuration: only [n/2] ips cells are required.
— Two delays between each cell for a-slow and one delay for b.
— The initial delay is 2 * [n/2]; the overall operation time is
— n + 2 * [n/2].
proc system =
  seq
    if
      (n\2) = 0
        s: = (n/2)
      true
        s: = ((n − 1)/2)
    del: = (2 * s) + 1
    time: = (n + del) + 1
    par
      source (af.c[0], a, n, time)
      square (af.c[0], af.c[1], as.c[0], b.c[0], time)
      par i = [0 for s]
        par
          ips (af.c[i + 1], as.c[(i * 3) + 2], b.c[(i * 2) + 1],
               af.c[i + 2], as.c[(i * 3) + 3], b.c[(i * 2) + 2], time)
          par j = [0 for 2]
            delay (as.c[(i * 3) + j], as.c[(i * 3) + (j + 1)], time)
          delay (b.c[i * 2], b.c[(i * 2) + 1], time)
      sink (af.c[s + 1], as.c[3 * s], b.c[2 * s], b, del, time):

proc getdata =
  seq
    get (n, "degree of polynomial")
```

```
      fp.get.n(a, (n+1), "of coefficients"):

proc putdata =
   seq
      fp.put.n(b, (n+1), "of coefficients"):

seq
   getdata
   system
   putdata
```

A RE-USABLE SYSTOLIC ARRAY FOR MATRIX-VECTOR MULTIPLICATION

D. J. EVANS and K. G. MARGARITIS*

Parallel Algorithms Research Centre
Loughborough University of Technology
Loughborough, Leicestershire, U.K.

This paper presents a systolic array for matrix-vector multiplication (*mvm*), with the additional feature that its output can be re-used as input for a consecutive *mvm*. This fact makes the array especially useful for iterative systolic algorithms based on successive *mvm* computations. Some of these applications are presented in this paper. The array is simulated in Occam.

KEY WORDS: Systolic array, matrix vector multiplication, iterative systolic algorithm
C.R. CATEGORIES: C.1.2., G.1.3

1. INTRODUCTION

In [10] a re-usable systolic array for matrix multiplication is presented. This array has the important feature that the matrix product $C = AB$ can be fed back immediately as input for another matrix multiplication. The same array, augmented with a row and/or column of boundary cells (multiplexers-adders) can be used to implement combinations of matrix products and sums, e.g. $C = AX + B$ (see Figure 1.1).

The re-usable matrix multiplication array forms the basis for a series of systolic iterative algorithms. These algorithms use successive matrix operations, in the form of matrix squarings, matrix products, or matrix products and sums. All these operations can be expressed in a generic computation of the form

$$C = AX + B \tag{1.1}$$

and is termed the "matrix–matrix inner product step" (*mmips*).

Systolic algorithms implemented include: Iterative solution of systems of linear equations, based on Jacobi and Gauss–Seidel methods [8], [9]. Calculation of matrix eigenvalues using Power and Matrix Squaring methods [8], [7]. Further, the calculation of matrix powers, matrix inverse, matrix square root, inverse matrix square root, matrix exponential, as well as other matrix functions and matrix polynomials [8], [6].

*Informatics Centre, University of Macedonia, Thessaloniki, Greece.

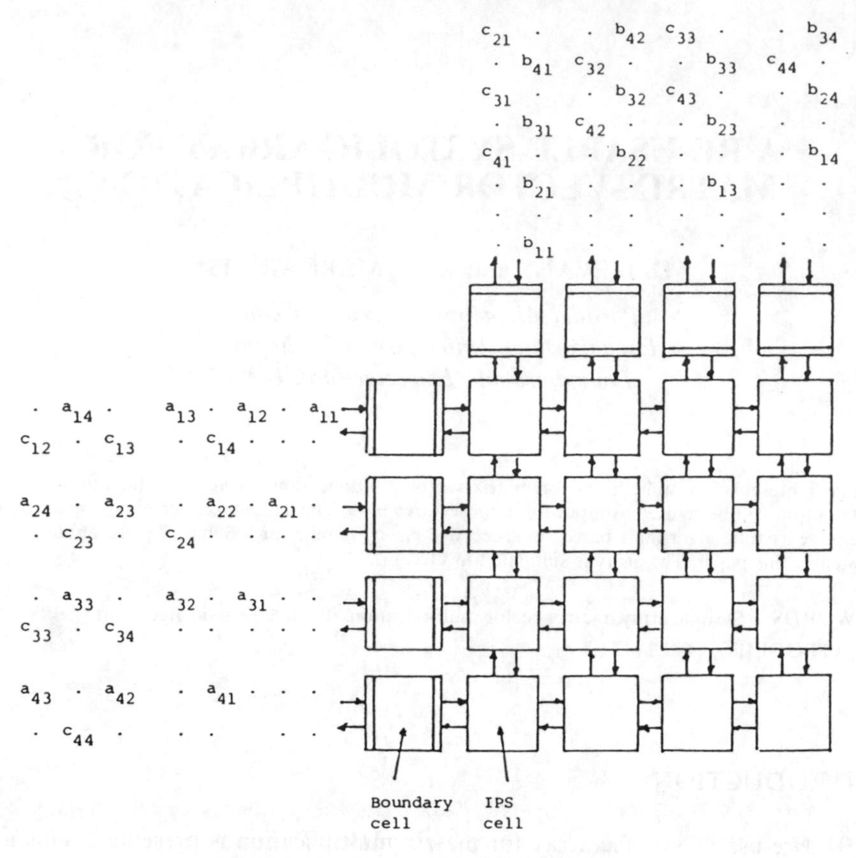

Figure 1.1 Re-usable *mmm* systolic array. ($n = 4$).

2. SYSTOLIC DESIGN

In this paper we propose a re-usable systolic array for matrix-vector multiplication (*mvm*), which can be seen as a single row or column implementation of the matrix multiplication array. It consists of n Inner Product Step (IPS) processors (cells), where n is the size of the matrix. This new array computes successive *mvm*'s, of the form $\mathbf{y} = A\mathbf{x}$. Further, the array can be augmented with a single boundary cell (e.g. multiplexer-adder), so that it can perform combinations of matrix-vector products and sums. The generic matrix-vector operation

$$\mathbf{y} = A\mathbf{x} + \mathbf{b} \tag{2.1}$$

is termed the "matrix-vector inner product step" (*mv ips*).

Each one of the IPS cells perform a computation similar to that of the IPS cell of the re-usable *mmm* array, i.e. it accumulates successive multiplications-additions, it forms an element of the resulting vector, and passes its result towards the output (see Figure 2.1).

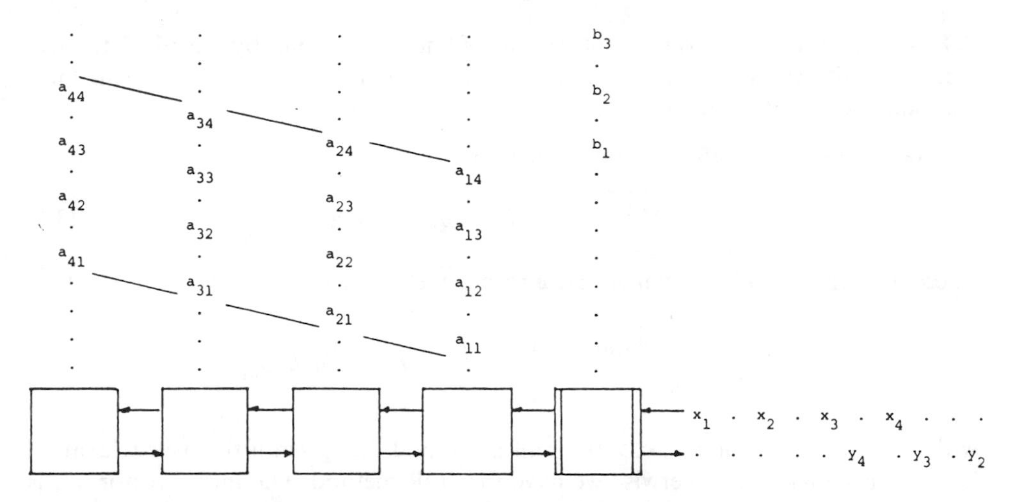

Figure 2.1 Re-usable *mvm* systolic array. ($n = 4$).

The only difference is that each one of the IPS cells keeps in local memory a number of elements of the coefficient matrix A. Depending on the algorithm formulation it can be row, column or diagonal elements of A. Thus the need of communication channels for the components of matrix A is eliminated at the expense of local storage of n elements, where n is the size of A. Similarly, the multiplexer-adder cell keeps in local memory the addition vector **b**, if this computation is required.

Therefore, the re-usable *mvm* array has area requirements of $(n + 1)$ IPS cells, where n is the problem size. The computation time for a single *mv ips* operation is calculated as follows: it takes $2(n + 1)$ IPS cycles for the first result to come out, and another $2(n - 1)$ IPS cycles for the whole output vector to be produced, giving a total computation time of $4n$ IPS cycles. However, it is important that after $2(n + 1)$ IPS cycles, and thereafter every $2n$ IPS cycles, a new (*mv ips*) operation can start, using as input the output vector that is currently being produced. Therefore, for k successive (*mv ips*) computations the time needed is $2(k + 1)n - 1$ IPS cycles.

Note that for the re-usable *mmm* systolic array, the corresponding area requirements are $(n + 1)^2 - 1$ IPS cells, while the computation time is $5n - 1$ IPS cycles, for a single *mmips*, and $2(k + 1)(n + 1) - 1$ IPS cycles, for k successive *mmips* operations. Therefore, the computation time in both cases is approximately the same, but the *mv ips* array uses only $(n + 1)$ cells, instead of $(n + 1)^2 - 1$ at the expense of about $(n + 1)^2$ local memory locations.

The systolic designs of Figure 1.1 and 2.1 have been simulated soft-systolically using Occam (see Appendix, [8]).

3. APPLICATIONS: ITERATIVE SYSTEM SOLUTION

There is a large number of important computations in numerical linear algebra and digital signal processing that can be seen as a series of *mv ips* computations.

First, we consider the solution of systems of linear equations, by means of iterative methods. Given a system of equations, of the general form $A\mathbf{x} = \mathbf{b}$, we can form the following iterative methods [12], [14]:

i) Jacobi Over-relaxation methods (J, JOR).

$$\mathbf{x}^{(k+1)} = (M + (1 - \omega)I)\mathbf{x}^{(k)} + \mathbf{g} \tag{3.1}$$

where M, can be derived from A and $\mathbf{b}$ respectively, as follows

$$m_{ij} = \begin{cases} -\omega(a_{ij}/a_{ii}), & i <> j \\ 0, & i = j \end{cases} \qquad g_i = -\omega(b_i/a_{ii}) \tag{3.2}$$

with $i, j = 1, 2, \ldots, n$ and $a_{ii} <> 0$ for all i. Method J is given if the over-relaxation factor ω is equal to 1; otherwise we have the JOR method. The initial vector $\mathbf{x}^{(0)}$ is either a known approximation of the solution vector, or an arbitrary choice.

ii) Gauss–Seidel, Successive Over-relaxation method (GS, SOR).

$$\mathbf{x}^{(k+1)} = L\mathbf{x}^{(k+1)} + (U + (1 - \omega)I)\mathbf{x}^{(k)} + \mathbf{g} \tag{3.3}$$

with $L(U)$ strictly lower (upper) triangular and $L + U = M$, where M, $\mathbf{g}$ are defined as in (3.2). Again, the GS method is derived from SOR method, for $\omega = 1$.

From (3.2) it can be seen that $(M + (1 - \omega)I)$, L, $(U + (1 - \omega)I$ and $\mathbf{g}$ can be computed once from A, $\mathbf{b}$, ω, at the beginning of the computation, as they remain unchanged throughout the iterative process. Simple systolic pre-processors for these calculations can be found in [8].

From (3.1), (3.3) it is clear that J, JOR methods perform essentially a series of successive *mv ips* operations as in (2.1), i.e.

$$\mathbf{y}^{(k+1)} = C\mathbf{x}^{(k)}, \qquad \text{with} \quad C = (M + (1 - \omega)I)$$
$$\mathbf{x}^{(k+1)} = \mathbf{y}^{(k+1)} + \mathbf{g} \tag{3.4}$$

A systolic algorithm for the J, JOR methods, based on the re-usable *mvm* systolic array of Figure 2.1, is shown in Figure 3.1, for matrix size $n = 4$. Notice the initial time-skewed delay that is required for the local memory elements, especially for vector $\mathbf{b}$. This delay can be implemented either by means of dummy elements or by attaching control signals to the first and last elements of vector $\mathbf{x}$.

Notice also the overlapping in the calculation of the elements of successive vectors. Thus, while the elements of $\mathbf{y}^{(k+1)}$ are calculated, the components of $\mathbf{y}^{(k)}$ travel towards the output.

The area and time requirements for J, JOR methods are exactly the same as in the general case of successive *mv ips*. Thus, we need $(n + 1)$ IPS cells and $2(k + 1)n - 1$ IPS cycles, where n is the size of matrix A and k is the number of iterations.

A systolic algorithm for GS, SOR methods is shown in Figure 3.2, and Figure 3.3. In this case the arrangement of the array is somewhat different, since we require two

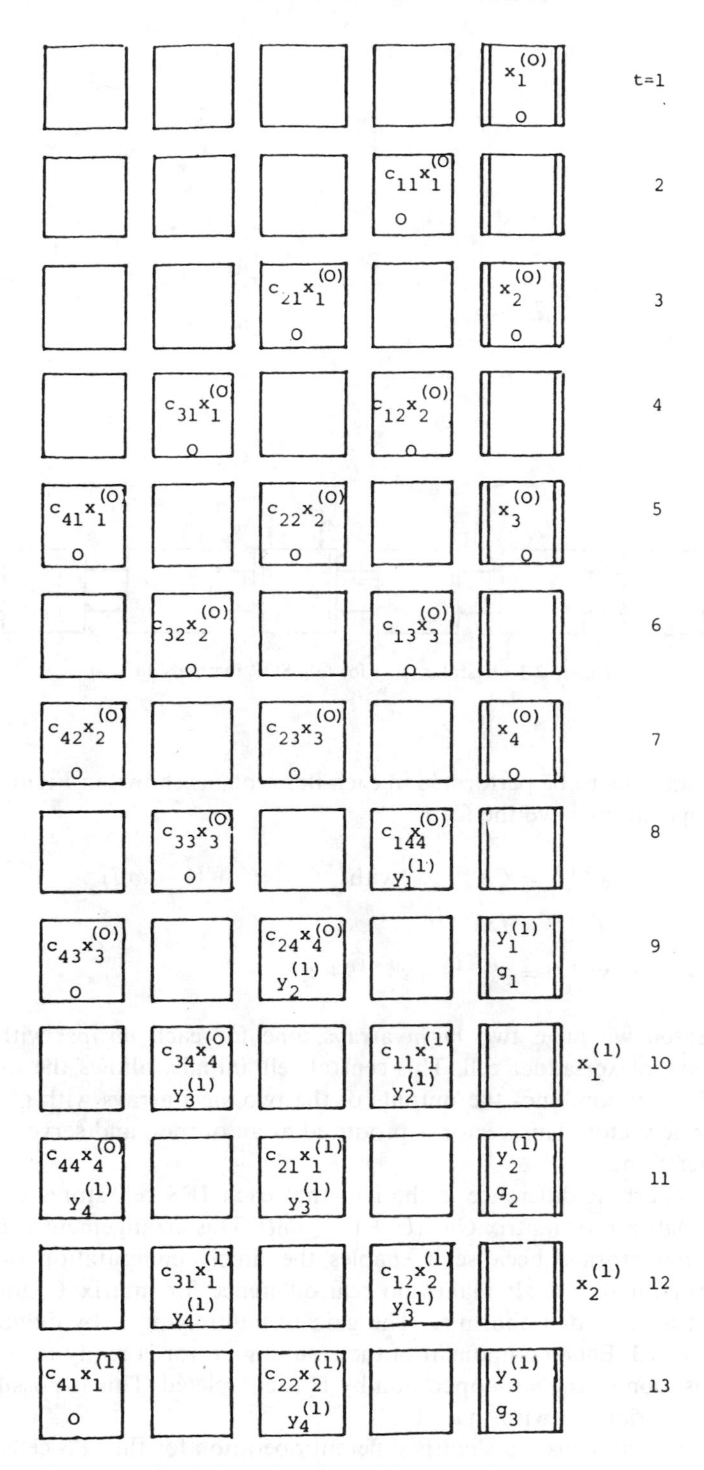

Figure 3.1 Systolic J, JOR methods. ($n = 4$).

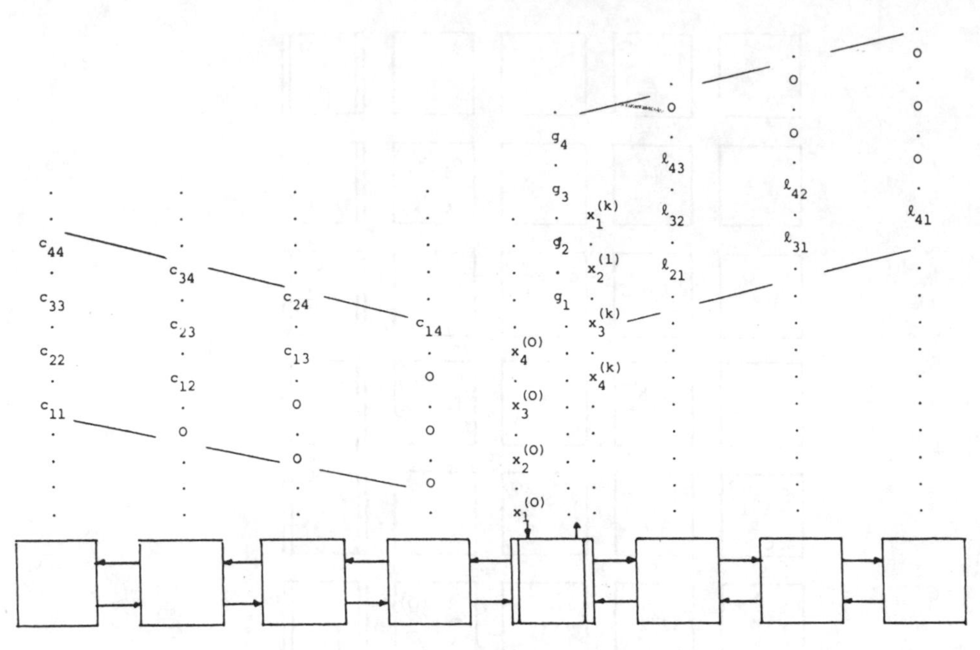

Figure 3.2 Systolic array for GS, SOR methods. ($n = 4$).

mv ips computations to be performed in each iteration, as shown in Figure 3.2. These two *mv ips* operations have the form

$$\begin{aligned}\mathbf{y}^{(k+1)} &= C\mathbf{x}^{(k)}, \qquad \text{with} \quad C = (U + (1 - \omega)I) \\ \mathbf{z}^{(k+1)} &= L\mathbf{x}^{(k+1)} \\ \mathbf{x}^{(k+1)} &= \mathbf{y}^{(k+1)} + \mathbf{z}^{(k+1)} + \mathbf{g}\end{aligned} \tag{3.5}$$

For this reason we have two *mv m* arrays, one for each *mv ips*, with a central three-way multiplexer-adder cell. This central cell, initially allows the input of $\mathbf{x}^{(0)}$, and from then on combines the outputs of the two *mvm* arrays with $\mathbf{g}$, to produce the next iterate vector. This vector is produced as an output, and serves as input for the next interation.

Another important difference is the fact that each IPS cell stores one diagonal from either matrix L or matrix $C = (U + (1 - \omega)I)$. This arrangement is preferred to column or row storage because it enables the timely computation of the result produced using matrix L. It makes no real difference for matrix U, but the same arrangement is kept for uniformity. The gain of multiplying L by diagonal can be seen in Figure 3.3. Each component of the resulting vector is ready to be produced, essentially as soon as the *mvm* operation by U is completed. This is possible because *mvm* by L is overlapped with that of U.

This arrangement causes a slightly different operation for the IPS cells: instead of accumulating inner products for a single output component in a single IPS cell, now

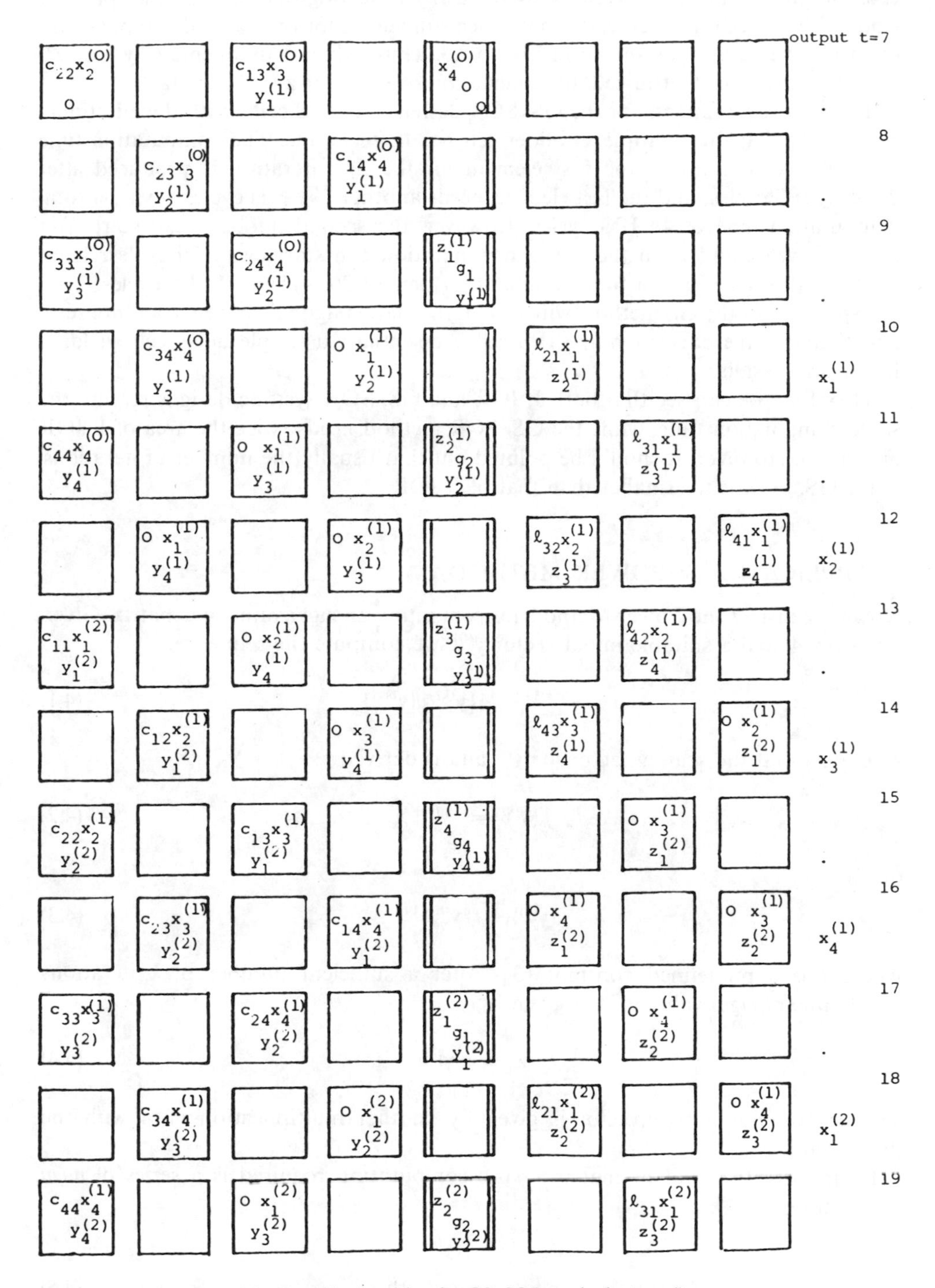

Figure 3.3 Systolic algorithm for GS, SOR methods. ($n = 4$).

each output component travels along the array collecting one partial inner product from each IPS cell. However, this difference still allows for overlapped computations of successive iterates, as shown in Figure 3.3. Thus, while components of $\mathbf{y}^{(k)}$ or $\mathbf{z}^{(k)}$ travel towards the central cell, the calculation of $\mathbf{y}^{(k+1)}$ or $\mathbf{z}^{(k+1)}$ can start.

The area requirements for the GS, SOR design is $2n$ IPS cells, with the additional complexity that the multiplexer-adder cell has three inputs. The computation time is calculated as follows: the first element for the first iteration is produced after $2(n+1)$ IPS cycles, and the last element needs another $2(n+1)$ cycles, giving a total computation time of $4n$ IPS cycles. However, the second iteration can start after $2(n+1)$ cycles, and from then on a new iteration can start every $2n$ cycles. Thus, for k iterations we have a total computation time of $2(k+1)n-1$ IPS cycles.

Especially for the GS method, with $\omega = 1$, the main diagonal elements of C are zero, and therefore the corresponding IPS cell is essentially a simple delay cell, yielding better area results.

Therefore, we can see that both J, JOR and GS, SOR systolic designs require the same computation time, while the GS, SOR method needs twice the area of J, JOR algorithm. However, it should be pointed out that usually the number of iterations, k, for GS, SOR is far smaller than that of J, JOR.

4. APPLICATIONS: POWER METHOD

Now we consider the Power Method, for computing an eigenvalue of a matrix. Given a matrix A, and a suitable initial vector $\mathbf{y}^{(0)}$, we compute the sequence:

$$\mathbf{y}^{(k+1)} = A\{\mathbf{y}^{(k)}/\mu(\mathbf{y}^{(k)})\} \tag{4.1}$$

where $\mu(\mathbf{y}^{(k)})$ is the scaling factor for $\mathbf{y}^{(k)}$ and is derived as

$$\mu(\mathbf{y}^{(k)}) = \|\mathbf{y}^{(k)}\|_\infty \tag{4.2}$$

or

$$\mu(\mathbf{y}^{(k)}) = t\|\mathbf{y}^{(k-1)}\|_\infty \tag{4.3}$$

where t is a predefined constant [3]. After a sufficient number of k iterations the dominant eigenvalue of A is given by

$$\lambda_1 = \mu(\mathbf{y}^{(k)}) \tag{4.4}$$

The corresponding eigenvector is given by another multiplication of A with the final vector.

From (4.1) it is evident that the basic computation required is a series of *mvm* calculations, i.e.

$$\begin{aligned} \mathbf{x}^{(k+1)} &= A\mathbf{y}^{(k)} \\ \mathbf{v}^{(k+1)} &= \mu(\mathbf{x}^{(k+1)}) \end{aligned} \tag{4.5}$$

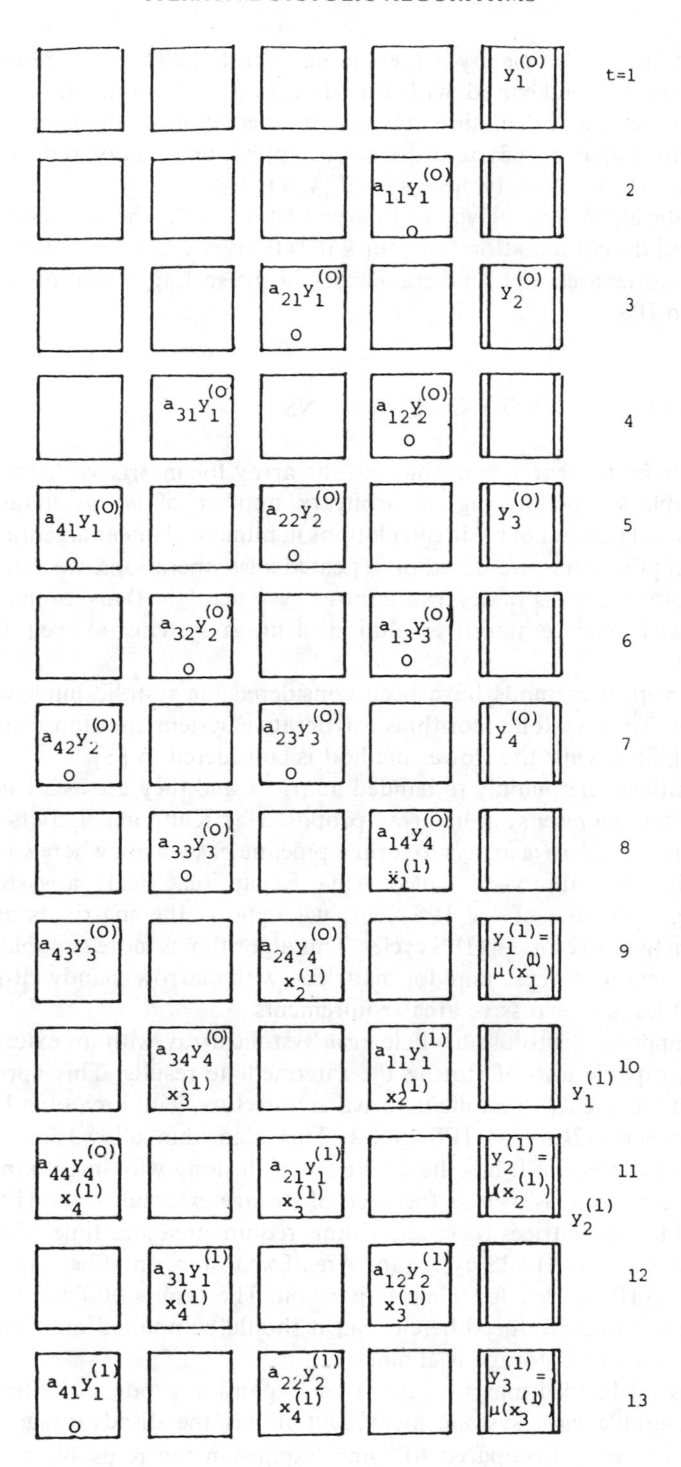

Figure 4.1 Systolic Power Method. ($n = 4$).

The only additional complexity is the scaling operation. Thus, the re-usable systolic array of Figure 2.1 can be used, with a modification of the boundary cell, which now performs the scaling calculation, instead of the addition of *mv ips*. The scaling factor is computed as in (4.3), in order not to impose an additional delay of $2n$ cycles per iteration, which would be necessary if (4.2) is used.

The systolic algorithm is given in Figure 4.1 for $n = 4$. The area needed is $(n + 1)$ IPS cells, and the computation time, for k iterations is $2(k + 1)n - 1$ IPS cycles. It is supposed that the area and time complexity of the scaling operation is comparable to that of an IPS.

5. COMPARISONS AND CONCLUSIONS

In this paper we present a re-usable systolic array for matrix-vector multiplication (*mvm*), suitable for performing an arbitrary number of *mv ips* iterations. Given that a significant number of basic calculations in numerical linear algebra, as well as in digital signal processing, are based on repeated *mvm* operations, we can see that this array can form the basis of several iterative systolic algorithms. Some applications are presented herein, i.e. iterative solution of linear systems, as well as eigenvalue computation.

These numerical methods have been considered for systolic implementations in several cases. Thus, systolic alorithms for iterative system solution can be found in [4], [2], [1], [11] while the Power method is considered in [3].

The algorithms refer mainly to banded matrices, and they use as a basic computational structure the *mvm* systolic array proposed by Kung and Leiserson [15]. The first option is to use k *mvm* arrays to form a pipeline of k stages, where k is the number of iterations. This approach, called Area Expansion, yields a systolic pipeline architecture, with area of kw IPS cells, where w is the matrix bandwidth. The computation time is $(2n + kw)$ IPS cycles. This algorithm is more suitable for a limited number of iteration steps, and for matrices with narrow bandwidth, i.e. $w \ll n$. Otherwise it leads to excessive area requirements.

Another approach is to use a single *mvm* systolic array with an external feedback mechanism, capable also of storing the intermediate results. This approach called Time Expansion, yields a systolic iterative architecture, with area of ω IPS cells, and computation time $(2kn + w)$ IPS cycles. This algorithm allows for an unlimited number of iterations, but it has the drawback of the long-wire interconnection in the feed-back mechanism, as well as the need of storage external to the IPS cells.

If applied to full matrices these algorithms require area and time of $k(2n - 1)$ IPS cells, and $2n + k(2n - 1)$ IPS cycles, for Area Expansion, and $(2n - 1)$ IPS cells and $2kn + (2n - 1)$ IPS cycles, for Time Expansion. The results of Time Expansion are comparable to those produced herein, but it should be pointed out that the need of external feedback and storage is eliminated.

In the case of banded matrices, the Area Expansion produces better time results than the re-usable *mvm* systolic array, but it has the disadvantage of a limited number of iterations. Compared to Time Expansion the re-usable array gives the same results, if the redundant IPS cells are replaced by simple delay elements. This is straightforward for the GS, SOR methods, since matrices are stored per diagonal. A similar arrangement can be achieved for the J, JOR methods, at the expense of $(n - 1)$ additional IPS cells or delay elements. Again, the main benefit is the elimination of external storage and feedback mechanism.

References

[1] O. Brudaru, Systolic algorithms to solve linear systems by iteration methods, Computer Centre, Polytechnical Institute, Iasi, Romania, 1985.
[2] M. Berzins, T. F. Buckley and P. M. Dew Systolic matrix iterative algorithms, Proc. Parallel Computing 83, North-Holland, 1984, pp. 483–488.
[3] H. J. Caulfield, J. H. Grunninger and W. K. Cheng, Using optical processors for linear algebra, Proc. SPIE, Optical Information Processing, 1983 (308), pp. 190–196.
[4] P. M. Dew, VLSI architectures for problems in numerical computation, in Supercomputers and parallel computation, Paddon, D. J. (ed.), Oxford Univ. Press, 1986, pp. 1–24.
[5] D. J. Evans and K. G. Margaritis, Improved systolic designs for iterative solution of linear systems, Proc. NUMETA 87, Swansea, 1987, pp. S18. 1–S18.17.
[6] D. J. Evans and K. G. Margaritis Systolic computation of matrix exponential and other matrix functions, *Int. J. Computer Math.*, 1988 **25**, pp. 345–358.
[7] D. J. Evans and K. G. Margaritis, Systolic designs for eigenvalue-eigenvector computations, Proc. PARCELLA 88, Berlin 1989.
[8] K. G. Margaritis, A study of systolic algorithms for VLSI processor arrays and optical computing, *Ph.D. Thesis*, Loughborough Univ., UK., 1988.
[9] K. G. Margaritis and D. J. Evans, A re-usable systolic optical matrix multiplication processor with applications, Proc. SPIE, Real Time Signal Processing X, (1987) vol (827).
[10] P. Quinton, B. Jannault and P. Gachet, A new matrix multiplication systolic array, in Parallel algorithms and architectures, Cosnard M. *et al.* (eds.), North-Holland, 1986, pp. 259–268.
[11] Y. G. Saridakis, Parallelism, applicability and optimality of modern iterative methods. *Ph.D. Thesis*, Clarkson Univ., USA, 1985.
[12] R. S. Varga, Matrix iterative analysis, Prentice-Hall, 1962.
[13] J. H. Wilkinson, The algebraic eigenvalue problem, Clarendon Press, 1971.
[14] D. Young, Iterative solution of large linear systems, Academic Press, 1971.
[15] H. T. Kung and C. E. Lerserson, Systolic Arrays for VLSI, in Proc. of Symp. on Sparse Matrix Computations. Edits. I. S. Duff and G. W. Stewart. 1978, 256–282.

SYSTOLIC MATRIX OPERATIONS

Systolic Givens Factorization of Dense Rectangular Matrices

MICHEL COSNARD and YVES ROBERT

CNRS, Laboratoire TIM3, Institut National Polytechnique de Grenoble, 38031 Grenoble Cedex, France and Ecole Normale Supérieure de Lyon, France

Given an m by n dense matrix $A(m \geqq n)$, we consider parallel algorithms to compute its orthogonal factorization via Givens rotations. First we describe an algorithm which is executed in $m+n-2$ steps on a linear array of $\lfloor m/2 \rfloor$ processors, a step being the time necessary to achieve a Givens rotation. The pipelined version of the new algorithm leads to a systolic implementation whose area-time performances overcome those of the arrays of Bojanczyk, Brent and Kung [1] and Gentleman and Kung [5].

KEY WORDS: Systolic algorithms, VLSI architectures, Givens rotations, QR factorization, linear least squares problems.

C.R. CATEGORIES: F.1.1 [Computation by abstract devices]: Models of Computation—*Systolic array*; F.2.1 [Analysis of algorithms and problem complexity]: Numerical algorithms and problems—*QR decomposition of a dense matrix*; B.7.1 [Integrated circuits]: Types and design styles—*Algorithms implemented in hardware, VLSI.*

1. INTRODUCTION

Given an m by n dense matrix $A(m \geqq n)$, we consider parallel algorithms to compute its orthogonal factorization via Givens rotations. That is, we want to compute an orthogonal n by n matrix

Q such that

$$QA = \begin{bmatrix} R \\ 0 \end{bmatrix}$$

where R is n by n upper triangular. Q is formed as the product of plane rotations. The sequential algorithm is well-known (see e.g. [7]) and has unconditional numerical stability [9]. We let $R(i, j, k)$, $i \neq j$, $1 \leqq i, j \leqq n$ and $1 \leqq k \leqq n$, denote the rotation in plane (i, j) which annihilates the element a_{ik}. For instance, $R(i, j, 1)$ combines row i and row j so that a_{i1} is annihilated, as depicted below:

$$\begin{bmatrix} a'_{j1}\, a'_{j2} \ldots a'_{jn} \\ 0 \quad a'_{i2} \ldots a'_{in} \end{bmatrix} = \begin{bmatrix} C & S \\ -S & C \end{bmatrix} \begin{bmatrix} a_{j1}\, a_{j2} \ldots a_{jn} \\ a_{i1}\, a_{i2} \ldots a_{in} \end{bmatrix}$$

When A is a general square matrix, computing the QR factorization is the major step to solve a linear system of eqs. $Ax = b$. When A is rectangular, this is also a key-step in computing unconstrained linear least square solutions [5, 14, 19]. Orthogonal triangularization is computationally expensive, however. Sequential algorithms for performing it typically require $0(mn^2)$ operations on general m by n matrices. In some applications, the size of the problem can be large [6], and in others the solution is needed in the shortest time possible as in real-time signal processing [10]. As a result, triangularization has become a bottleneck in some real-time applications [23]. As pointed out in [19], in either case, the use of multiprocessors offers definite advantages.

Indeed, parallel versions for SIMD or MIMD computers [4, 11, 22] of the Givens factorization algorithm have been introduced in the literature: see Lord *et al.* [16], Sameh and Kuck [21], Modi and Clarke [17], Cosnard *et al.* [2], Cosnard and Robert [3], and the survey papers of Heller [8] or Sameh [18, 20]. Assuming that enough processors are available to perform at each step up to $\lfloor m/2 \rfloor$ independent rotations simultaneously (where $\lfloor \cdot \rfloor$ denotes the floor function), Sameh and Kuck [21] propose a scheme whose number of steps is $m+n-2$ if $m>n$ and $2n-3$ if $m=n$. Sameh [19] also considers an implementation of the Givens factorization algorithm on a linear array of processors in the form of a ring. Finally, several systolic arrays have been introduced in the literature [1, 5] to compute the Givens factorization of a dense matrix in linear time using a quadratic number of elementary processors. We refer to [12], [13] for the systolic model.

We propose in this paper a new implementation of the Givens factorization algorithm on a linear array of processors. More precisely, let A be a dense m by n rectangular matrix ($m \geqq n$). We compute the Givens factorization of A in $m+n-2$ steps on a linear array of $\lfloor m/2 \rfloor$ processors, a step being the time necessary to achieve a Givens rotation. The pipelined version of the new algorithm leads to a systolic implementation whose area-time performances overcome those of the aforementioned arrays of [1, 5].

For the sake of clarity, we restrict ourselves to the square case $m=n$ until Section 4, where we consider the general case $m \geqq n$. Section 2 is devoted to the implementation of Sameh and Kuck's algorithm on a linear array of processors, and in Section 3 we deduce a new systolic array to perform the Givens factorization of a dense matrix.

2. SAMEH AND KUCK'S SCHEME

To introduce Sameh and Kuck's algorithm, let us fix $n=8$ for the purpose of illustration. The table of Figure 1 describes Sameh and Kuck's annihilation scheme [21]. An integer r is enetered when zeros are created at the rth step.

*							
7	*						
6	8	*					
5	7	9	*				
4	6	8	10	*			
3	5	7	9	11	*		
2	4	6	8	10	12	*	
1	3	5	7	9	11	13	*

Figure 1 Sameh and Kuck's scheme.

We do not specify completely each rotation in the table. In fact, when a rotation $R(i, j, k)$ is performed, the choice $j=i-1$ is systematically made. For instance at step 3, we perform simultaneously the rotations $R(6, 5, 1)$ and $R(8, 7, 2)$. Sameh and Kuck's scheme is easy to program and to analyze. Clearly, the total number of steps is $2n-3$ (13 in the example).

It can be easily seen that at step t we perform rotations $R(i, i-1, k)$ such that $n+2k=i+t+1$, $1 \leqq i \leqq n$.

We propose an implementation of Sameh and Kuck's scheme on a linear array of processors. Each processor has some local memory large enough to store two rows of the matrix A. As in Sameh and Kuck's scheme, we assume that a Givens rotation can be performed in one unit of time, irrespectively of the position k where the new zero is introduced.

There is an obvious solution which makes use of an array of $n-1$ processors, each processor P_k performing all the rotations $R(i, i-1, k)$, $k+1 \leqq i \leqq n$. This repartition of the rotations among the processors can be represented by the following table ($n=8$):

$$
\begin{array}{llllllll}
* \\
P_1 & * \\
P_1 & P_2 & * \\
P_1 & P_2 & P_3 & * \\
P_1 & P_2 & P_3 & P_4 & * \\
P_1 & P_2 & P_3 & P_4 & P_5 & * \\
P_1 & P_2 & P_3 & P_4 & P_5 & P_6 & * \\
P_1 & P_2 & P_3 & P_4 & P_5 & P_6 & P_7 & *
\end{array}
$$

Figure 2 Annihilation scheme with $n-1$ processors.

Rather than using $n-1$ processors, we can succeed with half this number by letting the rows of the matrix move backward as soon as all the rotations of the first column are completed, that is at time $n-1$. The repartition of the rotations among the processors is now the following ($n=8$):

$$
\begin{array}{llllllll}
* \\
P_1 & * \\
P_1 & P_1 & * \\
P_1 & P_2 & P_1 & * \\
P_1 & P_2 & P_2 & P_1 & * \\
P_1 & P_2 & P_3 & P_2 & P_1 & * \\
P_1 & P_2 & P_3 & P_3 & P_2 & P_1 & * \\
P_1 & P_2 & P_3 & P_4 & P_3 & P_2 & P_1 & *
\end{array}
$$

Figure 3 Annihilation scheme with $\lfloor n/2 \rfloor$ processors.

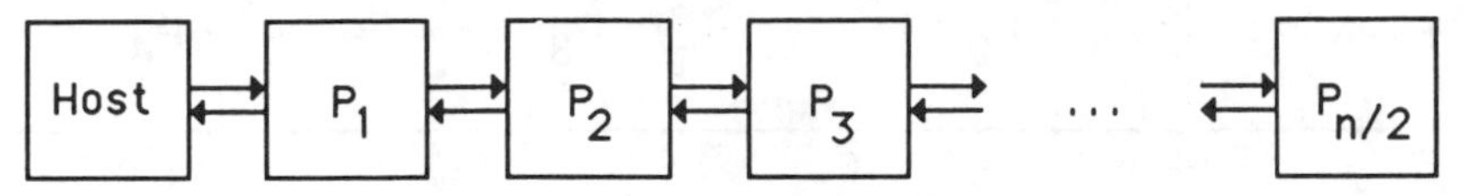

Figure 4 The linear array of processors.

The linear array of processors is depicted in Figure 4. There are two phases in the algorithm. At each time step $t \leqq n-1$ (7 in the example), data flow rightwards and each processor P_i operates as indicated in Figure 5. From step $t=n$ up to $t=2n-2$, the flow of data is inversed and the processors operate in a similar fashion (see Figure 6).

Let us describe phase 1 in more details. The rows of A are delivered by the host, a new row being input to the first processor P_1 at each time-step. The rows move rightwards, reaching a new processor every second step. The operation of the processors is the following:

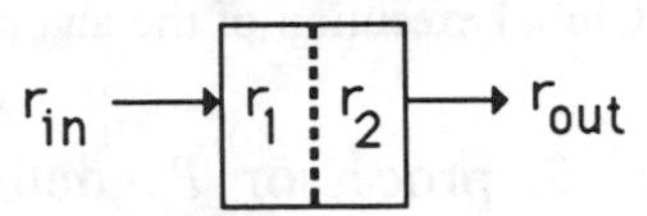

- perform a rotation between rows r_1 and r_2: $R(r_2, r_1, k)$; k is chosen to annihilate the leftmost non-zero element of r_2
- send r_2 to the right: $r_{out} := r_2$
- store r_1 and r_{in} in the local memory: $r_2 := r_1; r_1 := r_{in}$

Figure 5 Operation of the processors during phase 1.

At step $t=n$, the flow of data is reversed. We can either assume a global control or that each processor knows its number. In the latter case, processor P_i reverses the flow when it has performed $n-2i+1$ rotations. In the second phase, the processors operate as follows:

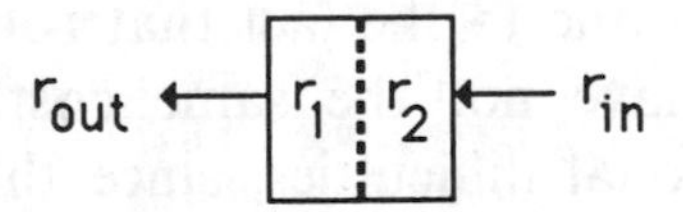

- perform a rotation between rows r_1 and r_2: $R(r_2, r_1, k)$; k is chosen to annihilate the leftmost non-zero element of r_2.
- send r_1 to the left: $r_{out} := r_1$
- store r_2 and r_{in} in the local memory: $r_1 := r_2; r_2 := r_{in}$

Figure 6 Operation of the processors during phase 2.

	P_1		P_2		P_3		P_4	
t=0	8							
t=1	78	R(8,7,1)						
t=2	67	R(7,6,1)	8					
t=3	56	R(6,5,1)	78	R(8,7,2)				
t=4	45	R(5,4,1)	67	R(7,6,2)	8			
t=5	34	R(4,3,1)	56	R(6,5,2)	78	R(8,7,3)		
t=6	23	R(3,2,1)	45	R(5,4,2)	67	R(7,6,3)	8	
t=7	12	R(2,1,1)	34	R(4,3,2)	56	R(6,5,3)	78	R(8,7,4)
t=8	23	R(3,2,2)	45	R(5,4,3)	67	R(7,6,4)	8	
t=9	34	R(4,3,3)	56	R(6,5,4)	78	R(8,7,5)		
t=10	45	R(5,4,4)	67	R(7,6,5)	8			
t=11	56	R(6,5,5)	78	R(8,7,6)				
t=12	67	R(7,6,6)	8					
t=13	78	R(8,7,7)						
t=14	8							

Figure 7 Global execution of the algorithm ($n=8$).

From $t=n$ to $t=2n-2$, processor P_1 delivers to the host a new row of the resulting matrix R every time step. The execution of the algorithm is globally described in Figure 7.

3. THE SYSTOLIC ARRAY

We propose in this section a systolic implementation of the previous algorithm. The key-idea is to pipeline in a systolic fashion the Givens rotations which were generated during the execution of the preceding algorithm. Therefore there will still be two main phases in the execution, the data input from the host moving first rightward in the array and being sent backward just as before. But now, the timing will be different, due to the fact that rotations creating a zero in different columns have not the same cost. Moreover, we shall encounter some additional difficulties, since the processing elements can no longer recognize their index, according to the generic criteria in systolic design [24], and global control must be avoided to reverse the direction of data flow.

Let A be a dense square matrix of order n, n even. The systolic array is a trapezoidal array of elementary processors, it is composed of $n/2$ columns and n rows (see Figure 8). Processors in row i are

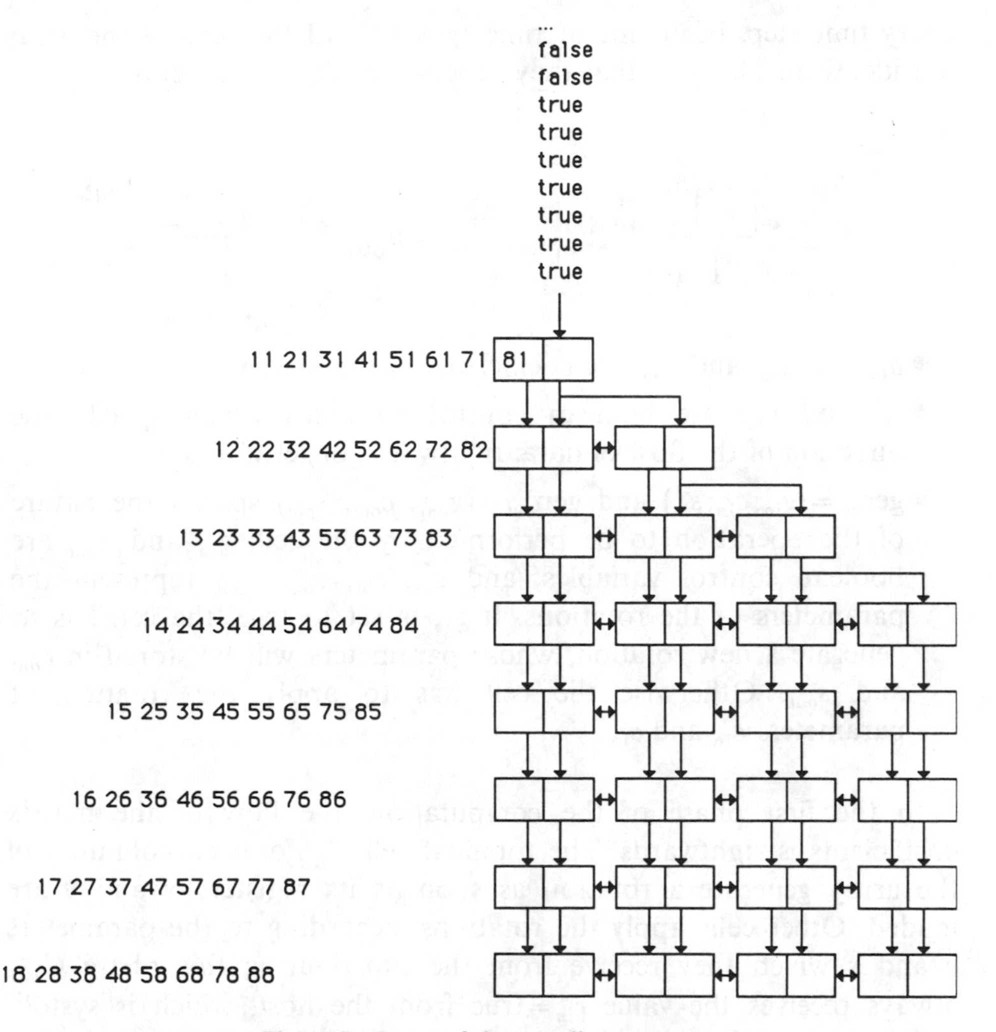

Figure 8 Input of the systolic array.

numbered from left to right $C_{i,1}, C_{i,2}, \ldots, C_{i,q}$ where $q=\min(i, n/2)$. Each processor is able to generate and apply a rotation. Hence two kinds of operations are required: (i) determination of the rotations parameters c and s (see [7] for details) and (ii) application of the rotation, which is equivalent to $x := cx + sy$ and $y := -sx + cy$.

To avoid a formal proof, we describe informally the operation of the array. The input format is shown in Figure 8. The matrix A is fed into the array column by column. More specifically, column k of the matrix is input in the array in the reverse order, one new element every time-step, beginning at time $t=k-1$. All the cells of the array are identical. The data that they process are described below:

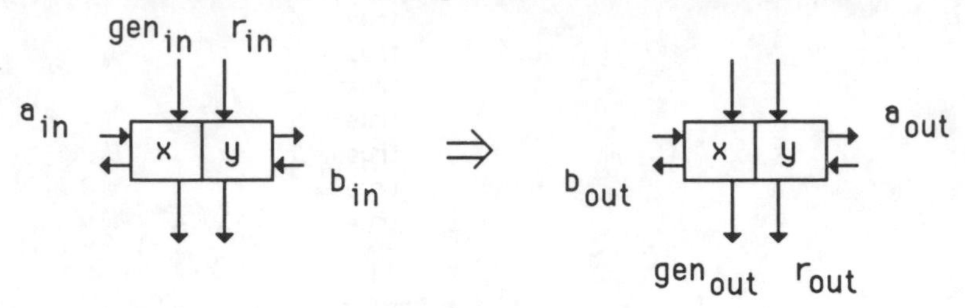

- a_{in}, b_{in}, a_{out} and b_{out} are coefficients of the matrix
- r_{in} and r_{out} are boolean control variables which specify the direction of the flow of data: rightwards or leftwards.
- $\text{gen}_{in}=(g_{in}, c_{in}, s_{in})$ and $\text{gen}_{out}=(g_{out}, c_{out}, s_{out})$ specify the nature of the operation to be performed by the cell: g_{in} and g_{out} are boolean control variables, and c_{in}, s_{in}, c_{out}, s_{out} represent the parameters of the rotations. If g_{in} is set to true, the cell has to generate a new rotation, whose parameters will be stored in c_{out} and s_{out}. Otherwise the cell has to apply the rotation of parameters c_{in} and s_{in}.

In the first phase of the computation, the flow of the matrix coefficients is rightwards. The topmost cell $C_{j,j}$ of each column j of the array generate a rotation as soon as its registers x and y are loaded. Other cells apply the rotations, according to the parameters c and s which they receive from the top. During this phase $C_{1,1}$ always receives the value $r_{in}=\text{true}$ from the host, which is systolically propagated to the whole array.

Phase 2 begins when $C_{1,1}$ receives the value $r_{in}=\text{false}$ from the host: this happens at step $n-1$ if a_{n1} is input to C at step 0. The instruction to reverse the direction of the flow is then systolically propagated to the whole array, a new row of cells reversing the flow every time-step. When a cell has its register y empty ($y=\text{nil}$), it sends the instruction to generate a rotation to its southern neighbour. Hence in a given column j, the cells which generate a rotation are successively $C_{j,j}, C_{j+1,j}, \ldots, C_{n,j}$. As in phase 1, at a given time-step, there is a single cell per column which generates a rotation, all the cells below apply previous rotations.

The operation of the cells is detailed in Figure 9. We assume that the topmost cells $C_{j,j}$, of all the columns always receive the value $g_{in}=\text{true}$.

We can state our main result:

THEOREM *The systolic array of* $3n(n-2)/8$ *cells computes the Givens factorization of the dense square matrix* A *of order* n, n *even, in* $3n-1$ *steps.*

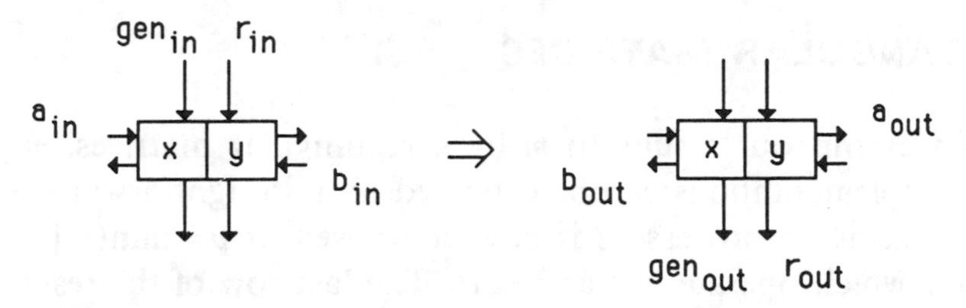


```
{ send the instruction to generate a rotation below iff register y is
empty }
if y = nil then g_out := true else g_out := false ;
{ select the operation of the cell according to g_in }
if g_in then GENERATE(x,y,c_out,s_out)
else APPLY(x,y,c_in,s_in) ;
{ select the direction of the flow according to r_in }
if r_in then RIGHTWARDS(x,y,a_in,a_out)
else LEFTWARDS(x,y,b_in,b_out) ;
r_out := r_in ;

Procedure GENERATE (x,y,c,s):
    { rotate (x,y) to annihilate y }
    if y = 0 then begin c := 1; s := 0 end
    else
      if |y| ≥ |x| then begin t := x/y ; s := 1/(1+t^2)^1/2 ; c := st end
      else begin t := y/x ; c := 1/(1+t^2)^1/2 ; s := ct end ;
    x := cx + sy ; y := nil ;

Procedure APPLY (x,y,c,s):
    { apply rotation (c,s) to (x,y) }
    u := x ; v := y ; x := cu + sv ; y := -su + cv ;

Procedure RIGHTWARDS(x,y,a,a1)
    a1 := y ; y := x ; x := a ;

Procedure LEFTWARDS(x,y,b,b1)
    b1 := x ; x := y ; y := b ;
```

Figure 9 Operation of the cells in the array.

We point out that our array performs the factorization within the same number of steps as the arrays of [1] and [5], although it requires only $3n^2/8$ cells instead of $n^2/2$. The small price we pay for our array is a little additional control (two booleans per cell).

4. RECTANGULAR MATRICES

We can extend our results to m by n rectangular matrices, $m \geqq n$. Our two implementations can be extended in a straightforward way. The linear array of processors is now composed of $p = \min(n, \lfloor m/2 \rfloor)$ processors, which operate just as before. The last row of the resulting matrix is now delivered to the host by the leftmost processor at step $m+n-2$. Figures 10 and 11 show the cases $m=8$, $n=3$ and $m=8$,

	P_1		P_2		P_3	
t=0	8					
t=1	78	R(8,7,1)				
t=2	67	R(7,6,1)	8			
t=3	56	R(6,5,1)	78	R(8,7,2)		
t=4	45	R(5,4,1)	67	R(7,6,2)	8	
t=5	34	R(4,3,1)	56	R(6,5,2)	78	R(8,7,3)
t=6	23	R(3,2,1)	45	R(5,4,2)	67	R(7,6,3)
t=7	12	R(2,1,1)	34	R(4,3,2)	56	R(6,5,3)
t=8	23	R(3,2,2)	45	R(5,4,3)		
t=9	34	R(4,3,3)				

Figure 10 Global execution of the algorithm ($m=8, n=3$).

	P_1		P_2		P_3		P_4	
t=0	8							
t=1	78	R(8,7,1)						
t=2	67	R(7,6,1)	8					
t=3	56	R(6,5,1)	78	R(8,7,2)				
t=4	45	R(5,4,1)	67	R(7,6,2)	8			
t=5	34	R(4,3,1)	56	R(6,5,2)	78	R(8,7,3)		
t=6	23	R(3,2,1)	45	R(5,4,2)	67	R(7,6,3)	8	
t=7	12	R(2,1,1)	34	R(4,3,2)	56	R(6,5,3)	78	R(8,7,4)
t=8	23	R(3,2,2)	45	R(5,4,3)	67	R(7,6,4)	8	
t=9	34	R(4,3,3)	56	R(6,5,4)	78	R(8,7,5)		
t=10	45	R(5,4,4)	67	R(7,6,5)				
t=11	56	R(6,5,5)						

Figure 11 Global execution of the algorithm ($m=8, n=5$).

$n=5$. Notice that we have simply deleted the operations with indices out of range from the original table (square case).

For the systolic implementation, the array consists now of p columns ($p=\min(n, \lfloor m/2 \rfloor)$) of respectively $n, n-1, \ldots, n-p+1$ cells. The new array can be obtained by suppressing data and cells relative to indices out of range.

The Givens factorization array in [1] has been described only in the square case but can be easily extended to handle rectangular matrices. We can derive the following table, where asymptotic values are given:

ARRAY	[1]	[5]	this paper
number of steps	$m+2n$	$m+2n$	$m+2n$
number of cells	$n(m-n/2)$	$n^2/2$	$n^2/2$ if $n<m/2$
			$m(n-m/4)/2$ otherwise

We see that the three systolic arrays operate in the same number of steps, but their cell requirements are different. We can distinguish two cases:

i) $n<m/2$. Our array has the same number of cells than that of [5], and less than that of [1].

ii) $n>m/2$. Our array has fewer cells than that of [5], which in turn has fewer cells than that of [1].

Let $n=\alpha m, \alpha \leqq 1$. The ratio R_α of the number of cells in our array over the number of cells in [5] is $R_\alpha=1$ if $\alpha \leqq 1/2$ and $R_\alpha=(\alpha-1/4)/\alpha^2$ otherwise. The maximum value of R_α for $\alpha \in [1/2, 1]$ is obtained when $\alpha=1$, with $R_1=3/4$.

References

[1] A. Bojanczyk, R. P. Brent and H. T. Kung, Numerically stable solution of dense systems of linear equations using mesh-connected processors, Technical Report, Carnegie-Mellon University (1981).

[2] M. Cosnard, J. M. Muller and Y. Robert, Parallel QR decomposition of a rectangular matrix, *Numerische Mathematik* **48** (1986), 239–249.

[3] M. Cosnard and Y. Robert, Complexity of QR factorization, *J.A.C.M.* **33**, 4 (1986), 712–723.

[4] M. J. Flynn, Very high-speed computing systems, *Proc. IEEE* **54** (1966), 1901–1909.

[5] W. M. Gentleman and H. T. Kung, Matrix triangularisation by systolic arrays, *Proc. SPIE* **298,** Real-time Signal Processing IV, San Diego, California (1981).

[6] G. H. Golub and R. Plemmons, Large-scale geodetic least-squares adjustment by dissection and orthogonal decomposition, *Linear Algebra and Appl.* **38,** (1980), 3–28.

[7] G. H. Golub and C. F. Van Loan, *Matrix Computations*, The Johns Hopkins University Press, 1983.

[8] D. Heller, A survey of parallel algorithms in numerical linear algebra, *Siam Review* **20,** (1978), 740–777.

[9] D. Heller and I. Ipsen, Systolic networks for orthogonal equivalence transformations and their applications, *Proc. 1982 Conf. Advanced Research in VLSI*, MIT 1982, pp. 113–122.

[10] S. Horwarth Jr, A new adaptative recursive LMS filter. In: *Proc. Digital Signal Processing* (Cappellini and Constantinides, eds.), Academic Press, 1980, pp. 21–26.

[11] K. Hwang and F. Briggs, *Parallel Processing and Computer Architecture*, MacGraw-Hill, 1984.

[12] H. T. Kung, Why systolic architectures, *IEEE Computer* **15,** 1 (1982), 37–46.

[13] H. T. Kung and C. E. Leiserson, Systolic arrays for (VLSI). In: *Proc. of the Symposium on Sparse Matrices Computations* (I. S. Duff and G. W. Stewart, eds.), Knoxville, Tenn, 1978, pp. 256–282.

[14] C. Lawson and R. Hanson, *Solving Least Squares Problems*, Prentice-Hall, 1974.

[15] C. E. Leiserson, Area-efficient VLSI computation, Ph.D Thesis, Carnegie-Mellon University, Pittsburgh, PA, USA, October 1981.

[16] R. E. Lord, J. S. Kowalik and S. P. Kumar, Solving linear algebraic equations on an MIMD computer, *J. ACM* **30,** 1 (1983), 103–117.

[17] J. J. Modi and M. R. B. Clarke, An alternative Givens ordering, *Numerische Mathematik* **43** (1984), 83–90.

[18] A. Sameh, Numerical parallel algorithms—a survey. In: *High Speed Computer and Algorithm Organization* (D. Kuck, D. Lawrie and A. Sameh, eds.), Academic Press, 1977, pp. 207–228.

[19] A. Sameh, Solving the linear least squares problem in a linear array of processors, Proc. Purdue Workshop on algorithmically-specialized computer organizations, W. Lafayette, Indiana, September 1982.

[20] A. Sameh, An overview of parallel algorithms, *Bull. EDF* **C1,** (1983), 129–134.

[21] A. Sameh and D. J. Kuck, On stable parallel linear system solvers, *J. ACM* **25,** 1 (1978), 81–91.

[22] U. Schendel, *Introduction to Numerical Methods for Parallel Computers*, Ellis Horwood Series, J. Wiley & Sons, New York, 1984.

[23] J. M. Speiser and M. J. Whitehouse, Architecture for real-time matrix operations, Proc. of Government Microcircuits Applications Conference, 1980.

[24] J. D. Ullman, *Computational Aspects of VLSI*, Computer Science Press, 1983.

Triangular Systolic Arrays for Matrix Product and Factorisation

G. M. MEGSON* and D. J. EVANS

Department of Computer Studies, Loughborough University of Technology, Loughborough, Leicestershire, UK

Triangular systolic arrays for performing the matrix product $C=AB$ and LU triangular matrix factorisation of 2 $n*n$ matrices A and B with bandwidths w_1, w_2 respectively, and $w=p+q-1$ are derived using the properties of reflection and refraction of systolic wavefronts. When compared with the hexagonally connected arrays of Kung and Leiserson [1] these new arrays have an efficiency of $e=2/3$ rather than 1/3, and for $w_1 \approx w_2$ (product), $p \approx q$ (factorisation) save approximately half the number of cells whilst maintaining the same computation time $T=3n+\min(a,b)$ where,

$$a=\begin{cases} w_1 \\ p \end{cases}, \quad b=\begin{cases} w_2 & \text{product} \\ q & \text{factorisation} \end{cases}$$

KEY WORDS: Ray diagram, triangular hex array, optical preprocessing, computational interference.

C.R. CATEGORIES: F.1.1, F.2.1, B.7.1.

1. INTRODUCTION

The development of systolic arrays can now be classified into two major activities:

1) The design of an initial array by algorithm synthesis; for example by mapping a synchronous algorithm into a systolic design (Saxe and Leiserson [4]).

*Oriel College, University of Oxford.

2) Retiming data and replacing cells to optimise the array (Rothe [2]).

In this paper we extend a technique developed by the authors in [9] which uses the principles of reflection, refraction, and wave interference from optics to model improvements to existing systolic arrays. The principle tool is the expression of the array as a ray diagram with related data inputs emanating from distinct and distant point sources. Different sources are coherent (or of the same type) and the interference of data wavefronts reflects the effectiveness of retiming by constructive, neutral, and destructive interferences. Introducing reflecting boundaries identifies virtual images of portions of the array allowing the replacement of cells and the simulation of virtual computations by retiming onto the real parts of the array.

The techniques used here are intended to aid the systolic array designer rather than replace him/her completely by the automatic array synthesis techniques currently under investigation for silicon compilers.

2. A TRIANGULAR MATRIX PRODUCT ARRAY

In Figure 1 the cell activation pattern is illustrated for a typical hexagonal connected array of Kung and Leiserson [1, 3]. Partitioning the array indicates that the order of computation in the left and right portions are identical. If we consider the input matrices A and B to emanate from two distant point sources and interpret the input data sequences as rays forming distinct wavefronts (acting according to the Huygens' principle). The array computation is mimicked by the interference of two waves.

Constructive interference being a true inner product step (ips) calculation,

Destructive interference an incorrect computation,

and Neutral interference a masked calculation (of the form $y=y+(0*a)$).

It follows that the original hex array generates a resultant wavefront C from constructive and neutral interference. Any modification of

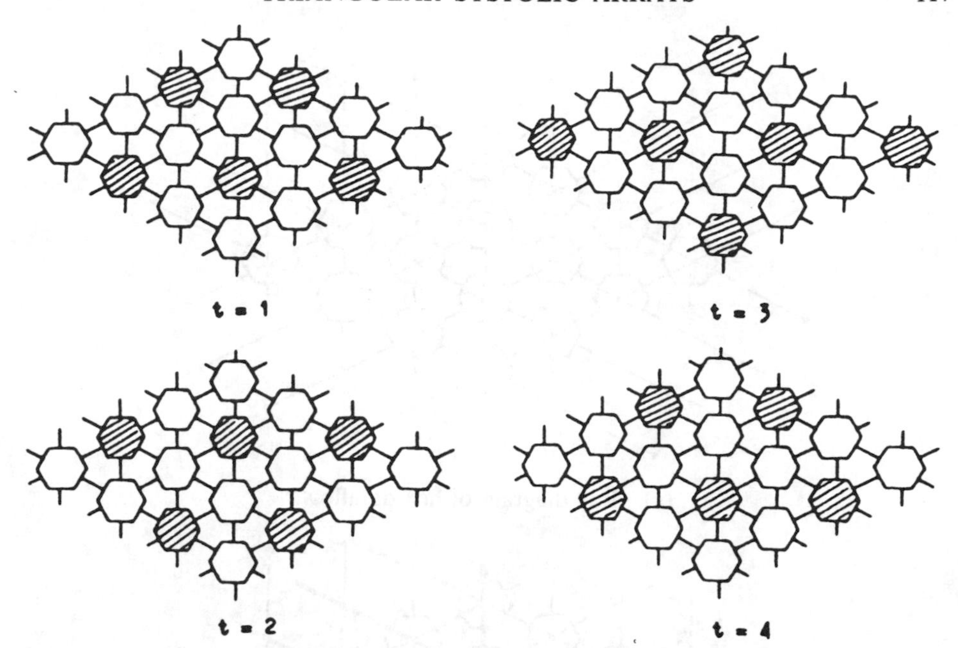

Figure 1 Activation of cells in a hexagonal array.

the existing arrays must therefore retain a constructive wavefront pattern by replacing and/or retiming data and cells.

This interference can be charted by the use of the ray diagram shown in Figure 2a, and the symmetry of the cell activation modelled by a "mirror" placed on the central cell column. Now, if we consider a single point source combining A and B, the array for the product $C=AB$ in Figure 2b consists of a triangular right half computing real calculations and a "mirror" section which defines virtual computations for the real left half of Figure 2a produced by mirror reflection. The trick is now to define new cells on the mirror boundary to reflect data and interleave the real computations on the right part with simulations of the virtual computations in the left half of the array.

A simple intuitive approach is to fold the virtual array part over onto the real part. This immediately reduces the number of cells by half mapping two computations to each active cell resulting in either doubling the cell cycle time or the cell hardware to maintain constructive interference and in the latter case computation time. A less intuitive approach which maintains cycle time and cell hardware is to re-time the virtual part of the array so that it operates one

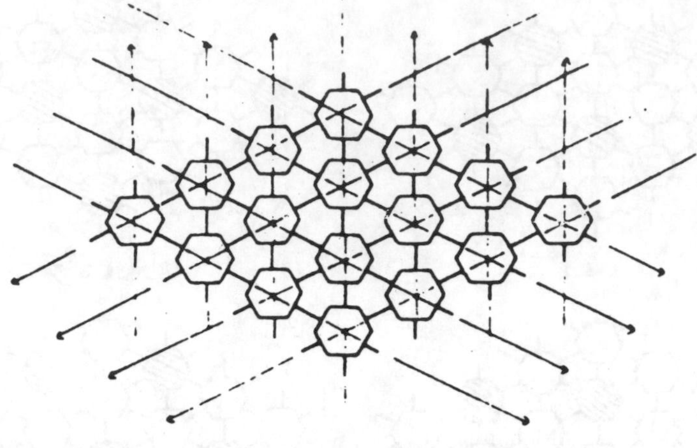

(a) Ray diagram of hex dataflow.

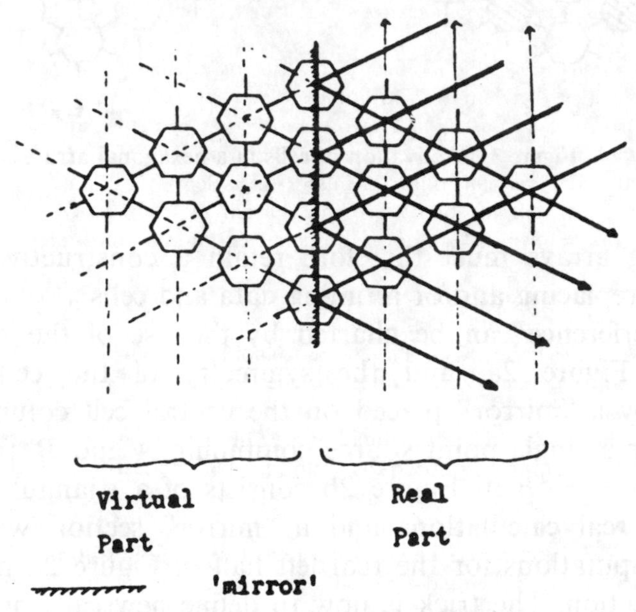

(b) Ray diagram with reflecting boundary.

Figure 2

cycle behind the real part. Now array folding allows the real and simulated virtual calculation to share the same cell in a mutually exclusive fashion. This is by essentially utilising the inactive cell pattern in Figure 1 produced by synchronising delay elements of the original Kung and Leiserson input data stream.

Snapshots of this new array format are shown in Figure 3, clearly the efficiency $e=1/3$ in the original hex connected scheme is im-

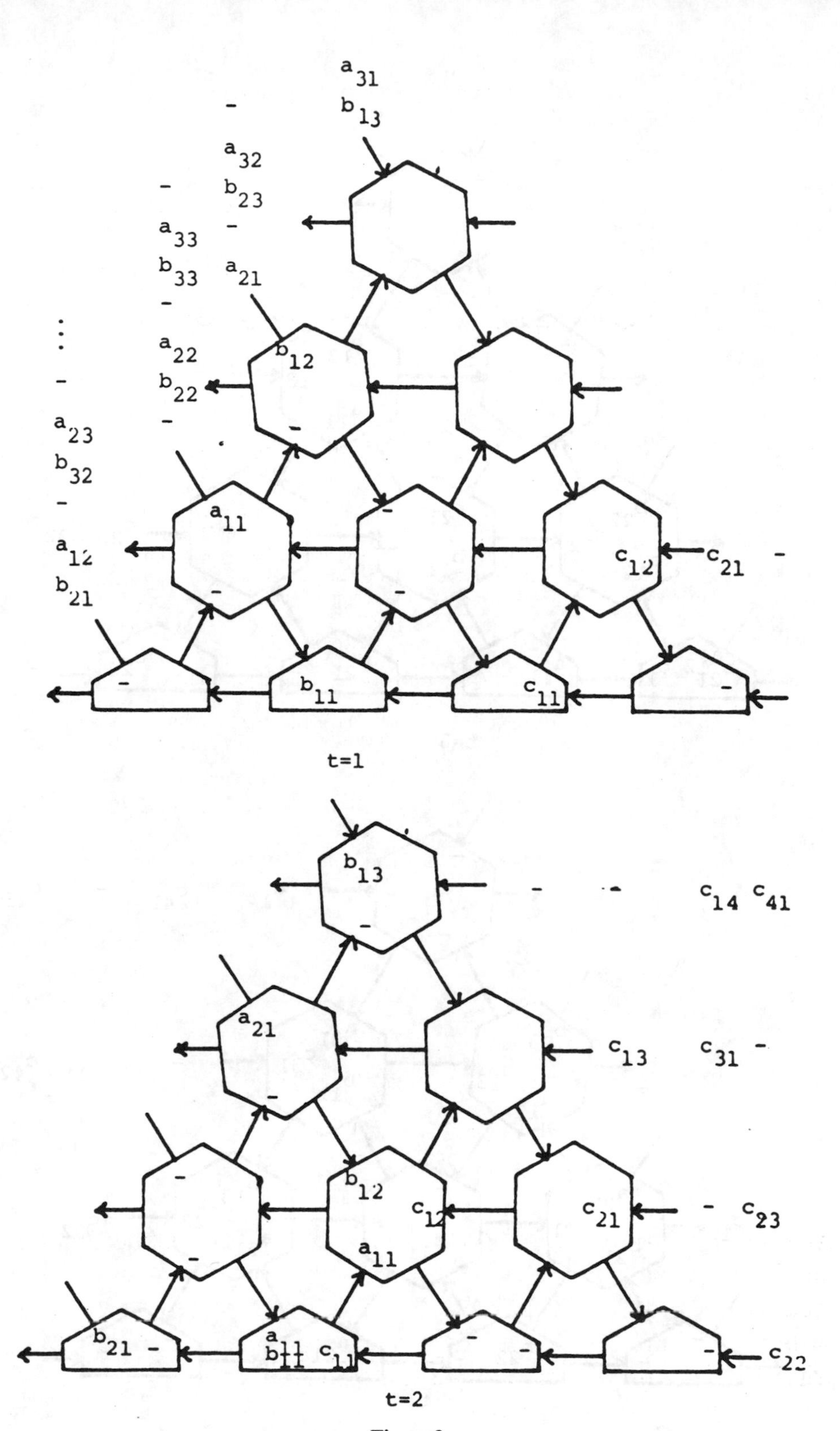

a_{31}
b_{13}
a_{32}
b_{23}
a_{33}
b_{33}
a_{21}
b_{12}
a_{22}
b_{22}
a_{23}
b_{32}
a_{11}
a_{12}
b_{21}
c_{12}
c_{21}
b_{11}
c_{11}
t=1
b_{13}
c_{14} c_{41}
a_{21}
c_{13} c_{31}
b_{12}
c_{12}
c_{21}
c_{23}
a_{11}
b_{21}
c_{11}
c_{22}
t=2

Figure 3

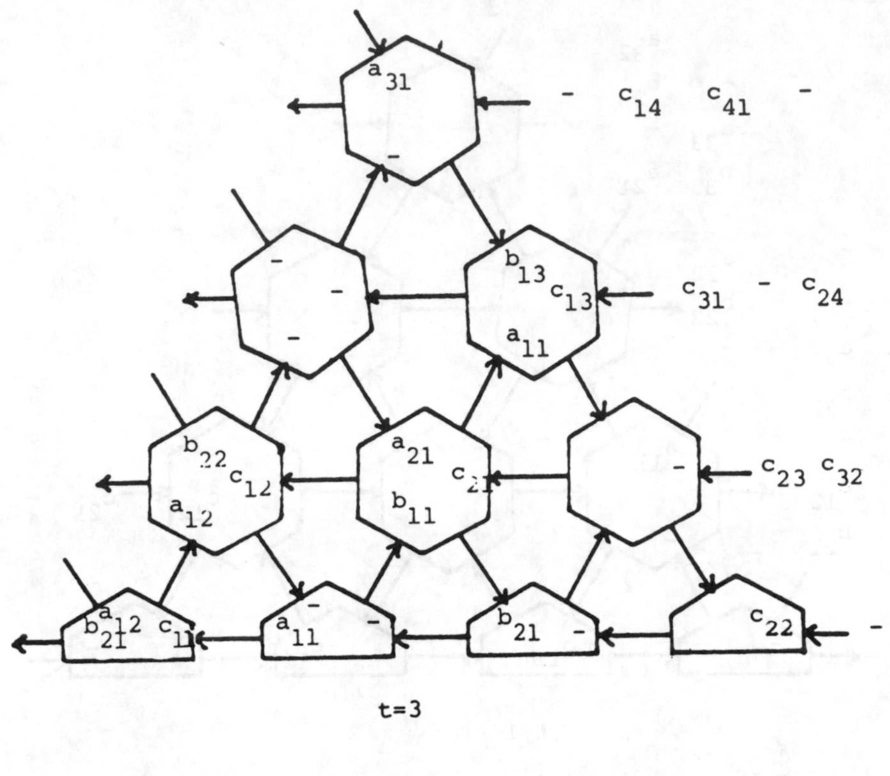

t=3

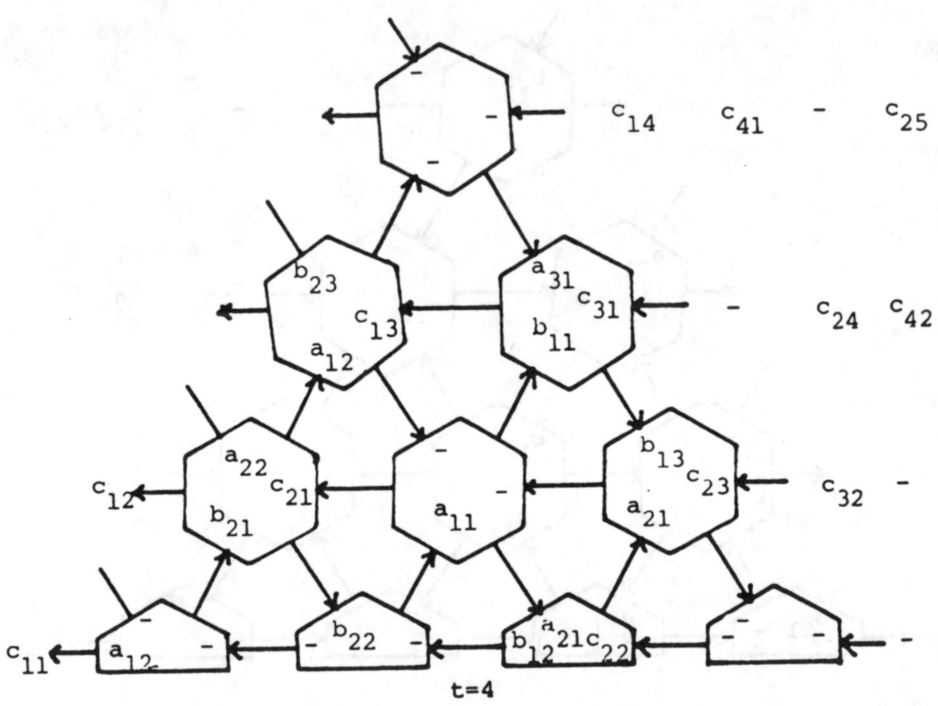

t=4

Figure 3

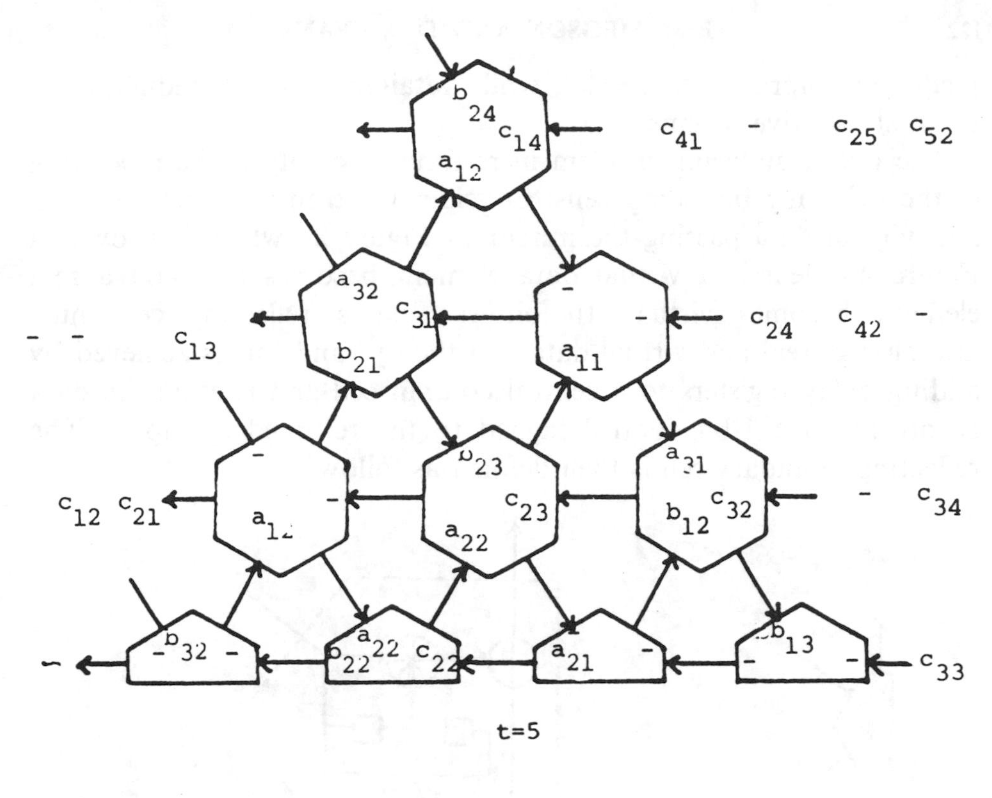

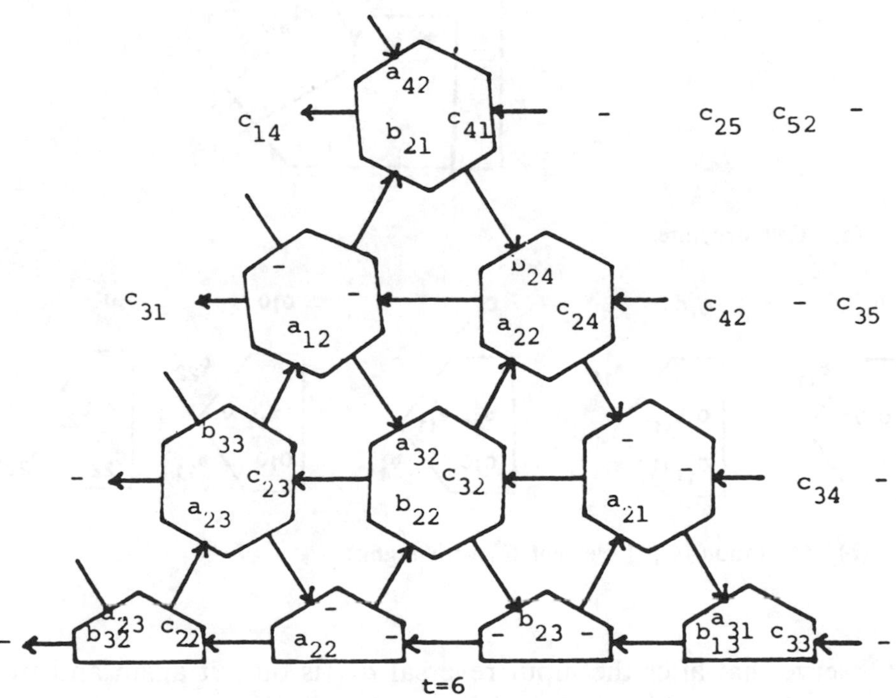

Figure 3 Snapshots of triangular matrix product.

mediately improved to $e=2/3$, while retaining the cell reduction of the first intuitive scheme.

The only significant problem in re-timing activity is the modelling of the reflecting boundary cells. Consider the transition between real and virtual data passing the mirror in Figure 2b which is shown in Figure 4. Clearly, a virtual data element becomes real and a real element becomes virtual. It follows that a reflecting cell must interchange real and virtual data "on-the-fly" and this is achieved by adding delay registers to a central column cell and routing the data according to a 1-bit control tagged to the re-timed c_{ii} inputs. The reflecting boundary cell is then defined as follows:

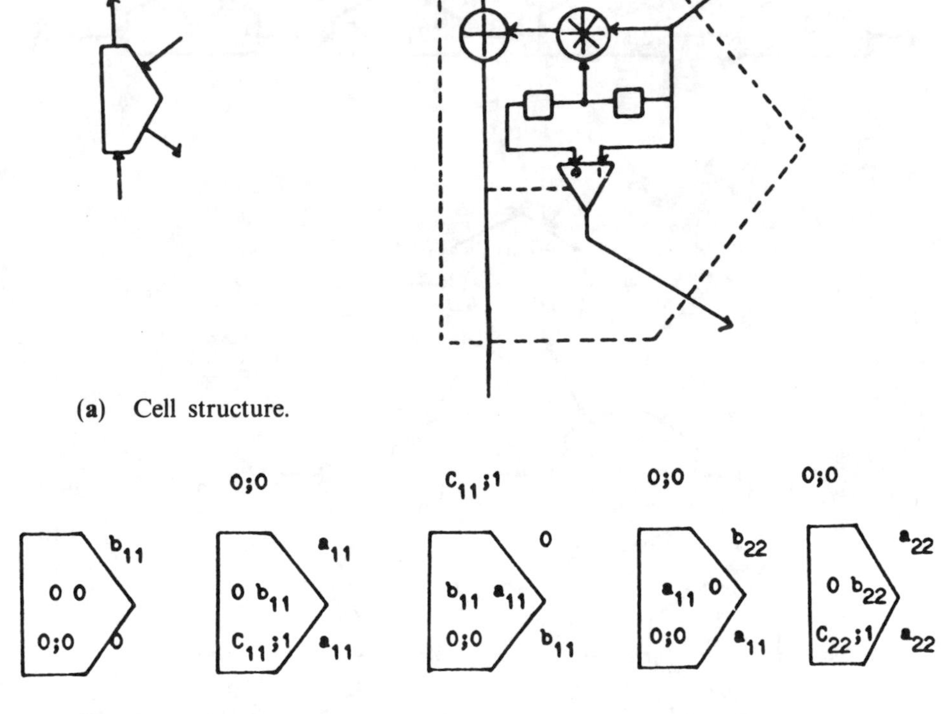

(a) Cell structure.

(b) Operation (C_{ii}; 1 element C_{ii} with tagbit).

Observe that after the input reversal a_{11} is output again and from the snapshots in Figure 3 it follows that only neutral interference occurs when this element enters a cell. Thus, in general no C_{ij} values

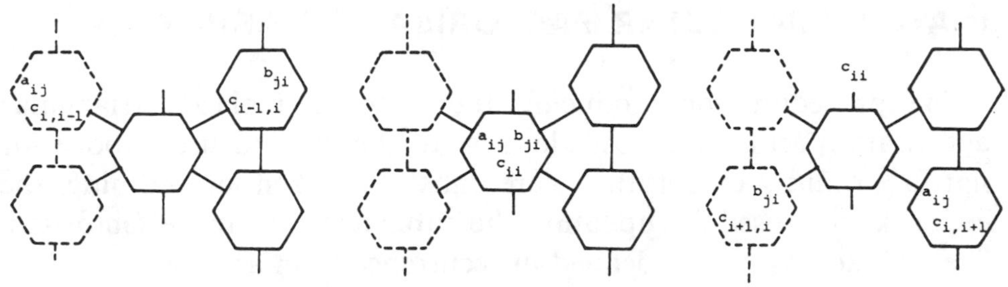

(a) Real to virtual scheme.

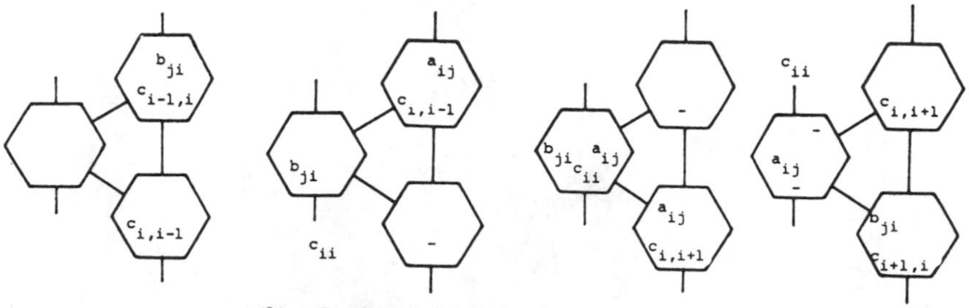

(b) Retimed folded scheme.

Figure 4 Boundary cell computation.

are incorrectly modified and computation is preserved using a combination of neutral and constructive interference.

Finally, the computation time of the original hex-connected array is unchanged because:

i) The number of boundary cells in the folded array is the same as the number of central cells in the original design, maintaining the output delay.

ii) The original central diagonal input has the form

$$c_{11}\ 0\ 0\quad c_{22}\ 0\ 0\quad c_{33}\ \ldots\quad \ldots\ 0\ 0\quad c_{nn}\ 0\ 0$$

of length $3n$, consequently the retimed input has the form

$$0\quad c_{11}\ 0\ 0\quad c_{22}\ 0\ 0\quad c_{33}\ \ldots\quad \ldots\ 0\ 0\quad c_{nn}\ 0$$

which is still of length $3n$.

Thus, folding reduces the hardware of the hex design but leaves the input data length unchanged.

3. A *LU* TRIANGULAR FACTORISATION ARRAY

The procedure for producing the re-timed replaced triangular factorisation array is almost identical to the method used above for matrix product, except that care must be taken in retiming the feedback loop used for updating the submatrix still to be factorised. The LU factorisation is defined in recurrence form as,

$$a_{ij}^{(1)}=a_{ij}$$

$$a_{ij}^{(k+1)}=a_{ij}^{(k)}+l_{ik}(-u_{kj})$$

$$l_{ik}=\begin{cases}0 & i<k\\ 1 & i=k\\ a_{1k}^{(i)}/u_{kk} & i>k\end{cases}$$

and

$$u_{kj}=\begin{cases}0 & k>j\\ a_{kj}^{(k)} & k\leqq j\end{cases}$$

where k denotes the kth submatrix. In the original array of Kung and Leiserson [1] the hex cells on the upper boundary were rotated right 120° on the left, and left 120° on the right. A natural correspondence between the matrix product array and factorisation array follows immediately by interpreting the modified cells as elements of a reflecting mirror producing two virtual sources as shown in Figure 5.

It is now evident that the same folding technique is applicable and providing the reciprocal cell outputs -1 for neutral inputs and the reciprocal for real inputs, the feedback calculations are preserved. Snapshots of the triangular array operation are shown in Figure 6 and as before the efficiency is $e=2/3$ while only half the number of cells are required compared with the original array.

Finally, consider the output of the L and U elements. In the original scheme, the l_{ij} values were produced from the south-east

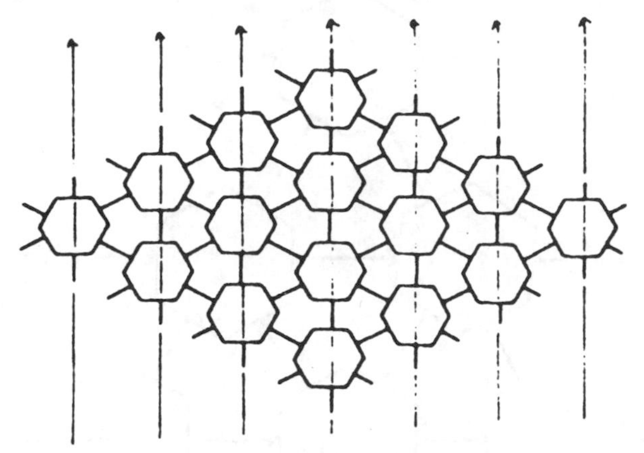

(**a**) Ray diagram of *LU* hex dataflow.

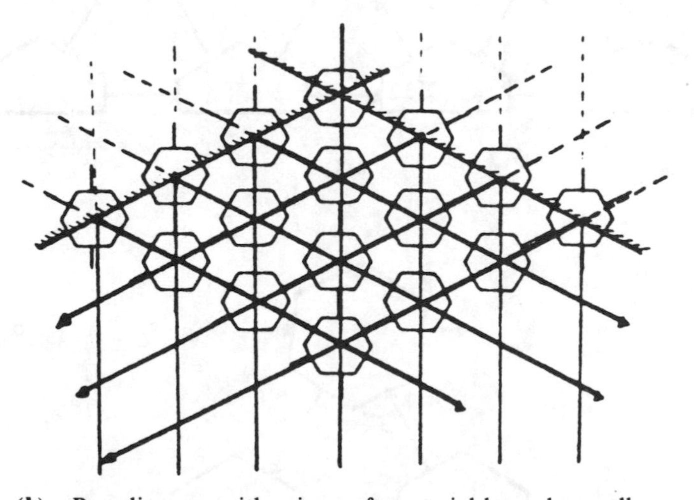

(**b**) Ray diagram with mirrors for special boundary cells.

Figure 5 Locating virtual sources of the *LU* hex array.

output of boundary cells on the left, and the u_{ij} by north outputs of boundary cells on the right. In the folded array, the u_{ij} and l_{ij} elements are output from the SW and N outputs of an upper boundary cell, indicating that folding does not reduce the array communication bandwidth as it did for the product array.

However, further observation of the data flow in Figure 6 shows that the results can be filtered out from the reflecting "mirror" boundary associated with the central column of unfolded cells. Now as $a_{nn}^{(n-1)}$ must be updated by the last u_{ij} and l_{ij} values produced on the cycle before reaching the reciprocal cell, it follows that the

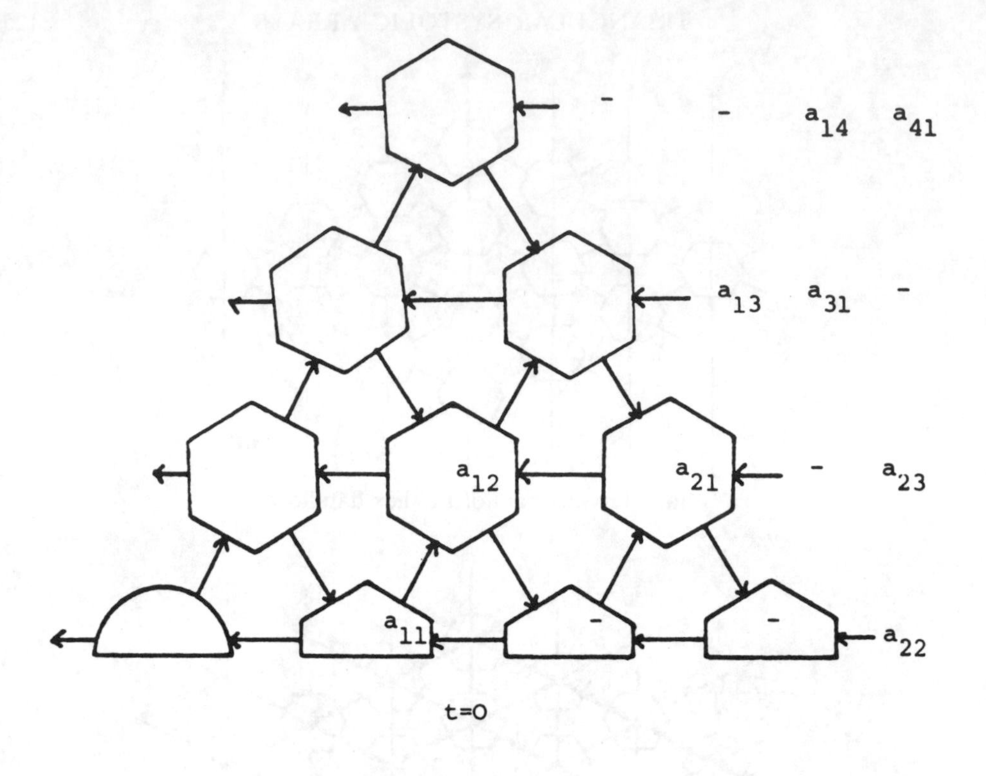

t=0

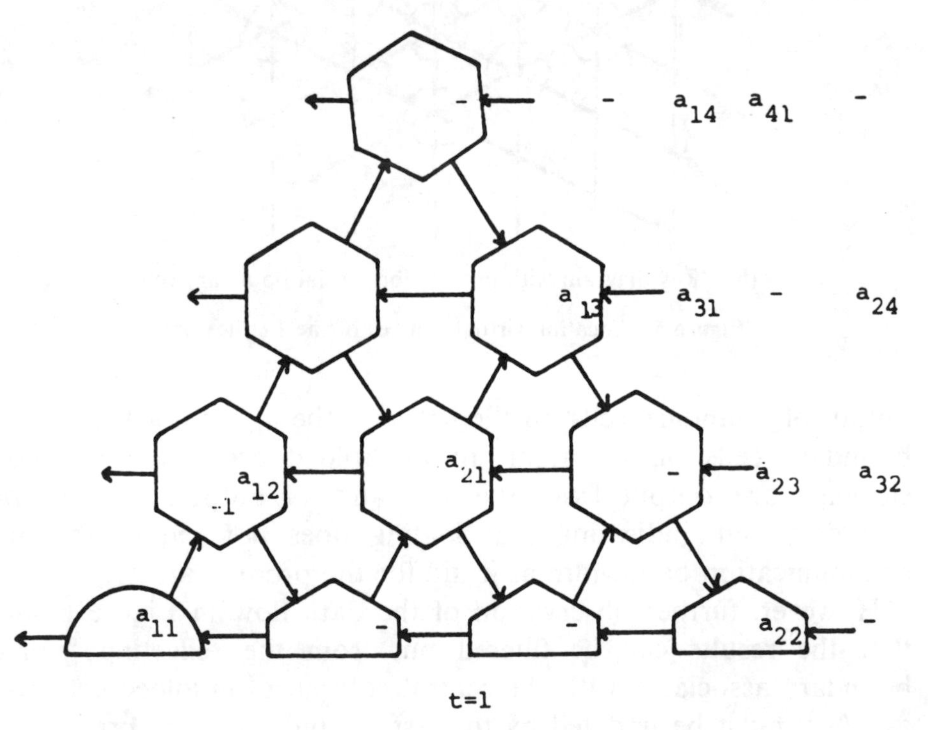

t=1

Figure 6

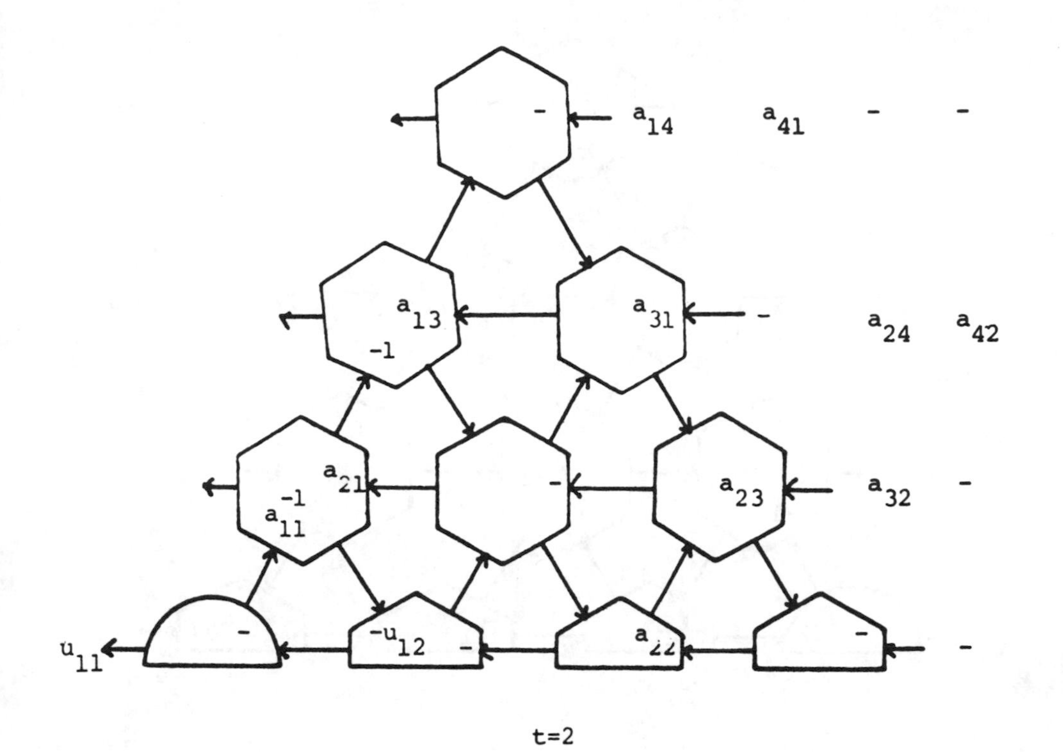

t=2

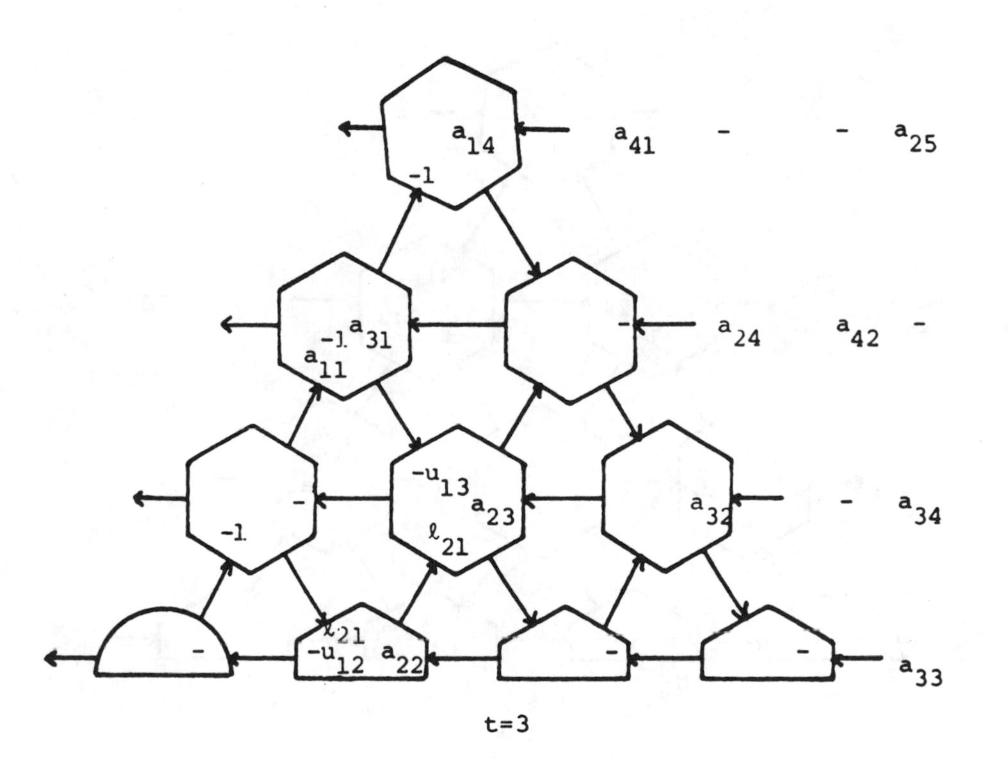

t=3

Figure 6

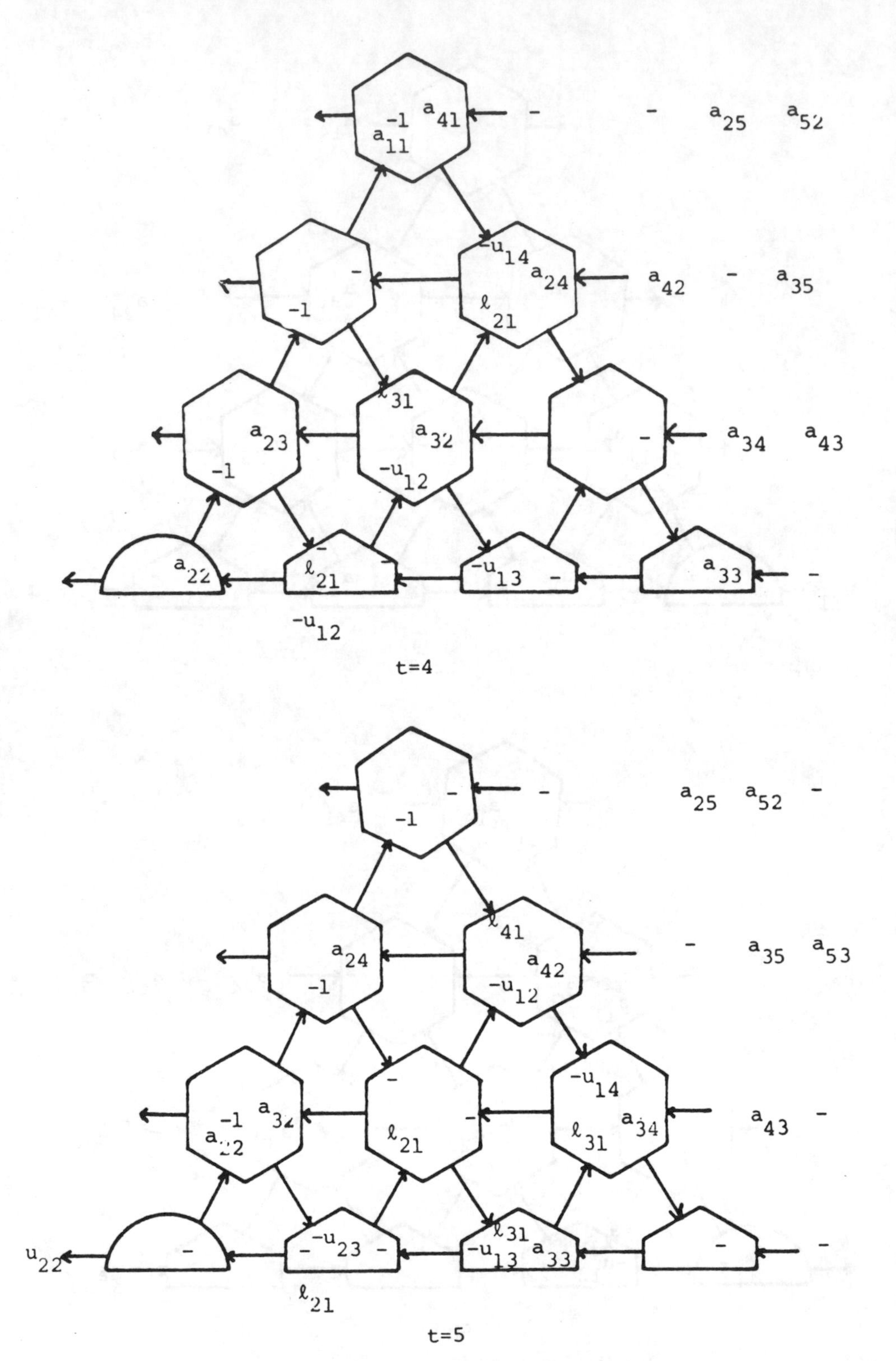

Figure 6 Snapshots of triangular *LU* factorisation.

computation time of the folded array is at most $t=3n+\min(p,q)+1$ with an additional cycle for the last l_{ij} to leave the reflecting boundary cell. Notice however that the triangular array is completely planar, whereas, the rotation of the left boundary cells in the unfolded scheme make it non-planar.

4. FOLDING SKEWED ARRAYS

It should be clear that the folding scheme proposed above is only applicable when the hexagonal array is symmetrical about the central column which implies the following:

i) $w_1=w_2$ for a matrix product, and $p=q$ for matrix factorisation.

ii) To preserve the computation pattern of Figure 1 we must have $w_1=p_1+q_1-1$ and $w_2=p_2+q_2-1$ for a product such that $p_1=q_2$ and $p_2=q_1$.

Notice that a matrix factorisation naturally satisfies this if (i) is true.

For arrays with a skewed structure (i.e. $w_1 \neq w_2, p \neq q$) the folding technique breaks down as the swapping of reflected elements becomes increasingly difficult as the interleaved b_{ij} and a_{ij} values become increasingly spatially and temporally separated. A simple solution to these problems is to pad out the array up to a symmetrical shape by adding extra cells, as illustrated by Figure 7.

Now, to make a skewed and folded array triangular results in the region A in Figure 7a where $w=|w_2-w_1|$. The folding of the original skewed hex actually saves,

$$\sum_{i=1}^{\min(w_1,w_2)} i=\tfrac{1}{2}\min(w_1,w_2)(\min(w_1,w_2)+1) \tag{4.1}$$

while the region A adds,

$$\sum_{i=1}^{w} i=\tfrac{1}{2}w(w+1) \tag{4.2}$$

consequently with the original unfolded array requiring w_1w_2 hex ips cells the folded and padded triangular array requires a total of,

$$A=w_1w_2-\tfrac{1}{2}\min(w_1,w_2)(\min(w_1,w_2)+1)+\tfrac{1}{2}w(w+1) \tag{4.3}$$

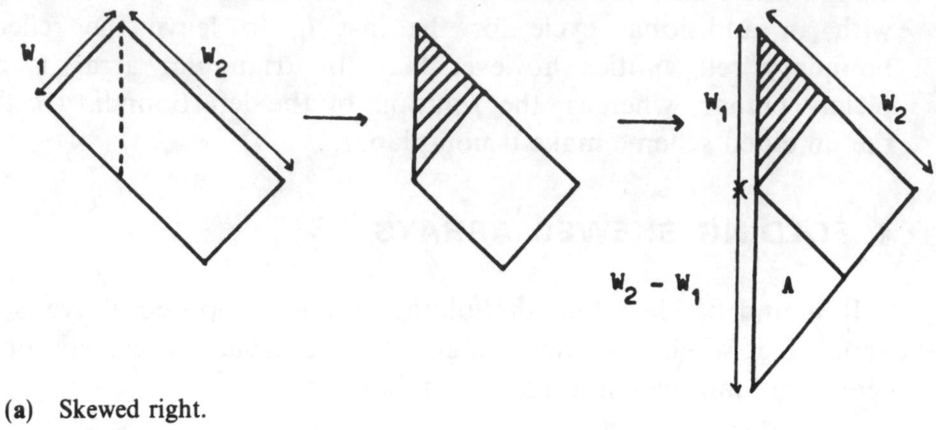

(a) Skewed right.

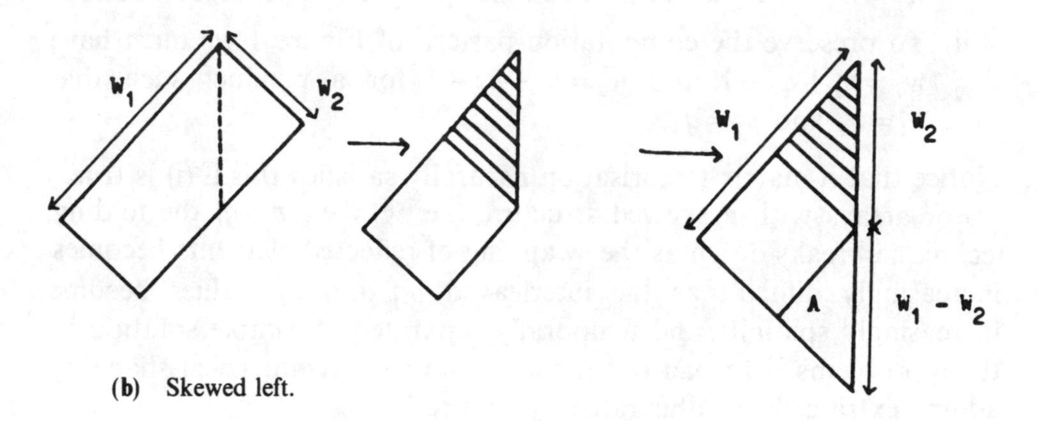

(b) Skewed left.

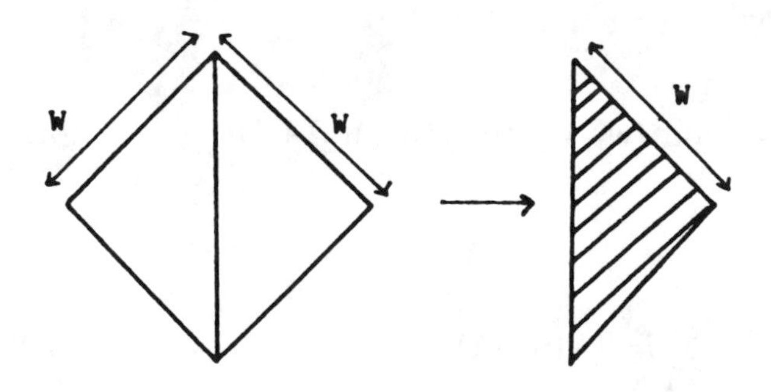

(c) Symmetrical.

Figure 7

ips cells. Thus as $w_1 \to w_2$, and $w \to 0$,

$$A \to w_1^2 - \tfrac{1}{2}w_1^2 - \tfrac{1}{2}w_1. \tag{4.4}$$

In fact we must add $\frac{1}{2}w_1$ to this total as the central column, even after folding, must always be full. Thus, for $w_1 = w_2$ the triangular array requires $A = \frac{1}{2}w_1 w_2$ saving half the cells and confirming our earlier statements.

Next let $w_1 < w_2$, if $w_2 = 2w_1, w = w_1$ and from (4.3),

$$A = w_1 w_2 - \tfrac{1}{2}w_1(w_1 + 1) + \tfrac{1}{2}w_1(w_1 + 1) \tag{4.5}$$

and it follows that provided $w_2 < 2w_1$ the triangular array always saves some hardware over the original unfolded scheme with computation time bounded by,

$$T = 3n + \max(w_1, w_2) \tag{4.6}$$

as the total delay of the folded array is increased to $w_1 + (w_2 - w_1)$ with our assumption that $w_2 > w_1$.

A further complication arises if $p_1 \nleq q_2$ and/or $q_1 \nleq p_2$ (which can occur even when $w_1 = w_2$) because padding w_1 up to w_2 creates a symmetrical array without a symmetrical computation of the form in Figure 1. The size of the array must then be further increased to produce both array and data flow symmetry by setting $\bar{w} = \max(p_1, q_1) + \max(p_2, q_2) - 1$. This new array has $\bar{w}^2$ cells which on folding reduces to,

$$A = \sum_{i=1}^{\bar{w}} i = \tfrac{1}{2}\bar{w}(\bar{w} + 1) \tag{4.7}$$

and requires a time,

$$T = 3n + \bar{w}. \tag{4.8}$$

A comparison of the cell counts of the unfolded and triangular array then demands that,

$$w_1 w_2 > \tfrac{1}{2}\bar{w}(\bar{w} + 1), \tag{4.9}$$

and if we let $p=\max(p_1,q_2)$, $q=\max(p_2,q_1)$ then,

$$(p_1+q_1-1)(p_2+q_2-1)>\tfrac{1}{2}(p+q)(p+q-1) \tag{4.10}$$

and the following four cases given general hardware bounds.

Case (i) $p=q_2, q=q_1$

We have,

$$(p_1+q_1-1)(p_2+q_2-1)>\tfrac{1}{2}(q_1+q_2)(q_1+q_2-1) \tag{4.11}$$

which after some manipulation yields,

$$q_1(2p_2-1)+q_2(2p_1-1)+2(E+1)>(q_1^2+q_2^2), \tag{4.12}$$

where,

$$E=p_1p_2-p_1-p_2.$$

Case (ii) $p=p_1, q=p_2$

We have,

$$(p_1+q_1-1)(p_2+q_2-1)>\tfrac{1}{2}(p_1+p_2)(p_1+p_2-1) \tag{4.13}$$

which similar to (4.11) produces,

$$p_1(2q_2-1)+p_2(2q_1-1)+2(E+1)>(p_1^2+p_2^2), \tag{4.14}$$

where $E=q_1q_2-q_1-q_2$.

Now provided $q_1,q_2>0$, and $E+1\geqq 0$ (4.14) is satisfied only if $p_1<(2q_2-1)$ and $p_2<(2q_1-1)$. The combination $E+1$ ensures that an even tighter bound on p_1,p_2 is actually true. The above condition is sufficient to show that for reasonably banded structures where $p_1=q_2+e_0$ and $p_2=q_1+e_1$ where $e_0,e_1\ll q_2,q_1$ ensures that the folded triangular array always saves some cells over the original array. A similar result applies to case (i).

Case (iii) $p=p_1, q=q_1$

We have,

$$(p_1+q_1-1)(p_2+q_2-1)>\tfrac{1}{2}(p_1+q_1)(p_1+q_1-1) \tag{4.15}$$

from which it trivially follows that cells are only saved iff

$$2w_2 > w_1 + 1, \tag{4.16}$$

implying

$$2p_2 \leqq p_1 \text{ and } 2q_2 \leqq q_1.$$

Case (iv) $p = q_2, q = p_2$
Finally we have

$$(p_1 + q_1 - 1)(p_2 + q_2 - 1) > \tfrac{1}{2}(q_2 + p_2)(q_2 + p_2 - 1) \tag{4.17}$$

and

$$2w_1 > w_2 + 1, \tag{4.18}$$

that is,

$$2p_1 \leqq p_2 \text{ and } 2q_1 \leqq q_2.$$

We conclude that provided the 2 matrices to be multiplied are of approximately the same bandwidth the triangular array will always reduce the total cell count. Notice, however, that the computation time can be increased by w-min(w_1, w_2) cycles, where the computational array symmetry must be forced onto the design before folding. Similar results hold for the factorisation array when $w_1 = p$ and $w_2 = q$, but generally less cells are added for $p = q$ as $q_1 = q_2 = 0$ can be assumed.

5. EFFECTS OF SPECIAL MATRIX FORMS

In the previous sections we have characterised the effectiveness of the array folding techniques. Now we can assess its applicability to some practical problems.

First, consider the problem of computing the product $C = AB$ where A, B and C are symmetric matrices. Clearly the bandwidths of A and B can be written as $w_1 = 2p_1 - 1$, and $w_2 = 2p_2 - 1$ as $p_1 = q_1$

and $p_2 = q_2$. Thus, from cases (i) and (ii) above, we have,

$$p_1(2p_2-1)+p_2(2p_1-1)+2(p_1p_2-p_1-p_2+1)>p_1^2+p_2^2$$

or

$$8p_1p_2-3(p_1+p_2)+2>p_1^2+p_2^2$$

and as $p_1p_2 \geqq p_1+p_2$ provided $p_1, p_2>1$ this becomes,

$$8p_1p_2-3(p_1+p_2)+2>8(p_1+p_2)-3(p_1+p_2)+2>(p_1+p_2)^2. \quad (5.1)$$

Thus some cells are saved if,

$$0>(p_1+p_2)^2-5(p_1+p_2)-2. \quad (5.2)$$

By solving the quadratic in (p_1+p_2) on the righthand side of (5.2) yields the range,

$$0 \leqq p_1+p_2 \leqq 5,$$

on the semi-bandwidths of A and B. Consequently only tridiagonal and quindiagonal matrices which occur widely in ordinary and partial differential equations can be multiplied with cell savings by the triangular array. Also observe that (5.2) is a very tight bound, because generally $p_1p_2>p_1+p_2$ for larger p_1 and p_2 indicating that the array has much wider applications. Indeed with Case (iii) and Case (iv) above we obtain,

$$\left.\begin{aligned} 2(2p_2-1)>(2p_1-1)+1 \Rightarrow 2p_2>p_1+1 \\ 2(2p_1-1)>(2p_2-1)+1 \Rightarrow 2p_1>p_2+1 \end{aligned}\right\} \quad (5.3)$$

for cell savings.

Finally for a symmetric matrix product observe the output pattern of Figure 3 $c_{ij}, c_{ji}, 0$, here we have a natural advantage which reduces the output communication giving an intuitive response to the fact that only a triangular portion of a symmetric product needs to be formed.

Next consider the problem of forming the matrix powers, $M^2, M^4, \ldots, M^{2^i}$ ... of a matrix M, which can be used to provide rapidly convergent approximations to truncated series expansions such as the Neumann and sin(M), cos(M), exp(M) power series of M. For a product $A*B=M$ and the bandwidth of M is w where $w=w_1=w_2$, giving an immediate symmetric structure to the product array, although additional padding may be required for symmetric dataflow. The key point is that the input format of Figure 3 i.e., $b_{ij}, a_{ji}, 0=m_{ij}, m_{ji}, 0$ can be used to halve the input bandwidth by use of a simple optical preprocessor using the concepts from [7, 8] as shown in Figure 8. Now both the input and output of the tri-

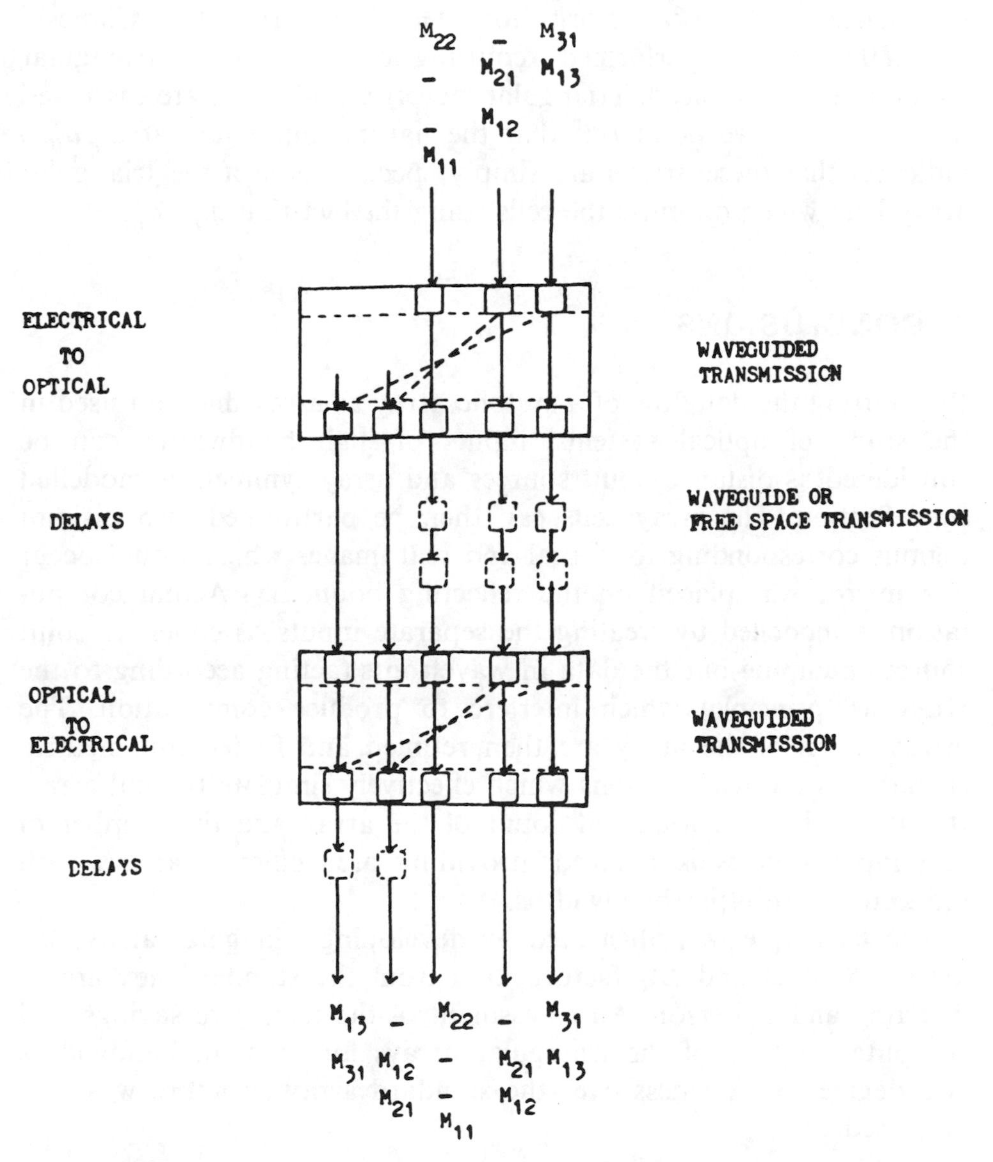

Figure 8 Optical preprocessor for expanding matrix input bandwidth.

angular array have the same format and hence the array can be cascaded to form a pipelined power generator as discussed in [9]. Where folded arrays use half the area of the original hex arrays two matrix squarings can be performed with the equivalent hardware (plus preprocessors) for a single matrix product on the unfolded array.

Finally, consider the matrix factorisation array. Here, the input and output bandwidths are already minimised and no preprocessor is required to modify the input. When the input matrix A is symmetric a Choleski factorisation $A=LL^t$ (or root free Choleski $A=LDL^t$) can be performed requiring only the lower triangular factor L to be produced. Triangular factorisation arrays are discussed in [6, 10], and we point out that the natural input format $a_{ij}, a_{ji}, 0$ indicates that these arrays are simply special cases of the triangular array here which optimise the cells using the fact that $a_{ij}=a_{ji}$.

6. CONCLUSIONS

By charting the dataflow of a systolic array as a ray diagram used in the study of optical systems, inputs of high bandwidths can be considered as distance point sources and array symmetries modelled by reflection. The array cells can then be partitioned into disjoint regions corresponding to virtual and real images which would occur if a mirror was placed on the reflecting boundary. Actual computation is modelled by treating the separate inputs as coherent point sources pumping out the data in wavefronts (acting according to the Huygens' principle) which interfere to produce computation. The virtual parts of the array are then retimed and folded into neutral elements of the real portions which effectively simulate the full array. It follows that the total cell count of the array and the number of real input sources is reduced, providing area efficient arrays with reduced input/output bandwidths.

The technique was illustrated by developing triangular arrays for matrix product and LU factorisation from the standard hex arrays of Kung and Leiserson. An assessment of the hardware savings and computation time of the triangular arrays for different bandwidths and degrees of skewness over the standard array structure was also discussed.

For symmetric matrices various advantages regarding the locally placed data illustrated the effectiveness of these arrays for certain problems. For matrix product the concepts of optical computing in the form of hybrid or electro-optical preprocessing elements to expand communication bandwidth and produce cascadable components was introduced. Such components can then be used for more complex tasks such as matrix squaring and power series generation.

To conclude the techniques here use optics on two levels. At the first level, the principles of reflection, refraction and interference can be used for the identification of replacement and retiming properties of an array to facilitate quick design improvements to existing systolic arrays. At the second level, the use of actual optical processing techniques is adopted as a method of overcoming communication interface problems. Both techniques together address the main problems encountered with VLSI implementations of arrays, area and pin restrictions. The design techniques discussed here should play an important role in developing future systolic arrays.

References

[1] H. T. Kung and C. E. Leiserson, *Introduction to VLSI Design*, Ch. 8, Mead and Conway, eds., Addison Wesley, 1979.

[2] G. Rothe On the connection between hexagonal and unidirectional rectangular systolic arrays, AWOC 86 2nd Workshop Parallel Computing and VLSI.

[3] C. E. Leiserson, *Area-Efficient VLSI Computation*, Ph.D Thesis, CMU, Pittsburgh, Oct. 1981.

[4] Saxe and Leiserson, Optimising synchronous systems, *J. VLSI and Computer Systems* **1,** 1 (1983), 41–67.

[5] W. M. Gentleman and H. T. Kung, Matrix triangularisation by systolic arrays, *SPIE* **298,** *Real Time Signal Processing IV* (1981), 19–26.

[6] R. Schrieber, *On Systolic Array Methods for Band Matrix Factorisations*, TRITA-NA-8316, Dept. Numerical Analysis & Computer Science Report, The Royal Institute of Technology, Stockholm, Sweden.

[7] Caulfield, Rhodes, Foster and Horvitz, Optical implementation of systolic array processing, *Optics Communications* **40,** 2 (1982), 86–90.

[8] Goodman, Leonberger, Kung and Athale, Optical interconnections for VLSI systems, *IEEE PROC.* **72,** 7, 850–866.

[9] G. M. Megson and D. J. Evans, Matrix power generation using an optical reduced bandwidth systolic array, J. Rose, ed., 7th International Congress of Cybernetics and Systems, Vol. 2, Thales Publishers, 1987, pp. 631–643.

[10] G. M. Megson, Novel algorithms for the soft-systolic paradigm, Ph.D Thesis, Dept. Computer Studies, Loughborough University of Technology, 1987.

Algorithmic fault tolerance for matrix operations on triangular arrays

G.M. MEGSON * and D.J. EVANS

Department of Computer Studies, Loughborough University of Technology, Loughborough, Leicestershire, United Kingdom LE11 3TU

Abstract. In this paper the technique of algorithm-based fault tolerance which is used to detect and correct transient or permanent hardware faults by checksum matrices is reconsidered for triangular systolic arrays. Linear error detecting arrays are developed for both matrix product and triangular factorisation and are shown to interface neatly with triangular schemes. The overheads associated with error detecting redundancy is offset by hardware reduction due to the folding of the array to produce triangular rather than the standard hex connected arrays. The result is shown to be improved efficiency and area efficient fault tolerant arrays.

Keywords. Triangular arrays, fault tolerance, matrix operations.

1. Introduction

The design of computing structures capable of surviving a number of random faults while maintaining robustness and reliability is the problem of fault tolerance. This increased reliability is paid for by the overheads of redundancy and increased algorithm computation time, which for practicality reasons must be kept low.

Since the requirements of many significant computational tasks (e.g. signal and image processing, solution of differential equations, etc.) can be reduced to a common set of matrix operations, the development of robust fault tolerant matrix calculations is of paramount importance. Consequently, it is highly desirable to produce a linear algebra system which achieves high performance and tolerates physical and transient failures but still produces correct results. An intuitive solution to both fault tolerance and fast computation is the use of parallel architectures where the total computation can be suitably partitioned and distributed among the processors. In the former case, each processor only affects a few elements of the total solution and so restricts the spread of errors due to such faults. In the latter case, the computation time is reduced by performing a number of operations simultaneously. Until the introduction of VLSI techniques the cost of additional processors to construct these architectures was prohibitively high, restricting their use to special one off designs mainly for research and high security applications. With VLSI it has become feasible to consider multiple processor networks, but the increased complexity of the architecture also introduces a higher probability of component failure. For these reasons the value of algorithmically specialized processors such as systolic arrays, where the fault tolerance can be tailored to specific algorithms, and consequently, optimised, has been emphasised (see [2,5]).

* Informatics Department, Rutherford Appleton Laboratory, Chilton, Oxon, United Kingdom OX11 0QX.

Reproduced with permission of Elsevier Science Publishers B.V. (North Holland)
Parallel Computing 10(207 – 219)
1989

Unfortunately, fault tolerance remains a subjective issue where completely different techniques become cost-effective under different constraints. Recent papers, [4,10] to name just two, have attempted to characterise the general approaches to fault tolerance and the common techniques employed. In particular, the following considerations have been identified:

(a) the types of failure which can be expected and their probability of occurrence;

(b) the appropriate recovery method;

(c) the cost of a failure or fault occurring and

(d) the additional hardware/time required for reliability.

Broadly speaking, fault tolerance schemes can then be further classified into two groups according to the type of failures they deal with, that is

(i) *production defects* where faults occur due to flaws in circuit elements introduced during manufacture. In the case of VLSI chips this means testing the device before packaging and discarding the faulty circuits. As the chances of a flaw during manufacture is relatively high with the current state of technology, this produces a low yield and the costs of production must be carried by the operational chips.

(ii) *operational defects* where faults develop during the working life of a device, and although the probability of failure is lower than the first mentioned item the effects of it occurring have much wider implications. For instance, it may not be possible to access the device easily to replace the faulty part, or even worse the error might go undetected (for a limited time period at least) and hence reducing reliability.

In [4] it was pointed out that flaws during manufacture were inevitable, and with the recent trends towards Wafer Scale Integration (WSI) such designs must implement fault tolerance which must be cost effective. Now, a common approach to fault tolerance for production defects involves the inclusion of redundant circuitry which avoids manufacturing flaws by reprogramming the interconnection network using the implicit assumptions that

(a) the connection links have a low probability of failure compared with the processors;

(b) the additional connections do not represent a significant overhead.

As circuits become smaller and more complex both these constraints may not be valid.

For regularly connected processor arrays such as systolic arrays, the above assumptions are generally valid and fault tolerance can be considered on two levels, the cell and array level. When faults are localised to cells, the cell becomes the smallest replaceable unit and the whole array can be made fault tolerant by utilising spare cells or routing around the faulty cells. Kung and Lam [5] have recently suggested new systolic arrays which avoid feedback loops and re-defined algorithms with unidirectional dataflow and systolic ring architectures. For these designs, the costs in terms of

(a) added connections,

(b) probability of survival for a given number of faults,

(c) added computation time

were assessed, and it was found that the designs tended to degrade gracefully under increasing faults.

Huang and Abraham [2], in contrast have introduced the concept of algorithmic fault tolerance for operational type failures in the form of permanent or transient hardware faults. The technique has three main characteristics:

(1) data used by the algorithm is encoded,

(2) the algorithm is redesigned to operate on the encoded data and produce encoded output, and

(3) redundancy in the encoding is used to detect and correct errors.

The encoding schemes must be re-designed so that the information part of the data is easily recovered, and also that any faulty processor does not mask the error during detection and correction. Reference [2] examined the application of checksum matrices to mesh connected and systolic arrays for detecting (but not correcting) errors in the matrix product, LU factorisation and matrix inversion algorithms. For the product of matrices, the hexagonal array discussed optimised the computation time by removing synchronising delays from the standard hexagonal array of [6], and the 2-D array while tree arrangements for error detection were added.

In contrast, the scheme presented in this paper utilises the delay elements of the original hex-connected array to produce cell savings which can compensate for the hardware overhead associated with error detection. In addition, only 1-D linear arrays are used for detection which interfaces naturally with the product array, and can easily be extended to the LU factorisation problem. The overall cell saving implies less area and consequently a lower probability of encountering production defects.

2. Checksum matrix encodings

Before describing the new fault tolerant arrays we briefly review the checksum technique for encoding matrix computations. We start with a few definitions. The row, column and full checksum matrices for an $n \times m$ matrix $A = (a)_{ij}$ are defined as follows:

(1) The column checksum matrix A_c of A is an $(n+1) \times m$ matrix with A in the first n rows and a column sum vector,

$$a_{n+1,j} = \sum_{i=1}^{n} a_{ij}, \quad j = 1(1)m$$

in the $(n+1)$st row.

(2) Similarly, the row checksum matrix A_r of A is an $n \times (m+1)$ matrix with A in the first m columns and a row sum vector,

$$a_{i,m+1} = \sum_{j=1}^{m} a_{ij}, \quad i = 1(1)n.$$

(3) A full checksum matrix A_f of A is an $(n+1) \times (m+1)$ matrix with A in the first n rows and m columns augmented with row and column sum vectors in the $(m+1)$st and $(n+1)$st column and row respectively.
With $e^T = (1, 1, \dots, 1)$, we have

$$A_c = \left[\frac{A}{e^Ta}\right], \qquad A_r = [A \mid Ae], \qquad A_f = \left[\begin{array}{c|c} A & Ae \\ \hline e^TA & e^TAe \end{array}\right].$$

Consequently, for the two $n \times n$ matrices A and B, the basic matrix manipulation operations yields the following relevant results:

- *Matrix product:*

$$A_c \times B_r = \left[\frac{A}{e^TA}\right][B \mid Be] = \left[\begin{array}{c|c} AB & ABe \\ \hline e^TAB & e^TABe \end{array}\right] = C_f, \tag{2.1}$$

- *LU factorisation:* Let $C \equiv LU$ denote the triangular factorisation of an $n \times n$ matrix C. Then

$$C_f = \left[\begin{array}{c|c} C & Ce \\ \hline e^TC & e^TCe \end{array}\right] = \left[\frac{L}{e^TL}\right][U \mid Ue] = L_cU_r. \tag{2.2}$$

Note that if pivoting is required, (2.2) can be formed but L and U are not necessarily triangular matrices.

Fault tolerance is then implemented as a three stage process given by:

(I) *Detect error:* Assume e^TA and Be are known and compute $c = e^TAB$, $r = ABe$ and A_cB_r. Denote by $\bar{c}$ the last row's first m elements and by $\bar{r}$ the last column's n elements of C_f respectively. An error is indicated in C if $\hat{c} = c - \bar{c} \neq 0$ and/or $\hat{r} = r - \bar{r} \neq 0$.

(II) *Error location:* A map of the error locations is given by the outer product matrix $(\hat{c}\hat{r})^T$ where

$$(\hat{c}\hat{r})_{ij} = \begin{cases} 1 & \text{iff } \hat{c}(i)\hat{r}(j) \neq 0, \\ 0 & \text{otherwise.} \end{cases}$$

(III) *Correct error:* The erroneous values are corrected by

(a)

$$c_{ij} = \begin{cases} c_{ij} \pm \hat{r}(j) \\ \text{or} \\ c_{ij} \pm \hat{c}(i) \end{cases}, \quad \text{iff } (\hat{c}\hat{r})^T_{ij} = 1,$$

(b) by replacing r with $\bar{r}$ and c with $\bar{c}$ in the case of a checksum vector error.

Notice that if the outer product form in (II) has the definition

$$(\hat{c}\hat{r})_{ij} = \begin{cases} \hat{c}(i) \text{ or } \hat{r}(j) & \text{iff } \hat{c}(i)\hat{r}(j) \neq 0, \\ 0 & \text{otherwise,} \end{cases}$$

the error map contains the error locations and the amounts by which the elements must be modified.

This simple scheme is suitable only for the detection and correction of errors in a matrix product operation. For (2.2), the matrix vector multiplication and matrix inversion operation errors can only be detected and more sophisticated schemes using weighted checksums as discussed in [3,7], must be used.

Finally, we point out the following features of algorithmic fault tolerance common to all techniques:

(i) Checksum methods can correct errors from a single fault in a full checksum matrix.

(ii) The formation of the summation vectors does not affect the wordlength of calculations significantly even with floating-point calculations.

(iii) A tolerance should be allowed when testing $\hat{c}(j) = 0$ or $\hat{r}(i) = 0$ so that rounding errors do not trap 'false' errors.

3. Checksum techniques and triangular arrays

Next we consider fault tolerant arrays for matrix product and factorisation using the triangular arrays of [9] whose input formats are shown in Fig. 1. These folded arrays frequenty use less cells than the standard hexagonal arrays and can be used to offset the additional hardware costs for fault tolerance against total cell savings.

The initial step for fault tolerance is to suitably partition the computation between any host machine and array components. Since the simple checksum technique above only detects the errors for many algorithms, a natural partitioning appears to be the detection and location of errors in the array and their correction by the host machine is apparent. This approach also has the added flexibility of allowing designs to extend easily to weighted checksum schemes.

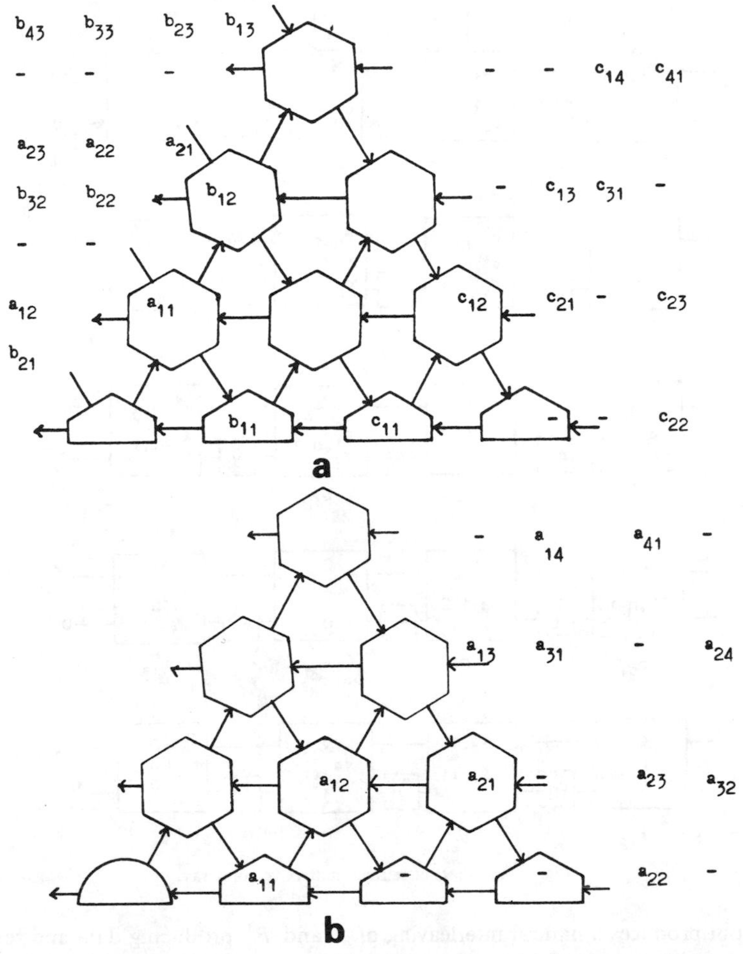

Fig. 1. Input formats for triangular arrays; (a) Triangular matrix product array; (b) Triangular LU factorisation array.

For the array components the calculations are written as follows:

	Product	Factorisation
Pre-processing:	$c^{\mathrm{T}} = B\bar{a}$, $\bar{a} = A^{\mathrm{T}}e^{\mathrm{T}}$	$c^{\mathrm{T}} = C^{\mathrm{T}}e$
	$r = A\bar{b}$, $\bar{b} = Be$	$r = Ce$
Array:	$C_{\mathrm{f}} = A_{\mathrm{c}}B_{\mathrm{r}}$	$C_{\mathrm{f}} = L_{\mathrm{c}}U_{\mathrm{r}}$
Post-processing:	$\bar{c} = e^{\mathrm{T}}C_{\mathrm{f}}$, $\hat{c} = c - \bar{c}$	$\bar{c} = e^{\mathrm{T}}L$
	$\bar{r} = C_{\mathrm{f}}e$, $\hat{r} = r - \bar{r}$	$\bar{r} = Ue$

When $\bar{a}$, $\bar{b}$ are known in advance, pre-(post-)processing can be performed by simple matrix vector and summation arrays.

Figure 2 illustrates a retimed matrix vector array which accepts the same matrix input format as the triangular array given in Fig. 1(a) for product pre-processing. Observe that the calculation of c^{T} and r is interleaved on the array (consisting of inner product and delay cells) with the vector input $\bar{a}(i) = a_{n+1,i}$ and $\bar{b}(j) = b_{j,n+1}$, i, $j = 1(1)n$. The folding of the triangu-

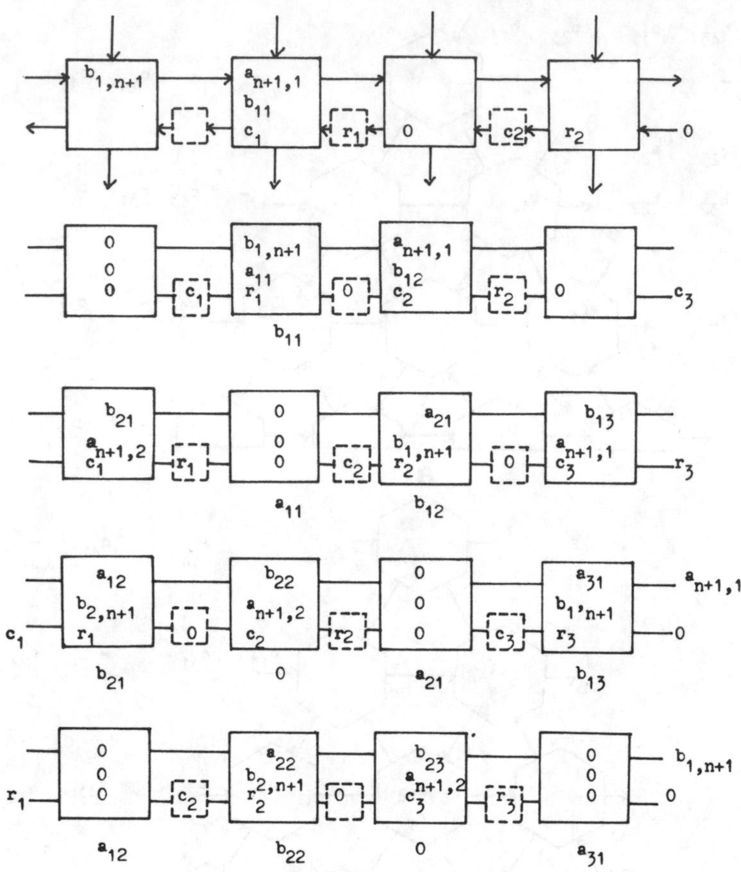

Fig. 2. Snapshots of retimed matrix vector array.

lar array input produces a natural interleaving of A and B^T producing data and results of the two calculations in nearest neighbour positions and (by the properties used in [1]) ensures that the results move systolically in the same direction.

A more involved task is the generation of the row and column summation $\bar{c}$, $\bar{r}$ during post-processing using the output C of Fig. 1(a). The accumulation of both sums is again interleaved as shown in Fig. 3 and demands a linear array with the structure and operation of Fig. 4.

The boundary cell acts like a mirror reflecting the partially accumulated $\bar{c}$ and $\bar{r}$ vectors back along the array to complete their summations from left to right, and takes three cycles to compute

$$\begin{aligned} t:&\quad \text{read } \bar{c}_i \\ t+1:&\quad \text{read } \bar{r}_i,\ r_i \\ &\quad \text{compute } \hat{r}_i = r_i - \bar{r}_i - C_{ii} \\ t+2:&\quad \text{read } -,- \\ &\quad \text{compute } \hat{c}_i = c_i - \bar{c}_i - C_{ii} \end{aligned} \tag{3.1}$$

The basic cell consists of an adder and subtracter and performs two tasks simultaneously. First, the partial column or row sum moving left is updated and second the error location

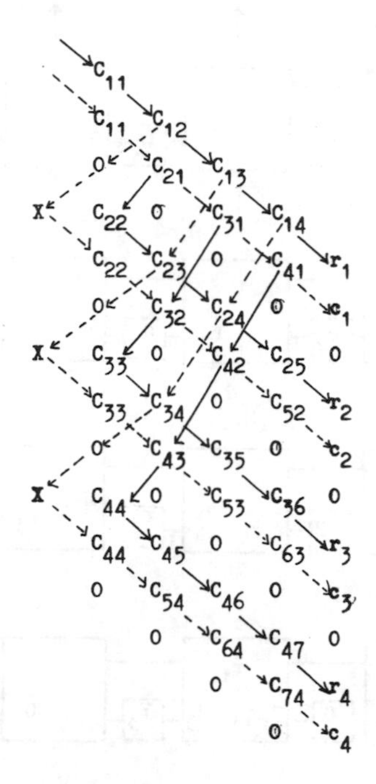

Fig. 3. Summation of row and column checksums.

vectors ($\hat{r}$, $\hat{c}$) are updated. It follows from the dataflow of the array that valid results emerging from the right-most end cell are zero only if $r_i = \bar{r}_i$ or $c_i = \bar{c}_i$ and that non-zeros indicate possible errors and the amount to be added or subtracted from the incorrect element of C_f. Also, observe that if $r_i = c_i = 0$ is assumed in (3.1), then Fig. 4 produces the negated row and column summations. Consequently, the preprocessing for factorisation, which has the same input format as the product array output, can use the same array. For the factorisation post-processing, $\bar{c}$ is the sum of columns taken over the lower triangular elements, and $\bar{r}$ the same for rows over the elements in the upper triangular part of U, yielding the dataflow if Fig. 5 and as no reflection is required, an array of simple adders is sufficient.

Finally, we must consider the placement of the error detecting arrays for synchronisation and interfacing between the triangular array and the host machine. In the case of the product, Fig. 6 illustrates the connections for a two-tier array combining both pre- and post-processing which forms a natural interface for the side QS of the triangular array represented in Fig. 7.

Suppose that the a_{ii} and b_{ii} elements of A and B are input at P. An input travels along PR contributing to the off-diagonal calculations of C until reaching R where the C_{ii} value is picked up and accumulates its remaining partial product terms along SR. Now suppose $QS = QT = ST = W$, where W is the number of cells and hence data delays along a boundary. If $PS = p$ from [9] and the hex array structure $PR = SR = PS$, then $2p$ ips cycles is the delay from the input of a_{ii}, b_{ii} to the output of C_{ii}. Likewise from Fig. 6 there is exactly one cell from the pre/post processing tiers for every cell along QS and from Fig. 2, the c_i value associated with the a_{ii} (and r_i value for b_{ii}) inputs leave the preprocessor tier after $2p - 2$ cycles. Thus, by delaying the pre-processor output a single cycle, c_i, $\bar{c}_i$, r_i, r_i and c_{ii} synchronise in the boundary cell of the post-processor.

Now the original hexagonal array requires W^2 cells and the triangular array only $\frac{1}{2}W(W+1)$ cells. The pre(post) processor of Fig. 6 contributes W ips cells and $2W$ adder/subtracters and

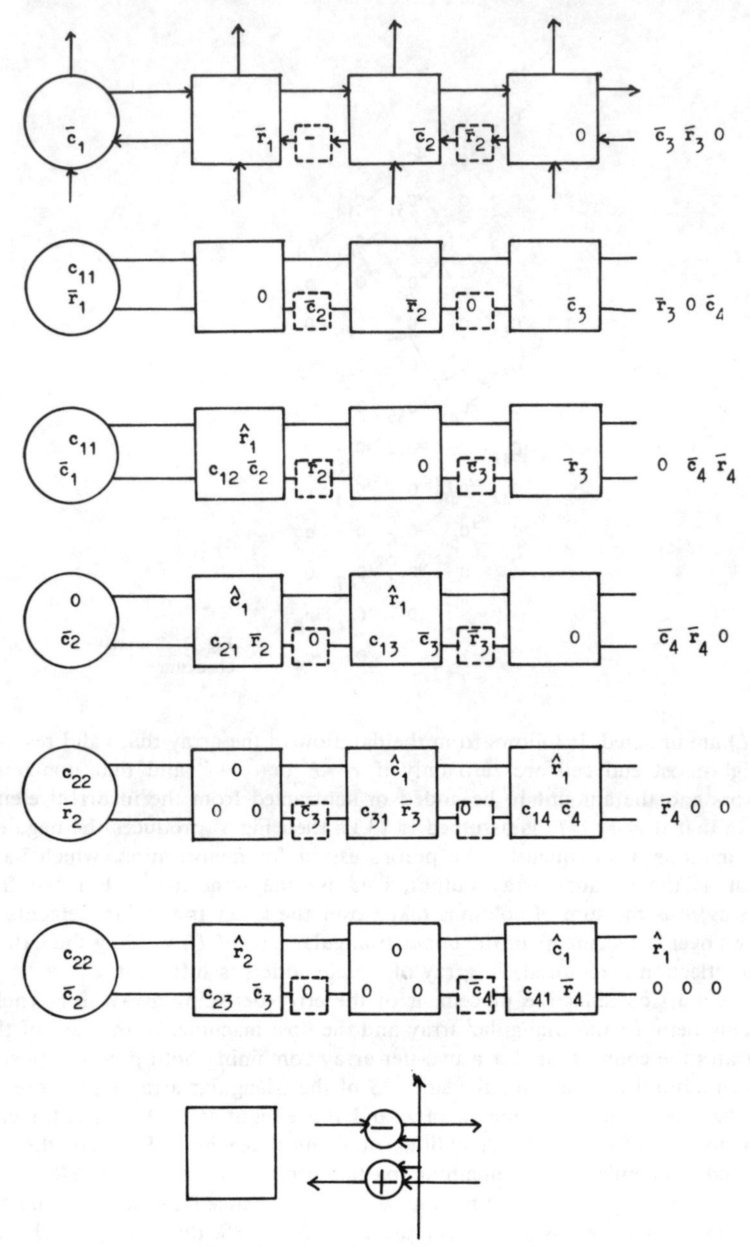

Fig. 4. Snapshots of checksum accumulation.

$2W$ delays producing an overall net cell reduction. The timing of the array for the $n \times n$ matrix product is

$$T = 3n + W + (W + p), \tag{3.2}$$

using the timings of the triangular array in [6] and adding p cycles for the initial synchronisa-

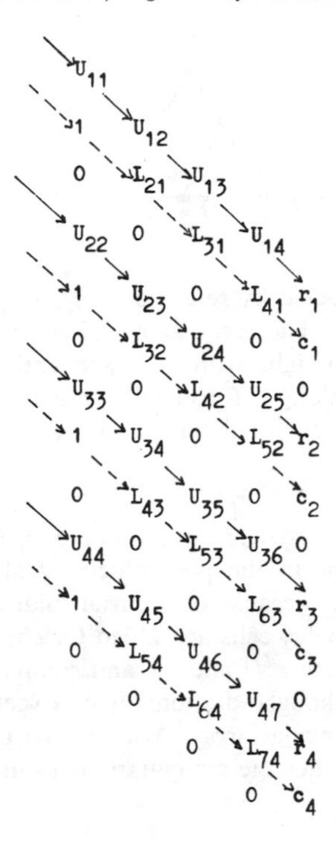

Fig. 5. Summation of L and U checksum vectors.

tion of the pre-processor tier along SP (using $\bar{a}$ and $\bar{b}$) and W cycles to filter the last $\hat{r}$, $\hat{c}$ components along SQ.

In the case of LU factorization, a two-tier arrangement is not possible and the array in Fig. 3 is placed along boundary QT and the modified adder version for accumulating $\bar{c}$, and $\bar{r}$ along

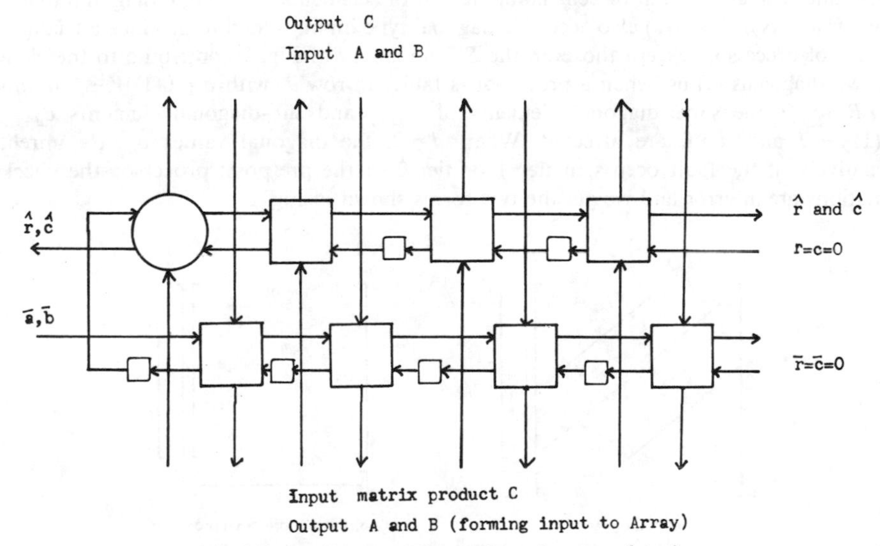

Fig. 6. Pre(post)-processor for matrix product error detection.

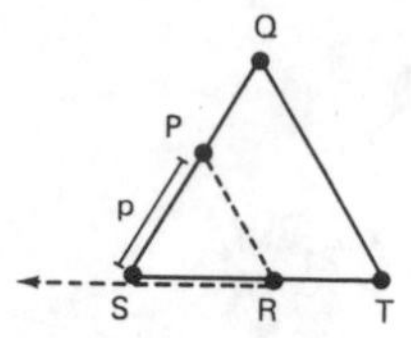

Fig. 7.

ST. Now vectors c and r are constructed by reflection at T and accumulation along TQ and begin to emerge at Q after W cycles. The array along ST receives its first term after $W+1$ cycles and accumulates from left to right with results $\bar{c}$ and $\bar{r}$ emerging after $2W$ cycles. Consequently, adding W delay cells along QT to feedback c and r allows c, r, $\bar{c}$ and $\bar{r}$ to be output simultaneously from the corner T of the array in Fig. 1(b). The timing of the array is

$$T = 3n + W + W, \tag{3.3}$$

where no time is required for the initial synchronisation ($c = r = 0$) and only W cycles must be added for the accumulation of terms in the post processor along ST. Again, hardware is reduced from W^2 to $\frac{1}{2}W(W+1)$ cells because of the triangular array used and a total of $3W$ adders/subtracters and at most $2W$ delay cells are added for checksum computations. Further work is required in the host to compute $\bar{c}^{\mathrm{T}}U$ and $L\bar{r}$ and compared with c and r to yield the quantities $\hat{c} = c - \bar{c}$ and $\hat{r} = r - \bar{r}$. Although additional matrix vector arrays could be used, they are not easily interfaced with the triangular array. Also, if used they require at least W and at most $2W$ ips cells depending on whether the computations using L and U are interleaved or not.

4. Tolerance of errors

The standard hexagonal arrays for the matrix product operation accepts a diagonal input pattern, and a single column of cells computes all the elements of a single diagonal of C. The triangular arrays in Fig. 1(a) also accept a diagonal type input, folded to produce a triangle, but each row of processors except those on the ST boundary of Fig. 7 contribute to the elements from two diagonals. Thus, when a processor is faulty in row d, with $d = 0(1)W-1$ numbered from R to Q, the super-diagonal elements of $c_{i,d+i}$ and sub-diagonal elements $c_{d+i,i}$ for $i = 1(1)n-d$ and $d > 0$ are affected. When $d = 0$, the diagonal values c_{ii} are unreliable. Alternatively, if the fault occurs in tier 1 or tier 2 of the pre(post) processor the checksum summations are in error and we get the two forms shown in Fig. 8.

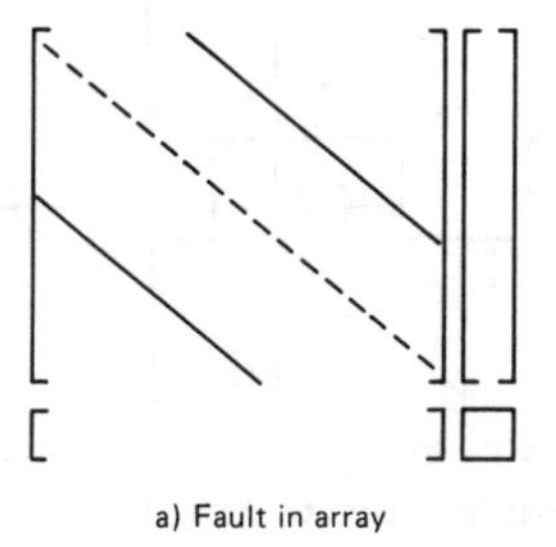

a) Fault in array

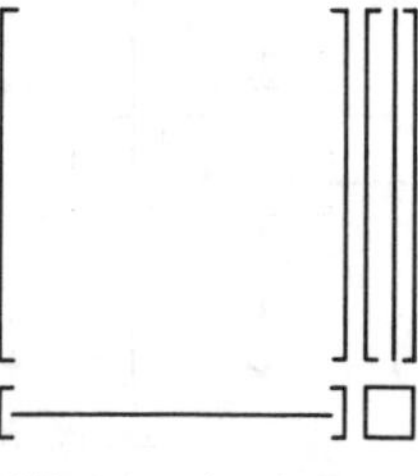

b) Fault in pre(post) processor

Fig. 8.

Thus if we assume that faults do not occur simultaneously in the array and pre(post) processor and that faults in the triangular array are also limited to a single column of cells errors can always be detected and corrected. For case (b) in Fig. 8 this is trivial because C is correct and the redundant checksum is simply discarded. For (a), we must update both the incorrect diagonals by a forward recursive type correction which ensures that at the time of modification the error is the only one in a row or column. This is achieved in a two-stage correction procedure:

(1) Correct the next superdiagonal element using the row checksum and update the corresponding column checksum.

(2) Correct the next subdiagonal element using the correct column checksum, and update the corresponding row checksum.

Finally, a significant result can be obtained from the theorems in [2] regarding error undetectable matrices. These theorems essentially prove that the error pattern formed from the outer product $\hat{c}\hat{r}$ can be connected in a closed loop if a single cell computes $2n$ or more elements of C. Now, from Fig. 1 row $d=0$ is the largest and computes the diagonal of C, that is it contributes to only n elements. Consequently row d of Fig. 1(a) computing two diagonals of C contributes to at most $2(n-d)$ elements, and it follows that row $d=1$ yields $2(n-1)$ elements of C. This result corroborates the recursive correction procedure as no loop can ever be formed in $\hat{c}\hat{r}$ as long as the assumptions on fault positioning remains true. More importantly, this result indicates that any further folding of the triangular array to reduce further hardware costs will permit the formation of loops and any checksum techniques will not be able to detect errors even for a single fault.

For the weighted checksum scheme, the forms corresponding to (2.1) and (2.2) are derived from the following checksum matrix definitions (see [3]):

$$A_c = \begin{bmatrix} A \\ e^T A \\ e_w^T A \end{bmatrix}, \qquad A_r = [A\ Af\ Af_w], \qquad A_f = \begin{bmatrix} A & Af & Af_w \\ e^T A & e^T Af & e^T Af_w \\ e_w^T A & e_w^T Af & e_w^T Af_w \end{bmatrix}$$

where $e_w^T = [1, 2, \dots, 2^{n-1}]$, $f^T = [1, 1, \dots, 1]$ and $f_w^T = [1, 2, \dots, 2^{m-1}]$, and for,

– *matrix product:*

$$C_f = A_c B_r = \begin{bmatrix} AB & ABf & ABf_w \\ e^T AB & e^T ABf & e^T ABf_w \\ e_w^T AB & e_w^T ABf & e^T Bf_w \end{bmatrix},$$

– *matrix factorisation:*

$$C_f = L_c U_r,$$

where,

$$\begin{bmatrix} C & Ce & Ce_w \\ e^T C & e^T Ce & e^T Ce_w \\ e_w^T C & e_w^T Ce & e_w^T Ce_w \end{bmatrix} = \begin{bmatrix} L \\ e^T L \\ e_w^T L \end{bmatrix} [U\ Ue\ Ue_w]$$

and with $\bar{a}(1) = e^T A$, $\bar{a}^{(2)} = e_w^T A$, $\bar{b}^{(1)} = Bf$, $\bar{b}^{(2)} = Bf_w$ known in advance. The pre- and post-processing for a product is achieved by pipelining two arrays like in Fig. 6 to compute the ordinary and weighted checksums. Likewise in a factorisation, arrays like Fig. 4 are pipelined. Consequently, the weighted checksum scheme requires an additional $k > 0$ cycles to account for propagation delays and extra synchronisation cycles in the extended pipeline, and double the hardware for errror detection.

5. Conclusions

In this paper triangular systolic arrays and linear error detecting arrays for fault tolerant checksum calculations in matrix product and LU factorisation systolic algorithms have been examined. The triangular fault tolerant schemes in most cases use approximately 50% of the hardware and produce an asymptotic cell efficiency of $e = \frac{2}{3}$ and computation time $O(3n)$ ips cycles, compared with $O(3n)$, efficiency $e = \frac{1}{3}$ and $O(n)$, $e = 1$ for the original [6] and modified [2] hexagonal arrays. These results serve to place our designs in context, as we can discern three types of strategy for improving the overheads associated with the fault tolerance arrays above:

(1) Maintain both area and time of the original array and accept a redundancy overhead in additional error correcting apparatus.

(2) Optimise the computation time of the array at the expense of additional area required for fault tolerance.

(3) Optimise the area of the array at the expense of computation time to calculate redundant data.

(4) Optimise both area and time of the array to offset increased redundancy and computation time.

We point out that (1) is not really practical as overheads may become too high, however linear arrays like those above can also be developed for the original hex arrays requiring $O(4W)$ additional cells. The pre(post) processor would cover the whole of the arrays upper boundary, and the resulting checksum vectors r and c move towards the central column, $\hat{r}$ and $\hat{c}$ away from the column to the right and left. Thus r and c would then emerge at the central column and $\hat{r}$ and $\hat{c}$ spatially separated at the left and right ends of the array. In additional, the pre(post) processor hardware and initial synchronisation are strict overheads in both area and time.

The modified hex array used in [2] is of type (2) and optimises the time offsetting the increased time for fault detection against the overall time reduction, to produce a net speed increase. However the global unidirectional flow of the A, B and C matrices in the modified array makes it difficult to interface the error detecting arrays neatly with the hex boundary. Consequenty 2-D fanin-trees and buffers are required to accumulate the checksum vectors and locate errors, effectively increasing the array bandwidth. The overhead in hardware and connection difficulties is offset by the factor of three speedup obtained by the dense packing of input data.

The triangular arrays of this paper are of type (3) above and reduce the product forming hardware by half, and offset this against error detecting hardware overheads to yield a net area saving. In addition, the array folding minimises the array bandwidth (halving it) while retaining the neat interfacing of the original hexagonal scheme with pre(post) processors. A direct result of the folding is a reduced hardware overhead of $O(2W)$ cells and the local placement of the r, c and $\hat{r}$, $\hat{c}$ vector outputs which are combined on just two outputs rather than four. The penalty for fault tolerance is the additional time overhead of $O(W)$ which is negligible as n grows when A, B are really banded.

Where time is a premium, the modified array with its factor of 3 speedup has a clear advantage. However, the triangular scheme is easily extended to weighted checksum calculations giving a wider range of applications, with the additional cell overhead absorbed by the initial reduction due to folding. We conclude that although the modified scheme is fast, the price paid is inflexibility. Likewise, the triangular array restricts the ability to detect errors as further folding produces undetectable checksum matrices. Alternatively, we can argue that the cell reduction for large triangular arrays may significantly affect the chances of producing a hardware or production defect and so implicitly improves tolerance.

To conclude, a desirable area of further study is the production of a scheme of type (4) possibly by the combination of (2) and (3) to produce a hybrid scheme using retiming, and replacement of cells.

References

[1] D.J. Evans and G.M. Megson, Matrix inversion by systolic rank annihilation, *Internat. J. Comput. Math.* **21** (1987) 319–358.

[2] Huang and Abraham, Algorithm based fault tolerance for matrix operations, *IEEE Trans. Comput.* **33** (6) (1984) 518–528.

[3] Jou and Abraham, Fault-tolerant matrix arithmetic and signal processing on highly concurrent computing structures, *Proc. IEEE* **74** (5) (1986) 732–741.

[4] Koren and Pradhan, Yield and performance enhancement through redundancy in VLSI and WSI multiprocessor systems, *Proc. IEEE* **74** (5) (1986) 699–711.

[5] H.T. Kung and Lam, Wafer-scale integration and two level pipelined implementations of systolic arrays, *J. Parallel Distrib. Comput.* **1** (1984) 32–63.

[6] C.E. Leiserson, Area efficient VLSI computation, Ph.D. Thesis. Carnegie-Mellon University, Pittsburg, 1981.

[7] F.T. Luk, An analysis of algorithm based fault tolerance techniques, Report EE-CEG-86-11, Computer Engineering Group, Cornell University, and to appear in: *Advanced Algorithms & Architectures for Signal Processing, Proc. SPIE*, **696**.

[8] G.M. Megson, Novel algorithms for the soft-systolic paradigm, Ph.D. Thesis, Loughborough University of Technology, 1987.

[9] G.M. Megson and D.J. Evans, Triangular arrays for matrix product and factorisation, *Internat. J. Comput. Math.* (1988).

[10] Sami and Stefanelli, Reconfigurable architectures for VLSI processing arrays, *Proc. IEEE* **74** (5) (1986) 712–711.

ON SYSTOLIC ARRAYS FOR COMPLEX MATRIX PROBLEMS

G. M. MEGSON

Computing Laboratory, University of Newcastle upon Tyne, Newcastle upon Tyne, UK

D. J. EVANS

Department of Computer Studies, Loughborough University of Technology, Loughborough, Leicestershire, UK

We consider systolic arrays for matrix computations involving complex elements, and show that in certain circumstances the complex calculations can be decomposed into a number of real matrix sub-problems which can be used to good effect in reducing computation times and increasing array cell efficiency. Computations involving matrix vector, matrix product and matrix factorisation are examined. It was found that matrix vector and product calculation produce arrays which have $e=1$ cell efficiency and the same computation time as their real counterparts, with only an increase in hardware related to the bandwidth of the systems.

Matrix factorisation is achieved by using $2*2$ block factorisation requiring four times the hardware and is only twice as slow as the real version of the factorisation.

KEY WORDS: Systolic array, complex matrix.

C.R. CATEGORIES: G1.3, C1.2.

1. INTRODUCTION

The rapid development of systolic array architectures for a wide range of numerical matrix problems has resulted in many area efficient and linear-time architecture structures for well-known problems which form the basis of almost all numerical algorithms. These basic problems are matrix multiplication (and the special case of matrix vector multiplication), solving triangular systems of linear equations, *LU* decomposition and Gaussian elimination procedures. All these arrays are based on the basic inner product step (*ips*).

$$y=\sum_{i=1}^{n} a_i x_i,$$

where a, x are n-component vectors. Apart from the inner product step all the current arrays have one other feature in common, they all deal explicitly with only real components or elements in the vectors and matrices. This report investigates the increases in cells and computation time when complex numbers are allowed in vectors and matrices.

As a starting point for the development of complex systolic arrays we must decide on a new matrix representation which is easily formed into an input for the systolic array. In systolic array design great emphasis is placed on the banded nature of a system which is solved. The bandwidth of a system affects both the number of cells in the design and contributes terms to the final computational complexity (often only minor terms due to the relatively small bandwidth compared to problem size).

Consider a $n*n$ matrix A with entries $a_{ij} \in C$, the real case is a special case of the more general complex formulation of the problem, where,

$$a_{ij} = \begin{cases} a_{ij}^{(1)} + i0 & \text{Real } a_{ij}^{(1)}, a_{ij}^{(2)} \in \mathbb{R} \\ a_{ij}^{(1)} + ia_{ij}^{(2)} & \text{Complex.} \end{cases} \tag{1.1}$$

It follows that an $n*n$ complex matrix of bandwidth W can be represented by a $2n*2n$ matrix of bandwidth W_1 as follows,

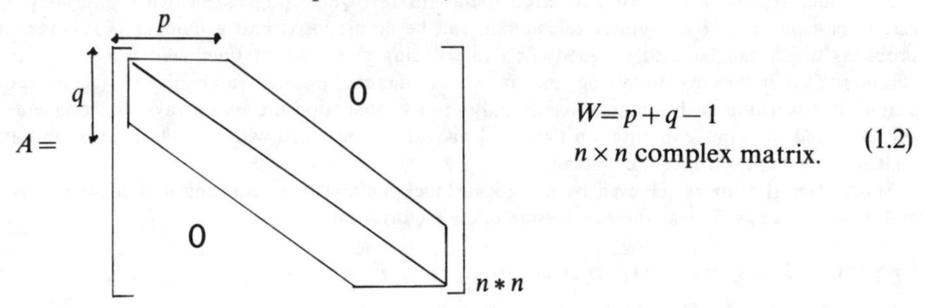

The real equivalent is given by a matrix of the form,

$$\bar{A} = \left[\begin{array}{c|c} A_R & -A_C \\ \hline A_C & A_R \end{array}\right]_{2n\times 2n} \tag{1.3}$$

with

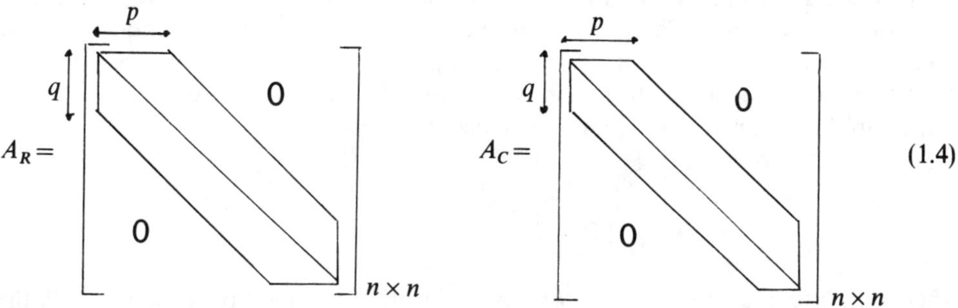

where

$$A_R(i,j) = a_{ij}^{(1)}, \; A_C(i,j) = a_{ij}^{(2)}.$$

Hence $\bar{A}$ has bandwidth $W_1 = 2n+p+q-1$. Thus the bandwidth of the extended real matrix $\bar{A}$ is related to the complex $n*n$ matrix by $W_1 = 2n+W$, which gives an intuitive feel to the complexity of complex systolic matrix problems. Thus we would expect a complex systolic array to be identical to the real version except that we require $0(n)$ more cells and the length of the input is doubled. Notice also that when A is a real matrix $\bar{A}$ forms a $2n*2n$ real matrix consisting of two decoupled $n*n$ matrices which are identical (i.e. A_r), we need only to solve a single $n*n$ problem with bandwidth W. These results are disappointing because the bandwidth of the extended real system is now related to n (the problem size) not just the bandwidth, indicating that the number of cells in the complex problem is dramatically increased. Notice that although the structures of A_r and A_c are banded the extra sparsity of the extended band cannot be utilised.

Solution of the matrix–matrix multiplication, matrix–vector multiplication and LU factorisation problems for the $n*n$ complex banded system can be performed on the standard architectures of [1] with the following cell/time tradeoffs.

THEOREM 1 (Matrix–vector multiplication) *The multiplication of an $n*n$ complex matrix and an n-component complex vector can be performed in time $T=4n+(2n+W)$ using a linear array of $2n+W$ ips cells where W is the bandwidth of the complex matrix.*

Proof Simply perform real matrix vector multiplication on the matrix $\bar{A}$, timings and cell counts follow from [1].

THEOREM 2 (Matrix–matrix multiplication) *The multiplication of two $n*n$ complex matrices of bandwidth W_1 and W_2 is performed in a time $T=6n+\min(2n+W_1, 2n+W_2)$ ips cycles using a hex connected array with $4n^2+2n(W_1+W_2)+W_1W_2$ ips cells.*

THEOREM 3 (LU factorisation) *The LU factorisation of an $n*n$ complex matrix A of bandwidth $W=p+q-1$ is performed in a time $T=6n+\min(n+p, n+q)$ ips cycles using $n^2+n(p+q)+pq$ ips cells in a hex connected array format.*

Proof Performing LU factorisation on the real matrix $\bar{A}$. See [1].

We show in the remainder of this paper that these results indicate only upper bounds on the time required to compute the complex problems involved and that the hardware/cell requirement in certain cases can be dramatically reduced.

2. COMPLEX MATRIX VECTOR MULTIPLICATION

The simplest problem is the complex matrix–vector multiplication. This is formulated as the problem,

$$\underbrace{\left[\begin{array}{c|c} A_R & -A_C \\ \hline A_C & A_R \end{array}\right]}_{\bar{A}} \underbrace{\left[\begin{array}{c} x_r \\ \hline x_c \end{array}\right]}_{\bar{x}} = \underbrace{\left[\begin{array}{c} y_r \\ \hline y_c \end{array}\right]}_{\bar{y}} \tag{2.0}$$

with $\bar{A}$ a $2n \times 2n$ extended real matrix $\bar{x}, \bar{y}$ extended real vectors of size $(2n \neq 1)$. Now put,

$$A_R x_r = t_1 \tag{2.1a}$$

$$A_C x_c = t_2 \tag{2.1b}$$

$$A_C x_r = t_3 \tag{2.1c}$$

$$A_R x_c = t_4 \tag{2.1d}$$

$$y_r = t_1 - t_2 \tag{2.1e}$$

$$y_c = t_3 + t_4 \tag{2.1f}$$

It follows that if the complex $n * n$ matrix has bandwidth W then the sub-problems (2.1a–d) are also $n * n$ real matrices of bandwidth W. Using the interleaving of separate instances of these problems on the same arrays reduces the matrix–vector timings as follows:

THEOREM 1a *The multiplication of an $n * n$ complex matrix and an n-component complex vector can be performed in time $T = 2n + (W + 2)$ using a bi-linear array of $2W$ ips cells, an adder and a subtracter.*

Proof See Figure 1.

3. MATRIX–MATRIX MULTIPLICATION

In the previous section we managed to reduce the time of the complex matrix–vector computation down to the same time a real computation would take by using only twice as much hardware. This is a significant saving over the upper-bound for complex computations given in Theorem 1. The reduction in cells and increase in speed were achieved by a simple observation that matrix–vector operations can be decomposed into a number of smaller related problems (in fact 1/4 the size of the original).

We now apply the same principles to the matrix–matrix multiplication of complex matrices. Consider the partitioned form of the extended real matrix multiplication

$$\left[\begin{array}{c|c} A & -B \\ \hline B & A \end{array}\right] \left[\begin{array}{c|c} C & -D \\ \hline D & C \end{array}\right] = \left[\begin{array}{c|c} AC-BD & -AD-BC \\ \hline BC+AD & AC-BD \end{array}\right] \tag{3.1}$$

from which we immediately notice that only the terms,

$$AC - BD, BC + AD \tag{3.2}$$

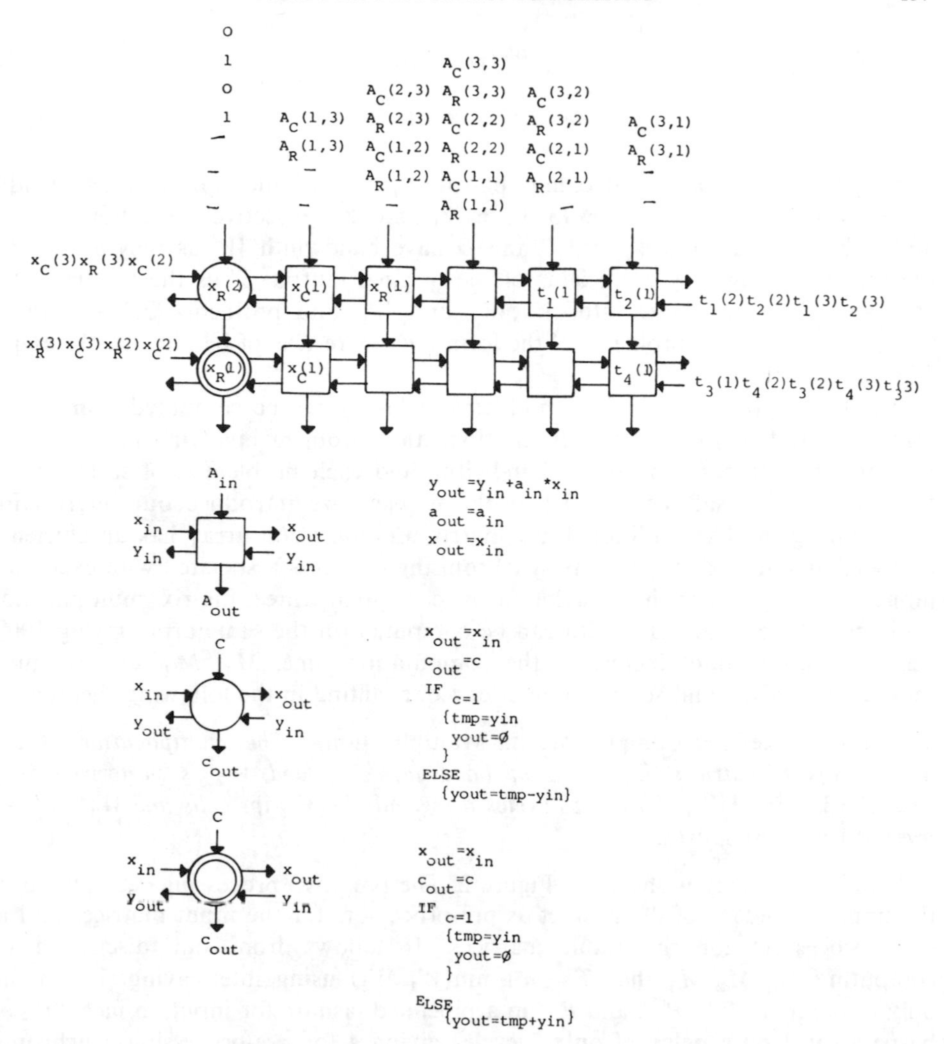

Figure 1 Bi-linear array for complex matrix vector.

are independent with the remaining terms produced by negating the result of the latter term. This can be computed in only three matrix multiplication problems,

$$M_1 = P^{(1)}C, \tag{3.3a}$$

$$M_2 = BP^{(2)}, \tag{3.3b}$$

$$M_3 = P^{(3)}D, \tag{3.3c}$$

where,

$$p^{(1)} = A - B, \tag{3.3d}$$

$$P^{(2)} = C - D, \tag{3.3e}$$

$$P^{(3)} = A + B. \tag{3.3f}$$

These problems are $n * n$ matrix multiplication problems and real. If the bandwidth of the two complex matrices in (3.1) are W_1 and W_2 respectively, it follows that A and B have bandwidth W_1 and C and D have bandwidth W_2 as they are simply the real and imaginary coefficients of the original matrix. Thus the matrices $P^{(1)}$, $P^{(2)}$, and $P^{(3)}$ have bandwidths W_1, W_2, and W_1, and problems (3.3a–c) are all matrix multiplication problems of the form EF where E is of bandwidth W_1 and F is of bandwidth W_2.

It follows that any one of the problems (3.3a − c) can be computed using a hex array identical to the real matrix multiplication problem involving matrices with the same structure EF (i.e. same bandwith), and each problem is of size n, hence the array in [1] suffices for each problem. Next we introduce our interleaving argument again. The traditional hex matrix multiplication array has an efficiency of $e = 1/3$ due to the two dummy synchronising elements associated with each data input. Using problem interleaving it is clear that three matrix multiplication problems of identical bandwidth can be computed on the same array giving 100% cell efficiency without increasing the computation time. M_1, M_2 and M_3 meet these requirements and so can be interleaved, resulting in the following theorem.

THEOREM 2a (Revised Complex Matrix Multiplication) *The multiplication of two* $n * n$ *complex matrices* E *and* F *of bandwidth* W_1 *and* W_2 *is performed in a time* $T = 3n + \min(W_1, W_2) + 4$ *ips cycles using only* $W_1 W_2$ *ips cells and* $W_1 + W_2 - 1$ *pre(post) processing cells.*

Proof The array is shown in Figure 2. The pre(post) processing cells placed on the upper boundary of the hex act as pre-processors for the input matrices and as post processors for the result matrices. It follows from our discussion on computing M_1, M_2, M_3 that $T = 3n + \min(W_1, W_2)$ using interleaving. It remains only to compute $P^{(1)}$, $P^{(2)}$, and $P^{(3)}$ in a pipelined fashion for input, in fact this can be performed by a delay of only 2 cycles giving 4 for pre-processing synchronisation and post-processing delays. The number of cells follows also from the fact that the ordinary real case requires $W_1 W_2$ cells and the perimeter of the upper boundary is $W_1 + W_2 - 1$ cells in length.

Now the computation proceeds as follows. Initially $P^{(1)} = A$ and $P^{(3)} = 0$ (zero matrix) while $P^{(2)} = D$. Dataflow is identical to that of the hex array of [1], after $\min(W_1, W_2) + 2$ cycles the first results begin to leave the array and correspond to the forms,

$$M_1 - M_2 = AC - BD, \text{ and } M_2 + M_3 = BC - AD$$

which are constructed by the pre(post) boundary cells.

It follows that complex matrix product can be computed with only $W_1 + W_2 - 1$ extra cells which contain adder/subtractor arrangements and an extra 4 *ips* cycles

BASIC IPS CELL

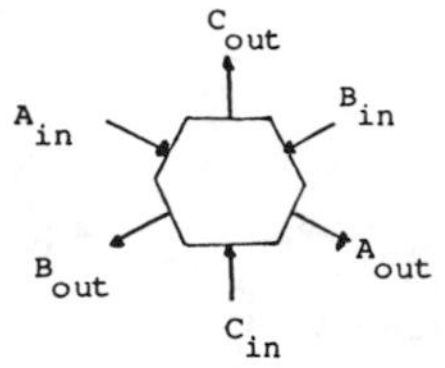

C_{out}, B_{in}, A_{in}, B_{out}, A_{out}, C_{in}

$C_{out} = C_{in} + A_{in}B_{in}$
$B_{out} = B_{in}$
$A_{out} = A_{in}$

CENTER PRE(POST) CELL

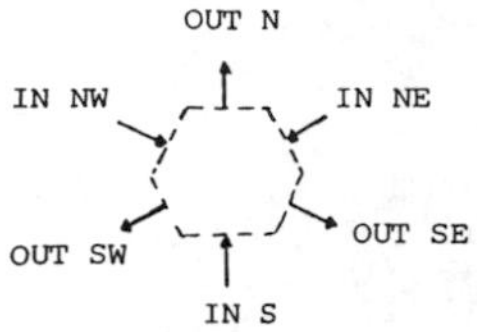

OUT N, IN NW, IN NE, OUT SW, OUT SE, IN S

t: IN S A
OUT N B+C
t+1: IN S B
OUT N ZERO
t+2: IN S C
OUT N A-B

LEFT PRE(POST) CELL

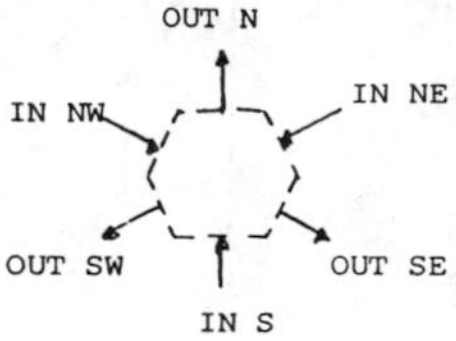

OUT N, IN NW, IN NE, OUT SW, OUT SE, IN S

t: IN NW A
IN S r_1
OUT SW $\bar{B}$
OUT N r_2+r_3
t+1: IN NW B
IN S r_2
OUT SW $\bar{A}-\bar{B}$
OUT N ZERO
t+2: IN S r_3
OUT SW A-B
OUT N r_1-r_2
$\bar{A}=A$ $\bar{B}=B$

RIGHT PRE(POST) CELL

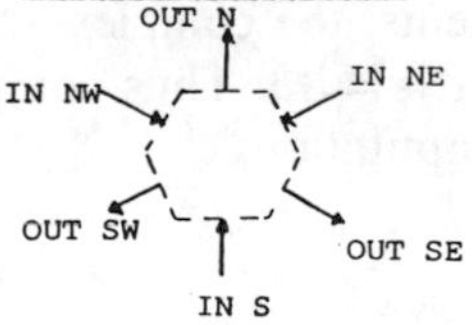

OUT N, IN NW, IN NE, OUT SW, OUT SE, IN S

t: IN NE C
IN S r_1
OUT SW $\bar{C}-\bar{D}$
OUT N r_2+r_3
t+1: IN NE D
IN S r_2
OUT SW $\bar{D}$
OUT N ZERO
t+2: IN S r_3
OUT SW C
OUT N r_1-r_2
$\bar{C}=C$, $\bar{D}=D$

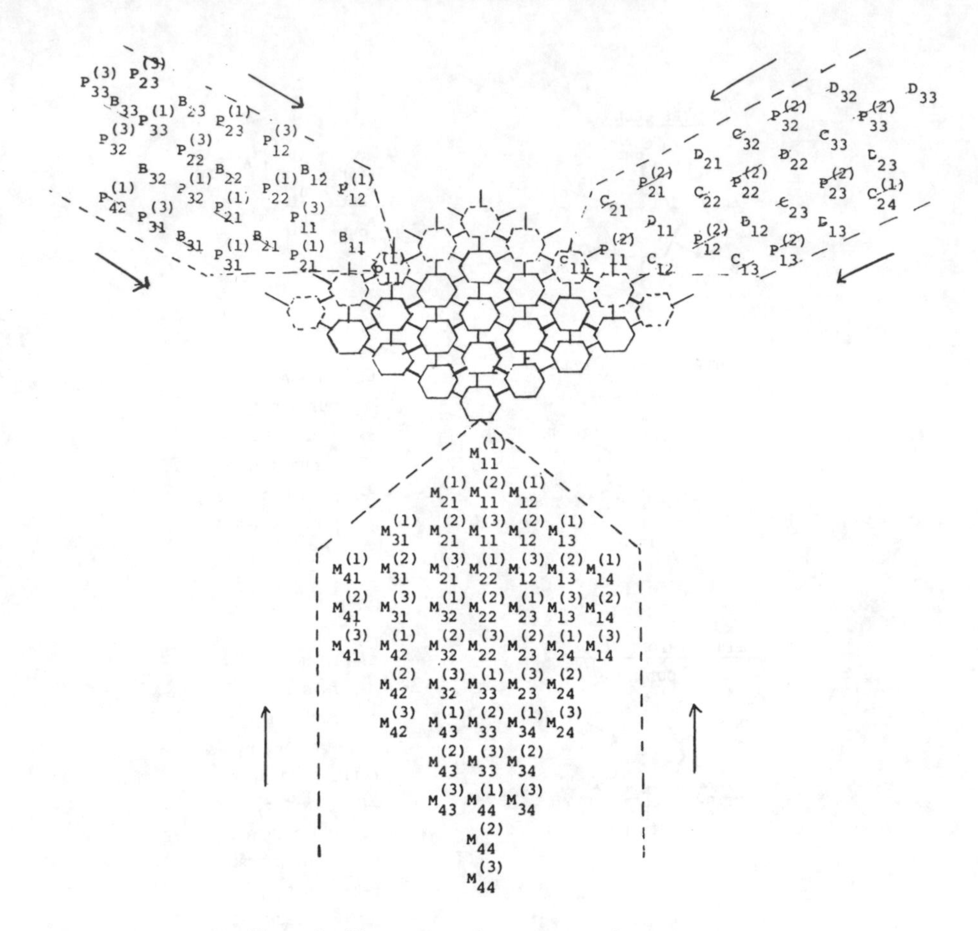

Figure 2 Hex connected complex matrix product.

than the real matrix product array. Hence, complex and real matrix computation are asymptotically the same in time requirements, the complex array computes at $e=1$ efficiency, while the real case computes at $e=1/3$. Thus complex calculations utilise the systolic concept better than real computations.

4. COMPLEX *LU* FACTORISATION

By continuing the theme of decomposing the extended real matrix computations into component problems resembling the original problem we now consider the solution of a linear system of equations of the form,

$$Ax=b, \tag{4.1}$$

wc rewrite (4.1) in extended real matrix form,

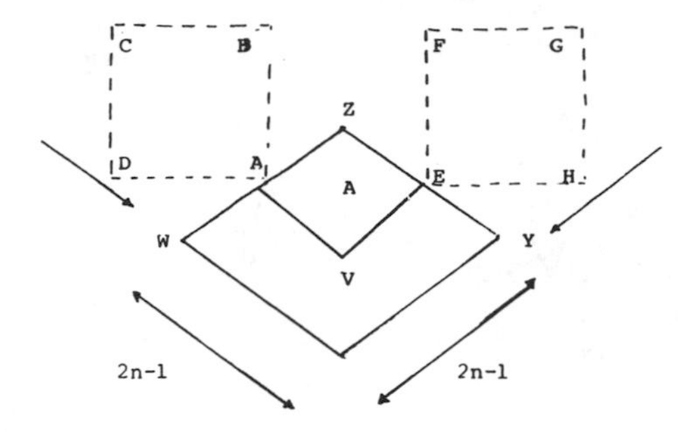

Figure 3 Accumulation of product terms in a hex array.

$$\begin{bmatrix} C & -D \\ D & C \end{bmatrix} \begin{bmatrix} y \\ z \end{bmatrix} = \begin{bmatrix} c \\ d \end{bmatrix} \tag{4.2}$$

where C and D are $n \times n$ real matrices, $x=(y+iz), b=(c+id)$, which can be rewritten as,

$$Zy = D^{-1}c + C^{-1}d, \tag{4.3a}$$

$$Zz = D^{-1}d + C^{-1}c, \tag{4.3b}$$

$$Z = D^{-1}c + C^{-1}d, \tag{4.3c}$$

Now we are required to construct Z and then factorise it, this can be performed in three steps:

(i) compute $D^{-1}c = K^{(1)}$, (4.4a)
(ii) compute $C^{-1}d = K^{(2)}$, (4.4b)
(iii) compute Z and factorise. (4.4c)

These three problems all involve $n * n$ matrices. The only drawback is the fact that we require the explicit knowledge of D^{-1} and C^{-1} these matrices will tend to be full when C and D are banded, implying a hex of dimension $W * (2n-1)$ where W is the bandwidth of A. For the moment we will assume two properties of A:

a) D^{-1} and C^{-1} are easily computed, and so are available at the start of the computation.

b) A is dense (i.e. full) giving a bandwidth $2n-1$.

Cases (a) and (b) may seem to contradict one another but later we will remove (b).

When C and D are full the products (4.4a, b) can be computed in an interleaved fashion on a hex array with $(2n-1)^2$ cells, this follows from Section 3. Now observe in Figure 3 which represents the multiplication hex and the associated

OUT N
IN NW
IN NE
OUT SE
OUT SW
IN S

```
INTERNAL REGION A CELL

IN S c
IN NW a;k
IN NE b
OUT N cout
OUT SE aout
OUT SW bout
IF
  K=O
  {cout=c+(a*b)
   aout=a
   bout=b
   r1=r2
   r2=cout
  }
  K=1
  {m=a*b      }
   c=r1+r2    }  *
   cout=c+m   }
  }
```

```
LEFT BOUNDARY REGION A CELL

IN S c
IN NW a,k
IN NE b
OUT N cout
OUT SE aout
OUT SW bout
IF
  Flag
    {aout=c*b,bout=b}
  K=O
    {cout=c+(a*b)
     aout=a
     bout=b
     r1=r2
     r2=cout
    }
  K=1
    {flag=true
     c=r1+r2
     aout=c*b
     bout=b
    }
```

```
RIGHT BOUNDARY REGION A CELL

IN S c
IN NW a,k
IN NE b
OUT N cout
OUT SE aout
OUT SW bout
IF
   flag
    {bout=O+a*c, aout=a}
   K=O
    {cout=c+(a*b)
     aout=a
     bout=b
     r1=r2
     r2=cout
    }
   K=1
    {c=r1+r2
     bout=O+a*c
     aout=a
     flag=true
    }
```

*computed in parallel with one adder and one multiplier

interleaved data input for C^{-1}, D^{-1}, C, and D. Now because C and D are dense, the data flow has a special structure not available when banded systems are computed. As the data shifts across the array, computations start when the vertex A and vertex E of the two data fields reach vertex V of the region A. It follows that the first result accumulates its last term when vertices B and F reach vertex Z of region A. Likewise when the vertices C and G reach vertex V the last matrix element z_{nn} is completed. It follows that region A consists of all the cells in the hex which will complete the construction of an element of the matrix Z. Further, as the region is $n*n$ each cell computes one and only one final result of a matrix element z_{ij} of Z (that is, it finishes the accumulation of partial products of z_{ij}). It also follows after a little thought that cells finish accumulation in an order which mimicks a wavefront passing across region A from vertex Z to V.

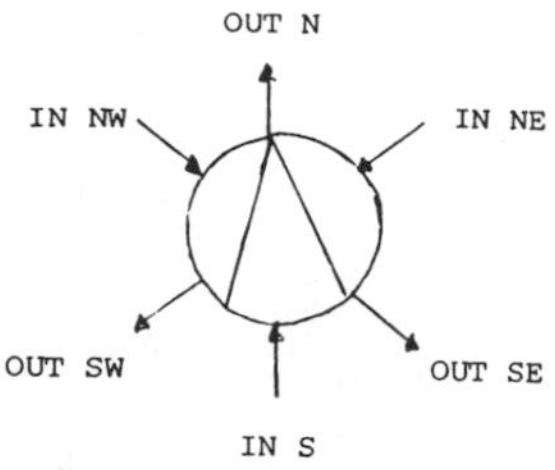

OUT N

IN NW IN NE

OUT SW OUT SE

IN S

```
CENTER/PROCESSOR (VERTEX Z) REGION A

IN S c
IN NW a,k
IN NE b
OUT N cout
OUT SW bout
OUT SE aout
IF
  Flag
    {cout=1/c, bout=cout, aout=-1}
  K=0
    {cout=c+(a*b)
     r1=r2
     r2=cout
     aout=a
     bout=b
    }
  K=1
    {cout=1/(r1+r2)
     bout=cout
     aout=-1
     flag=true
    }
```

REMARK: Initially flag=false, $r_1=r_2=0$

Now we compute (4.4) by interleaving the three steps in the same array, the data flow is shown in Figure 4. To explain the operation of the array, we start when the first matrix value reaches the boundary cell at vertex Z. On this cycle the element $K_{11}^{(1)}$ is produced. On the next cycle $K_{12}^{(1)}$ and $K_{21}^{(1)}$ and $K_{11}^{(2)}$ are constructed starting the wavefront from Z to V. On the third cycle $K_{13}^{(1)}$, $K_{31}^{(1)}$, $K_{22}^{(1)}$, $K_{12}^{(2)}$ and $K_{21}^{(2)}$ are completed but more importantly $z_{11}^{-1}=(K_{11}^{(1)}+K_{11}^{(2)})^{-1}$ is computed at vertex Z. The fourth cycle gives $K_{14}^{(1)}$, $K_{41}^{(1)}$, $K_{32}^{(1)}$, $K_{23}^{(1)}$ and $K_{13}^{(2)}$, $K_{31}^{(2)}$, $K_{22}^{(2)}$, and the multiplier for start of the factorisation is computed from $z_{12}=(K_{12}^{(1)}+K_{12}^{(2)})$ and $z_{21}=(K_{21}^{(1)}+K_{21}^{(?)})$ and $L_{21}=z_{21}^{*}z_{11}^{-1}, U_{12}=z_{12}$. The next cycle sees the computation of $z_{22}=(K_{22}^{(1)}+K_{22}^{(2)})$ and a further modification to give $z_{22}=z_{22}-(z_{21}^{*}z_{11}^{-1}{}^{*}z_{12})$. The factorisation proceeds in this interleaved manner producing the full factorisation of z. In order to achieve the correct computations the cells in region A are redefined as shown below, the essential point is that they can simulate either a matrix product cell or a LU hex cell from [1]. The switch is made by a simple 1-bit control value tagged to the input of the left upper boundary of the hex. The control bit is set only once on the cycles that immediately follow values on the line BC of the left data input of Figure 3.

This construction now gives the revised theorem for matrix factorisation.

THEOREM 3a (Complex matrix factorisation) *The LU factorisation of an $n*n$ complex matrix with bandwidth W is performed in time $T=3n+(2n-1)+2$ cycles and requires $(2n-1)^2$ ips cells.*

Proof The array in Figure 4.

Remark Note that the time of this array is decreased by $2n$ cycles at least but the number of *ips* cells is much larger than the corresponding hex in Theorem 3, except when the matrix is full.

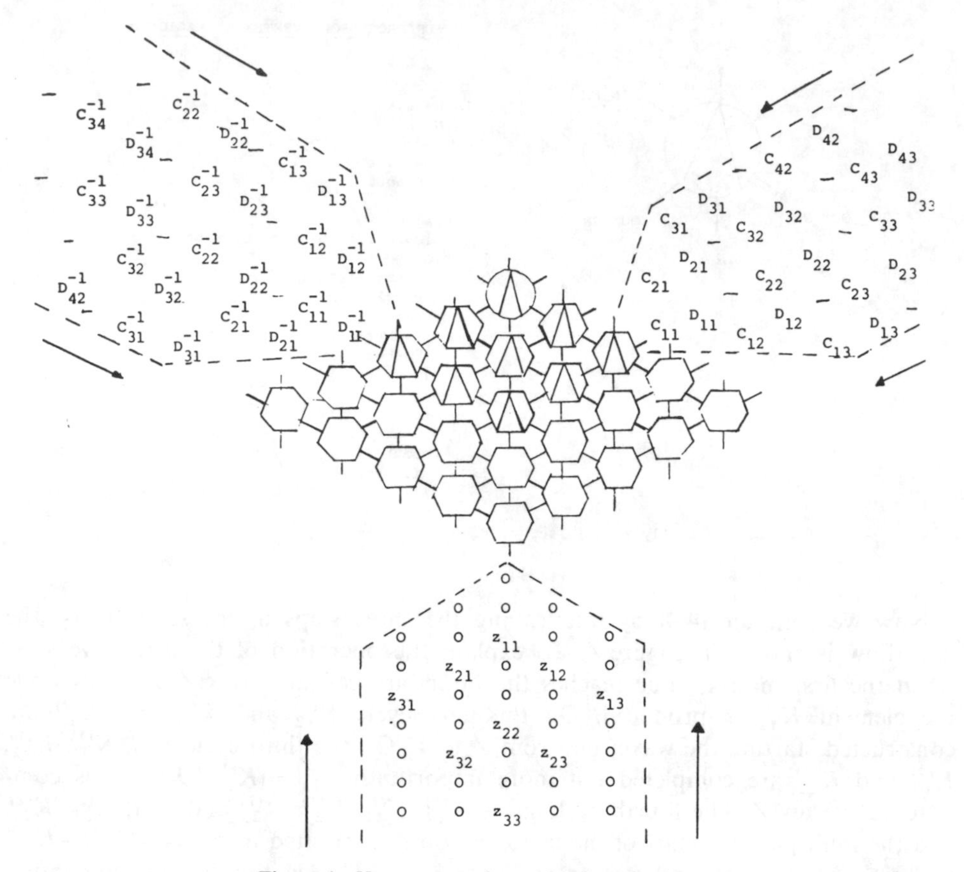

Figure 4 Hex connected complex factorisation.

Hence, when we have A full complex, Theorem 3a performs $3n$ cycles faster than the straightforward array in Theorem 3, with the same hardware requirement except for switching and extra registers in the region A cells. This result is somewhat disappointing, because a banded system will probably give the simplest inverse for C and D allowing easy computation, and this produces the worst cell tradeoff between the two designs. Finally, we note that the full solution of (4.1) can be obtained by using the matrix–vector array in Figure 1 to compute the right hand sides of (4.3) (a) and (b), while the principle of interleaving is easily extended to backsubstitution to solve the two systems in parallel on a single array.

4.1 2 ∗ 2 Block Form of Complex Factorisation

So far we have only considered the production of a real extended matrix from the complex form along the lines of (1.3) which for the matrix product and matrix vector problems was adequate. However for the LU array as just illustrated it was not very successful.

Now consider a single complex number $a+bi$, which can be written in the form of a $2*2$ matrix,

$$\begin{bmatrix} a & -b \\ b & a \end{bmatrix}. \tag{4.5}$$

It follows that every complex element can be expanded to produce the $2*2$ block form of a $2n*2n$ real matrix which is a permuted form of (1.3). Now for a factorisation process we must be able to perform a modification of each row and form the multipliers, which involves in the block case a $2*2$ block inner-product and a $2*2$ matrix inversion.

COMPLEX INNER PRODUCT STEP

$$\begin{bmatrix} e & -f \\ f & e \end{bmatrix} = \begin{bmatrix} e & -f \\ f & e \end{bmatrix} + \begin{bmatrix} a & -b \\ b & a \end{bmatrix} \begin{bmatrix} c & -d \\ d & c \end{bmatrix}$$

$$(e+fi) \quad = \quad (e+fi) \quad + \quad (a+bi)(c+di)$$

COMPLEX RECIPROCAL

$$\frac{1}{(a+bi)} = \begin{bmatrix} a & -b \\ b & b \end{bmatrix}^{-1} = \frac{1}{a^2-b^2} \begin{bmatrix} a & b \\ -b & a \end{bmatrix}.$$

It follows that if we perform a block $2*2$ factorisation on the permuted form of (1.3) we can factorise the complex matrix. Such an array already exists (see Robert [2]) and requires $T=2n+\min(p,q)$ for an $n*n$ real matrix of bandwidth $W=p+q-1$. We can use this on a $2n*2n$ matrix by noticing that the $2*2$ block form of the complex matrix only doubles the bandwith to give $W_1=2p+2q-1=2W+1$. Giving the theorem,

THEOREM 3b ($2*2$ block complex matrix factorisation) *An $n*n$ complex matrix of bandwidth W can be factorised on a hexagonally connected array in time $T=4n+\min(2p,2q)$ requiring approximately $4pq$ ips cells.*

Proof See Robert [2] for the block $2*2$ block form increases the number of cells proportional to the bandwidth of the original complex matrix and so is superior to that in Theorem 3a. However when the matrices are full Theorem 3a uses less hardware than Theorem 3b and the efficiency is slightly better than the block form.

5. CONCLUSIONS

We have exploited the systolic principle to produce arrays for complex matrix operations from existing arrays for real matrix operations. For matrix vector computations we have doubled the hardware using a bilinear array but adjusted the computation by only a constant from the real case.

For a matrix–matrix product we have showed that the complex matrix product can be represented by three real matrix products with exactly the same structure (bandwidth) as the complex matrices. Using an interleaving strategy of these three problems on the same array a real hexagonal array was used to multiply complex matrices in the same time and with the addition of only W_1+W_2-1 extra cells, as for real matrix product. The complex matrix product interleaved array operates with a cell efficiency of $e=1$, while the real matrix product has $e=1/3$.

The final problem considered was the LU factorisation of complex matrices. Two approaches were adopted for solving the complex factorisation. The first attempted to extend the ideas of interleaving and decomposition into smaller matrix problems in a similar manner to the complex product case. However the inverse of two $n*n$ real matrices was required and when they were easily invertible (e.g. really banded, symmetric, etc.) the matrices filled in giving a high processor count. It was only when the matrix to be factorised was fully dense did we get a satisfactory trade-off with array efficiency $e=2/3$. The best approach for the complex factorisation turned out to be the representation of the complex $n*n$ matrix by a $2*2$ block $2n*2n$ real matrix. We then apply a $2*2$ block factorisation requiring approximately four times the hardware than the array necessary for the corresponding real problem, with efficiency $e=1/2$, and twice as slow.

References

[1] H. T. Kung and C. L. Leiserson, Systolic arrays for VLSI, in: *Introduction to VLSI Systems*, C. A. Mead and L. A. Conway, eds., Addison-Wesley, 1980.

[2] Yves Robert, Block *LU* decomposition of a band matrix on a systolic array, *Int. Jour. Comp. Math.* **17** (1985), 295–316.

Systolic Computation of the Matrix Exponential and Other Matrix Functions

D. J. EVANS and K. MARGARITIS
Department of Computer Studies, Loughborough University of Technology, Loughborough, Leicestershire, UK

This paper discusses the systolic implementation of the computation of the exponential of a matrix by means of techniques involving "scaling and squaring" as applied to the Taylor series approximation. Further, it is shown that a number of other matrix functions, such as A^{-1}, $A^{1/2}$, $A^{-1/2}$, $\cos(A)$, $\sin(A)$, $\log(A)$, can be computed systolically using similar techniques.

KEY WORDS: Systolic design, matrix exponential, matrix functions, scaling and matrix squaring.

C.R. CATEGORIES: F1.1, F2.1, B7.1.

1. INTRODUCTION

Mathematical models of many scientific and engineering problems involve systems of linear, constant coefficient ordinary differential equations,

$$\dot{x} = Ax, \tag{1}$$

where A is a given $(n \times n)$ matrix. In principle, the solution is given by,

$$x = e^A x_0, \tag{2}$$

where x_0 is a given initial condition.

Now, e^A can be formally defined by the convergent power series,

$$e^A = I + A + \frac{A^2}{2!} + \frac{A^3}{3!} + \cdots \tag{3}$$

The systolic computation of this matrix function is discussed herein.

The definition of e^A in (3) can be the basis for a simple algorithm calculating the exponential of a matrix using Taylor series approximation techniques. Thus,

$$e^A \approx T_k(A) = \sum_{i=0}^{k} \left(\frac{1}{i!} A^i \right). \tag{4}$$

However, such an algorithm is known to be unsatisfactory, since k is usually very large, for a sufficiently small error tolerance. Furthermore the round-off errors and the computing costs of the Taylor approximation increase as $\|A\|$ increases, or as the spread of the eigenvalues of A increases.

These difficulties can be controlled by exploiting a fundamental property unique to the exponential function:

$$e^A = (e^{A/m})^m. \tag{5}$$

In the scaling and squaring method, m is chosen so that:

$$m = 2^j \quad \text{and} \quad \frac{\|A\|}{2^j} \leqq 1. \tag{6}$$

With this restriction, the Taylor approximation in (4) can be satisfactorily used, and then e^A is formed by j successive squarings, [4].

For a given error tolerance ε, and magnitude of $\|A\|$, Table 1 summarizes the optimum (k, j) associated with $[T_k(A/2^j)]^{2^j}$, [10].

Table 1

$\|A\|$ \ ε	10^{-3}	10^{-6}	10^{-9}	10^{-12}	10^{-15}
10^{-2}	1,0	2,1	3,1	4,1	5,1
10^{-1}	3,0	4,0	4,2	4,4	5,4
10^{0}	5,1	7,1	6,3	8,3	7,5
10^{1}	4,5	6,5	8,5	7,7	9,7
10^{2}	4,8	5,9	7,9	9,9	10,10
10^{3}	5,11	7,11	6,11	8,13	8,14

The algorithm for the calculation of the exponential of a matrix can be described as follows:

Step 1 Given A, ε obtain (k, j) from Table 1.

Step 2 Calculate $A/2^j$.

Step 3 Calculate $T_k(A/2^j)$ using (4) for k steps.

Step 4 Calculate $[T_k(A/2^j)]^{2^j}$ with j successive matrix squarings.

2. SYSTOLIC IMPLEMENTATION

Step 3 of the algorithm is a matrix polynomial computation in which Horner's scheme of nested multiplication can be used [4], [9], [13]:

$$S_0=\frac{1}{k!}I$$

$$S_i=AS_{i-1}+\frac{1}{(k-i)!}I, \quad \text{for} \quad i=1,2,\ldots,k, \tag{7}$$

or in a more general form,

$$S_0=X_k$$

$$S_i=AS_{i-1}+X_{k-i}, \quad \text{for} \quad i=1,2,\ldots,k, \tag{8}$$

where,

$$X_i = \frac{1}{i!} I, \quad S_k = T_k(A). \tag{9}$$

Thus, the evaluation of the Taylor series approximation has been reduced to a series of matrix-matrix inner product steps (mmips),

$$S = AS + X. \tag{10}$$

Step 4 of the algorithm is a successive matrix squaring given by the following recurrence:

$$S_0 = A$$

$$S_i = S_{i-1}^2, \quad \text{for} \quad i = 1, 2, \ldots, j, \tag{11}$$

where,

$$S_j = A^{2^j}.$$

Similarly with Step 3 the successive matrix squaring has been reduced to a series of matrix-matrix multiplications (mm), which is a simpler form of a mmips given in (10), for $A = S$ and $X = 0$.

The underlying computational structure for both the Taylor series evaluation and successive matrix powers, is a systolic network allowing for repeated mmips, where the input of each new computation can be the output of the previous computation, and, possibly, some new matrices.

Following the definitions introduced in [8], the processes of repeated mmips can be expanded either in time, yielding an iterative network, or in area, producing a pipeline, (see Figure 1). In general terms, there is an area-time tradeoff between the two approaches, i.e. the area expansion requires $\cong k$ times the area of the time expansion, while the time expansion requires $\cong k$ times the computation time of the area expansion, where k is the number of repetitions of the processes.

Now with the process in question being the mmips, the two approaches have been proposed in [13] and [2], [9], for time expansion and area expansion respectively. These two architectures are utilised in the following sections for the systolic implementation of the calculation of e^A.

3. TIME EXPANSION

A re-usable mmm array is proposed in [13], which, with some modifications can perform both the calculations in (8) and (11). The array is shown in Figure 2(a), for $n=3$, where n is the size of matrix A. In Figure 2(b) the input-output-feedback sequences required for the computation of e^A are given.

The area required is not more than $(n+1)^2$ IPS cells, and the computation time is $2n(k+j+2)+n-1$ IPS cycles. The area occupied and the delay caused by the control operations within the n^2 IPS cells of the main array are assumed negligible. The area-time complexity of the multiplexing-addition operations of the boundary row-column is taken to be not more than 1 IPS.

The time-expanded systolic system is mostly suitable for full matrices, since both its area and time complexities depend heavily on n. On the other hand, however, the same array can perform an unlimited number of iterations (i.e. for any k, j).

4. AREA EXPANSION

A pipeline of hexagonally-connected systolic arrays is proposed in [2] for the computation of successive powers as in (11); a similar pipeline for the calculation of a matrix polynomial is discussed in [9].

The hex-connected array used is shown in Figure 3(a), for $w=3$, where w is the bandwidth of matrix A. In Figure 3(b) a pipeline is illustrated for $k=2, j=1$ (as defined in Table 1).

The area required is $1+\sum_{i=0}^{k-1}(w(iw-(i-1)))+\sum_{i=0}^{j-1}(2^j w'-(2^i-1))^2$ IPS cells, and the computation time is $n+w+\sum_{i=1}^{k-1}(iw-(i-1))+\sum_{i=0}^{j-1}(2^j w'-(2^i-1))$ IPS cycles; where $w'=kw-(k-1)$.

The area occupied and the delay caused by reformatting delays and inter-connections are assumed to be negligible.

The area expanded systolic design is mostly suitable for banded matrices with $n \gg w$, since both its area and time complexities depend mainly on w. Notice, however, that a new pipeline configuration is necessary for a different pair of k, j.

In the remainder of this paper, the systolic matrix inversion and the systolic computation of several matrix functions are briefly discussed.

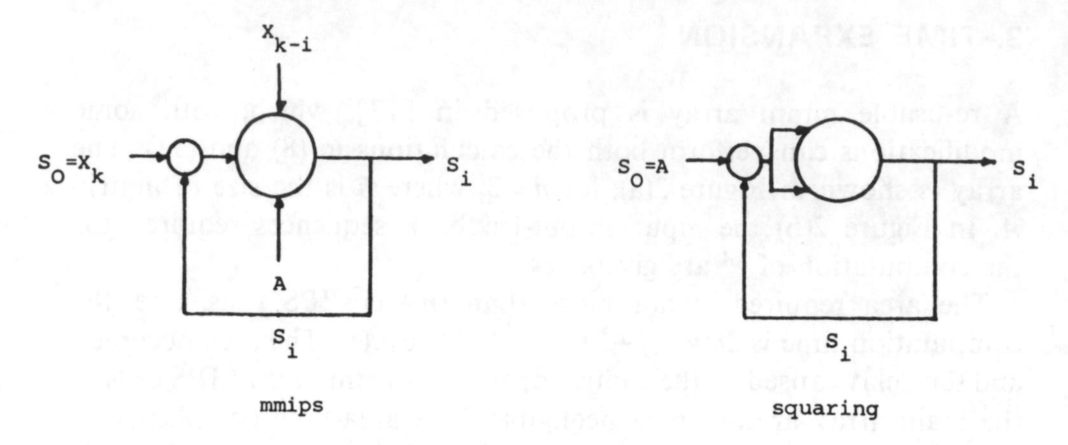

a) Time expansion

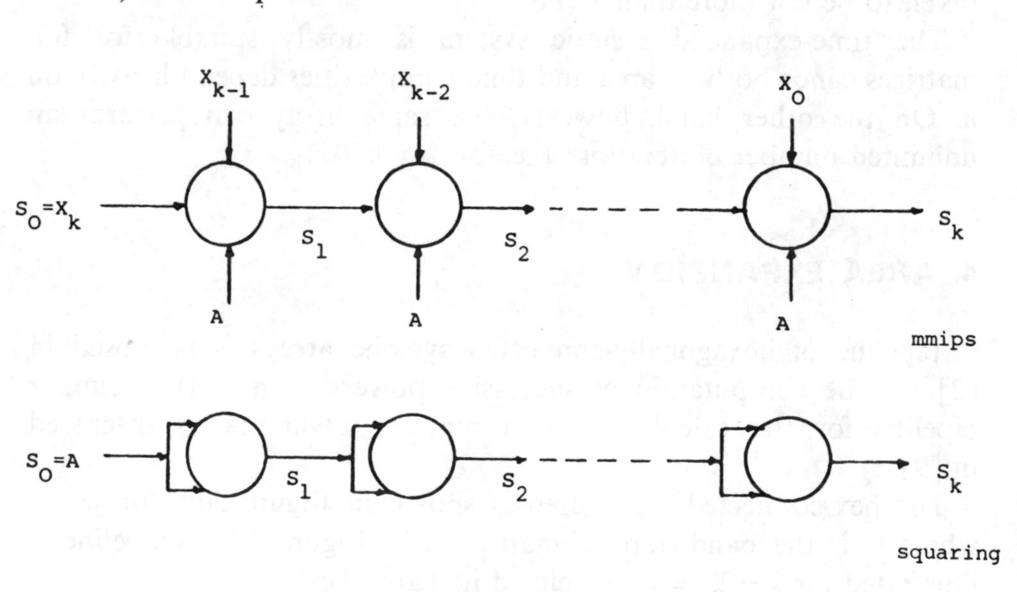

b) Area expansion

Figure 1

5. SYSTOLIC INVERSION USING MATRIX POWERS

Let A be a $(n \times n)$ real matrix and we want to compute A^{-1}; we assume that an approximate inverse B is known. Then, for

$$M = I - AB \tag{12}$$

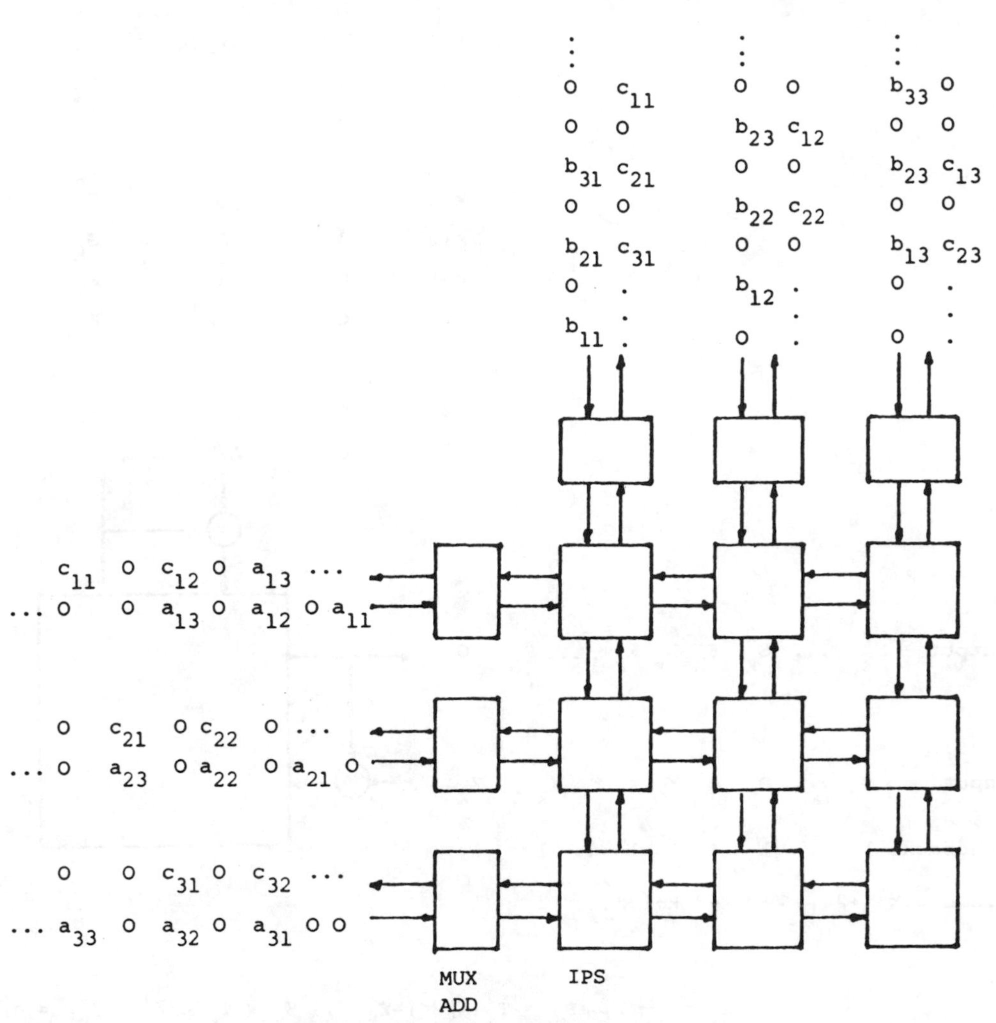

Figure 2(a) Reusable mmm array: $C = C + A \cdot B$.

the inverse is calculated as

$$A^{-1} = B(I - M)^{-1} = B(I + M + M^2 + M^3 + \cdots) \tag{13}$$

with $\|M\| < 1$ [3]. The efficient calculation of the approximate inverse B is investigated in [12], so that the convergence condition is satisfied.

Three alternative ways of evaluating the sum

$$S_k = I + M + M^2 + \cdots + M^{k-1} \tag{14}$$

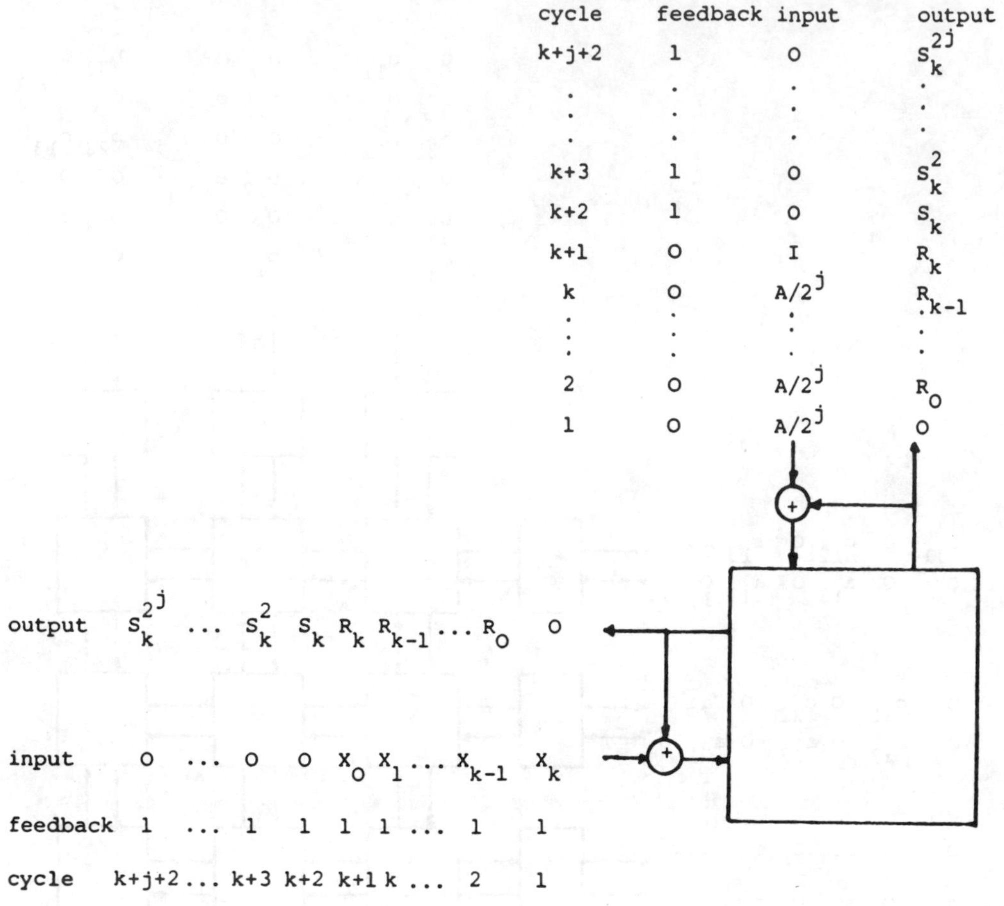

cycle	feedback	input	output
k+j+2	1	0	$S_k^{2^j}$
⋮	⋮	⋮	⋮
k+3	1	0	S_k^2
k+2	1	0	S_k
k+1	0	I	R_k
k	0	$A/2^j$	R_{k-1}
⋮	⋮	⋮	⋮
2	0	$A/2^j$	R_0
1	0	$A/2^j$	0

output	$S_k^{2^j}$	...	S_k^2	S_k	R_k	R_{k-1}	...	R_0	0
input	0	...	0	0	X_0	X_1	...	X_{k-1}	X_k
feedback	1	...	1	1	1	1	...	1	1
cycle	k+j+2	...	k+3	k+2	k+1	k	...	2	1

$$X_i = \frac{1}{i!} I, \quad R_i = T_i(A/2^j) - X_{k-i}, \quad S_k = T_k(A/2^j), \quad S_k^{2^j} = e^A$$

Figure 2(b)

are proposed in [1]:

i) Nested multiplication (Horner's scheme): $(k-1)$ mmips yield S_k:

$$S_k = I + M + (I + M(I + \cdots M(I + MI)\cdots)). \tag{15}$$

ii) Especially for $k = 2^{(l+1)}$, successive squares of M are computed, and S_k is calculated. Starting from $S_0 = I$, $S_2 = I + M$ one computes $S_4, S_8, \ldots, S_k$ in $l+1$ steps, using the recurrence:

$$S_{2^{(i+1)}} = S_{2^i} + S_{2^i} M^{2^i}, \qquad i = 0, 1, 2, \ldots, l \tag{16}$$

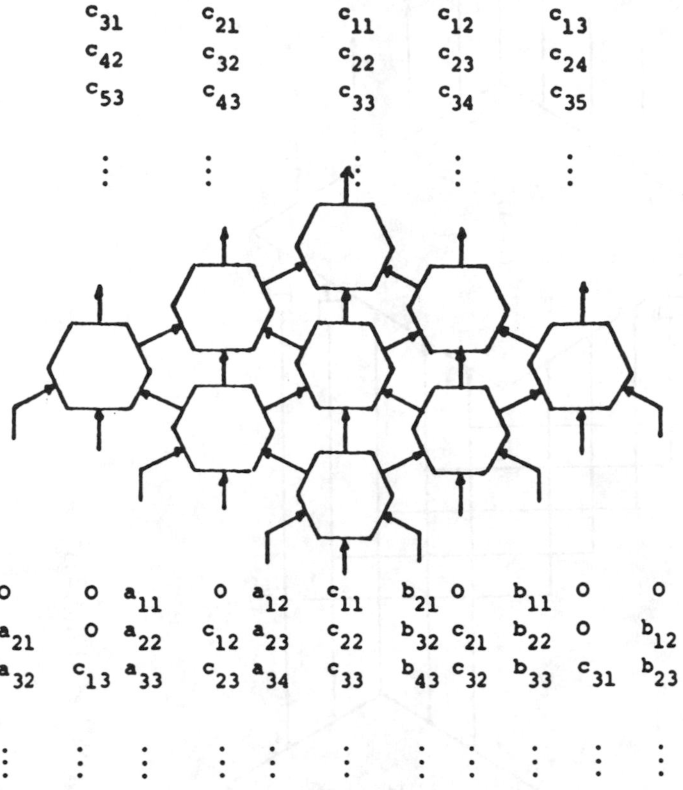

Figure 3(a) Hexagonally-connected mmips array, $C=C+A\cdot B$.

iii) Another procedure based on successive squarings can be derived from (16):

$$S_k=(I+M)(I+M^2)(I+M^4)\cdots(I+M^{2^l}). \tag{17}$$

Notice, that each step of (16, 17) consists of 1 mmips and a mmm for matrix squaring and has approximately twice more work than that of (15). On the other hand, only $l+1$ steps are required instead of 2^{l+1}. Further, if we truncate the power series in (13), after the first two terms, we obtain the Newton method for matrix inversion:

$$B_{k+1}=B_k(2I-AB_k). \tag{18}$$

Each step of the Newton method can be analysed into a mmm, i.e. $T_k=2I-AB_k$, and a mmips, i.e. $B_{k+1}=B_kT_k$.

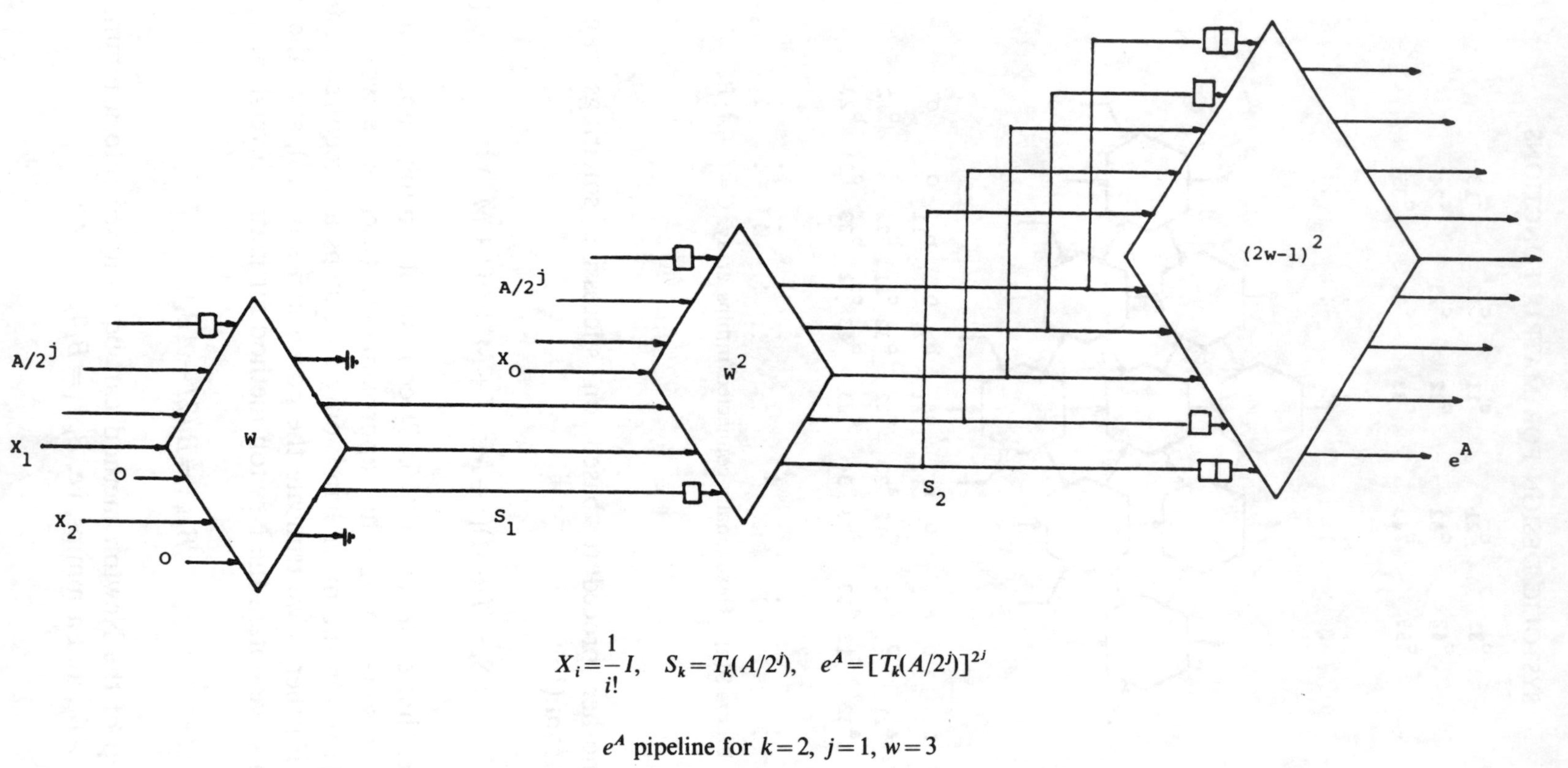

$$X_i = \frac{1}{i!} I, \quad S_k = T_k(A/2^j), \quad e^A = [T_k(A/2^j)]^{2^j}$$

e^A pipeline for $k=2$, $j=1$, $w=3$

Figure 3(b)

All these recurrences can be readily realised using the systolic pipeline or iterative designs proposed in the previous sections. Especially for the Newton method, the iterative structure seems more appropriate, since B_k can be stored in order to be used in both steps of the iteration.

6. SYSTOLIC COMPUTATION OF MATRIX FUNCTIONS

The calculation of $A^{1/2}$ has been studied in [6], [7], [4]. The Newton's method in the form

$$P_0 = A, \quad Q_0 = I$$

$$P_{k+1} = \tfrac{1}{2}(P_k + Q_k^{-1})$$

$$Q_{k+1} = \tfrac{1}{2}(Q_k + P_k^{-1}), \quad k = 0, 1, 2, \ldots \tag{19}$$

is recommended as the most stable, but its main disadvantage for a systolic implementation is the computation of matrix inverses in each iteration. It is interesting to note that the computation of $A^{-1/2}$ involves no matrix inversion during the iterations [3], [11]:

$$Z_k = I - T_k A T_k$$

$$T_{k+1} = T_k(I + \tfrac{1}{2}Z_k), \quad k = 0, 1, 2, \ldots \tag{20}$$

where T_0 is an initial approximation of $A^{-1/2}$. Thus, apart from the importance of $A^{-1/2}$ itself in numerical applications [11], it can also provide $A^{1/2}$, since $A^{1/2} = AA^{-1/2}$. The systolic implementation of (20) is based on iterative mmips and mmm operations, and therefore computational structures similar to that of Sections 3, 4 can be used. The iterative array seems especially more attractive since T_k must be temporarily stored, so that it can be used in both steps of an iteration.

In [4], [14] the calculation of $\cos(A)$, $\sin(A)$ is discussed, based on the double-angle method

$$\cos(2A) = 2\cos(A)^2 - I$$
$$\sin(2A) = 2\sin(A)\cos(A). \tag{21}$$

Thus, in a manner analogous to the matrix exponential computation, we can choose a scaling factor j, so that $\|A\|/2^j \leqq 1$. Then, the Taylor series approximations for $\cos(B)$, $\sin(B)$ are calculated, where $B = A/2^j$, for a given number k of steps.

$$\sin(B) = S_0 = B - \frac{B^3}{3!} + \frac{B^5}{5!} - \cdots$$

$$\cos(B) = C_0 = I - \frac{B^2}{2!} + \frac{B^4}{4!} - \cdots \tag{22}$$

Finally, the actual values of $\cos(A)$, $\sin(A)$ are given by the recurrence

$$S_k = 2S_{k-1}C_{k-1}$$
$$C_k = 2C_{k-1}^2 - I, \quad k = 1, 2, \ldots, j \tag{23}$$

Again the similarity of the calculations with that of Sections 3, 4 is evident. Two pipelines or two iterative arrays working in parallel can produce systolically $\cos(A)$, $\sin(A)$.

Finally, a similar computation is proposed in [4], [5], [15] and $\log(A)$, based on the Taylor series approximation

$$\log(A) = (A - I) - \frac{(A-I)^2}{2} + \frac{(A-I)^3}{3} - \cdots \tag{24}$$

However, no scaling method such as those given for the other matrix functions is generally available. Thus, the computation of (24) can be efficient only for $\|A\| < 1$.

7. CONCLUSION

This paper discusses the systolic implementation of the matrix exponential computation. The scaling and squaring method is used as applied to the Taylor series approximation. The systolic designs proposed have been implemented soft-systolically in OCCAM.

Two techniques for the realisation of a repetitive process are used: the time expansion, yielding the design of Figure 2, and the area expansion producing the design of Figure 3. In general terms, the area and time requirements for these systems are summarised in Table 2, where n is the size of the matrix, w its bandwidth, k the Taylor series terms calculated and j the successive squarings.

In summary, it can be observed that the time-expansion method yields a design with fixed area requirements but low throughput and considerably slower than the area-expansion design. On the other hand, the area-expansion method yields a pipeline with fast-increasing area requirements that can be suitable only for $w \ll n$ and for small k, j.

Computations similar to that required for e^A are met in the calculation of other matrix functions; subsequently similar designs can also be produced for the computation of A^{-1}, $A^{1/2}$, $A^{-1/2}$, $\cos(A)$, $\sin(A)$, $\log(A)$.

Table 2

	Time-expansion	*Area-expansion*
Area ≅	n^2	$\sum_{i=0}^{k-1} (iw^2) + \sum_{i=0}^{j-1} ((2^j w')^2)$
Time ≅	$2n(k+j)$	$n + \sum_{i=1}^{k} (iw) + \sum_{i=0}^{j-1} (2^i w')$

References

[1] E. Bobewig, *Matrix Calculus*, North-Holland, 1959.

[2] D. J. Evans, K. G. Margaritis and M. P. Bekakos, Systolic and holographic pyramidical soft-systolic designs for successive matrix powers, Report CS-284, Dept. of Computer Studies, Loughborough Univ. of Technology, June 1986; to appear in *Parallel Computing*, 1988.

[3] C. E. Froberg, *Introduction to Numerical Analysis*, Addison-Wesley, 1965.

[4] G. H. Golub, C. F. van Loan, *Matrix Computations*, North-Oxford Academic, 1983.

[5] B. W. Helton, Logarithms of matrices, *Proc. American Mathematical Society* **19** (1968), 733–736.

[6] H. J. Higham, Newton's method for the matrix square root, *Math. Comput.* **46** (1986), 537–549.

[7] W. D. Hoskins and D. J. Walton, A faster method of computing the square root of a matrix, *IEEE Trans. on Automatic Control* **AC-23** (1978), 494–495.

[8] L. Johnsson and D. Cohen, A mathematical approach to modelling the flow of data and control in computational networks. In: *VLSI Systems and Computations*, H. T. Kung *et al.* (eds.), Computer Science Press, 1981, pp. 213–225.

[9] K. G. Margaritis and D. J. Evans, Systolic designs for matrix inversion and solution of linear systems using matrix powers, Report CS-306, Dept. of Computer Studies, Loughborough Univ. of Technology, Sept. 1986; presented at Int. Conf. on High Performance Computer Systems, Paris, Dec. 1987.

[10] C. Moler and C. F. van Loan, Nineteen dubious ways to compute the exponential of a matrix, *SIAM Review* **20** (1978), 801–838.

[11] B. Philippe, Approximating the square root of the inverse of a matrix, Technical Report 508, Centre for Supercomputing R&D, Urbana, IL., 1985.

[12] V. Pan and J. Reif, Efficient parallel solution of linear systems, *Proc. 17th Annual Symposium on Theory of Computing*, 1985, pp. 143–152.

[13] P. Quinton, B. Jannault and P. Gachet, A new matrix multiplication systolic array. In: *Parallel Algorithms and Architectures*, M. Cosnard *et al.* (eds.), North-Holland, 1986, pp. 259–268.

[14] S. M. Serbin and S. A. Blalock, An algorithm for computing the matrix cosine, *SIAM J. Sci. Stat. Comput.* **1** (1980), 198–204.

[15] B. Singer and S. Spilerman, The representation of social processes by Markov models, *American J. of Sociology* **28** (1976), 1–55.

QUADRATURE AND DIFFERENTIAL EQUATIONS

Romberg integration using systolic arrays

D.J. EVANS and G.M. MEGSON

Department of Computer Studies, Loughborough University of Technology, Loughborough, Leicestershire, United Kingdom

Abstract. A systolic array is presented to improve numerical approximations to integrals using Richardson's extrapolation procedure in the form of Romberg integration. Two designs are presented, the first is an intuitive linear systolic array, the second, a systolic ring using approximately 1/3 of the cells of the first array. Both systolic arrays have a computation time of $3n$ cycles, which is a significant improvement on the $O(n^2)$ steps required to construct the extrapolation table sequentially.

Keywords. Systolic arrays, Richardson's extrapolation procedure, Romberg integration.

1. Introduction

The use of systolic arrays to date has largely been applied to fundamental problems in matrix and linear algebra. These problems, i.e. the solution of differential equations, as well as signal and image processing applications, account for approximately 70% of numeric computation. Apart from various sorting and searching systolic algorithms, little application has been made to other computational areas. Research in [6,7] has been aimed at developing new algorithm classifications such as hard, soft and hybrid systolic algorithms in order to define other areas capable of exploiting systolic design principles.

In this report we focus attention on the area of interpolation and extrapolation, where McKeown [5] has shown that Aitken's Iterated Interpolation algorithm can be performed by a systolic array in $O(n)$ time. Also by extension, Neville's Iterated Interpolation can also be solved by a systolic array in the same time. Here we discuss the extrapolation procedure in the Romberg Integration algorithm.

Interpolation and extrapolation techniques also have wide uses in numerical computation and often result in tables of a triangular form. The triangular form and the manner in which the table is formed indicates that systolic techniques used for matrix type algorithms such as the backsubstitution process and matrix vector multiplication [4], may be applicable to interpolation and extrapolation algorithms.

2. The Romberg integration algorithm

The Romberg algorithm is well-known, and is based on the Newton–Cotes formula (see [1,2] for definitions). We use the particular Newton–Cotes formula known as the trapezoidal method, which is one of the easiest to use, but which is usually not as accurate as required. The

Reproduced with permission of Elsevier Science Publishers B.V. (North Holland)
Parallel Computing 3(289 – 304)
1986

Romberg algorithm is widely applicable, using this easy-to-apply formula to obtain initial approximations, then Richardson's extrapolation to improve these approximations to gain a required accuracy.

Thus, to evaluate the integral $I = \int_a^b f(x)\,dx$, we select an integer $n > 0$ followed by the procedure

Input: endpoints a, b; integer n.
Output: an array R. (R_{nn} is the approximation to I. Computed by rows; only 2 rows are saved in storage.)
Step 1: Set $h = b - a$

$$R_{1,1} = \tfrac{1}{2}h\{f(a) + f(b)\}$$

Step 2: OUTPUT($R_{1,1}$)
Step 3: For $i = 2(1)n$ DO Steps 4–8
Step 4: Set

$$R_{2,1} = \tfrac{1}{2}[R_{1,1} + h \sum_{k=1}^{2^{i-2}} f(a + (k - 0.5)h)]$$

(Approximation using trapezoidal method)
Step 5: For $j = 2(1)i$ set

$$R_{2,j} = \frac{4^{j-1}R_{2,j-1} - R_{1,j-1}}{4^{j-1} - 1} \quad \text{(extrapolation)}$$

Step 6: OUTPUT($R_{2,j}$ for $j = 1, 2, \ldots, i$)
Step 7: Step $h = h/2$
Step 8: For $j = 1(1)i$
Set $R_{1,j} = R_{2,j}$ (update row 1 of R)
Step 9: STOP

As a start for our systolic algorithm we re-state these two basic steps:

(i) Approximate using the trapezoidal rule with $m_1 = 1$, $m_2 = 2$, $m_3 = 4, \ldots, m_n = 2^{n-1}$ for integer $n > 0$ and step size $h_k = (b - a)/m_k = (b - a)/2^{k-1}$ and use $R_{k,1}$ for the approximation $k = 1(1)n$, neglecting the error terms $O(h_k^2)$.

(ii) By using Richardson Extrapolation procedure we speed-up the convergence

$$R_{k,2} = \frac{4R_{k,1} - R_{k-1,1}}{3} \quad \text{for next approximation, } k = 2(1)n.$$

Generally,

$$R_{ij} = \frac{4^{j-1}R_{i,j-1} - R_{i-1,j-1}}{4^{j-1} - 1}, \quad i = 2(1)n,\ j = 2(1)i.$$

This results in the triangular table of approximations given by

$$\begin{array}{llllll}
R_{11} \\
R_{21} & R_{22} \\
R_{31} & R_{32} & R_{33} \\
R_{41} & R_{42} & R_{43} & R_{44} \\
\vdots & \vdots & \vdots & & \ddots \\
R_{n1} & R_{n2} & R_{n3} & & \ldots & R_{nn}
\end{array}$$

Romberg Extrapolation Table

where the elements for the first column $R_{i,1}$, $i = 1(1)n$ are produced by step (i), and the diagonal entries $R_{i,i}$, $i = 1(1)n$ are the terms that converge to the correct value of the integral. In general, the diagonal terms $\{R_{ii}\}_{i=1}^{\infty}$ of the array converge much faster than $\{R_{m,1}\}_{m=1}^{\infty}$ and we stop when $|R_{n-1,n-1}| <$ tol, tol = required accuracy.

In the above algorithm we have constructed a table of size n, and take R_{nn} as the approximation regardless of whether the method has converged (before R_{nn}) or not after the table has been constructed. For the systolic array this has important consequences, firstly we must have a fixed number of cells, in order to fabricate the design, and secondly we have to decide the number of cells to ensure that approximations are sufficiently accurate. As in the algorithm above, we construct a finite sized table of n rows, but allow the systolic array to close down if convergence has been achieved before the full table is constructed. In order to achieve accurate results we can choose n as large as possible subject to chip area restrictions.

Next we note the important point that although only the last row of the Romberg Triangle, and the $R_{i,1}$ value of the next row is required to construct row i, the parallel evaluation of the triangle terms as discussed in [5] can still be applied, i.e.,

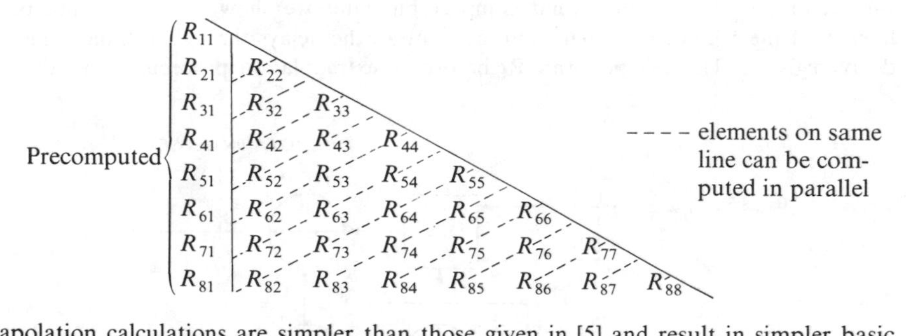

The extrapolation calculations are simpler than those given in [5] and result in simpler basic cells and input output techniques.

3. Systolic array computation

The two basic steps defined above provide us with a natural partitioning of the computation between the host machine and the systolic array.

(i) The host computes $R_{k,1}$, $k = 1(1)m$, for $m \leqslant n$, before waking the array. (Note n is the maximum sized table, but we can compute a smaller one on the size n array.)

(ii) The systolic array computes R_{ij}, $i = 2(1)m$, $j = 2(1)i$ (i.e. construct the table by extrapolation).

This division is natural because (i) involves evaluating the function $f(x)$, and since $f(x)$ can be arbitrary (within the constraints of integration, continuous, etc.) including it as part of the array would require an arbitrary number of complex basic array cells, which is not a desirable nor practical approach to the design.

For step (ii) we first consider a set of linearly connected cells, i.e., the basic cells implement the Richardson Extrapolation Procedure (REP), hence are called REP cells. The array consists of $n - 1$ REP cells (see Fig. 1), each with two inputs and three outputs, two outputs are inputs to the adjacent cell to the right, the third connects to a Fanin network, for filtering out the results.

The Fanin network is used as only the R_{ii}, $i = 2(1)n$ values are required to decide convergence, and as each REP cell will compute a column of the triangle, with cell i computing the $i + 2$th column. Each cell will output only one diagonal value, and if c is the latency of the

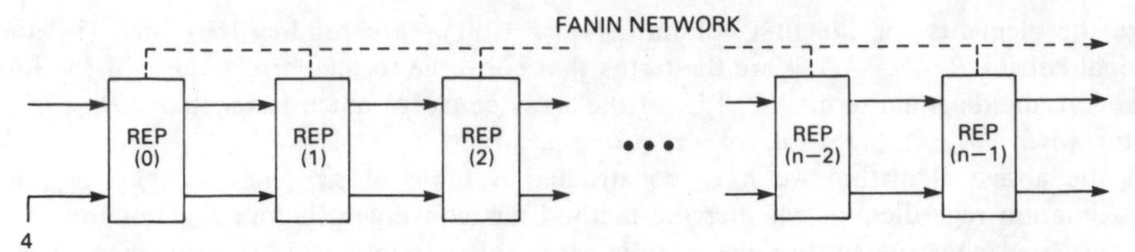

Fig. 1. Romberg linear systolic array.

REP cell, outputs only one diagonal value, and if c is the latency of the REP cell, outputs from cells will occur at c, $2c$, $3c$, respectively, on the Fanin network line, thus no values will be confused and only one output wire is sufficient for our needs.

3.1. The REP cell

The Richardson Extrapolation cell is shown in Fig. 2, it consists of 1 ips, a multiplier, subtractor and divider, this seems complex but later we show that this is acceptable. The latency of the cell is $c = 3$, taking into account all the delays through arithmetic elements and delay registers. The cell performs Richardson's extrapolation procedure, and that each cell

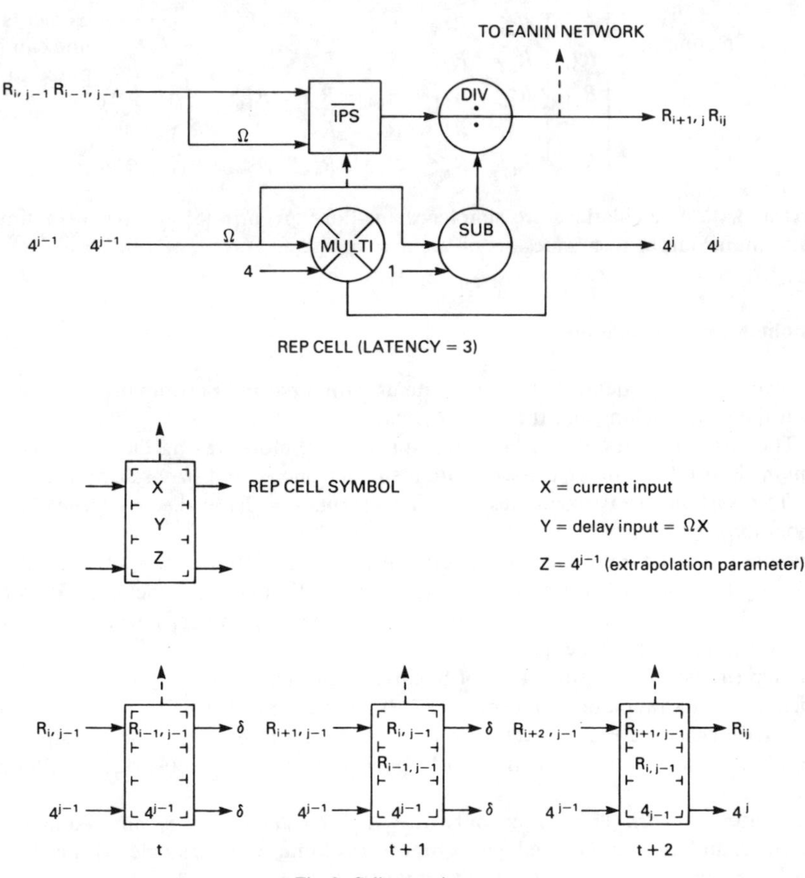

Fig. 2. Cell operation.

computes the power 4^j for the next cell, hence the two input and two output lines. The leftmost cell accepts the elements $R_{11}, R_{21}, \ldots, R_{n1}$ on one input line and the second input is hardwired to the value 4 (using a permanently register stored value). This allows the construction of the powers of 4 systolically, rather than sequentially by the host, although the multiplier seems extravagant because we could precompute and then load them into each cell before starting the array (i.e. replace the multiplier by a register). Later we show that the computation of the powers of 4 systolically using the extra multiplier in the REP cell can be justified.

The operation of the REP cell is simple to understand, and a one-bit control line is added which is associated with the R_{ii} values, $i = 1(1)m$, such that,

$$\text{control} = \begin{cases} 1\text{–send result of cell on Fanin network} \\ \quad \text{as well as normal output,} \\ 0\text{–normal output only.} \end{cases}$$

The control value moves systolically with the results from cell to cell, ensuring that all the R_{ii} values will be placed on the Fanin network back to the host, and no other values will use it (i.e. only true approximations). Figure 3 shows the first 7 steps of an array with $n = 6$. The total time for the array is thus given by,

$$T = (\text{cell latency}) * (\text{number of cells}) = 3(n-1) = 3n - 3$$

where T is the time to construct the complete table. If we converge before all rows of the table are computed we can just stop inputting and reading output results, essentially closing down the array and no explicit commands are needed as the cells do not retain any values. Likewise if we run out of R_{k1} values, $k = 1(1)m$ for $m < n$ inputting a suitable dummy element (δ), zero for instance, will not affect the construction of the first m rows of the triangle, already being computed along the array, hence smaller triangles can be constructed with time

$$T = 3(m-1) < 3(n-1).$$

Finally, remark that a time cycle is equivalent to a single ips step, and that each REP cell has area equivalent to 2.5 ips cells.

3.2. Systolic ring

From the given example, the last input R_{61} enters the array just as the 2nd cell is about to output a result to the third cell. Generally an output is about to leave the $\frac{1}{3}n$th cell when the last value of column 1 of the triangle R_{n1} is already in the first cell. It follows, that once the last input has entered the first cell successive inputs will be dummy elements, and consequently only $\frac{1}{3}n$ REP cells will be doing useful computations at any particular time. Notice that the REP cell as described implements the concept of two-level pipelining, a values can be using the ips cell while the divider is working with the previous ips result, allowing input and output within the cell to be overlapped with computation of R_{ij} values.

Now we reduce the size of the array to contain $m' = \lceil \frac{1}{3}n \rceil$ cells. When the last input R_{n1} has entered the 1st cell, the last cell will be computing the output in the divider, on the next cycle the output of the last cell is output and a dummy value is inserted into the 1st cell, to keep data moving systolically. Now we wrap the results around the array connecting the last cell output to the 1st cell input, the dummy value input replaced by the last cell output. The result computed in the ips of the 1st cell at this time is a garbage result, but would have been anyway because of the dummy input and does not affect results further down the array, it will not be output on the Fanin network because the control signal has not reached the divider (it will do so in another two cycles). The next cycle will now start to compute the same computation that would have occurred in the $(\frac{1}{3}n + 1)$th cell of the linear array, it follows that the systolic ring

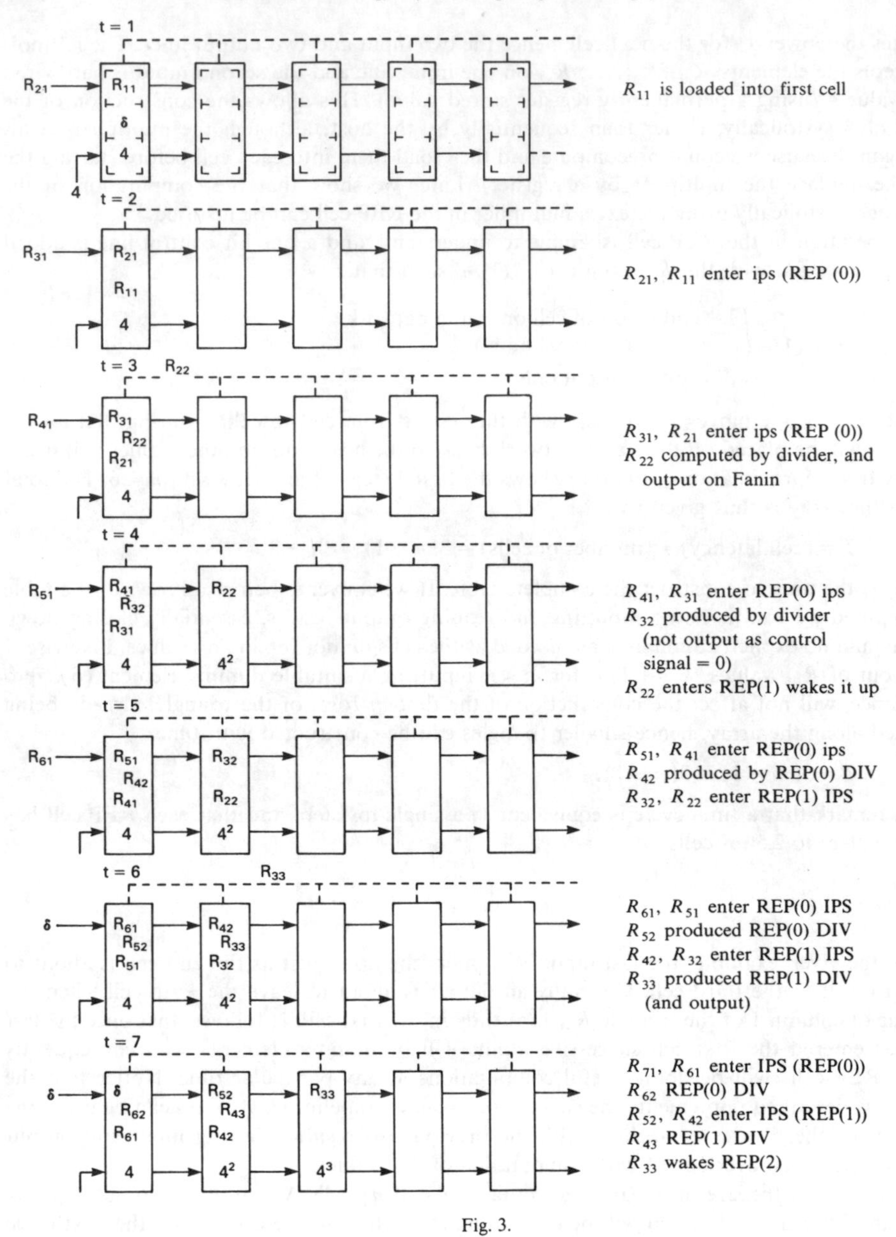

Fig. 3.

given in Fig. 4 will compute the same table as the linear array, but will require only $\frac{1}{3}$ of the cells, and take the same amount of time. This justifies the more complex REP cell, as each cell is equivalent to 2.5 ips cells, so the systolic ring requires the equivalent of at most n ips cells to implement it.

From [3] we note that this is an efficient layout, and if $m' = \lceil \frac{1}{3}n \rceil$, the ring can be constructed in a box of side $\sqrt{m'}$. We also remark that the design is superior to those in [4] for tasks like backsubstitution because the Fanin line avoids the awkward problems of outputs

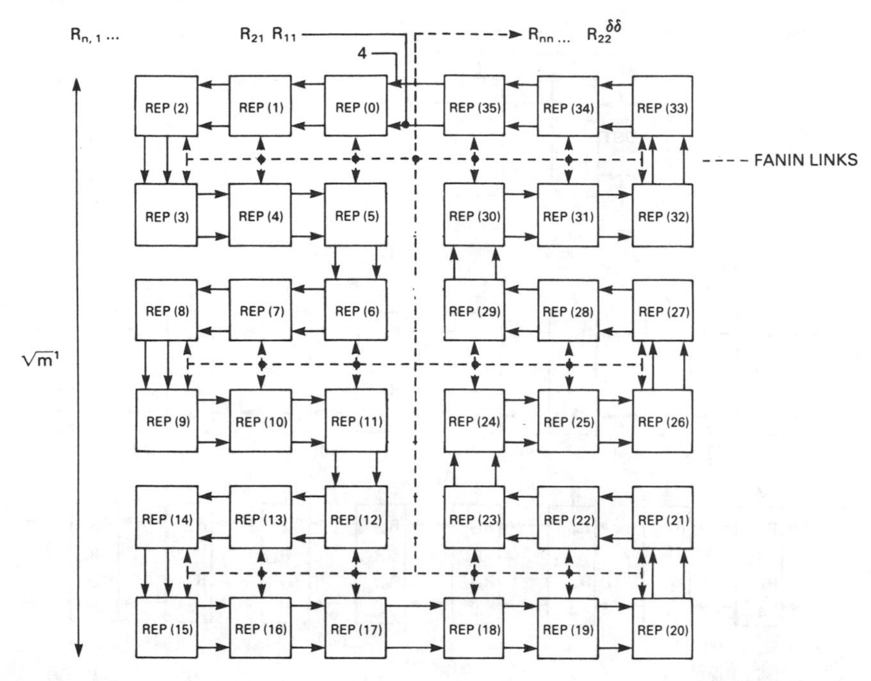

Notes:

(i) The input lines to the ring are switched in by a toggle switch.
The switch is controlled by the 1-bit line associated with input data.

Control	0	....	0	0	1
Data	R_{nn}	....	R_{31}	R_{21}	R_{11}

(ii) Initially the toggle is off and inputs from REP(0) come from REP(35). On receiving the control singal (high) associated with R_{11}, the toggle switches on, and REP(0) takes input from the input lines.

(iii) The control signal is pipelined around the ring, on each 3rd cycle an output is sent down the Fanin line, switch on by the control signal.

(iv) On returning to REP(35) the control is output from REP(35) switching the toggle off, making REP(0) receive input from REP(35).

(v) Total Fanin distance $\alpha\ \frac{3}{2}\sqrt{m'}$.

Fig. 4. Systolic ring for Romberg integration ($n = 3m$, $m' = 36$ cells).

from every cell. The Fanin network itself is also efficient in area, being embedded inside the ring to minimise its length which is proportional to $\frac{3}{2}\sqrt{m'}$. Minimising the length of the Fanin connection is important because for large values of n, the time for a value to reach the host along the line may be long enough to cause latch mistimings, this is especially true of the linear systolic array (where the distance $= n$). Here the systolic ring greatly reduces the length of the Fanin line.

The control line commanding the output onto the Fanin network can be utilised to the full with the systolic ring. Firstly, we can justify the use of the extra multiplier to compute the 4^j powers in each cell systolically. A systolic ring clearly will not allow us to preload these values, as they would be incorrect on the 2nd cycle around the ring. The host could continually pump the new values into the array but this significantly complicates the host array interface and the control of the array. Here we use the REP as described, and use the control signal which is high only with the first value to enter an unused cell (in the linear array) to cause loading of the 4^j value computed by the previous cell. The loaded value is retained until the control signal

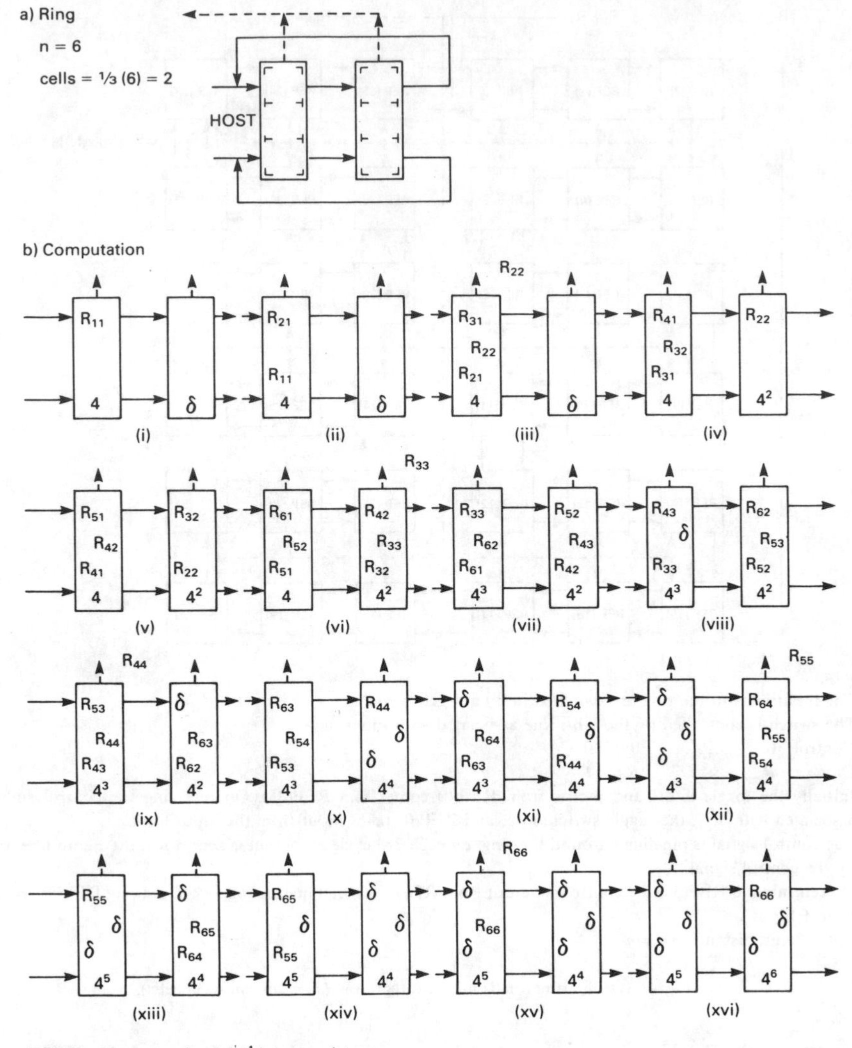

Fig. 5. Systolic ring for Romberg integration table ($n = 6$), (a) array, (b) ring computation.

completes a circuit of the ring triggering the overwrite again. The control signal is also used to control the connections between the host input and the ring connections from m'th to 1st cell. On startup the 1st cell REP(0) takes data from the host input connection, which includes the control. When the control signal has completed a circuit and is leading the m'th cell, by the ring property that the last host input has entered the array it can be used to switch out the host input, and switch in the connections from the m'th cell to the 1st cell. On further cycles the switches are disabled, this is easy to control using the host, note also that the Fanin connection is not affected by this switching. As an example, Fig. 5 illustrates the use of the systolic ring that results from the linear array Fig. 3. Notice we only require 2 REP cells.

Finally notice that Fig. 4 consists of 36 cells allowing a table of size $n = 108$ which would probably be adequate for most applications, and any $m < 108$ sized table can also be constructed. The systolic ring also has the useful property that only 1 host input and 1 host output is required, the input of 4 can be hardwired to a register on the chip. Thus the design is suitable for single chip implementation.

4. Conclusion

We have presented a linear systolic array and systolic ring which will compute the Romberg integral table in a time $3n$, this is significantly faster than the $O(n^2)$ operations required by algorithm performed on a sequential machine for n large enough.

With the REP cell equivalent to 2.5 ips cells in area terms, the systolic ring requires the equivalent of only n ips cells, i.e. $\lceil \frac{1}{3}n \rceil$ REP cells, and the efficient ring layout Fig. 4 requires a square area with side $\sqrt{\lceil \frac{1}{3}m \rceil}$ for its fabrication. Further the ring requires only 1 input and 1 output line which will allow floating point numbers to be represented with the available pins on a chip.

The Fanin connection is a significant restriction for the linear array but the systolic ring allows its length to be reduced by two thirds at least, reducing problems of latch mistimings for large n.

Finally, it should be noted that the systolic ring could not be used if the entire table were to be output, hence the reason for developing the linear systolic array. For a full table output, the Fanin connection would be replaced by one output for each REP and values would be output every cycle see [5], on each wire. Here however the host need only compare the diagonal elements of the Romberg triangle to decide on convergence, and both the Fanin network and ring are applicable.

Appendix A

Test example:

$$\int_0^{\pi} \sin x \, dx, \quad n = 6.$$

Romberg table:

0					
1.57079633	2.09439511				
1.89611890	2.00455976	1.99857073			
1.97423160	2.00026971	1.99998313	2.00000555		
1.99357034	2.00001659	1.99999975	2.00000001	1.99999999	
1.99839336	2.00000103	2.00000000	2.00000000	2.00000000	2.0

Linear systolic array (no. of REP cells = 5):

Starting values	*Diagonal entries* (*Fanin*)
$R[0] = 0.000000$	2.094395
$R[1] = 1.570796$	1.998571
$R[2] = 1.896119$	2.000005
$R[3] = 1.974232$	2.000000
$R[4] = 1.993570$	2.000000
$R[5] = 1.998393$	

N.B. Systolic ring produced same results with

(no. of cells) $m = 2$ (size of input) $n = 6$.

Appendix B. Linear systolic array

```
-- program 9

-- Systolic Array : To construct the extrapolation table in
--                  Romberg's Integration algorithm.

-- The basic cell is the REP or Richardson's Extrapolation cell
-- which computes the extrapolation values

-- Table size
def n = 5:

-- library routines
EXTERNAL Proc str.to.screen(value s[]) :
EXTERNAL Proc num.to.screen(value n) :
EXTERNAL Proc fp.num.to.screen(Value float f):
EXTERNAL Proc fp.num.from.keyboard(Var float f) :

Proc REP(Chan in1, out1, in2, out2, out3, cntrlin, cntrlout ) =
  --
  -- Richardsons extrapolation cell
  --
  var float t1, t2, p.4, p.res, rnew, rold, res :
  var switch, running, toggle, c.fifo[4] :
  seq
    -- intialisation
    p.res := 0.0
    rold := 0.0
    res := 0.0
    t1 := 0.0
    t2 := 1.0
    seq i =[0 for 3 ]
      c.fifo[i] := 0
    switch := true
    running := true
    toggle  := true
    -- cell
    while running
      seq
        -- control input
        if
          switch
            cntrlin?c.fifo[0]
        -- decide on input/output
        if
          (c.fifo[0] = 6 ) and switch
            -- close input
            switch := false
          c.fifo[3] = 6
            --destroy cell
            running := false
        cntrlout!c.fifo[3]
        -- i/0
        if
          switch
            par
              in1?rnew
              in2?p.4
        if
          running
            par
              out1!res
              out2!(p.res)
              -- output to fanin network
```

```
            if
              c.fifo[3] = 1
                seq
                  toggle := false
                  out3!res
              toggle
                out3!0
        -- extrapolation formula
        res := t1/t2
        t1 := (p.4*rnew)- rold
        t2 := p.4- 1.0
        rold := rnew
        p.res := p.4 * 4.0
        -- shift control fifo
        seq i =[ 0 for 3]
          c.fifo[3-i] := c.fifo[(3-i)-1]
        if
          c.fifo[0] = 6
            c.fifo[0] := 0 :

proc fnet(chan gather[], var float vec[], var k ) =
  --
  --  fanin network: primitive routine to collect values
  --
  seq
    par j =[0 for n]
      -- check all processes
      seq
        if
          j > k
            -- those still to output
            gather[j]?any
          j = k
            var float tmp :
            -- current output
            seq
              gather[k]?tmp
              if
                tmp <> 0.0
                  seq
                    vec[k] := tmp
                    k := k + 1 :

proc getdata( chan out1,out2, cntrl ) =
  --
  -- read starting values , and pump into
  -- array then close down array systolically
  --
  var float four, vec[n+1] :
  seq
    four := 4.0
    str.to.screen("*nEnter Romberg Starting values")
    seq i=[0 for (n+1)]
      seq
        str.to.screen("*nR[")
        num.to.screen(i)
        str.to.screen("] =")
        fp.num.from.keyboard(vec[i])
        fp.num.to.screen(vec[i])
    str.to.screen("*n*n*n")
    -- start pumping
    seq i=[0 for (n+1)]
      par
        if
          i = 0
            cntrl!1
          true
            cntrl!0
```

```
          out1!vec[i]
          out2!four
      -- close down
      cntrl!6 :

  proc putdata( chan in1, in2, fanin[], cntrl ) =
    --
    -- collect garbage falling off array, and
    -- call fanin to collect next result and print out
    -- diagonal entries
    --
    var float vec[n] :
    var running, c1, k :
    seq
      k := 0
      running := true
      -- collect results until stopped
      while running
        seq
          cntrl?c1
          if
            c1 = 6
              running := false
            true
              par
                in1?any
                in2?any
                fnet(fanin, vec, k)
      -- output diagonal approximations.
      str.to.screen("*n*n Diagonal Table Entries")
      seq i =[0 for n]
      seq
        str.to.screen("*n")
        fp.num.to.screen(vec[i]) :

-- main
--
-- The Romberg array
chan in1[n+1], in2[n+1], fanin[n], cntrl[n+1] :

par
  getdata(in1[0], in2[0], cntrl[0])
  par i =[1 for n]
    REP(in1[i-1], in1[i], in2[i-1], in2[i], fanin[i-1], cntrl[i-1], cntrl[i])
  putdata(in2[n], in1[n], fanin, cntrl[n])
```

Appendix C. Systolic ring

```
  -- program 10

  -- Systolic Array : A Systolic Ring implementation of the Romberg
  --                  table construction.
  --

  -- Ring size and Table size respectively
  def n = 2, m= 6 :

  -- library routines
  EXTERNAL Proc str.to.screen(value s[]) :
  EXTERNAL Proc num.to.screen(value n) :
  EXTERNAL Proc fp.num.to.screen(Value float f):
  EXTERNAL Proc fp.num.from.keyboard(Var float f) :
```

```
Proc REP(Chan in1, out1, in2, out2, out3, cntrlin, cntrlout ) =
  --
  -- Modified Extrapolation cell (see report)
  --
  var float t1, t2, p.4, p.res, rnew, rold, res :
  var switch, running, toggle, c.fifo[4] :
  seq
    -- intialisation
    p.res := 0.0
    rold := 0.0
    res := 0.0
    t1 := 0.0
    t2 := 1.0
    seq i =[0 for 3 ]
      c.fifo[i] := 0
    switch := true
    running := true
    toggle  := false
    -- cell
    while running
      seq
        -- control i/o
        par
          if
            switch
              cntrlin?c.fifo[0]
          cntrlout!c.fifo[3]
        -- decide on data i/o
        if
          (c.fifo[0] = 6 ) and switch
             switch := false
          c.fifo[3] = 6
             running := false
        -- switch on fanin output line
        if
          c.fifo[0] = 1
            toggle := true
        -- i/O
        par
          if
            switch
              par
                in1?rnew
                if
                  c.fifo[0] = 1
                    in2?p.4
          if
            running
              par
                out1!res
                if
                  c.fifo[3] = 1
                    seq
                      toggle := false
                      par
                        out3!res
                        out2!p.res
                  toggle
                    out3!0
        -- computation
        res := t1/t2
        t1 := (p.4*rnew)- rold
        t2 := p.4- 1.0
        rold := rnew
        p.res := p.4 * 4.0
        -- shift control fifo
        seq i =[ 0 for 3]
          c.fifo[3-i] := c.fifo[(3-i)-1]
```

```
        if
          c.fifo[0] = 6
            c.fifo[0] := 0 :

proc fnet(chan gather[], var float vec[], var k,z ) =
  --
  -- Modified fanin simulator.
  --      Sequentially poll ring cells looking for outputs
  --      and accept them if non-zero
  --
  -- Note : z = index of diagonal entry next output;
  --          k = index of next ring cell expected to output diagonal.
  --
  seq j =[0 for n]
    seq
      if
        j = k
          var float tmp :
          seq
            gather[k]?tmp
            if
              tmp <> 0.0
                seq
                  vec[z] := tmp
                  z := z + 1
                  k := k + 1 :

proc host( chan out1,out2, cntrlin, in1,in2, fanin[], cntrlout ) =
  --
  -- Combined getdata and putdata to act as ring arbiter, to switch
  -- from Host input to ring input and collect fanin results
  --
  chan link :
  par
    -- equivalent process to getdata, uses link to
    -- create switch from Host to ring input
    var float four, vec[m] :
    seq
      four := 4.0
      str.to.screen("*nEnter Romberg Starting values")
      seq i=[0 for m]
        seq
          str.to.screen("*nR[")
          num.to.screen(i)
          str.to.screen("] =")
          fp.num.from.keyboard(vec[i])
          fp.num.to.screen(vec[i])
      str.to.screen("*n*n*n")
      -- pump host inputs into ring
      seq i=[0 for m]
        par
          link!0
          if
            i = 0
              par
                cntrlin!1
                out2!four
            true
              cntrlin!0
          out1!vec[i]
      -- switch to ring
      link!1

    -- equivalent to putdata, but augmented with
    -- control to wrap around ends of ring
    -- when link = 1
    var float vec[m] :
    var running, switch, c1, k, z, l1, r1, r2 :
```

```
var rs1, rs2, cs1 :
seq
  -- intilaise
  z := 0
  k := 0
  cs1 := 0
  running := true
  switch := false
  -- collect and pump till all values
  -- received
  while running
    seq
      -- switch ?
      if
        not switch
          link?l1
      cntrlout?c1
      -- ring wrap around
      if
        l1 = 1
          seq
            -- first value
            l1 := 0
            switch := true
            cntrlin!cs1
        switch and (z <= (m-1))
          seq
            -- rest
            cntrlin!cs1
        switch and (l1 = 0)
          seq
            -- close down ring
            l1 := 2
            cntrlin!6
      if
        c1 = 6
          -- kill this process
          running := false
        true
          seq
            -- collect garbage and results
            seq
              in1?r1
              if
                c1 = 1
                  in2?r2
              fnet(fanin, vec, k, z)
            -- ring i/o
            if
              switch and (l1 <> 2)
                seq
                  out1!rs1
                  if
                    cs1 = 1
                      out2!rs2
      -- re-sync ring data and control
      rs1 := r1
      rs2 := r2
      cs1 := c1
      if
        (cs1 = 1) and (z <= (m-1))
          k := 0
  -- print results for user, vec = memory in Host
  str.to.screen("*n*n Diagonal Table Entries")
  seq i =[0 for (m-1)]
  seq
    str.to.screen("*n")
    fp.num.to.screen(vec[i]) :
```

```
-- main

-- Systolic Ring definition
chan in1[n+1], in2[n+1], fanin[n], cntrl[n+1] :

par
  host(in1[0], in2[0], cntrl[0], in1[n], in2[n], fanin, cntrl[n])
  par i =[1 for n]
    REP(in1[i-1], in1[i], in2[i-1], in2[i], fanin[i-1], cntrl[i-1], cntrl[i] )
```

References

[1] Burden, Faires and Reynolds, *Numerical Analysis* (Prindle, Weber & Schmidt, Boston, 2nd ed., 1981) Ch. 3 and 4.

[2] L.W. Johnson and R.D. Reiss, *Numerical Analysis* (Addison-reading, MA, 1977) ch. 5.

[3] H.T. Kung and M.S. Lam, Wafer-scale integration and two-level pipelined implementation of systolic arrays, *J. Parallel and Dist. Computing* **1** (1984) 32–63.

[4] Mead and Conway (eds.), *Introduction to VLSI Design* (Addison-Wesley, Reading, MA, 1980) 271–292.

[5] G.P. McKeown, Iterated interpolation using a systolic array, Internal Report CSA/21/1984, UEA, Norwich, 1984.

[6] G.M. Megson and D.J. Evans, Design and simulation of systolic arrays, Internal Report 230, Computer Studies Department, Loughborough University of Technology, 1985.

[7] G.M. Megson and D.J. Evans, BATS: (Banded and Toeplitz systems) systolic array, Internal Report, Computer Studies Department, Loughborough University of Technology, 1985.

Construction of extrapolation tables by systolic arrays for solving ordinary differential equations

D.J. EVANS and G.M. MEGSON

Department of Computer Studies, Loughborough University of Technology, Loughborough, Leicestershire, United Kingdom

Abstract. We consider here the systolic array construction of extrapolation tables used in the solution of Ordinary Differential Equations (ODE's) associated with initial value type problems. The technique is examined first for a low order formula, i.e. Euler's method which is combined with extrapolation to improve the estimates of the solution. We also show how this is extended to the Bulirsch and Stoer algorithm and hence a generic form can be given to systolic arrays for the construction of extrapolation tables.

Keywords. Systolic arrays, solving ordinary differential equations, extrapolation tables

Introduction

The construction of extrapolation and difference tables has a form very similar to the operations in certain matrix equations. This similarity can be exploited to produce systolic arrays to construct the tables. In [5] Romberg integration was solved using a systolic array with these matrix properties, while [4] solved Aitken's iterated interpolation algorithm on a systolic array. In this report, certain similarities to matrix computations and extrapolation tables are indicated giving a generic definition of an extrapolation systolic array. This correspondence is developed and combined with the solution of initial-value problems associated with ODE's of the form

$$y' = f(t, y), \quad a \leqslant t \leqslant b \text{ with initial condition } y(a) = \alpha. \tag{1}$$

Before the development of systolic arrays for any application two fundamental points must be considered:

(i) The systolic array can only be applied to existing initial value ODE's algorithms by constructing the extrapolation tables;

(ii) As the array must be of fixed size, a limit to the number of levels or step divisions must be made, so that the table size is of fixed size and manageable.

The first point indicates that the evaluation of $f(t, y)$ or any part of the algorithm that involves the evaluation of $f(t, y)$ must be placed outside the systolic array. The evaluation of $f(t, y)$ in general can be arbitrary as long as it is integrable and a systolic array would also become arbitrarily complex if it was included in the array. Together with this is the fact that all the function values which are used to estimate a point $y(t)$ at every level must be evaluated before the array can be used. This appears to defeat the object of the extrapolation idea as we will

Reproduced with permission of Elsevier Science Publishers B.V. (North Holland)
Parallel Computing 4(33 – 48)
1987

normally stop when convergence is reached, ignoring the computation at a lower level of step division, and moving on to the next point.

Extrapolation algorithms can be applied in two situations, where the function f is approximated by either polynomials or rational functions. The latter are usually better as they allow larger stepsizes to be used. Also, the use of extrapolation in ODE's is beneficial as the ODE solution is related to finding an indefinite integral and so extrapolation techniques allow the ODE methods to compute with algorithms like Romberg integration.

1. Extrapolation methods

We now show how an extrapolation procedure can be incorporated into the solution to ODE's using an algorithm attributed to Gragg as this will provide a suitable vehicle by which to illustrate the points (i) and (ii) above and also relate extrapolation techniques to matrix computations.

Consider the initial value problem (1) using Euler's method and step size $h > 0$, then

$$w_0 = \alpha,$$
$$w_{i+1} = w_i + hf(t_i, w_i), \quad i = 0(1)n-1. \tag{1.1}$$

Put

$$N = (b-a)/h \quad \text{and} \quad t_i = a + ih, \qquad i = 0(1)n-1. \tag{1.2}$$

The error $y(t_i) - w_i$ leads to a function $\delta(t)$ such that,

$$y(t_i) = w_i + h\delta(t_i) + O(h^2), \quad i = 1(1)n. \tag{1.3}$$

To apply extrapolation we need approximations of $y(t_i)$ for different step sizes. Put $w(t, h) \equiv$ approximation of $y(t)$ with step size h. As an example choose two step levels h_0 and h_1 $(< h_0)$. Now consider evaluating $y(b)$ and put

$$q_0 = (b-a)/h_0 \quad \text{and} \quad q_1 = (b-a)/h_1.$$

Now apply (1.1)–(1.3) twice once with $h = h_0$ and once with $h = h_1$ giving

$$y(b) = w(b, h_0) + h_0\delta(b) + O(h_0^2), \tag{1.4a}$$
$$y(b) = w(b, h_1) + h_1\delta(b) + O(h_1^2). \tag{1.4b}$$

Elementary manipulation of (1.4a) and (1.4b) results in,

$$y(b) = \frac{h_0 w(b, h_1) - h_1 w(b, h_0)}{h_0 - h_1} + O(h_0^2). \tag{1.5}$$

If the difference method has a particular type of error expansion [3], it can be generalised to construct an extrapolation table, with diagonal elements converging to a good (accurate) approximation to $y(t)$, e.g. using three levels (or step sizes) h_0, h_1, h_2 we obtain

$$y_{1,1} = w(t, h_0),$$
$$y_{2,1} = w(t, h_1), \quad y_{2,2} = \frac{h_0^2 y_{2,1} - h_1^2 y_{1,1}}{h_0^2 - h_1^2},$$
$$y_{3,1} = w(t, h_2), \quad y_{3,2} = \frac{h_1^2 y_{3,1} - h_2^2 y_{2,1}}{h_1^2 - h_2^2}, \quad y_{3,3} = \frac{h_0^2 y_{3,2} - h_2^2 y_{2,2}}{h_0^2 - h_2^2}.$$

We begin to relate this to matrix computation, as the extrapolation table is a lower triangular matrix of values for the above example,

$$Y = \begin{bmatrix} y_{11} & & \mathbf{0} \\ y_{21} & y_{22} & \\ y_{31} & y_{22} & y_{23} \end{bmatrix}. \tag{1.6}$$

From [3] we define Gragg's extrapolation algorithm for solving ODE's:

Approximate the initial value problem

$$y' = f(t, y), \quad a \leqslant t \leqslant b, \quad y(a) = \alpha$$

local array with given tolerance

Input: endpoints a, b, initial condition α, tolerence TOL, level limit $p \leqslant 8$, maximum step size ($h_{\max}$), minimum step size ($h_{\min}$).

Output: T, W, h where W approximates $y(t)$ and stepsize h was used or a message when the minimum step size exceeded.

Step 1: Initialise array NK = (2, 3, 4, 6, 8, 12, 16, 18).

Step 2: Set $T_0 = a$, $W_0 = \alpha$, $h = h_{\max}$.

Step 3: For $i = 1(1)7$
  For $j = 1(1)i$
    $Q_{i,j} = (\mathrm{NK}_{i+1}/\mathrm{NK}_j)^2 \qquad (Q_{ij} = h_j^2/h_{i+1}^2)$

Step 4: While (T_0 < b) do
  Step 5: $K = 1$
    FLAG = 0 (Boolean accuracy test)
  Step 6: While ($K \leqslant P$ and FLAG = 0) do
    Step 7: $\mathrm{HK} = h/\mathrm{NK}_k$
      $T = t_0$
      $w_2 = w_0$
      $w_3 = w_2 + \mathrm{HK} * f(T, w_2)$ (Euler step)
      $T = t_0 + \mathrm{HK}$
    Step 8: For $j = 1(1)N_k - 1$
      $w_1 = w_2$
      $w_2 = w_3$
      $w_3 = w_1 + 2 * H_k * f(T, w_2)$
      $T = T_0 + (j+1) * \mathrm{HK}$ (mid-point method)
    Step 9: $y_k = [w_3 + w_2 + \mathrm{HK} * f(T, w_3)]/2$ (smooth for $y_{k,1}$)
    Step 10: If $K \geqslant 2$ then
      Step 11: $j = k$
        $v = y_1$ (save y_{k-1}, y_{k-1})
      Step 12: While ($j \geqslant 2$) do
        $$y_{j-1} = y_j + \frac{y_j - y_{j-1}}{Q_{k-1,j-1} - 1} \quad \text{(extrapolation)}$$
        $j = j - 1$
      Step 13: If $|y_1 - v| \leqslant \mathrm{TOL}$ then FLAG = 1
        (accept y_1 as new w)
    Step 14: $K = K + 1$
  Step 15: $K = K - 1$

Step 16: If FLAG = 0 then
$h = h/2$ (w rejected decrease h)
If $h < h_{min}$ then OUTPUT 'minimum h exceeded fail'
STOP
else
$w_0 = y_1$
$T_0 = T_0 + h$
OUTPUT (t_0, w_0, h)
If ($K \leqslant 3$) and ($h < h_{max}/2$) then $h = 2h$ (increase h if possible)

Step 17: STOP

Now Step 3, provides the next matrix link, the general version of (1.5), can be simplified to give the form in Step 12, with $Q_{k-1,\,j-1}$ the ratio of the two step sizes squared. Step 3 defines a constant matrix for the algorithm of dimension $p * p$ of the form

$$Q = \begin{bmatrix} Q_{1,1} & & \\ Q_{2,1} & & \\ \vdots & \ddots & \\ Q_{p,1} & \cdots & Q_{p,p} \end{bmatrix}. \tag{1.7}$$

Now define E_p as the extrapolation operator, then the extrapolation table is given by,

$$Q(E_p)y_1 = Y, \qquad y_1 = (y_{1,1}, y_{2,1}, \ldots, y_{p,1})^{\mathrm{T}}, \tag{1.8}$$

e.g.

$$(Q_{1,1}, \ldots, 0)\begin{bmatrix} y_{1,1} \\ y_{2,1} \\ \vdots \\ y_{p,1} \end{bmatrix} = \begin{bmatrix} y'_{1,1} \\ y_{2,1} \\ \vdots \\ y_{p,1} \end{bmatrix} \quad \text{or} \quad y'_{1,1} = \frac{h_0^2 y_{2,1} - h_1^2 y_{1,1}}{h_0^2 - h_1^2} = y_{2,1} + \frac{y_{2,1} - y_{1,1}}{Q_{1,1} - 1}$$

and

$$(Q_{2,1}, Q_{2,2}, \ldots, 0)\begin{bmatrix} y'_{1,1} \\ y_{2,1} \\ \vdots \\ y_{p,1} \end{bmatrix} = \begin{bmatrix} y''_{1,1} \\ y'_{2,1} \\ \vdots \\ y_{p,1} \end{bmatrix} \quad \text{or}$$

$$y'_{2,1} = y_{3,1} + \frac{y_{3,1} - y_{2,1}}{Q_{2,2} - 1} \quad \text{and} \quad y''_{1,1} = y'_{2,1} + \frac{y'_{2,1} - y'_{1,1}}{Q_{2,1} - 1},$$

...etc.

This can be interpreted as a matrix-vector type computation:

$$Q_1 y_1^{(1)} = y_1^{(2)},\ Q_2 y_1^{(2)} = y_1^{(3)}, \ldots, Q_p y_1^{(p)} = y_1^{(p+1)}, \tag{1.9}$$

with Q_i = row i of Q and

$$y_1^{(1)} = y_1,\ y_1^{(2)} = (y'_{1,1} y_{2,1}, \ldots, y_{p,1})^{\mathrm{T}},\ y_1^{(3)} = (y''_{1,1}, y'_{2,1}, \ldots, y_{p,1}). \tag{1.10}$$

In the derivation of (1.9) and (1.10) we have used only the operations used in setting up the extrapolation table, and not on relating the step size in (1.5). Thus, we have a generic recurrence structure, where E_p is the extrapolation formula used and Q calculates the multiplicative factors relating the estimates together. The recurrence structure leads to the array in Fig. 1.

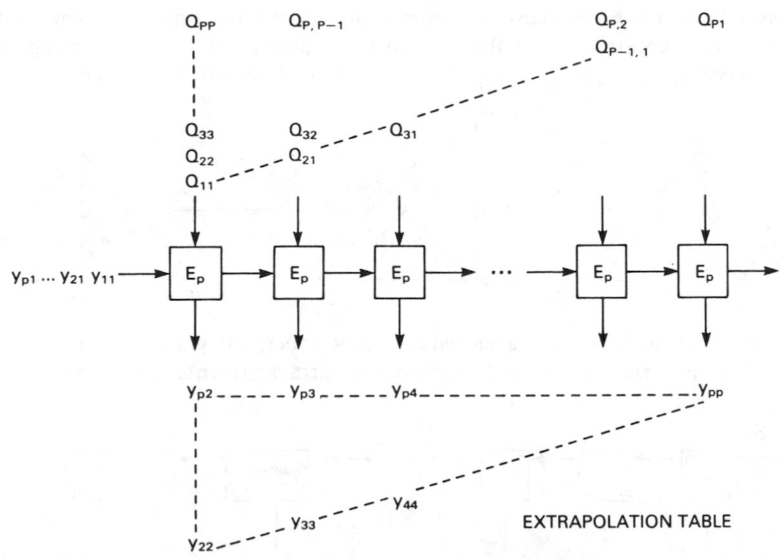

Fig. 1. Generic extrapolation array ($E_p \equiv$ extrapolation cell, computes extrapolation procedure, Time unit $\equiv$ latency (cost of) extrapolation computation for cell, Total time $= (p-1)C$, $p =$ number of step levels, $C =$ latency of E_p cell).

Now taking the p step levels $h_0, h_1, \ldots, h_{p-1}$ such that

$$h_1 = \frac{h_0}{2},\ h_2 = \frac{h_0}{4},\ h_3 = \frac{h_0}{8}, \ldots, h_{p-1} = \frac{h_0}{2^{p-1}},$$

then Q becomes,

$$Q = \begin{bmatrix} (2)^2 & & & \\ (4)^2 & & & \mathbf{0} \\ \vdots & \ddots & \ddots & \\ 2^{2(p-2)} & & & \\ 2^{2(p-1)} & 2^{2(p-2)} & \ldots & (4)^2(2)^2 \end{bmatrix}. \tag{1.11}$$

The array in Fig. 1 can now be simplified to give Fig. 2(a), where the multipliers are preloaded into the cells before the computation starts and must be calculated by the host machine. To save preloading and host computation of the $Q_{i,j}$ factors we can feed them systolically around the cells, computing the 'factors on the fly' and augmenting the cells with extra hardware to form the power required by the next cell. The number of cells is reduced to $m = \lceil p/c \rceil$ and by using two-level pipelining it is possible to input a value to each cell every cycle even though the latency of the cells is c. This produces the systolic ring of Fig. 2(b), after mc cycles the last (mth cell) outputs a value and the first cell reads in $Y_{p,1}$:

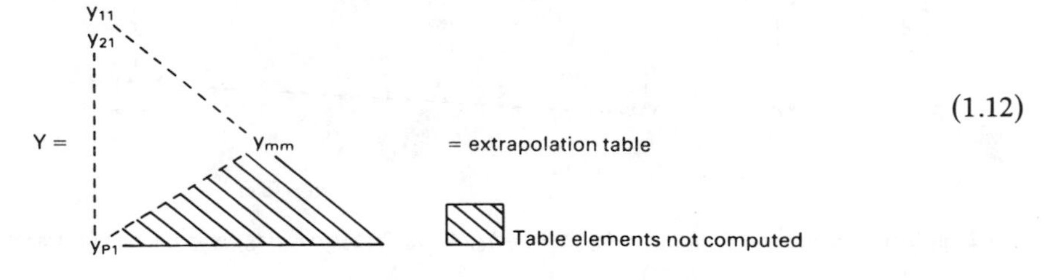

(1.12)

Thus the systolic ring now overlaps the computation of the remaining elements, with cell 1 starting the systolic construction of the extrapolation table from $y_{m+1,\ m+1}$ using the values from the mth cell $y_{m,\ m}$, $y_{m+1,\ m}, \ldots, y_{p,\ m}$, this is repeated on successive cycles:

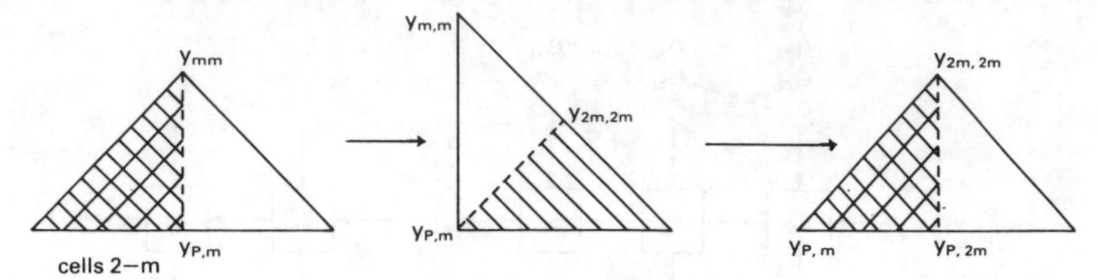

In [5] a very efficient layout was achieved for the Romberg integration algorithm and it can now be seen that this array can be easily derived from arrays presented here. In [5] we assumed

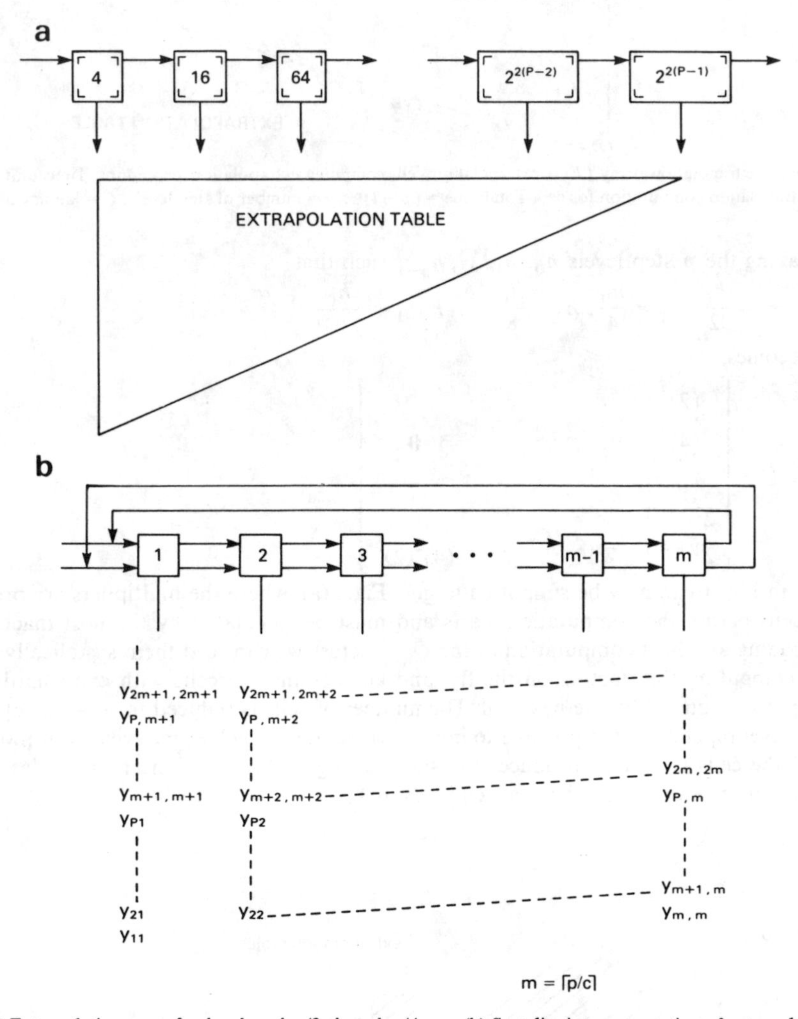

Fig. 2. (a) Extrapolation array for h_0, $h_1 = h_0/2$, $h_2 = h_0/4, \ldots,$. (b) Systolic ring computation of extrapolation table.

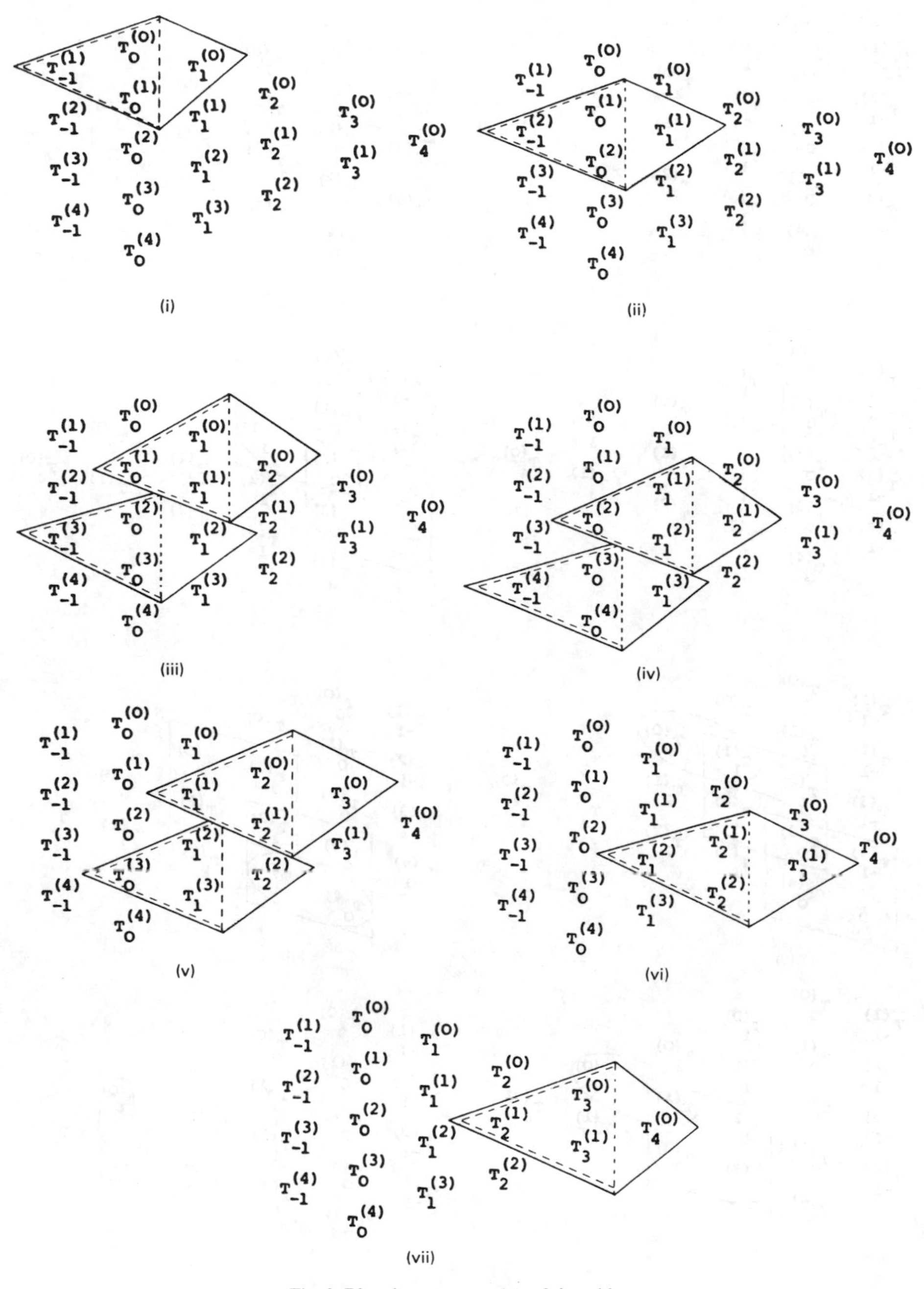

Fig. 3. Rhombus computation of the tableau.

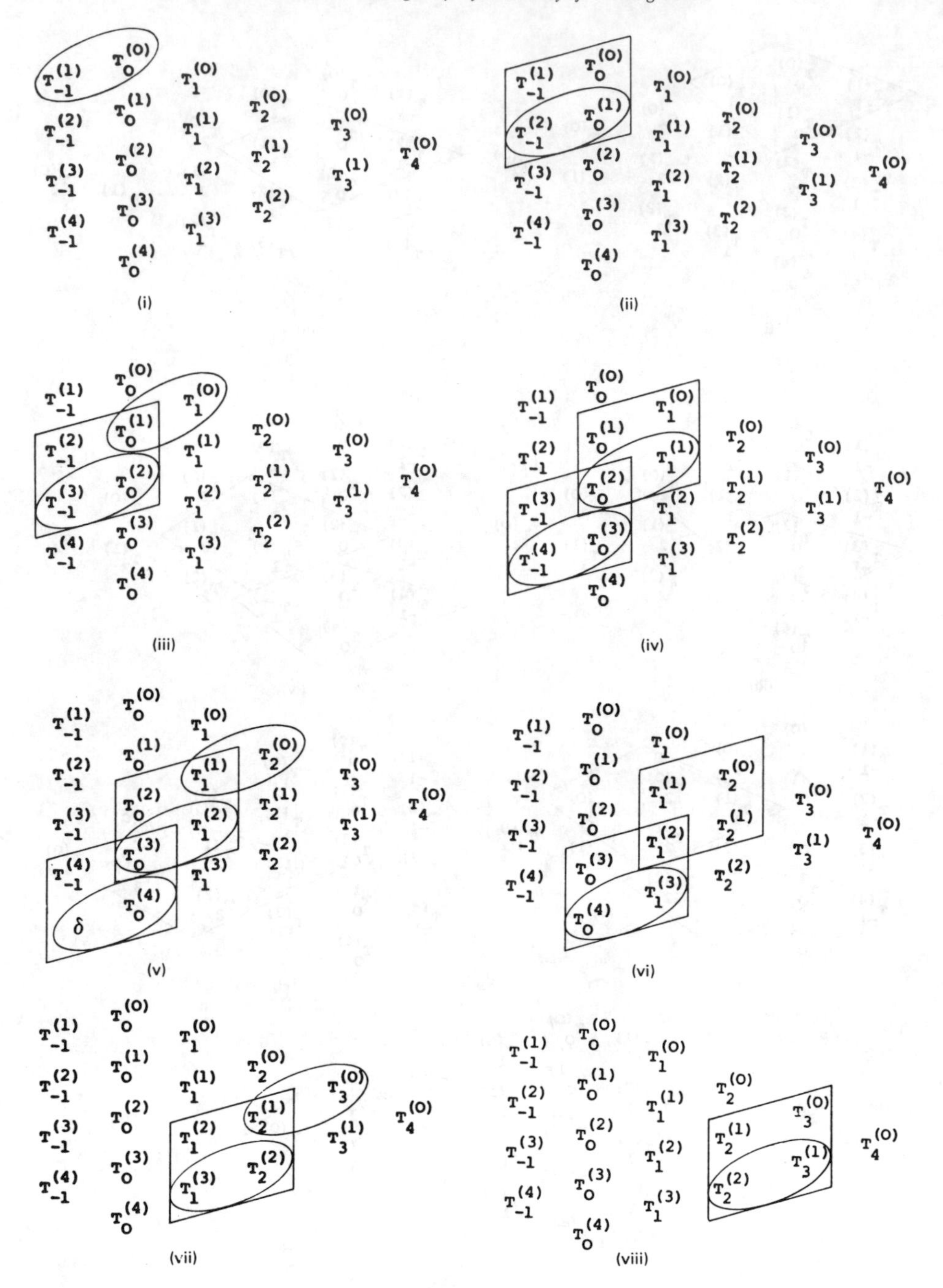

Fig. 4. Skew-rhombus computation of the tableau in parallel.

that only the diagonal table entries were required (as they are the ones used to decide convergence). Figure 2(b) can also be made into an efficient layout if the output links are made into Fanin links and only diagonal values are read in by the host. Thus, we can always produce an array with $O(p)$ E_p-cell area, where p = size of table, as we need only p cells, and with a systolic ring of m cells. If the latency c of each cell is equivalent to the number of ips equivalent elements used to implement it then the systolic ring requires area proportional to $O(p)$ ips cells.

2. E_p-cells

The overall structure of an extrapolation array has been given and for different extrapolation procedures, the E_p-cell changes complexity. In [5] the cell was defined for the Romberg integration algorithm using the Richardson's extrapolation procedure. For the algorithm presented above we define our E_p-cell to compute the Step 12 while loop body. The E_p-cell is given as

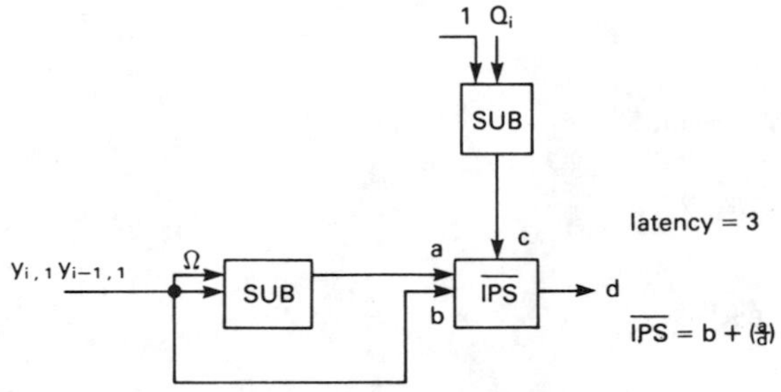

Thus, $c = 3$ and counting sub-cells as equivalent to $\overline{IPS}$ the systolic ring requires $O(p)$ $\overline{IPS}$ area. Computation time of array $T = 3(p-1)$.

Next, we consider the extrapolation table constructed by the Bulirsch and Stoer extrapolation method, the rational function approximation technique [1]. The Bulirsch and Stoer algorithm leads to the extrapolation table shown in Fig. 3 and relates $T_{k-2}^{(i+1)}$, $T_{k-1}^{(i)}$, $T_{k-1}^{(i+1)}$ and $T_k^{(i)}$ together by a Rhombus-rule. Figure 3 indicates the parallel order of computation to construct the table.

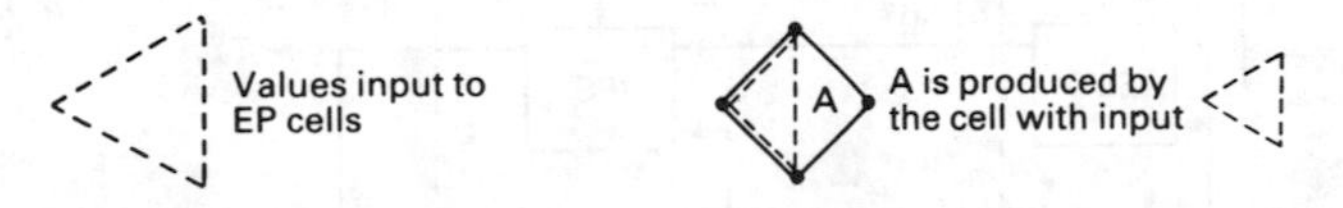

The cell that results from the three inputs as shown is unnecessarily complex involving delays and non-planarity to achieve the correct output sequence to the next cell. A less intuitive order of computation and input is the skew-rhombus computation rule which allows only two inputs and two outputs to each cell and reduces the bandwidth of the array. With the condition that $T_{-1}^{(i)} = 0$ from [1] the chip interface is 1 input and 1 output (Fig. 6, ignoring the Q values). In Fig. 4

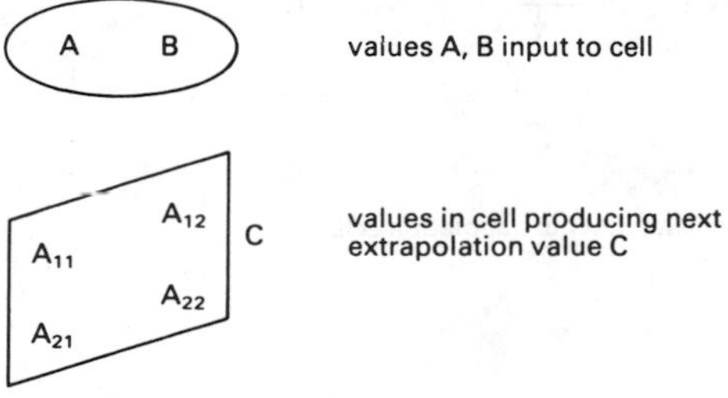

To compute

$$T_k^{(i)} = T_{k-1}^{(i+1)} + \frac{T_{k-1}^{(i+1)} - T_{k-1}^{(i)}}{\left(\frac{h_i}{h_{i+k}}\right)^2 \left[1 - \frac{T_{k-1}^{(i+1)} - T_{k-1}^{(i)}}{T_{k-1}^{(i+1)} - T_{k-2}^{(i+1)}}\right] - 1}$$

Put $A = T_{k-1}^{(i+1)} - T_{k-2}^{(i+1)}$, $B = T_{k-1}^{(i+1)} - T_{k-1}^{(i)}$, $Q = \left(\frac{h_i}{h_{i+k}}\right)^2$

$$T_k^{(i)} = T_{k-1}^{(i+1)} + \frac{B}{Q\left[1 - \frac{B}{A}\right] - 1}$$

STEP 1: read $T_{k-1}^{(i+1)}$, $T_k^{(i)}$, then $T_{k-2}^{(i+1)}$

compute B, A

STEP 2: $z_0 = 1 - B/A$

STEP 3: $z_1 = Qz_0 - 1$

STEP 4: $T_k^{(i)} = T_{k-1}^{(i+1)} + B/z_1$

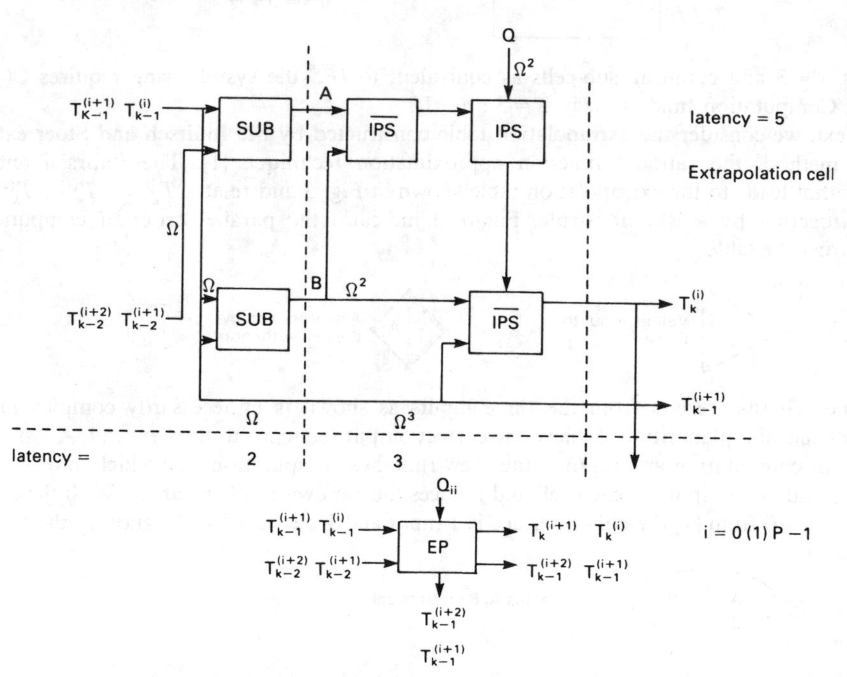

Fig. 5. Bulirsch and Stoer extrapolation cell.

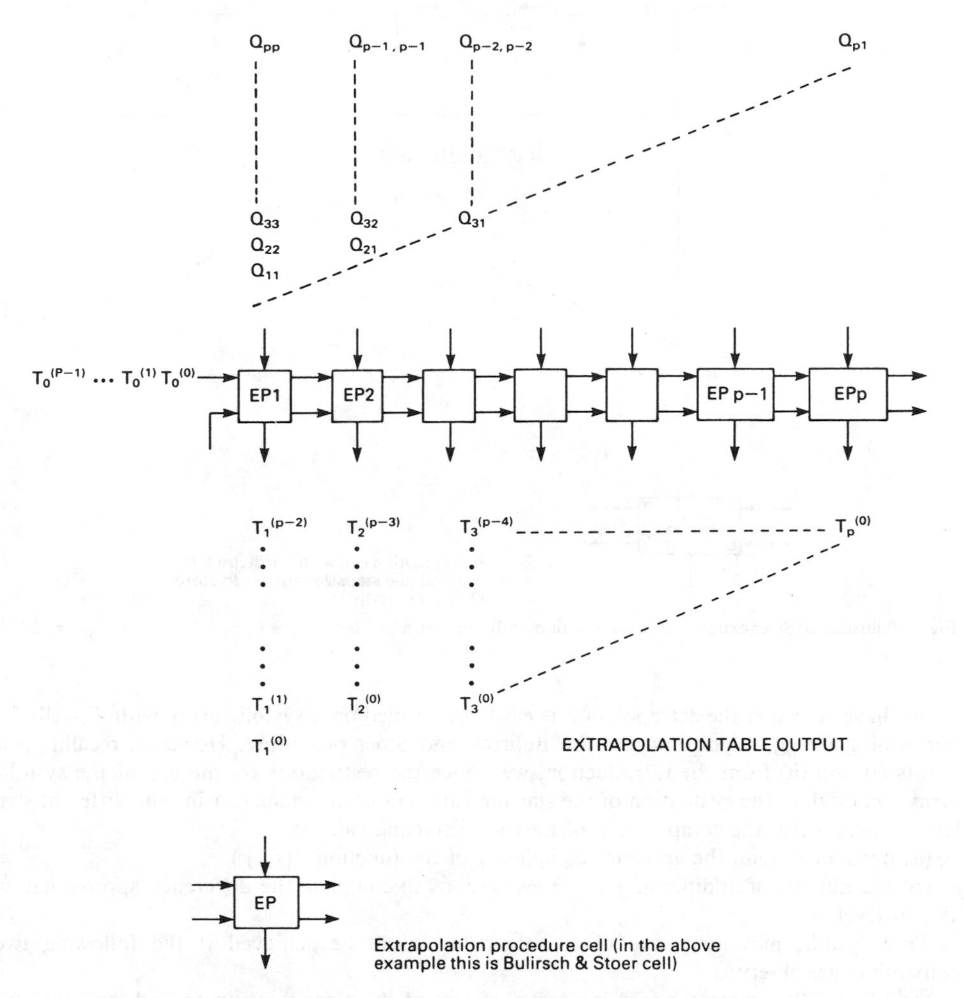

Fig. 6. General structure of systolic array to generate the Bulirsch and Stoer extrapolation table.

Figure 5 indicates the extrapolation cell formula and its implementation as a basic E_p-cell. Notice that the cell latency $c = 5$ and that the hardware requirement is bounded by 5 ips cells, hence if a systolic ring was used we would have $m = \lceil p/5 \rceil$ E_p-cells, and the hardware requirement would be $O(p)$ ips cells as for other designs.

The latency of the Bulirsch and Stoer extrapolation cell follows from Fig. 5 and the use of two-level pipelining. A new value is input to the array every time cycle = ips computation time and values are pipelined through the stages of the cell as well as the pipelining from E_p-cell to E_p-cell.

Figure 6 shows the generic extrapolation array amended for the Bulirsch and Stoer scheme. Note the two horizontal channels between cells. The case when the step size is halved at each level is indicated in Fig. 7.

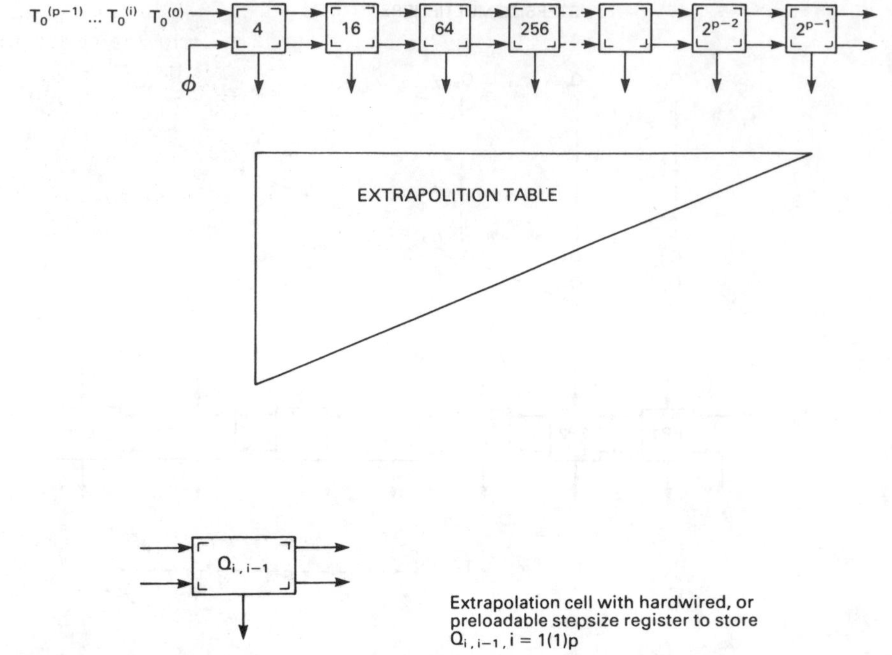

Fig. 7. Bulirsch and Stoer extrapolation array with overall step size h and levels $h_0 = h$, $h_1 = h/2, \dots, h_{p-1} = h/2^{p-1}$.

We have seen that the extrapolation is easily performed on a systolic array with E_p-cells for performing Gragg's algorithm and the Bulirsch and Stoer procedure. However, recalling the points (i) and (ii) from the introduction, we notice the restrictions on the use of the systolic arrays on ODE's. The evaluation of the starting values in both schemes, using the different step levels, ensures that the computation of next level starting value is

(a) dependent upon the arbitrary complexity of the function $f(t, y)$,

(b) the number of additional $f(t, y)$ evaluations to complete the difference approximation at that level.

True systolic movement overall the algorithm could be achieved if the following two constraints are observed:

(a) All the $Q_{i,j}$ were pre-calculated at the start of the algorithm, up to and including the deepest step level;

(b) The cost of successive level starting values could be computed in the interval between inputs to the systolic array (i.e. the periodicity).

Clearly, these conditions cannot be satisfied except under contrived situations, for instance with trivial ODE functions and a very small number of levels. All the $Q_{i,j}$ can be precomputed as can be trivially observed from the algorithm above. However if we consider an arbitrary ODE function, the periodicity of the array is equivalent to the cycle time of one $\overline{IPS}$ cell = (mult time + add time), which is the periodicity of the array. Clearly our ODE function must be computable within this time period, if not we must precompute all the starting values then pump them into the array.

If the ODE function is simple enough to be computed within the periodicity time, successive lower levels will require more and more function evaluations and ultimately the number of function evaluation to compute the next starting value will require a time greater than the periodicity, and the array would not function correctly. As a result of the arbitrary function and the changing levels, we are forced to evaluate the starting values before using the array, resulting in the construction of a finite table of predetermined size. If the array contains p

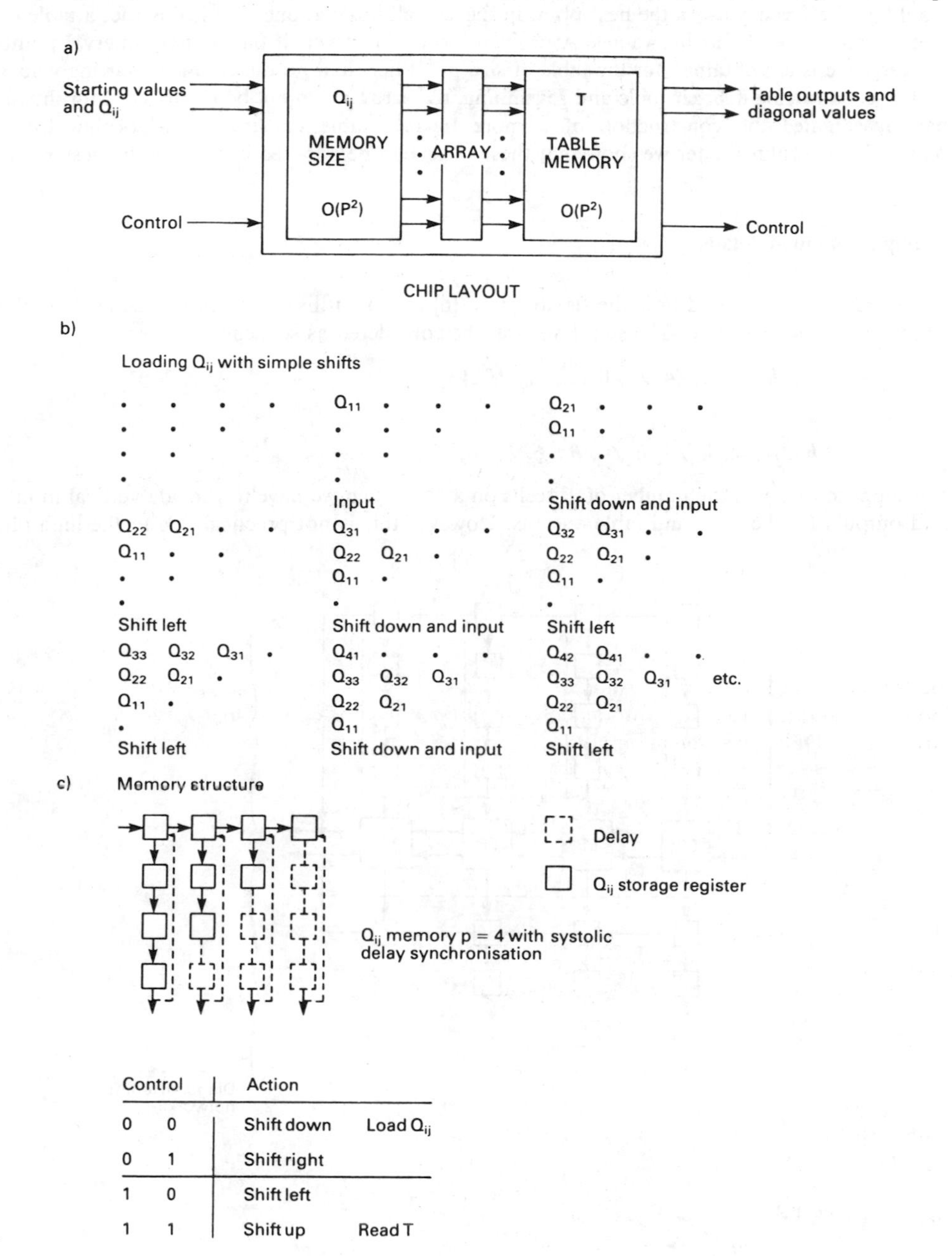

Control		Action	
0	0	Shift down	Load Q_{ij}
0	1	Shift right	
1	0	Shift left	
1	1	Shift up	Read T

Fig. 8.

E_p-cells, then any table of size $p' < p$ can also be constructed. However we do not know how large the table has to be for the convergence of the extrapolated result, and so we would have to precompute p starting values even if the table turned out to be only of level size p'. Clearly we would gain efficiency as on the next point in the overall interval one could construct a table of size p' with only p' starting values. A problem occurs however, if on the next interval point, convergence is not obtained for the table of size p'. Other than producing more starting values and re-constructing a larger table and re-running, the array p' cannot be improved. Murphy [6] has investigated the construction of a more flexible table construction algorithm for a sequential algorithm. Later we show that these ideas can be adopted in the systolic design.

3. Implementation details

In order to understand how the flexibility in [6] can be utilised we must consider how the array will be implemented. The step levels can be considered as sequences

$$\{h_0, h_0/2, h_0/4, h_0/8, \ldots, h_0/2^{p-1}\}$$

or

$$\{h_0, h_0/2, h_0/3, h_0/4, h_0/6, h_0/8, \ldots\}.$$

We suppose that p is the number of E_p-cells on a chip, then we have to provide vertical inputs and outputs for the $Q_{i,j}$ and table outputs. However this is not practical due to the high pin

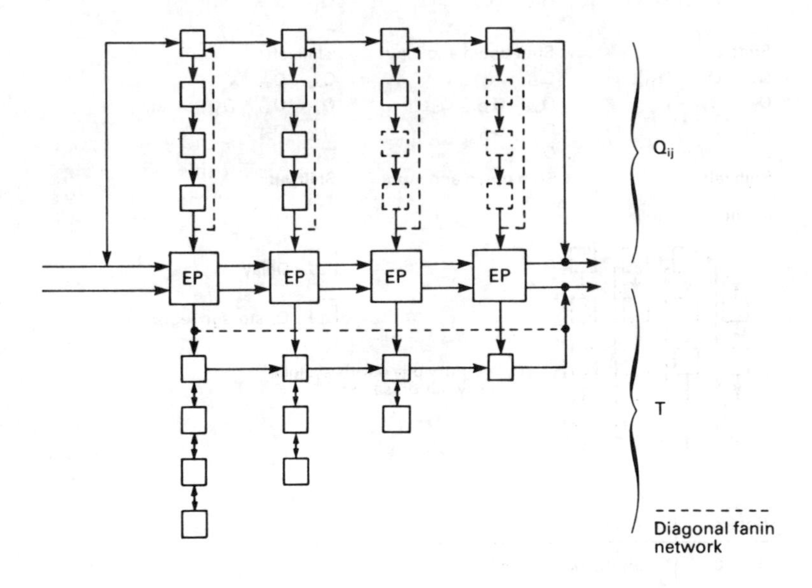

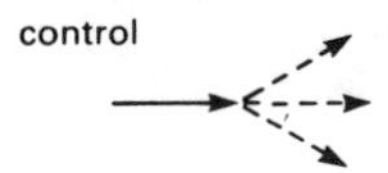

Fig. 9. Internal memory extrapolation array.

count. However precomputing the $Q_{i,j}$ and control the chip as:

(i) Load $Q_{i,j}$ into the chip (using chip storage);

(ii) Compute the table using the array (using chip storage to hold the table);

(iii) Output the tableau.This clearly requires extra storage and control in addition to the systolic array.

Figure 8(a) shows a possible arrangement for the chip using the systolic array connections to preload the $Q_{i,j}$ input to the chip memory. The memory is also shown in Fig. 8(b) with the control signals required. The feedback loops are added to allow the $Q_{i,j}$ to circulate once they are loaded. Thus, the array can be used for successive points over the interval $h = h_0$, without having to preload $Q_{i,j}$ for every table. Notice that the table has the same structure as the $Q_{i,j}$ and so the same method can be used for reading out the table at the end of computation, too, see Fig. 9. Clearly we should try to make p small to fit the design on a single chip, and to ensure that latch mistimings, etc., cannot occur with the feedback connection.

4. Adaptive systolic extrapolation array

The adaptive systolic array attempts to reduce the number of function evaluations needed to compute the starting values required, so that a convergent table can be constructed. We noted earlier that all the function values for a certain size table must be computed before the array is used, changing to a larger table while it is being computed using the current array is not possible. Also, a prediction of a smaller table would mean that some starting values were calculated and not used. For choice of a large table this is wasteful. We describe now how to implement the flexibility of the algorithm in [6], to produce what we shall call the 'adaptive array'. The array is based on the idea of the systolic priority queue for keeping real-time order statistics such as max, min see [3]. The idea is that the host inputs values to the array at intermittent intervals, with the array having no knowledge of when the next input will arrive. On the cycles between inputs, the cells of the systolic array continue to compute the remaining order of the list, with the max or min value always remaining in the host array interface cell. We suggest a modified form of an array like the priority queue for this. The extrapolation array performs no computation between inputs delayed by a significant time. We define two measures of time, array time and host time. If we compute p' starting values we can compute a table of size p' and array time identical to host time. Now if we compute $p'' < p'$ starting values we can compute the table of size p''. Then if we realise that the table will not be big enough to satisfy convergence to a certain accuracy, then before all the p'' inputs have entered the array, we can freeze the array, compute the remaining $p' - p''$ starting values and start up the array continuing as normal. Now we have enough values to compute the table of size p'. The array total time $= 3(p' - 1)$ while the host time = (array time) + (freeze time). Essentially the systolic array is insensitive to the time elapsed by the host to compute extra starting values (i.e. while it is frozen). Assuming that all the registers in the cells and memory of the chip described allow a value to be stored indefinitely, the freeze command can be achieved simply by gating the array clock signal with a convergence condition (which signifies more values are required). Normally, the practice of gating the clock signal is bad practice, but as it is gated only once before it restarts the rest of the array it will have a global effect and can be accepted, see Fig. 10.

Since $\overline{\text{FREEZE}}$ = (CLOCK) and (CONVERGENCE), convergence can be computed using Murphy's algorithm [6], which defines whether in the table the diagonal values are smoothly converging or not. The fact that we output diagonal results while also constructing the table can be used by the host to decide convergence criteria and:

(i) The detection of smooth convergence which by decreasing errors can decide:

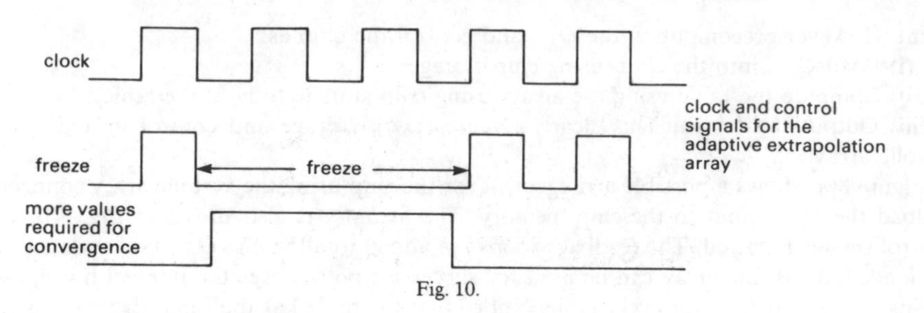

Fig. 10.

(a) whether to abandon the table because it is not converging,
(b) that the array will converge using only the already computed starting values,
(c) the size of the table required to provide convergence using extra starting values.

(ii) If (i) provides an incorrect prediction to the convergence we decide:
(a) whether to change the stepsize h_0 (increase or decrease),
(b) to re-run the array with the same stepsize, but with more starting values.

(iii) Whether to close down the array because convergence has already been achieved with a table smaller than p' (= no. of starting values).

(iv) Raise an exception that no more cells can be used as the maximum number in the array are already in use, i.e., no more starting values can be included.

It should be noted that the freeze command will allow the larger tables to be constructed, and that the original array allowed any table smaller than the available starting values to be constructed. This leaves only (iia) to be described, clearly if the step size is altered the $Q_{i,j}$ must be recalculated and loaded into the chip again, which of course requires that the array be closed down and re-started.

Clearly for large tables the results can be constructed by alternating the freeze/computation phases, so that we compute the minimum number of starting values and do not waste function evaluations to fit the restrictions of the non-adaptive array. A final warning to the practicality of the adaptive array is given as follows, we must be able to collect enough diagonal estimations to perform a prediction. The Bulirsch and Stoer E_p-cell has a latency $c = 5$, thus for two diagonal results using two-level pipelined cells requires at least 10 starting values. As a result we may have to compute more values than necessary before we can construct an adequate prediction. A way to reduce the throughput of the array is to build a cell without two-level pipelining (then two inputs would be required for each diagonal element). This was avoided previously as the cost of the overall computation is increased due to reduced throughput.

References

[1] R. Bulirsch and J. Stoer, Numerical treatment of ordinary differential equations by extrapolation methods, *Numer. Math.* **8** (1966) 1–13.
[2] Burden, Fiares and Reynolds, *Numerical Analysis* (Prindle, Weber & Schmidt, 2nd ed., 1978).
[3] C.E. Leiserson, Area-efficient VLSI computation, Ph.D. Thesis, Carnegie-Mellon University, 1981.
[4] G.P. Mckeown, Iterated interpolation using systolic arrays, Internal Report CSA/21/1984, UEA Norwich, 1984.
[5] D.J. Evans and G.M. Megson, Romberg integration using systolic arrays, *Parallel Comput.* **3** (1986) 289–304.
[6] C.P. Murphy, Numerical methods for solving ordinary and partial differential equations, Ph.D. Thesis, Loughborough University of Technology, 1978.
[7] J.K. Steadman, Rational approximation of the solution to a stiff differential equation using extrapolation, Ph.D. Thesis, Stevens Institute of Technology, Hoboken, NJ, 1981.

Systolic arrays for group explicit methods for solving first order hyperbolic equations

G.M. MEGSON * and D.J. EVANS

Parallel Algorithms Research Centre, Loughborough University of Technology, Loughborough, Leicestershire, UK

Abstract. In this paper, systolic techniques for the simulation of the Group Explicit method for parabolic equations by systolic array data structures using the Soft Systolic algorithm paradigm have been extended to include hyperbolic equations of first order. In particular, the Group Explicit Complete (GEC) strategy can be used as generic array construction for other simple schemes and for computations on large intervals. The easy form of the finite difference approximation to the hyperbolic equation results in a simpler and hence faster and more area efficient basic cell.

Keywords. Group explicit (GE) methods, Hyperbolic equation, Systolic array.

The possible configuration of systolic arrays for solving parabolic equations using the Group Explicit methods of Evans and Abdullah [4] has been fully discussed in [1–3]. In the above we have developed two lines of approach, firstly the *one time level, one cycle approach* in which a basic cell was used for each group of points, and a single cycle of the array produced a complete time level step of integration. The alternative approach to save cells was *the wavefront design* which proved complex when more general equations were used. The former design was emphasised due to its speed, the cell count for large intervals or small stepsizes being high. The Hopscotch schemes [3] included simpler non-group explicit finite difference formulae, and also the fast AGE scheme [2] was introduced to compress the number of cells and simplify internal cell structure.

Recent developments by Sahimi [5] have extended the Group Explicit (GE) principle to first order hyperbolic equations, and indicate more flexible array designs which are briefly outlined below.

The basic hyperbolic equation we shall consider is of the form,

$$\frac{\partial u}{\partial t} + \frac{\partial u}{\partial x} = 0, 0 \leqslant x \leqslant 1, \quad t \geqslant 0. \tag{1}$$

As before, we consider the solution in the infinite rectangular strip bounded by $0 \leqslant x \leqslant 1$, with finite difference approximations made at the intersecting points on a grid superimposed on the region with spacing $\Delta x = h$ along the x-axis, and $l = \Delta t$ along the t axis. We shall only briefly discuss the derivation of the new GE form for hyperbolic equations, the interested reader is referred to Sahimi [5].

* Computing Laboratory, University of Newcastle Upon Tyne.

Reproduced with permission of Elsevier Science Publishers B.V. (North Holland)
Parallel Computing 16(191 – 205)
1990

The hyperbolic form (1) is expressed as a weighted finite difference analogue equation at a particular grid point, say $(x_i, t_j + \Theta) = (i\Delta x, (j + \Theta)\Delta t)$ $0 \leqslant \Theta \leqslant 1$ as follows,

$$-\lambda\left[\Theta\left\{(1-w)u_{i+1,j+1} + (2w-1)u_{i,j+1} - wu_{i-1,j+1}\right\} + (1-\Theta)\left\{(1-w)u_{i+1,j} + (2w-1)u_{ij} - wu_{i-1,j}\right\}\right] = u_{i,j+1} - u_{ij}, \quad (2)$$

with $\lambda = \Delta t/\Delta x = l/h$.

When $w = 1$ we produce the equation,

$$(1+\lambda\Theta)u_{i,j+1} - \lambda\Theta u_{i-1,j+1} = (1 - \lambda(1-\Theta))u_{ij} + \lambda(1-\Theta)u_{i-1,j} \quad (3)$$

and with $w = 0$

$$(1-\lambda\Theta)u_{i,j+1} + \lambda\Theta u_{i+1,j+1} = (1 + \lambda(1-\Theta))u_{ij} - \lambda(1-\Theta)u_{i+1,j}, \quad (4)$$

At the point $((i-1)\Delta x, (j+\Theta)\Delta t)$ (4) becomes,

$$\lambda\Theta u_{i,j+1} + (1-\lambda\Theta)u_{i-1,j+1} = -\lambda(1-\Theta)u_{ij} + (1+\lambda(1-\Theta))u_{i-1,j} \quad (5)$$

coupling Equations (3) and (5) in groups of two adjacent points $(i-1,\ j+1)$ and $(i,\ j+1)$, etc., leads to the group explicit form,

$$Au_{j+1} = Bu_j, \quad (6)$$

where

$$A = \begin{bmatrix} -\lambda\Theta & (1+\lambda\Theta) \\ (1-\lambda\Theta) & \lambda\Theta \end{bmatrix}, \quad B = \begin{bmatrix} \lambda(1-\Theta) & 1-\lambda(1-\Theta) \\ 1+\lambda(1-\Theta) & -\lambda(1-\Theta) \end{bmatrix}$$

$$u_{j+1} = \begin{bmatrix} u_{i-1,j+1} \\ u_{i,j+1} \end{bmatrix}, \quad u_j = \begin{bmatrix} u_{i-1,j} \\ u_{i,j} \end{bmatrix}$$

which in explicit form yields,

$$u_{j+1} = A^{-1}Bu_j, \quad (7)$$

where

$$A^{-1}B = \begin{bmatrix} (1+\lambda) & -\lambda \\ \lambda & (1-\lambda) \end{bmatrix}$$

The computation can be represented by the molecule

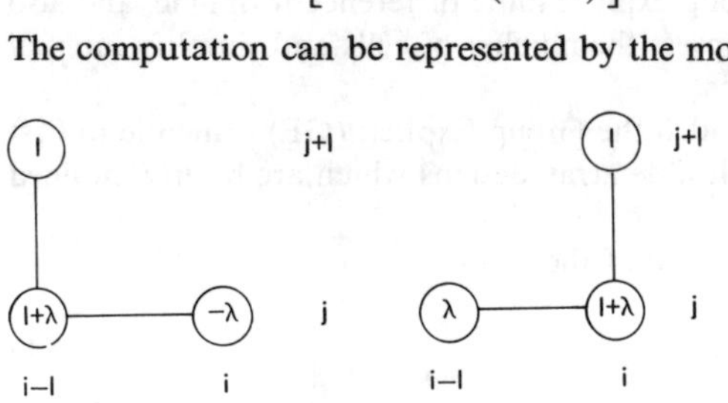

1 j+1 1 j+1

(8)

1+λ −λ j λ 1+λ j

i−1 i i−1 i

and the equations:

$$\left.\begin{aligned} (a)\quad & u_{i-1,j+1} = (1+\lambda)u_{i-1,j} - \lambda u_{ij} \\ (b)\quad & u_{i,j+1} = \lambda u_{i-1,j} + (1-\lambda)u_{ij} \end{aligned}\right\}. \quad (9)$$

Thus, in a similar manner to Evans and Abdullah [4] we can define various GE schemes. If there are m points in the region along grouped in pairs then x include the right boundary (or left boundary) ungrouped points occurring in the $(m-1)$th or 1st positions are computed by,

$$\begin{aligned}(a) \quad & u_{m-1,j+1} = \left[(1+\lambda(1-\Theta))u_{m-1,j} - \lambda(1-\Theta)u_{m,j} - \lambda\Theta u_{m,j+1}\right]/(1-\lambda\Theta) \\ (b) \quad & u_{1,j+1} = \left[\lambda(1-\Theta)u_{0,j} + (1-\lambda(1-\Theta))u_{1,j} + \lambda\Theta u_{0,j+1}\right]/(1-\lambda\Theta)\end{aligned} \tag{10}$$

Next define the $(m-1)*(m-1)$ matrices,

$$G_1 = \begin{bmatrix} G^{(1)} & & & & \\ & G^{(2)} & & & 0 \\ & & \ddots & & \\ & & & G^{(\frac{1}{2}(m-2))} & \\ & 0 & & & -1 \end{bmatrix}, \quad G_2 = \begin{bmatrix} 1 & & & \\ & G^{(1)} & & 0 \\ & & \ddots & \\ & C & & G^{(\frac{1}{2}(m-2))} \end{bmatrix}$$

$$\hat{G}_1 = \begin{bmatrix} 1 & & & \\ & G^{(1)} & & C \\ & & \ddots & \\ & & & G^{(\frac{1}{2}(m-3))} \\ & 0 & & & -1 \end{bmatrix}, \quad \hat{G}_2 = \begin{bmatrix} G^{(1)} & & & \\ & G^{(2)} & & C \\ & & \ddots & \\ & 0 & & \ddots \\ & & & & G^{(\frac{1}{2}(m-1))} \end{bmatrix}$$

with

$$G^{(i)} = \begin{bmatrix} -1 & 1 \\ -1 & 1 \end{bmatrix}$$

then the following schemes can be derived:

m even:

There are $(m-1)$ internal points – which is an odd number requiring boundary cells hence we have,

(i) Group Explicit ungrouped Right point (GER) scheme

$$\begin{aligned}&(I+\lambda\Theta G_1)u_{j+1} = (I-\lambda(1-\Theta)G_1)u_j + b_1, \\ &b_1 = \left(0, 0, \ldots, -\lambda(1-\Theta)u_{m,j} - \lambda\Theta u_{m,j+1}\right)^{\mathrm{T}}\end{aligned} \tag{11}$$

(ii) Group Explicit ungrouped Left point (GEL) scheme

$$\begin{aligned}&(I+\lambda\Theta G_2)u_{j+1} = (I-\lambda(1-\Theta)G_2)u_j + b_2, \\ &b_2 = \left(\lambda(1-\Theta)u_{0,j} + \lambda\Theta u_{0,j+1}, 0, \ldots, 0\right)^{\mathrm{T}}\end{aligned} \tag{12}$$

(iii) Single Alternating Group Explicit (SAGE) scheme

$$\left.\begin{aligned}(I+\lambda\Theta G_1)u_{j+1} &= (I-\lambda(1-\Theta)G_1)u_j + b_1 \\ (I+\lambda\Theta G_2)u_{j+2} &= (I-\lambda(1-\Theta)G_2)u_{j+1} + b_2\end{aligned}\right\} j = 0(2), \ldots \tag{13}$$

(iv) Double Alternating Group Explicit (DAGE) scheme

$$\left.\begin{aligned}(I+\lambda\Theta G_1)u_{j+1} &= (I-\lambda(1-\Theta)G_1)u_j + b_1 \\ (I+\lambda\Theta G_2)u_{j+2} &= (I-\lambda(1-\Theta)G_2)u_{j+1} + b_2 \\ (I+\lambda\Theta G_2)u_{j+3} &= (I-\lambda(1-\Theta)G_2)u_{j+2} + b_2 \\ (I+\lambda\Theta G_1)u_{j+4} &= (I-\lambda(1-\Theta)G_1)u_{j+3} + b_1\end{aligned}\right\} j = 0(4)\ldots \tag{14}$$

m odd:

There are $(m-1)$ internal points which this time is even requiring no boundary cells yielding,

(i) Group Explicit Ungrouped points (GEU) scheme

$$\left.\begin{aligned}&(I+\lambda\Theta\hat{G}_1)u_{j+1}=(I-\lambda(1-\Theta)\hat{G}_1)u_j+b_3\\&b_3=(\lambda(1-\Theta)u_{0,j}+\lambda\Theta u_{0,j+1},0,\ldots,0,-\lambda(1-\Theta)u_{m,j}-\lambda\Theta u_{m,j+1})^{\mathrm{T}}\end{aligned}\right\} \tag{15}$$

(ii) Group Explicit Complete (GEC)

$$(I+\lambda\Theta\hat{G}_2)u_{j+1}=(I-\lambda(1-\Theta)\hat{G}_2)u_j \tag{16}$$

(iii) SAGE

$$\left.\begin{aligned}&(I+\lambda\Theta\hat{G}_1)u_{j+1}=(I-\lambda(1-\Theta)\hat{G}_1)u_j+b_3\\&(I+\lambda\Theta\hat{G}_2)u_{j+2}=(I-\lambda(1-\Theta)\hat{G}_2)u_{j+1}\end{aligned}\right\} j=0(2)\ldots \tag{17}$$

(iv) DAGE

$$\left.\begin{aligned}&(I+\lambda\Theta\hat{G}_1)u_{j+1}=(I-\lambda(1-\Theta)\hat{G}_1)u_j+b_3\\&(I+\lambda\Theta\hat{G}_2)u_{j+2}=(I-\lambda(1-\Theta)\hat{G}_2)u_{j+1}\\&(I+\lambda\Theta\hat{G}_2)u_{j+3}=(I-\lambda(1-\Theta)\hat{G}_2)u_{j+2}\\&(I+\lambda\Theta\hat{G}_1)u_{j+4}=(I-\lambda(1-\Theta)\hat{G}_1)u_{j+3}+b_3\end{aligned}\right\} j=0(4)\ldots \tag{18}$$

An example of the group computation ordering is given in *Fig. 1*, illustrating the independent nature of non-alternating GE schemes, which is attractive from a systolic viewpoint.

First we develop a basic cell for our design, based like the parabolic scheme on a single level single cycle implementation using a linearly connected array of $\frac{1}{2}(m)$ GE cells. Clearly (9) can be expressed as,

$$\left.\begin{aligned}u_{i-1,j+1}&=u_{i-1,j}+\lambda\Gamma\\u_{i,j+1}&=u_{ij}+\lambda\Gamma\\\Gamma&=u_{i-1,j}-u_{ij}\end{aligned}\right\} \tag{19}$$

resulting in the simple arrangement of *Fig. 2a* from which a number of immediate benefits over the parabolic GE-cell are apparent:

(i) the GE cell requires a single ips (inner product step) cell and two adders (rather than two ips used in the parabolic scheme)
(ii) no control program is required for queueing up the cell operands
(iii) only a single register is required to hold coefficient data
(iv) no points are borrowed from adjacent GE cells.

The cell modifications are due mainly to simplifications in the finite difference formula (hyperbolic rather than parabolic) and save both hardware and time. For example, the hyperbolic scheme requires only 2 ips cycles to compute a group compared with at most 6 ips for the parabolic case yielding the immediate speedup $S_p=3$.

Remark. Remember that these comparisons indicate only the suitability of the GE problem to systolic arrays, not the choice of hyperbolic or parabolic equations for a problem.

Thus the hyperbolic GE method seems better suited to systolic computation. The amount of area consumed by the hyperbolic cell for data routing is also less than that of the parabolic cell. Consider the GER, GEL, GEU, and GEC schemes for both hyperbolic and parabolic

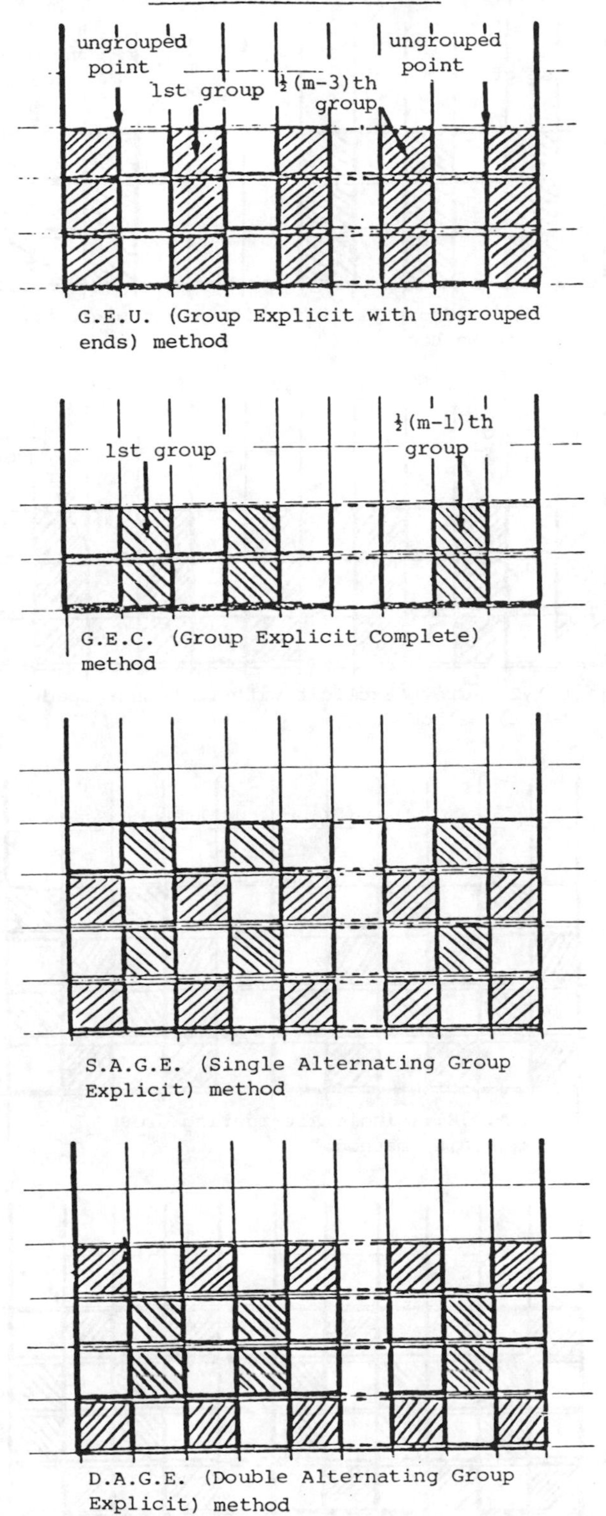

Fig. 1. GE computation for hyperbolic equations. (*Continued on next page.*)

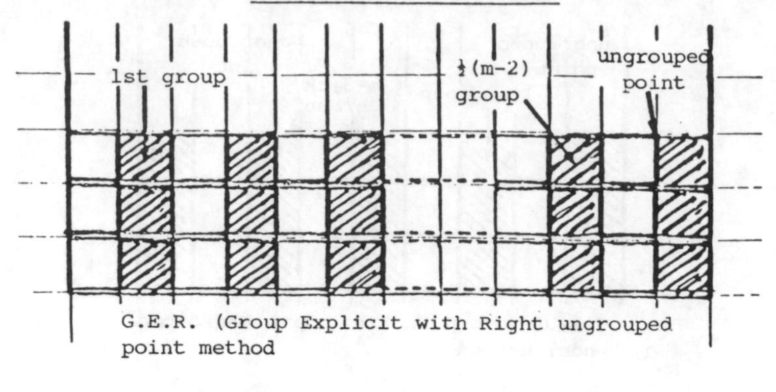

G.E.R. (Group Explicit with Right ungrouped point method

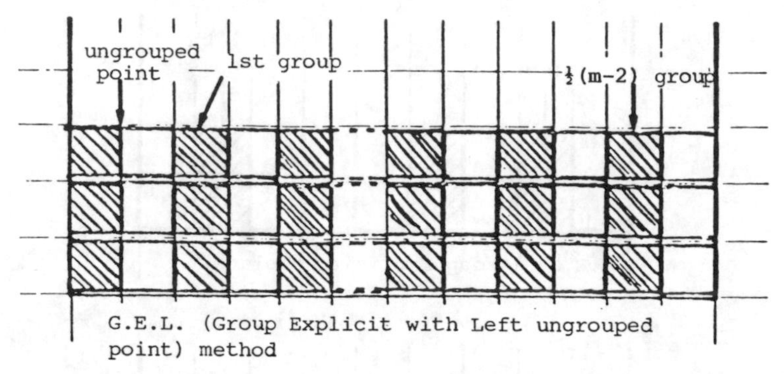

G.E.L. (Group Explicit with Left ungrouped point) method

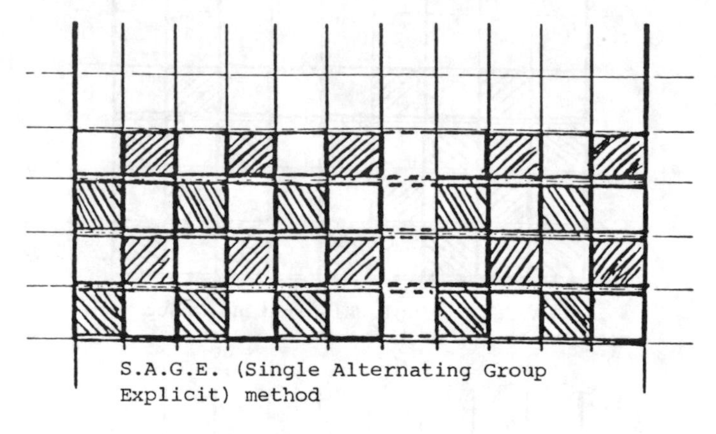

S.A.G.E. (Single Alternating Group Explicit) method

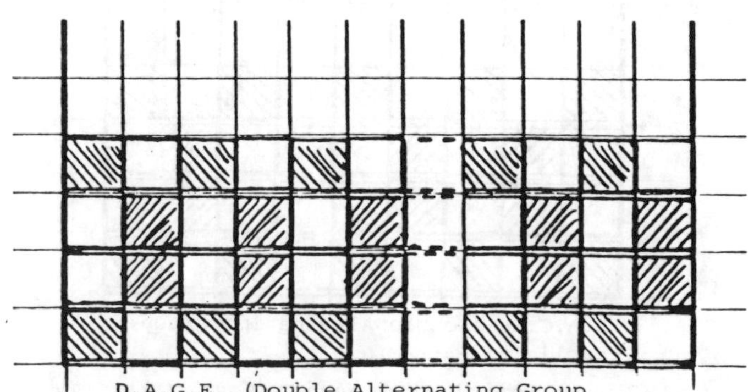

D.A.G.E. (Double Alternating Group Explicit) method

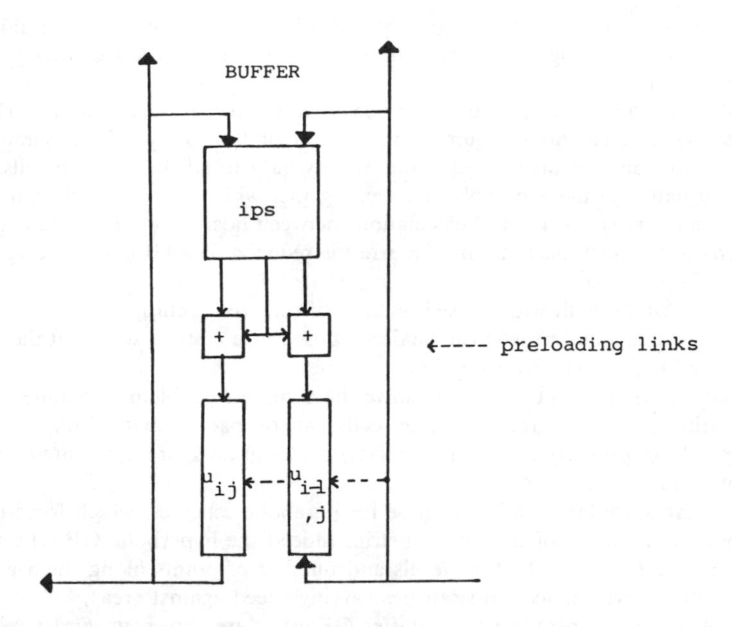

a) GER, GEL, GEU, GEC cell

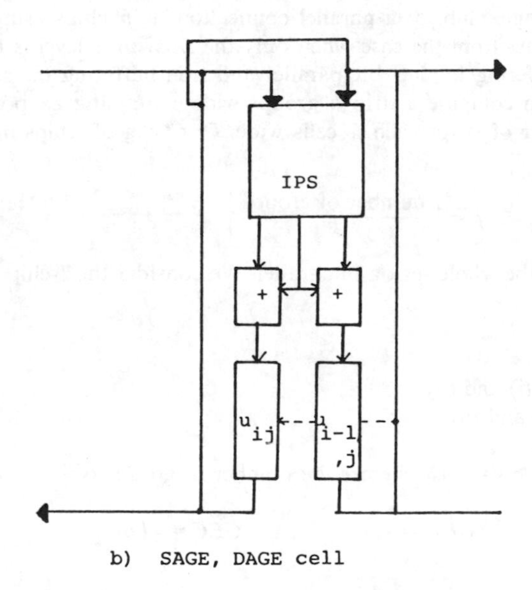

Fig. 2. Hyperbolic GE-cell.

problems. In the hyperbolic schemes the individual groups are disjoint, i.e. there are no overlapping points in adjacent groups, whereas the parabolic schemes computation can only proceed with point sharing. Consequently the hyperbolic array requires only single uni-directional connections which are activated only in loading (starting values) and unloading results. The removal of left and right data shuffling also removes all dataflow control yielding a simple and compact cell.

We still have the problem of boundary cells, and a trivial observation of (10) shows that ungrouped point calculations require more time than full groups. In a parabolic array this degrades performance as ungrouped points supply data to adjacent group cells. However, the independent nature of the hyperbolic scheme, together with the concept of an incomplete array suggests an alternative partition of calculations between host/array. By assigning the boundary calculations to the host machine the GE-array is reduced to a GEC scheme again simplifying the design. In particular:

(i) the array can be built with a fixed number of cells (on a chip)
(ii) the rectangular solution strip can be decomposed into k strips which fit the array in (i) and computed sequentially by a multipass scheme
(iii) we can linearly connect k chips to solve the complete problem in parallel
(iv) if a strip has $k' < k$ groups, unused cells can be padded with dummy values, and the independence property ensures that adjacent (true groups) are not contaminated by invalid computations.

These advantages are far superior to those for parabolic schemes, which force the array to be proportional to the width of the solution strip. Indeed the hyperbolic GE schemes produce an array independent of both the time levels and number of points along the x-axis spawning a number of alternative connection strategies varying speed against area.

Snapshots of array operation for a buffer GE-array are shown in *Fig. 3*. We consider the computation of simple hyperbolic schemes using a 'bag-of' approach. Essentially we suppose a finite number p of GEC chips each implementing a k cell array. We examine the computations associated with applying a single chip, or a parallel connection of p chips using buffered or unbuffered array designs. Apart from the case when only the final time level is required (and buffers can be removed) buffering implies bit parallel and non-buffering bit serial schemes respectively. A k cell chip can compute a strip k groups wide containing $2k$ points, a region wider than this uses a number of strips each k cells wide. Our 'bag of' chips must therefore contain at least,

$$p = \left\lceil \frac{\text{number of groups}}{k} \right\rceil$$

identical GEC chips, to solve the whole region in parallel. We consider the 1-chip versus p-chip bag under five cases:

(i) GEC array with buffers
(ii) GEC array without buffers
(iii) one chip arrangement of (i) and (ii)
(iv) p-chip arrangement of (i) and (ii)
(v) single time level approach.

We further suppose that for the various schemes the number of groups is

$$\left.\begin{matrix} GER \\ GEL \end{matrix}\right\} = \tfrac{1}{2}(m-1), \qquad GEU = \tfrac{1}{2}(m-3), \qquad GEC = \tfrac{1}{2}(m-2)$$

Buffered GEC:

(a) *1-chip sequential (multipass) scheme*

The cost of the array is derived as follows:

loading λ parameter into k cell chip	$2k$ cycles
loading initial values into k cells	$2k$ cycles
time to fill buffer with $2b$ registers	b cycles

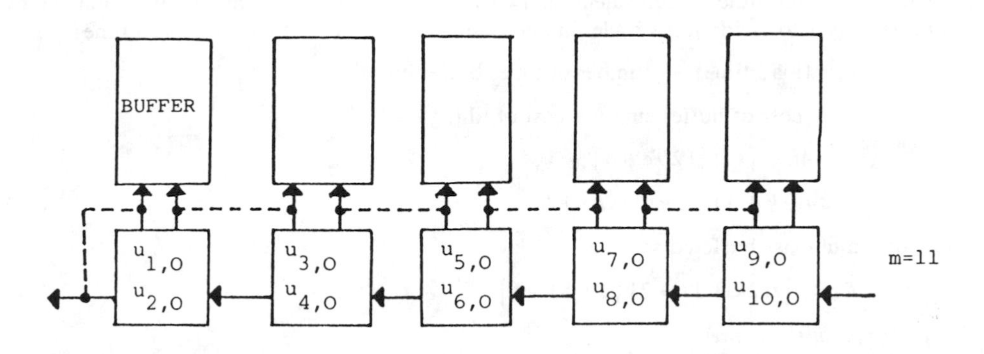

--- buffer out links

a) GEC array (with starting values) for producing all results

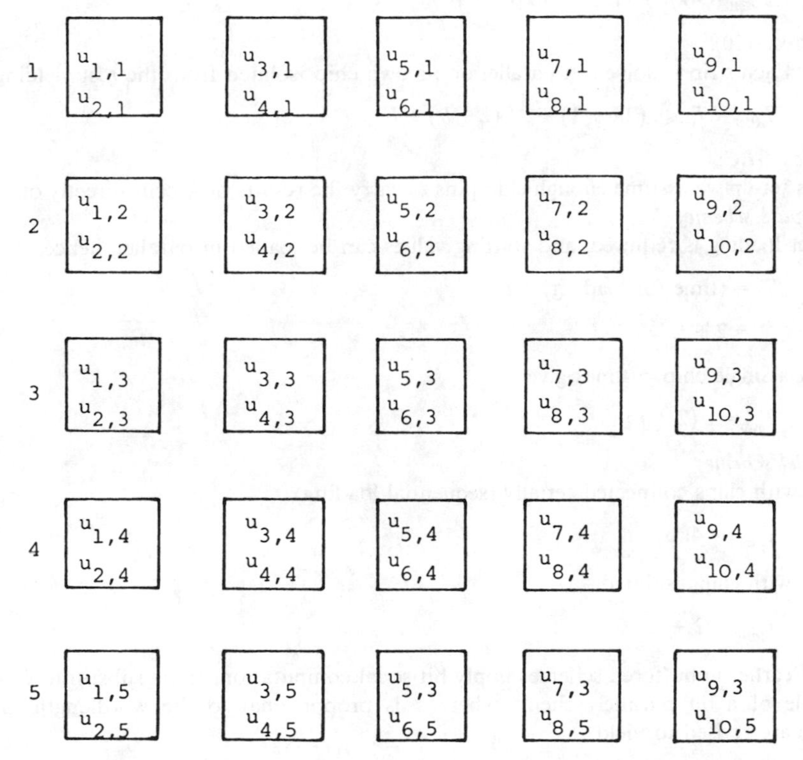

b) Five cycles of the hyperbolic GEC scheme

REMARK: Notice the independence of the computation

Fig. 3. Operating hyperbolic GE array.

The strip is computed to time level $t = t_z$ by filling the buffer $\lceil t/b \rceil$ times and consequently emptying this many times, while the buffer segment of each cell uses $2b$ cycles to empty. Thus, for k cells the time is $2bk$ for buffer load/empty.

Remark. Cost/unit time is equivalent to two ips cycles, thus ips timings are obtained by multiplying by 2 to yield an ips cycle timing. A single group strip therefore has a time

$$\begin{aligned} T_0 &= (\text{load time}) + (\text{number of times buffer empties})^* \\ &\quad [\text{cost of buffer empty} + \text{cost of filling buffer}] \\ &= 4k + \lceil t/b \rceil \{2bk + b\} \\ &\leqslant t(2k+1) + 2k(2+b) + b. \end{aligned} \tag{20}$$

Thus for a multipass buffered scheme,

$$T_{sb} \leqslant p\left[t(2k+1) + 2k(2+b) + b\right]. \tag{21}$$

(b) *p-chip parallel scheme*

There are two ways to connect the p chips in parallel.

Case (a): p-chips are chained together creating a multi-chip GEC array. This is equivalent to a single chip with kp cells. Thus,

$$T_{\mathrm{PCB}} = 4kp + \lceil t/b \rceil \{2\mathrm{bkp} + b\} \tag{22}$$

follows from (20).

Case (b): Each strip is solved in parallel on its own chip isolated from the rest yielding,

$$T_{\mathrm{PIS}} = T_0 \leqslant t(2k+1) + 2k(2+b) + b. \tag{23}$$

Unbuffered GEC:

In this set-up we assume enough chip pins to carry the results of k cells directly off the chip.

(a) Multipass scheme:

No buffer loading is required, and starting values can be loaded in parallel, hence,

$$\begin{aligned} T_1 &= (\text{time for loading}) + t \\ &= 2 + t. \end{aligned} \tag{24}$$

Applying a single chip p times gives,

$$T_{\mathrm{sub}} = p(2+t). \tag{25}$$

(b) Parallel scheme:

Case (i): with chips connected serially (sequential loading)

$$T_{\mathrm{pcub}} = 4\mathrm{kp} + t \tag{26}$$

Case (ii): with chips isolated

$$T_{\mathrm{piub}} = 2 + t. \tag{27}$$

As noted earlier unbuffered schemes imply bit serial computation. If we substitute s cycles for each cycle of a bit parallel scheme where s is proportional to the wordlength then Eqs. (25)–(27) are revised to yield

$$T_{\mathrm{sub}} = sp(2+t), \quad T_{\mathrm{pcub}} = s(4kp+t) \quad \text{and} \quad T_{\mathrm{piub}} = s(2+t).$$

Single time level:

In this case no buffers are required for the bit parallel scheme and a bit serial scheme uses its output pins only once. By shifting results sequentially left or right off the array the computations require

$$T_2 = (\text{time for load}) + (\text{computation for } t \text{ levels}) + (\text{time of unload})$$
$$= 4k + t + 2k = 6k + t.$$

Hence multipass (T_{sol}), parallel (T_{pcol}) and parallel but independent (T_{piol}) are given by,

$$\left.\begin{aligned} T_{sol} &= p(t + 6k) \\ T_{pcol} &= t + 6pk \\ T_{piol} &= t + 2 \end{aligned}\right\} \qquad (28)$$

cycles respectively.

The isolation of groups computing columns up the solution region admits many strategies for interleaving different hyperbolic problems on the same hardware. The p-chip 'bag of' is the simplest approach which allows p problems to be solved simultaneously with different λ parameters in each chip – and solving problems with greater than k groups in a multipass scheme. A more interesting problem is interleaving on a single chip or serially connected group of chips, as indicated by *Fig. 4*. A straightforward approach is simply to queue problems one behind the other in multipass, or along the cells, filling as many cells as possible. The problem here is that the last user (problem) has to wait for other problems in front to be filtered through. A more flexible approach would be a simple interleaving strategy to give all users a reasonable response time. Clearly the time of array operation is computed by substituting $k =$ (sum of all groups) in the above timings.

The hyperbolic scheme is also well suited to fault tolerance as *Fig. 5* demonstrates. For buffers and serial loading schemes there is only a single unidirectional line used only in preloading. This single line makes re-routing around faulty cells simple, and allows chip performance in conjunction with multipass computation to degrade gracefully. For a bit serial or unbuffered approach where loading/unloading proceeds in parallel no re-routing is required, and faulty cells can simply be discarded.

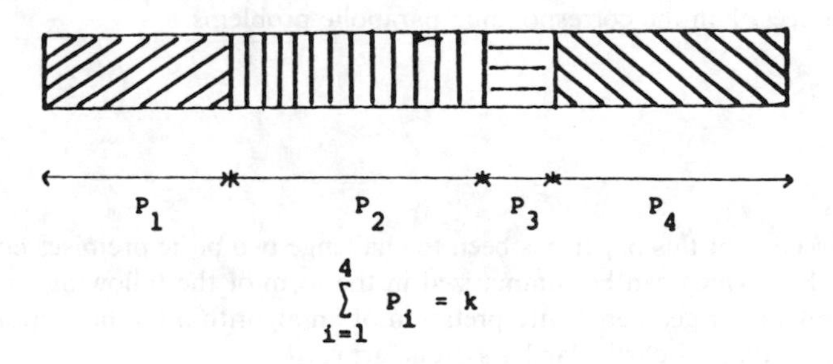

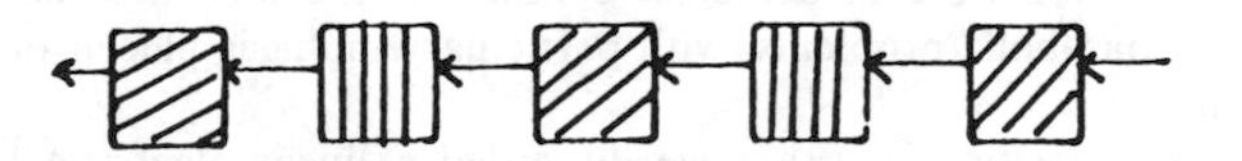

Fig. 4. Problem interleaving.

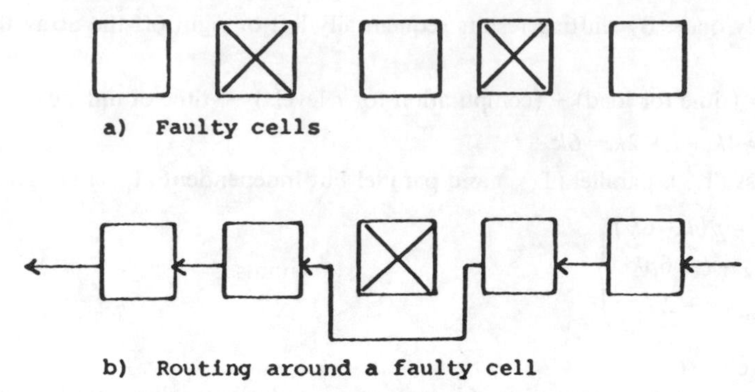

Fig. 5. Fault tolerance.

Finally we consider the Alternating Group Explicit forms of the hyperbolic GE method. Like the parabolic AGE the hyperbolic forms can be implemented with linear arrays that shift group data left or right, overwriting points in adjacent cells to provide alternation. A simple cell for hyperbolic AGE is shown in *Fig. 2b*. Clearly, the alternation of GER and GEL schemes provides a patchwork pattern across the region (see *Fig. 1*). This re-establishes the group dependency relationships demanding that:
(i) all level t is computed before level $t+1$ can start
(ii) boundary calculations must be included in the array (requiring a larger cell cycle time)
and preventing:
(i) interleaving of problems
(ii) multipass computation.
The latter features an extremely attractive implementation characteristic. A final design must balance the use of flexible, fast, and fault tolerant non-alternating GE arrays against the restrictive AGE method with unconditional stability (larger stepsizes) hence reduced cells and level calculations. We conclude that hyperbolic GE methods offer greater opportunities for hard-systolic devices than the corresponding parabolic problems.

Summary

The main objective of this paper has been to challenge two basic premises currently adopted for solving P.D.E.'s, which can be summarized in the form of the following two questions:
(i) Is the algorithmic or geometric interpretation of an algorithm the best approach to solving a P.D.E. in parallel (in particular by systolic arrays)?
(ii) Can a fast, area efficient parallel and conditionally stable design outperform slower unconditionally stable alternatives?
To illustrate the discussion we considered the solution of the 1-D (heat conduction) and 2-D (unsteady diffusion) parabolic problems, with particular emphasis placed on the 1-D case due to its simpler form.

In order to answer the first question we discarded a linear algebraic formulation of the P.D.E. problem interpreting the grid points of the solution region as a tableau of elements to be generated or modified by use of computational molecules relating the grid elements. In the case of the 1-D problem an open ended table was apparent, implying table generating techniques were applicable and gave rise to three main types of array:

(i) Non-stationary array: a design using a cascaded linear array based on the intuitive iterative algorithmic solution of linear systems, and which acted as a benchmark for other 1-D designs. The principle attribute of the array being the non-stationary movement of successive time-level approximation from the same grid column through the array.
(ii) Column-by-column array: here column i of grid point approximations remained stationary and tied to cell i of the linear array. The problem was similar to the generation of an open ended trapezium type table. Basic cells used the asymmetric molecules of Saul'ev [6] to implement a systolic marching technique.
(iii) Row-by-row array: again a stationary array with columns $2i-1$ and $2i$, $i = 1(1)\frac{1}{2}(m)$ tied to cell i of the array. The problem used a rectangular table generation pattern producing a single table row (or time level) every cell cycle. Basic cells utilised the unified Group Explicit (GE) molecule of Evans and Abdullah [4].

Similarly, for the 2-D case three types of design were considered for manipulating the regions grid points:
(i) Non-stationary array: another cascaded iteration array derived from the algorithmic formulation of an iterative matrix problem derived from the 2-D finite difference approximation and used for benchmarking 2-D systolic designs.
(ii) Wavefront processor: using the 2-D asymmetric approximations and occurring in two forms:
(a) a 2-D mesh with highly efficient pipelining of wavefronts, and useful for fast calculation up to a certain level t_z with no intermediate output.
(b) a 1-D linear asymmetric marching processor (LAMP) which reduced hardware by multipass simulation of (a) with each pass a single wavefront. This array had the natural capability of outputting all intermediate time levels up to and including t_z.
(iii) Mesh scheme: which adapted the Group Explicit molecules to achieve full parallel operation of processors, and appeared in two forms:
(a) a 2-D mesh of reduced instruction set processors, evaluating a single table update in one cell cycle.
(b) a 1-D array of macro GE-cells (incorporating a systolic ring) and computing molecules according to a generalised molecule template. Table updates were achieved by multipass with one update per pass.

For t_z levels and m divisions along the x and y-directions the 1-D case had $m-1$ and the 2-D case had $(m-1)^2$ initial values. The cascaded schemes required $O(m+t_z)$ ips cycles and $O(t_z)$ basic ips cells in the 1-D case, and $O(m^2+t_z)$ ips cycles and $O(mt_z)$ basic ips cells (neglecting synchronising delay cells) for the 2-D case. In contrast, the asymmetric molecule arrays required $O(m)$ and $O(m^2)$ basic ips cells for the 1-D, 2-D (LAMP) and 2-D wavefront scheme respectively (and apart from the LAMP array with $O(mt_z)$ time) had the same order of magnitude in timings. Thus, the algorithmic schemes favoured wide regions with few time levels, the geometric forms narrow regions with many levels to optimise the array speed/area tradeoff. Consequently, by fixed m, for any substantial calculations with a significant number of time levels and adopting a geometric approach area savings followed immediately. The actual cell savings depended upon the number of iteration arrays included in the cascaded array. For a fixed number of iteration arrays ($\bar{t}_z > 0$) the algorithmic approach times become $O(\lfloor t_z/\bar{t}_z \rfloor(m+\bar{t}_z))$ and $O(\lfloor t_z/\bar{t}_z \rfloor(m^2+\bar{t}_z))$ for the 1-D and 2-D cases. It follows that for finite hardware the geometric schemes run faster and seriously challenge the intuitive algorithmic arrays.

Now having established the answer to our first question above the next logical step was to produce the 'best' geometric array. For purposes of argument we define the 'best' array to be one which:

(1) improves the accuracy of grid point approximations
(2) reduces computation time and array area further.

(1) is controlled by the truncation error terms associated with the finite difference approximation used to model the P.D.E. For instance, the asymmetric forms had truncation errors $O(\alpha/h + h^2 + l)$ and $O(\alpha h + \beta h + l + h^2)$ for the 1-D and 2-D problems respectively, where h = stepsize in the x-direction and l the time step, α, β suitably chosen parameters. (2) is dictated by the simplicity of the computational molecule and the sizes of t_z and m. Clearly reducing t_z and m to speedup and compact the arrays requires increases in l and h which in turn increases the approximation error of the asymmetric computations. Likewise reducing h and l to provide more accurate results increases array time and area. Hence (1) and (2) provide conflicting goals. We resolved the problem by alternating the application of asymmetric formulas which retained the simple molecule-cell structures and improved accuracy because of truncation error term cancellations (due to opposite signs). For the 1-D case, alternation had the unfortunate effect of sequentialising array computations yielding low processor efficiency and making the algorithmic scheme attractive again. To overcome this difficulty the Group Explicit (GE) methods of Evans and Abdullah [4] were adopted and new arrays developed, using a hybrid molecule consisting of unified asymmetric molecules. The loosely coupled structure of the GE methods allowed parallel operation of array cells yielding high efficiency while retaining truncation term cancellations to maintain accuracy, at the expense of a more complex basic cell. Various types of arrays corresponding to positioning of grouped and ungrouped mesh points were developed, and alternating Group Explicit (AGE) arrays devised implementing an unconditionally stable method. Simple data shuffling and cycling operations were then introduced and the principle of cell unification applied to derive generic 1-D and unified 2-D arrays, implementing all the GE techniques. The former 1-D scheme adopted simple left/right data shifts, the 2-D method adopted a universal molecule template evaluated by accumulating terms of a systolic ring. As the 1-D molecule fitted the 2-D molecule the unified array could also be used to simulate the 1-D computation.

Next we produced FAST arrays based on restricted choices of l and h, with $r = l/h^2$, the above asymmetric formulae can be shown to be unconditionally stable for $r \geqslant 0$. Likewise the simple GE schemes are conditionally stable for $r \leqslant 1$, and the AGE methods again unconditionally stable. By restricting $r = 1$ approximating equations and hence molecules are simplified with terms disappearing altogether and coefficients involving r becoming constant. The FAST AGE array followed naturally, yielding a speedup $S_p = 6$ over the general arrays, from simplifying cell computation. However, fixing r also fixed the truncation error. Consequently the array speed increase was used to offset the larger stepsizes achievable by the general scheme by constructing more accurate approximations to a number of intermediate levels. It followed that for $r \leqslant 5$ the FAST AGE outperformed the general AGE array.

Fixed r values, where molecule terms disappeared also gave rise to incomplete versions of the P.D.E. solvers, using the well-known hopscotch techniques, in which only part of the whole solution region was produced. Omitted grid-points being placed in positions where they could be easily derived later from array results. The technique was discussed and arrays described for the ODD-EVEN, 1-point and 2-point (FAST AGE) hopscotch schemes. The ODD-EVEN method produced an unconditionally stable array by using a unified 2-point molecule derived from sequential application of the classical explicit and implicit formulae. The array computed at the same speed as the AGE array but had truncation error of $O(l + h^2)$ making it less accurate. The 1-point scheme adopted the classical explicit molecule (conditionally stable for $r \leqslant \frac{1}{2}$) with $r = \frac{1}{2}$ and truncation error $O(l + h^2)$, and produced a speedup of $S_p = 12$ over the ODD-EVEN and AGE arrays while using only $\frac{1}{2}$m inner product cells. The speedup was again interpreted as a method of computing more accurate intermediate levels allowing the conditionally stable scheme in some instances to outperform the unconditionally stable schemes. The same array compaction technique was then applied to the 2-point or fast AGE scheme also saving half the hardware.

Next the simple form of the 1-point hopscotch scheme was exploited to derive proposals for a hard-systolic implementation of the parabolic solver by bit parallel and bit serial computation strategies using fixed point arithmetic. The former scheme required buffering of input and output. The latter scheme producing an area efficient unbuffered cell structure – suggesting a FAST chip based design was possible.

Finally we considered the extension of the method to a simple first order hyperbolic equation which exhibited attractive VLSI design features. The derived group explicit molecule produced a fully decoupled approach to computation where pairs of individual table columns could be evaluated independently. This resulted in a decoupled collection of GE cells requiring communication only for the loading of initial values. It followed that a hyperbolic equation could be solved by multipass on a fixed sized architecture (independent of both t_z and m), and that a collection of problems could be solved in parallel by interleaving group columns of different instances on the same array. These attributes together with a uni-directional loading strategy combined to indicate a fault tolerant design which would degrade gracefully as individual cells became faulty.

We conclude that the geometric approach to solving P.D.E.'s is not only suited to soft-systolic frames but produces genuine proposals for hard-systolic implementations.

References

[1] D.J. Evans and G.M. Megson, Systolic arrays for group explicit methods for solving parabolic equations, in: K. Bromley, ed., *Systolic Arrays – SPIE* (1988) 159–179.

[2] D.J. Evans and G.M. Megson, A systolic array for the fast AGE method for parabolic equations, Int. Rep. No. 261, L.U.T., 1986.

[3] D.J. Evans and G.M. Megson, Hopscotch schemes for the solution of parabolic equations on area efficient systolic arrays, in: McKenny, McWhirter and Swartzlander, eds., *Systolic Array Processors* (Prentice Hall, Englewood Cliffs, NJ, 1989) 62–72.

[4] D.J. Evans and A.R.B. Abdullah, Group explicit methods for parabolic equations, *Internat. J. Comp. Math.* 14 (1983) 73–105.

[5] M.S. Sahimi, Numerical methods for solving hyperbolic and parabolic partial differential equations, Ph.D. Thesis, Loughborough University, 1986.

[6] V.K. Saul'yev, *Integration of Equations of Parabolic Type by the Method of Nets* (Macmillan, New York, 1964).

SOLUTION OF LINEAR SYSTEMS

Systolic block LU decompositions

D.J. Evans and K. Margaritis

Department of Computer Studies, Loughborough University of Technology, Loughborough, Leicestershire, U.K.

Abstract. This paper applies the "rotate and fold" (RF) concept to the block (2×2) LU/LDU decomposition of a banded matrix **A**, on the systolic array proposed by Y. Robert, Int. J. Comput. Math., Vol. 17 (1985). Although the hardware requirements are the same as proposed by Robert, the computation time is approximately halved. The solution of block (2×2) triangular linear systems of equations as produced by the LU/LDU decomposition methods is also discussed.

Keywords. LU decompositions, 2×2 block, triangular system solver, rotate and fold, systolic array.

1. Introduction

The importance of the LU decomposition algorithm for the solution of linear systems of equations is well known, and for this reason this method was amongst the first to be considered for systolic implementation [25,29].

Several similar architectures or modifications and extensions of the systolic algorithm have been proposed. For example, in [24] a wavefront LU decomposition algorithm is discussed. In [23,30], the decomposition procedure is used for the introduction of methodologies for the formal derivation of systolic algorithms and in [30] a new array is proposed. The partitioning of the LU decomposition method is addressed in [19,20] so that small VLSI arithmetic modules can accommodate bigger problems. The incorporation of the LU decomposition and the forward and backward substitution arrays is discussed in [1,35]. The block LU decomposition is investigated in [31] as a means for the improvement of the efficiency of the original array. For the same purpose the RF method has been applied in [6,12].

Reproduced with permission of Elsevier Science Publishers B.V. (North Holland)
VLSI 8(65 – 90)
1989

The original implementation uses Gaussian elimination without pivoting, a fact that makes the method suitable only for a specific but wide subset of applications [29,31]. Alternative methods for the solution of linear systems of equations have also been proposed. The Cholesky factorization method is investigated, amongst others, in [4], where some problem partitioning techniques are introduced. The QR decomposition, as well as other methods based on similarity transformations using rotation matrices, are discussed in [2,11,17,21]. Notice that similar methods can be used for the matrix eigenproblem solution. Some implementation aspects of QR decomposition are investigated in [33], and a problem partitioning method is described in [16].

In addition to QR decomposition, the Gauss elimination procedure with neighbour pivoting is introduced in [15] for the triangularization of a matrix.

The numerical properties of the pairwise (or neighbour) pivoting are investigated in [34] while a linear array for the same method is described in [5]. In [32] all the above mentioned methods are unified as four alternative methods, i.e. LU or QR decomposition with or without pivoting.

The optical implementation of LU, QR decompositions and generally the direct solution of linear systems of equations is addressed in [8–10]. Further, an LU factorization algorithm, based on a series of optical matrix multiplications is discussed in [7]; finally the optical LU decomposition using outer products is introduced in [3,13,14].

Another important aspect of matrix decomposition is the updating of the LU factors when a change in the original matrix occurs. The algorithm based fault-tolerant techniques for LU decomposition described in [18,22], have been expressed in terms of updating the corresponding LU factors in [26,27].

David J. Evans is Professor of Computing at Loughborough University of Technology and has been associated with Parallel Processing for many years. He was a member of the early Manchester University design team on ATLAS. More recently he has led a team researching into both Architectures and Algorithms for Parallel Processing and has successfully constructed 2 parallel MIMD systems. His current interests are in the design of Parallel Algorithms for VLSI Processor Arrays.
As European Editor of 'Parallel Computing' and Vice-Chairman of the International Parallel Computing Society he is actively involved in all aspects of Parallel Processing.

K.G. Margaritis. After initial studies at the Aristotle University of Thessalonika Mr. Margaritis has recently successfully completed his Ph.D. degree on VLSI Systolic Algorithms at Loughborough University of Technology under the supervision of Professor D.J. Evans and is currently completing his national service duties in Greece.

Some aspects of the implementation and various other applications of the LU decomposition method are addressed here. In Section 2 the efficiency of the original LU decomposition array is increased to 1 by combining block (2×2) and RF methods. A similar array for the triangular system solution is also described.
Finally, some extensions and further research topics in systolic LU decomposition are briefly addressed.

2. The RF method on systolic block LU decomposition

The use of block (2×2) LU decomposition of a banded matrix A is introduced in [31], so that the efficiency of the hex-connected array of Leiserson [25] is improved from $e = 1/3$ to $e = 1/2$. The term efficiency denotes the processor utilisation and the overall computation time.
More specifically the LU decomposition of an $n \times n$ banded matrix A with bandwidth $w = p + q - 1$ requires a hex-connected array with no more than pq processors and takes a time of $3n + \min(p, q)$ IPS cycles.
In [31] the same computation requires a hex-connected array with no more than pq processors and takes a time of $2n + \min(p, q)$ IPS cycles. It should be noted that the complexity of the systolic network is increased and some preprocessing of the input data streams is necessary. On the other hand, however, a considerable speed-up is achieved and the processor utilisation is also improved. The systolic array for the block (2×2) LU decomposition is shown in Fig. 1, where its operation is also explained. In Fig. 2 a preprocessing array for the formulation of the input data sequence is described.
The RF method is also used for the improvement of the efficiency of the systolic LU decomposition.
Thus, the LU decomposition can be performed on a hex-array with no more than pq processors in time $\lceil 3n/2 \rceil + \min(p, q)$ IPS cycles. The combination of these two strategies is investigated herein, i.e. the application of the RF concept on the block (2×2) LU decomposition. For simplicity the case of a block-tridiagonal matrix is examined in detail; again it is assumed that no pivoting is required.

2.1. Block RF LU decomposition

Suppose the linear set of equations,

$$A\mathbf{x} = \mathbf{b},$$

where A is an $n \times n$ block (2×2) tridiagonal matrix, with n even (without any loss of generality), i.e. $n = 2k$, $p = q = 4$, $w = 7$. Then

$$\begin{bmatrix} A_{11} & A_{12} & & & \\ A_{21} & A_{22} & A_{23} & & \\ & A_{32} & A_{33} & A_{34} & 0 \\ & 0 & \ddots & \ddots & A_{k-1,k} \\ & & & A_{k,k-1} & A_{kk} \end{bmatrix} \begin{bmatrix} \mathbf{x}_1 \\ \mathbf{x}_2 \\ \mathbf{x}_3 \\ | \\ \mathbf{x}_k \end{bmatrix} = \begin{bmatrix} \mathbf{b}_1 \\ \mathbf{b}_2 \\ \mathbf{b}_3 \\ | \\ \mathbf{b}_k \end{bmatrix}, \tag{2}$$

with

$$A_{ij} = \begin{bmatrix} a_{lm} & a_{l,m+1} \\ a_{l+1,m} & a_{l+1,m+1} \end{bmatrix}, \quad \mathbf{x}_i = \begin{bmatrix} x_l \\ x_{l+1} \end{bmatrix}, \quad \mathbf{b}_i = \begin{bmatrix} b_l \\ b_{l+1} \end{bmatrix}, \tag{3}$$

where $l = 2i - 1$, and $m = 2j - 1$.

The solution of the linear system in (1) can be effected by means of a LU or LDU decomposition, i.e.,

$$A = LU \text{ and } L\mathbf{y} = \mathbf{b}, \quad U\mathbf{x} = \mathbf{y}, \tag{4}$$

or

$$A = LDU' \text{ and } L\mathbf{y} = \mathbf{b}, \quad D\mathbf{y} = \mathbf{z}, \quad U'\mathbf{x} = \mathbf{z} \tag{5}$$

with

$$L = \begin{bmatrix} I & & & & \\ L_{21} & I & & 0 & \\ & L_{32} & I & & \\ & 0 & \ddots & \ddots & \\ & & & L_{k,k-1} & I \end{bmatrix}, \quad U = \begin{bmatrix} U_{11} & U_{12} & & & \\ & U_{22} & U_{23} & & 0 \\ & & U_{33} & U_{34} & \\ & 0 & & \ddots & U_{k-1,k} \\ & & & & U_{k,k} \end{bmatrix}, \tag{6}$$

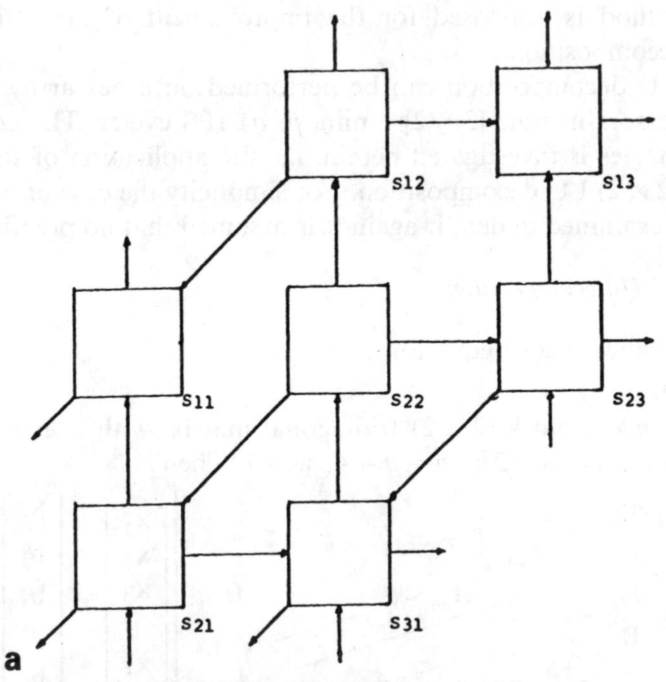

Fig. 1. (a) Block (2×2) LU, LDU decomposition array.

or

$$D = \begin{bmatrix} U_{11} & & & & \\ & U_{22} & & & \\ & & U_{33} & & 0 \\ & 0 & & \ddots & \\ & & & & U_{kk} \end{bmatrix},$$

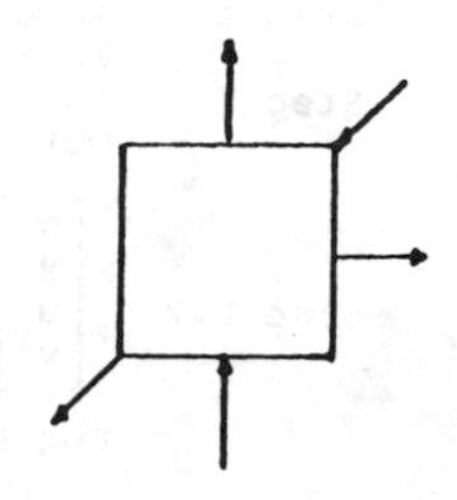

```
processor S21

Step t     In-S = l,m,n,p
           In-NE = a,b
Step t+1   In-NE = c,d
           Out-N = a*p-n*b,a*m-l*b
           Out-E = a*p-n*b,a*m-l*b
           Out-SW = a,b
Step t+2   Out-N = d*n-c*p,d*l-c*m
           Out-E = d*n-c*p,d*l-c*m
           Out-SW = c,d
```

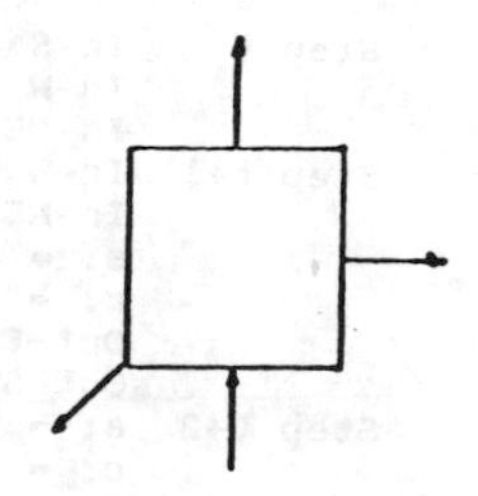

```
processor S22

Step t     In-S = a,b,c,d
Step t+1   Out-N = a,b,c,d,a*d-b*c
           Out-E = a*d-b*c
           Out-SW = a,b
Step t+2   Out-SW = c,d
```

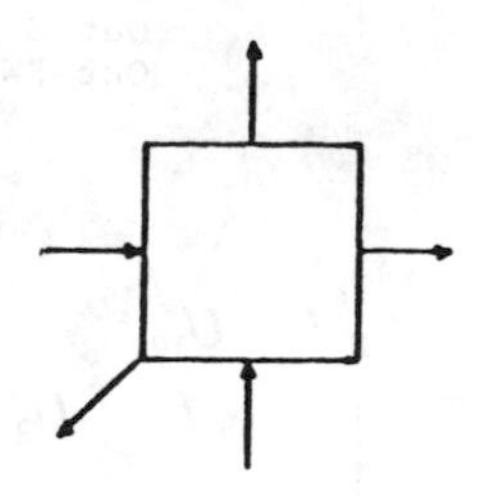

```
processor S23

Step t     In-S = u,v,w,x
           In-W = D
Step t+1   Out-N = w/D,x/D
           Out-E = D
           Out-SW = w/D,x/D
Step t+2   Out-N = u/D,v/D
           Out-SW = u/D,v/D
```

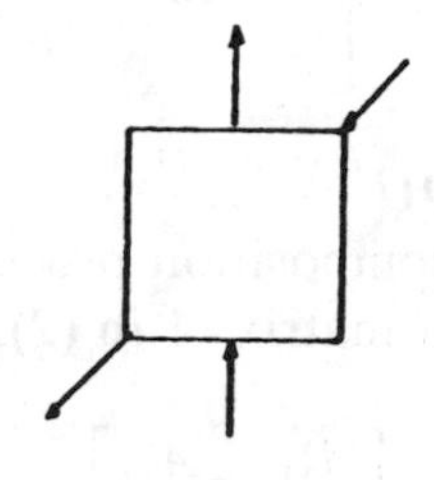

```
processor S11

Step t     In-S = n,l
           In-NE = D
Step t+1   In-S = p,m
           l: = l/D, n: = n/D
           Out-SW = D
Step t+2   m: = m/D, p: = p/D
           Out-N = l,m,n,p
```

Fig. 1. (b) Call definitions.

```
processor S12

Step t      In-S = a,b,c,d,D
Step t+1    Out-E = a,b,c,d
            Out-SW = D
Step t+2    Out-N = a,b,c,d
```

```
processor S13

Step t      In-S = w,x
            In-W = a,b,c,d
Step t+1    In-S = u,v
            Out-E = a,b,c,d
```

$$\text{Step } t+2 \quad \begin{bmatrix} w \\ x \\ u \\ v \end{bmatrix} := \begin{bmatrix} wb-ud \\ xb-vd \\ uc-wa \\ vc-xa \end{bmatrix}$$

```
            Out-N = u,v,w,x
```

```
processor S31

Step t      In-S = a,b,c,d
            In-W = n,l
            In-NE = w,x
Step t+1    In-W = p,m
            In-NE = u,v
            a: = a+l*w, b: = b+l*x
            c: = c+n*w, d: = d+n*x
            Out-E = n,l
            Out-SW = w,x
Step t+2    a: = a+m*u, b: = b+m*v
            c: = c+p*u, d: = d+n*v
            Out-N = a,b,c,d
            Out-E = p,m
            Out-SW = u,v
```

Fig. 1. (b) (continued).

$$U' = \begin{bmatrix} I & U'_{12} & & & \\ & I & U'_{23} & & \\ & & I & U'_{34} & 0 \\ & 0 & & \ddots & U'_{k-1,k} \\ & & & & I \end{bmatrix}, \tag{7}$$

and $U = DU'$.

The decomposition procedure is exemplified by taking the first four rows and columns of matrix A in (2),

$$A = \begin{bmatrix} A_{11} & A_{12} \\ A_{21} & A_{22} \end{bmatrix}. \tag{8}$$

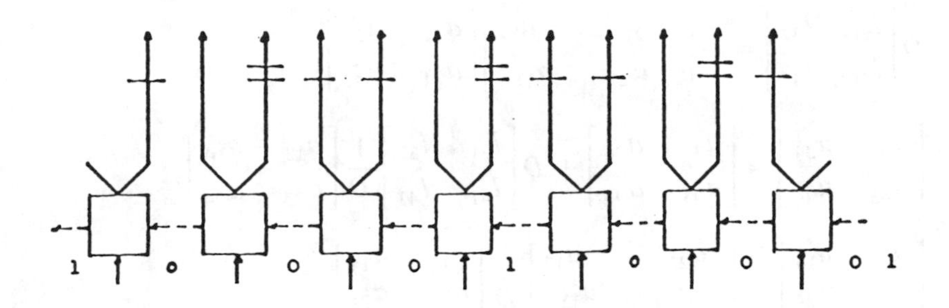

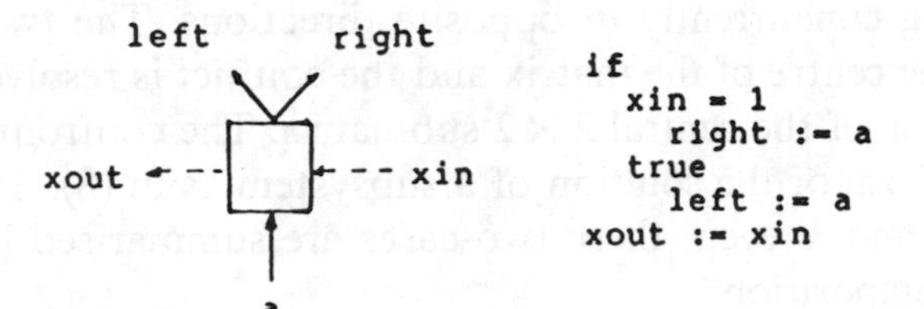

Fig. 2. Preprocessor array.

Then, we have,

$$\begin{bmatrix} I & 0 \\ L_{21} & I \end{bmatrix}^{-1} \begin{bmatrix} A_{11} & A_{12} \\ A_{21} & A_{22} \end{bmatrix} = \begin{bmatrix} U_{11} & U_{12} \\ 0 & U_{22} \end{bmatrix}, \tag{9}$$

with

$$U_{11} = A_{11}, \quad U_{12} = A_{12};$$

$$L_{21} = -(A_{11})^{-1} A_{21} = -(U_{11})^{-1} A_{21}, \quad U_{22} = A_{22} + L_{21} A_{12} \tag{10}$$

or

$$\begin{bmatrix} I & 0 \\ L_{21} & I \end{bmatrix}^{-1} \begin{bmatrix} A_{11} & A_{12} \\ A_{21} & A_{22} \end{bmatrix} = \begin{bmatrix} U_{11} & 0 \\ 0 & U_{22} \end{bmatrix} \begin{bmatrix} I & U'_{12} \\ 0 & I \end{bmatrix}, \tag{11}$$

with

$$U'_{12} = (U_{11})^{-1} A_{12} = (A_{11})^{-1} A_{12}. \tag{12}$$

Now, since A_{ij} is a 2×2 sub-matrix as in (3),

$$(A_{11})^{-1} = \frac{1}{D} \begin{bmatrix} a_{22} & -a_{12} \\ -a_{21} & a_{11} \end{bmatrix}, \tag{13}$$

with $D = a_{11} a_{22} - a_{12} a_{21}$, the determinant of A_{11}.

Therefore from (10) and (12):

$$D\begin{bmatrix} l_{31} & l_{32} \\ l_{41} & l_{42} \end{bmatrix} = -\begin{bmatrix} a_{22} & -a_{12} \\ -a_{21} & a_{11} \end{bmatrix}\begin{bmatrix} a_{31} & a_{32} \\ a_{41} & a_{42} \end{bmatrix},$$

$$\begin{bmatrix} u_{33} & u_{34} \\ u_{43} & u_{44} \end{bmatrix} = \begin{bmatrix} a_{33} & a_{34} \\ a_{43} & a_{44} \end{bmatrix} + D\begin{bmatrix} l_{31} & l_{32} \\ l_{41} & l_{42} \end{bmatrix}\frac{1}{D}\begin{bmatrix} a_{13} & a_{14} \\ a_{23} & a_{24} \end{bmatrix}, \tag{14}$$

$$\begin{bmatrix} u'_{13} & u'_{14} \\ u'_{23} & u'_{24} \end{bmatrix} = \begin{bmatrix} a_{22} & -a_{12} \\ -a_{21} & a_{11} \end{bmatrix}\frac{1}{D}\begin{bmatrix} a_{13} & a_{14} \\ a_{23} & a_{24} \end{bmatrix}.$$

As is shown in [31], the computation of U_{22} can be performed in 4 IPS cycles and at the fifth step U_{22} will be taken as pivot to begin the second elimination step, as in (6) and (7).

If we apply the RF method on (4) and (5) we proceed from the top and the bottom of matrix A simultaneously and obtain two LU decomposition streams functioning concurrently in opposite directions. The two streams confront each other in the centre of the matrix and the conflict is resolved by means of a double modification of the central 2×2 submatrix. The confrontation of the two streams takes the form of the solution of a subsystem as in (8). Two cases are considered, for k odd and k even; these two cases are summarised in Fig. 3 for the LU and LDU decomposition.

LU for $k = 5$ (odd):

$$\begin{bmatrix} A_{11} & A_{12} & & & \\ A_{21} & A_{22} & A_{23} & 0 & \\ & A_{32} & A_{33} & A_{34} & \\ 0 & & A_{43} & A_{44} & A_{45} \\ & & & A_{54} & A_{55} \end{bmatrix} = \begin{bmatrix} I & & & & \\ L_{21} & I & & 0 & \\ & L_{32} & I & L_{34} & \\ & 0 & & I & L_{45} \\ & & & & I \end{bmatrix}\begin{bmatrix} U_{11} & U_{12} & & & \\ & U_{22} & U_{23} & 0 & \\ & & U_{33} & & \\ & 0 & U_{43} & U_{44} & \\ & & & U_{54} & U_{55} \end{bmatrix}.$$

LU for $k = 6$ (even):

$$\begin{bmatrix} A_{11} & A_{12} & & & & \\ A_{21} & A_{22} & A_{23} & & 0 & \\ & A_{32} & A_{33} & A_{34} & & \\ & & A_{43} & A_{44} & A_{45} & \\ & 0 & & A_{54} & A_{55} & A_{56} \\ & & & & A_{65} & A_{66} \end{bmatrix} = \begin{bmatrix} I & & & & & \\ L_{21} & I & & & 0 & \\ & L_{32} & I & & & \\ & & L_{43} & I & L_{45} & \\ & 0 & & & I & L_{56} \\ & & & & & I \end{bmatrix}$$

$$\times \begin{bmatrix} U_{11} & U_{12} & & & & \\ & U_{22} & U_{23} & & 0 & \\ & & U_{33} & U_{34} & & \\ & 0 & & U_{44} & & \\ & & & U_{54} & U_{55} & \\ & & & & U_{65} & U_{66} \end{bmatrix}.$$

Fig. 3(a). Block (2×2) RF LU decomposition.

LDU for $k=5$ (odd):

$$
\begin{bmatrix} A_{11} & A_{12} & & & \\ A_{21} & A_{22} & A_{23} & 0 & \\ & A_{32} & A_{33} & A_{34} & \\ 0 & & A_{43} & A_{44} & A_{45} \\ & & & A_{54} & A_{55} \end{bmatrix} = \begin{bmatrix} I & & & & \\ L_{21} & I & & 0 & \\ & L_{32} & I & L_{34} & \\ & & & I & L_{45} \\ & 0 & & & I \end{bmatrix} \begin{bmatrix} U_{11} & & & & \\ & U_{22} & & 0 & \\ & & U_{33} & & \\ & & & U_{44} & \\ 0 & & & & U_{55} \end{bmatrix}
$$

$$
\times \begin{bmatrix} I & U'_{12} & & & \\ & I & U'_{23} & 0 & \\ & & I & & \\ & & U'_{43} & I & \\ 0 & & & U'_{54} & I \end{bmatrix}.
$$

LDU for $k=6$ (even):

$$
\begin{bmatrix} A_{11} & A_{12} & & & & \\ A_{21} & A_{22} & A_{23} & & 0 & \\ & A_{32} & A_{33} & A_{34} & & \\ & & A_{43} & A_{44} & A_{45} & \\ 0 & & & A_{54} & A_{55} & A_{56} \\ & & & & A_{65} & A_{66} \end{bmatrix} = \begin{bmatrix} I & & & & & \\ L_{21} & I & & 0 & & \\ & L_{32} & I & & & \\ & & L_{43} & I & L_{45} & \\ & 0 & & & I & L_{56} \\ & & & & & I \end{bmatrix}
$$

$$
\times \begin{bmatrix} U_{11} & & & & & \\ & U_{22} & & 0 & & \\ & & U_{33} & & & \\ & & & U_{44} & & \\ & 0 & & & U_{55} & \\ & & & & & U_{66} \end{bmatrix}
$$

$$
\times \begin{bmatrix} I & U'_{12} & & & & \\ & I & U'_{23} & & 0 & \\ & & I & U'_{34} & & \\ & & & I & & \\ & 0 & & U'_{54} & I & \\ & & & & U'_{65} & I \end{bmatrix}.
$$

Fig. 3(b). Block (2×2) RF LDU decomposition.

Now, the central submatrix has the form

$$
A = \begin{bmatrix} U_{m,m} & U_{m,m+1} \\ U_{m+1,m} & U_{m+1,m+1} \end{bmatrix}, \quad m = \begin{cases} \dfrac{k+1}{2}, & k = \text{odd} \\ \dfrac{k}{2}, & k = \text{even} \end{cases} \tag{15}
$$

which is reduced, for $k=$ odd, to

$$\begin{bmatrix} I & L_{m,m+1} \\ 0 & I \end{bmatrix} \begin{bmatrix} U^{+}_{m,m} & 0 \\ U_{m+1,m} & U_{m+1,m+1} \end{bmatrix}, \tag{16}$$

or for $k=$ even, to

$$\begin{bmatrix} I & 0 \\ L_{m+1,m} & I \end{bmatrix} \begin{bmatrix} U_{m,m} & U_{m,m+1} \\ 0 & U^{+}_{m+1,m+1} \end{bmatrix}. \tag{17}$$

In the case of LDU decomposition, (15) is reduced, for $k=$ odd, to

$$\begin{bmatrix} I & L_{m,m+1} \\ 0 & I \end{bmatrix} \begin{bmatrix} U^{+}_{m,m} & 0 \\ 0 & U_{m+1,m+1} \end{bmatrix} \begin{bmatrix} I & 0 \\ U'_{m+1,m} & I \end{bmatrix} \tag{18}$$

or, for $k=$ even, to

$$\begin{bmatrix} I & 0 \\ L_{m+1,m} & I \end{bmatrix} \begin{bmatrix} U_{m,m} & 0 \\ 0 & U^{+}_{m+1,m+1} \end{bmatrix} \begin{bmatrix} I & U'_{m,m+1} \\ 0 & I \end{bmatrix}. \tag{19}$$

The symbol + indicates a double modification; thus, the conflict is resolved by an additional LU or LDU decomposition step.

The solution of the resulting triangular systems in (4), (5) can be effected by means of the same technique, i.e. block RF forward and backward substitution. The matrices have the form given in Fig. 3. The discussion is concentrated firstly on a lower triangular system of the general form

$$\begin{bmatrix} A_{11} & & & & \\ A_{21} & A_{22} & & 0 & \\ & A_{32} & A_{33} & A_{34} & \\ & & & A_{44} & A_{54} \\ & 0 & & & A_{55} \end{bmatrix} \begin{bmatrix} \mathbf{x}_1 \\ \mathbf{x}_2 \\ \mathbf{x}_3 \\ \mathbf{x}_4 \\ \mathbf{x}_5 \end{bmatrix} = \begin{bmatrix} \mathbf{b}_1 \\ \mathbf{b}_2 \\ \mathbf{b}_3 \\ \mathbf{b}_4 \\ \mathbf{b}_5 \end{bmatrix} \tag{20}$$

The solution of the system can commence from both ends and proceed concurrently towards the centre. Again, the confrontation of the two streams is resolved by means of a double modification of the central 2×1 subvector of $\mathbf{x}$, here $\mathbf{x}_3$; in general $\mathbf{x}_m$ with $m=(k+1)/2$ for k odd and $m=(k/2)+1$ for k even. Thus

$$\mathbf{x}_m = (A_{mm})^{-1}(\mathbf{b}_m - A_{m,m-1}\mathbf{x}_{m-1} - A_{m,m+1}\mathbf{x}_{m+1}) \tag{21}$$

In the case of LDU decomposition, the lower triangular system has the form

$$\begin{bmatrix} I & & & & \\ A_{21} & I & & 0 & \\ & A_{32} & I & A_{34} & \\ & & & I & A_{54} \\ & 0 & & & I \end{bmatrix} \begin{bmatrix} \mathbf{x}_1 \\ \mathbf{x}_2 \\ \mathbf{x}_3 \\ \mathbf{x}_4 \\ \mathbf{x}_5 \end{bmatrix} = \begin{bmatrix} \mathbf{b}_1 \\ \mathbf{b}_2 \\ \mathbf{b}_3 \\ \mathbf{b}_4 \\ \mathbf{b}_5 \end{bmatrix} \tag{22}$$

and a diagonal system must also be solved.

3. Systolic implementation of block RF LU decomposition

The data sequence format necessary for the block (2 × 2) RF LU decomposition can be derived using the preprocessor illustrated in Fig. 4. The input of the array consists of the two streams of the LU decomposition overlapped: while in Fig. 2, there should be a dummy element between each two successive data items, here the dummy elements are replaced by the entries of the second LU decomposition stream.

Each one of the w demultiplexers produces the data item it accepts either on the left-hand or on the right-hand side output, according to the control signal C_1:

if
$C_1 = 1$
output to the right
true
output to the left.

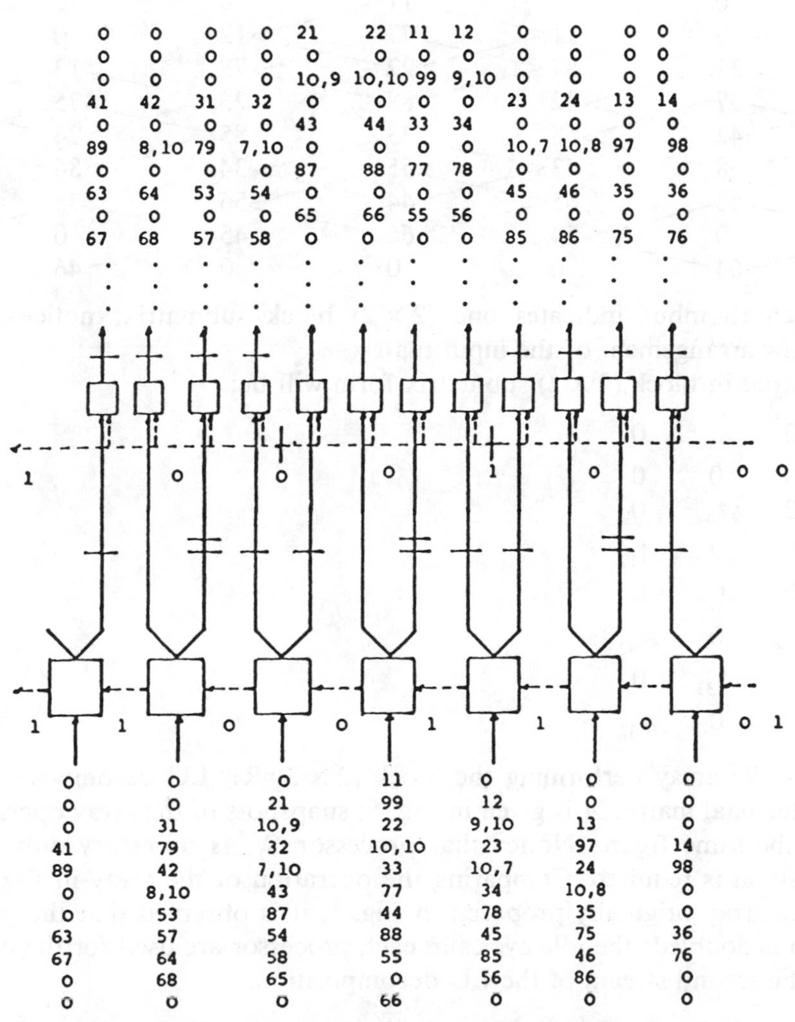

Fig. 4. Preprocessor array and input-output format for $k = 5$.

While in Fig. 2, C_1 has the form 1000, C_1 is 1100 for the RF method. The reformatting delays align all the entries of a block (2×2) submatrix in one line. In order to achieve one dummy cycle between successive submatrices a second control signal, C_2 is used. This signal is broadcast in a group of four delay cells and operates as follows:

if

$C_2 = 1$

delay output for one cycle; overwrite next output

true

no delay or overwrite occurs.

In order for the output of the preprocessing array to be the input of the main computational array, an additional delay is required for the off-diagonal blocks.

The case of $k = 5$ (odd), i.e. $n = 10$ is shown in Fig. 3 and this example will be used hereafter. For $k =$ even the results are similar; for example, the input format for the preprocessing array for $k = 4$, i.e. $n = 8$, is

0	0	0	11	0	0	0
0	0	21	77	12	0	0
0	31	87	22	78	13	0
41	57	32	88	23	75	14
67	42	58	33	85	24	76
0	68	43	55	34	86	0
0	53	65	44	56	35	0
63	0	54	66	45	0	36
0	64	0	0	0	46	0

where each rhombus indicates one (2×2) block submatrix; notice also the column-row arrangement of the input matrix.

The output in block (2×2) submatrix form will be:

0	A_{11}	0
0	0	0
0	A_{44}	0
A_{21}	0	A_{12}
0	A_{22}	0
A_{34}	0	A_{43}
0	A_{33}	0
A_{23}	0	A_{32}

The systolic array performing the block (2×2) RF LU decomposition for a block tridiagonal matrix A is given in Fig. 5; snapshots of the array operation are given in the same figure. Notice that processor S_{13} is necessary only if LDU decomposition is required. Comparing the operation of the array in Fig. 5, with that of the array originally proposed in Fig. 1, it is observed that the processor utilisation is doubled: the idle cycles in each processor are used for the computations of the second stream of the LU decomposition.

The only difference in the processor specifications of the two arrays is that in order for the double modification of the central (2×2) block submatrix to be achieved, this submatrix has to be kept in processor S_{31} for two additional IPS cycles (see Fig. 5). The central block (2×2) submatrix enters the array after $2k - 1 = n - 1$ IPS cycles; it is kept in processor S_{31} for 4 cycles and needs 3 more cycles to reach the output. Thus, the LU decomposition of a block (2×2) tridiagonal matrix ($w = 7$, $p = q = 4$) is computed on a systolic array having no more than pq processors in $n + \min(p, q) + 2$ IPS cycles.

A systolic array for the solution of the triangular systems using block RF methods is given in Fig. 6. The processors used are similar to those of the LU decomposition array; some snapshots of the computation are also given in the same figure. The inversion of a block (2×2) submatrix is achieved by processors S_{22} and S_{23}. Processors SD and MA calculate

$$\mathbf{x} = A^{-1}(\mathbf{b} - \mathbf{y}) \text{ and } \mathbf{y} = \mathbf{y} + A\mathbf{x} \tag{23}$$

respectively, where A is a block (2×2) submatrix and $\mathbf{x}$, $\mathbf{y}$, $\mathbf{b}$ are 2×1 subvectors. As is obvious from their specifications in Fig. 6, they can be regarded as "half" of processor S_{31} in terms of area requirements, while the computation retains the same format. However the calculation of SD requires a time unit equal to 1 IPS + 1 ADD to be completed, a fact that imposes a longer cycle for the array.

The confrontation of the two streams is resolved by keeping the middle 2×1 subvector for two additional cycles in processor SD and the corresponding inverse submatrix is delayed accordingly. Thus the solution of the block (2×2) triangular system in its general form (20) requires an array of no more than 8 processors and a computation time of $n + 5$ cycles.

Soft-systolic simulation programs in OCCAM, for the designs presented in this paper, are given in [28].

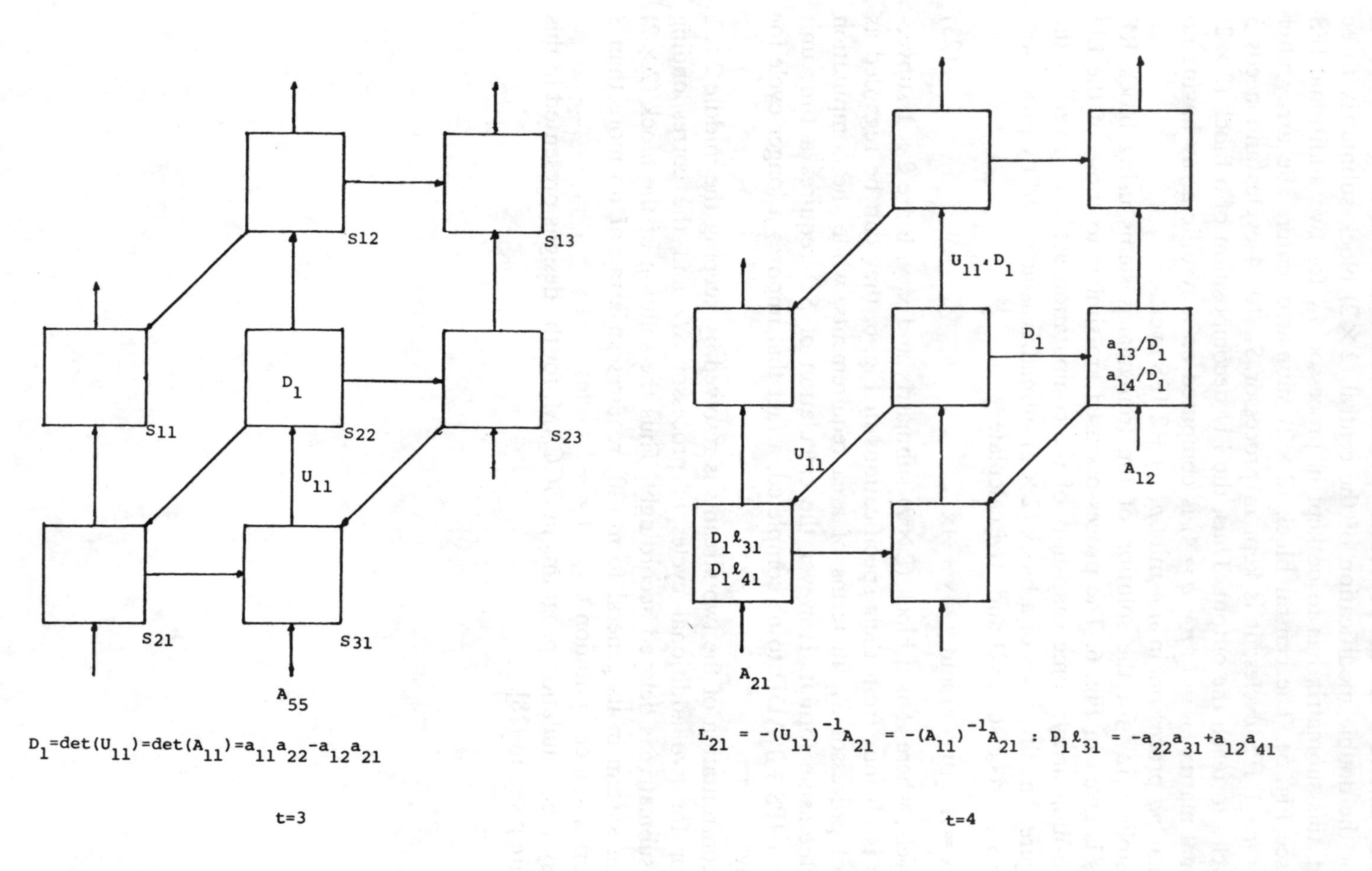

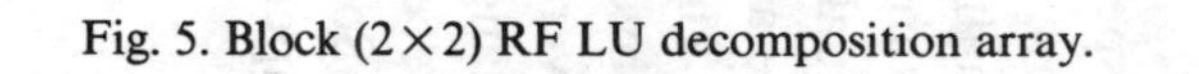
Fig. 5. Block (2×2) RF LU decomposition array.

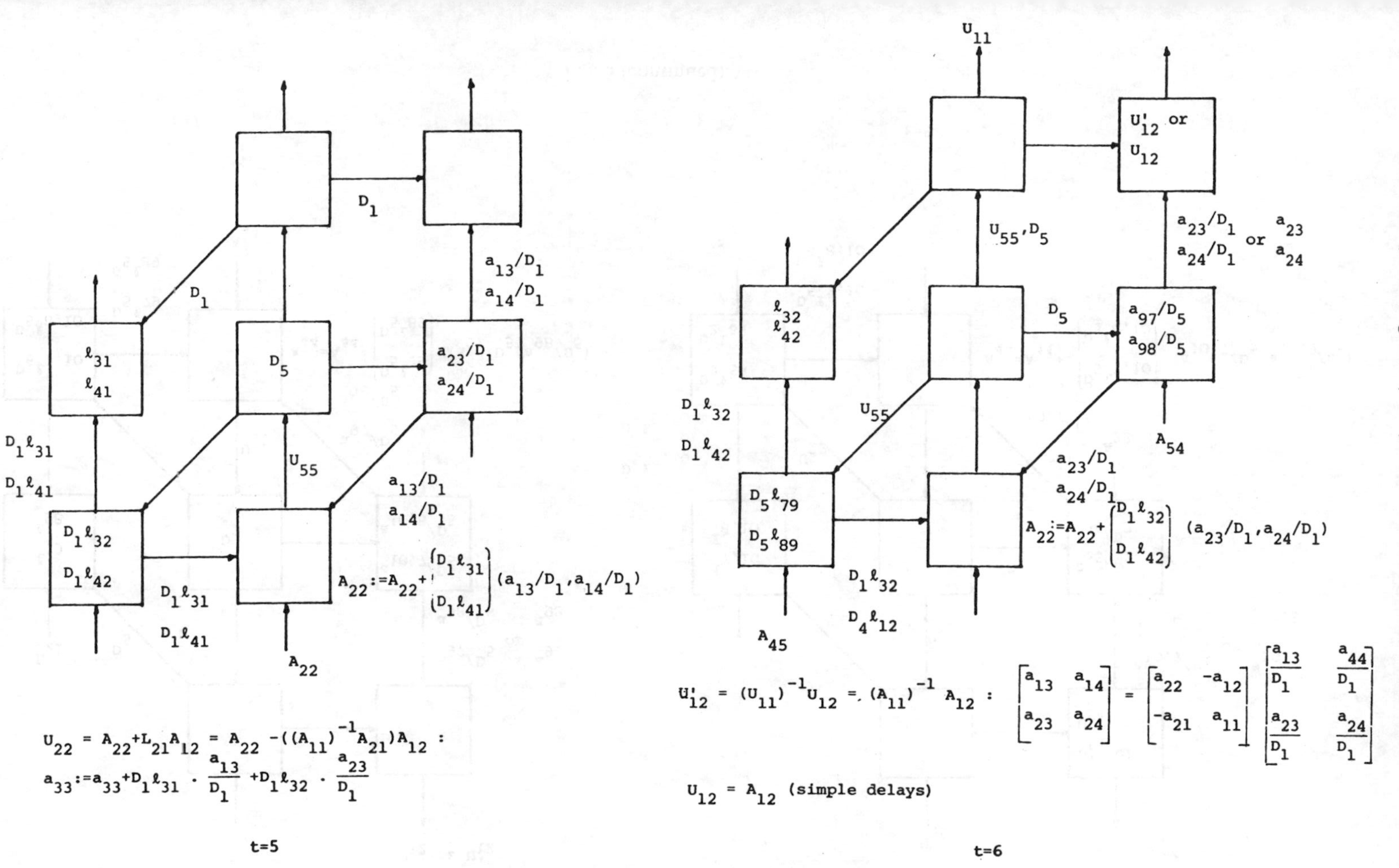

Fig. 5 (continued).

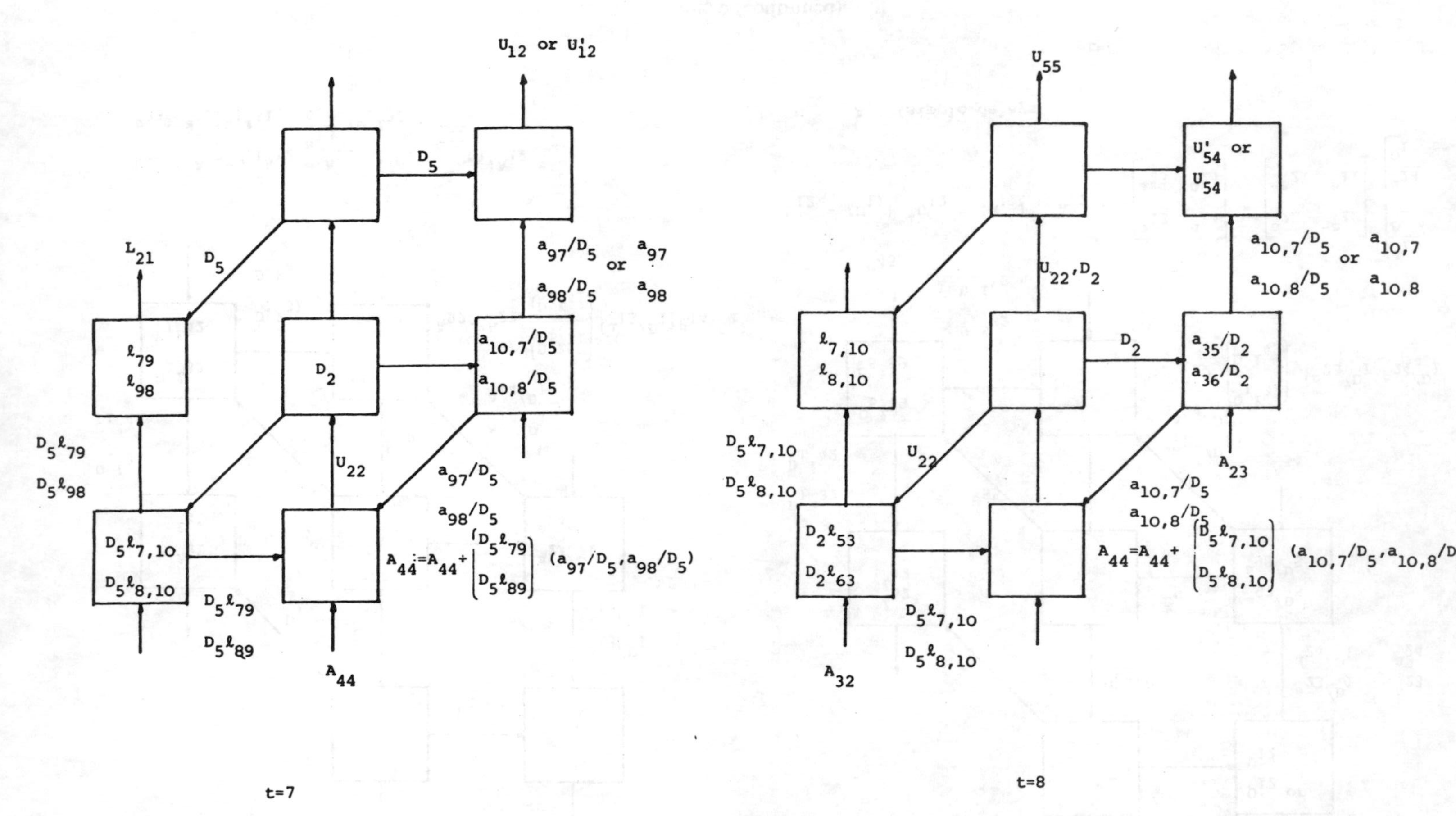

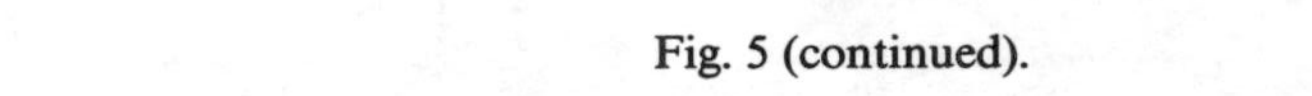
Fig. 5 (continued).

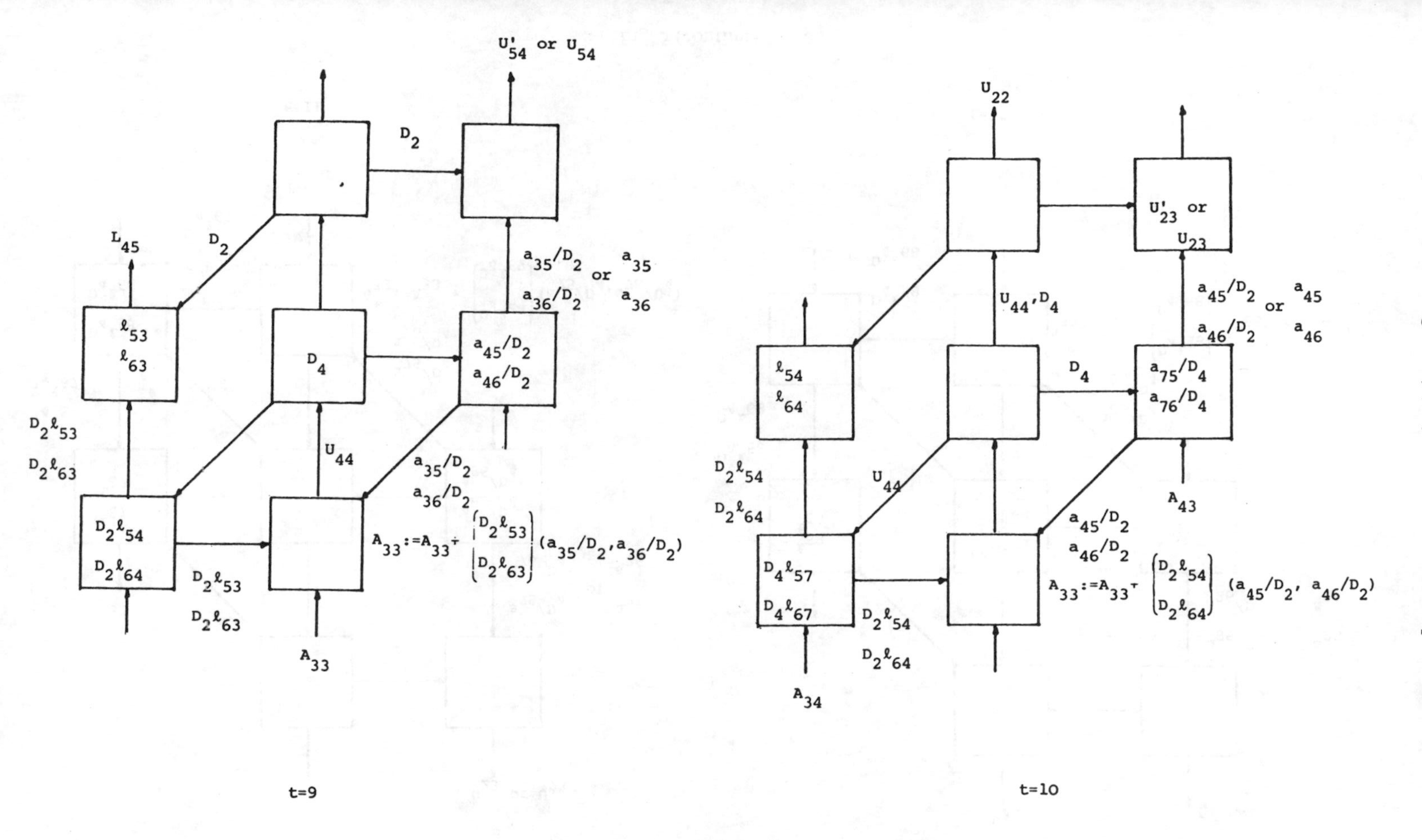

Fig. 5 (continued).

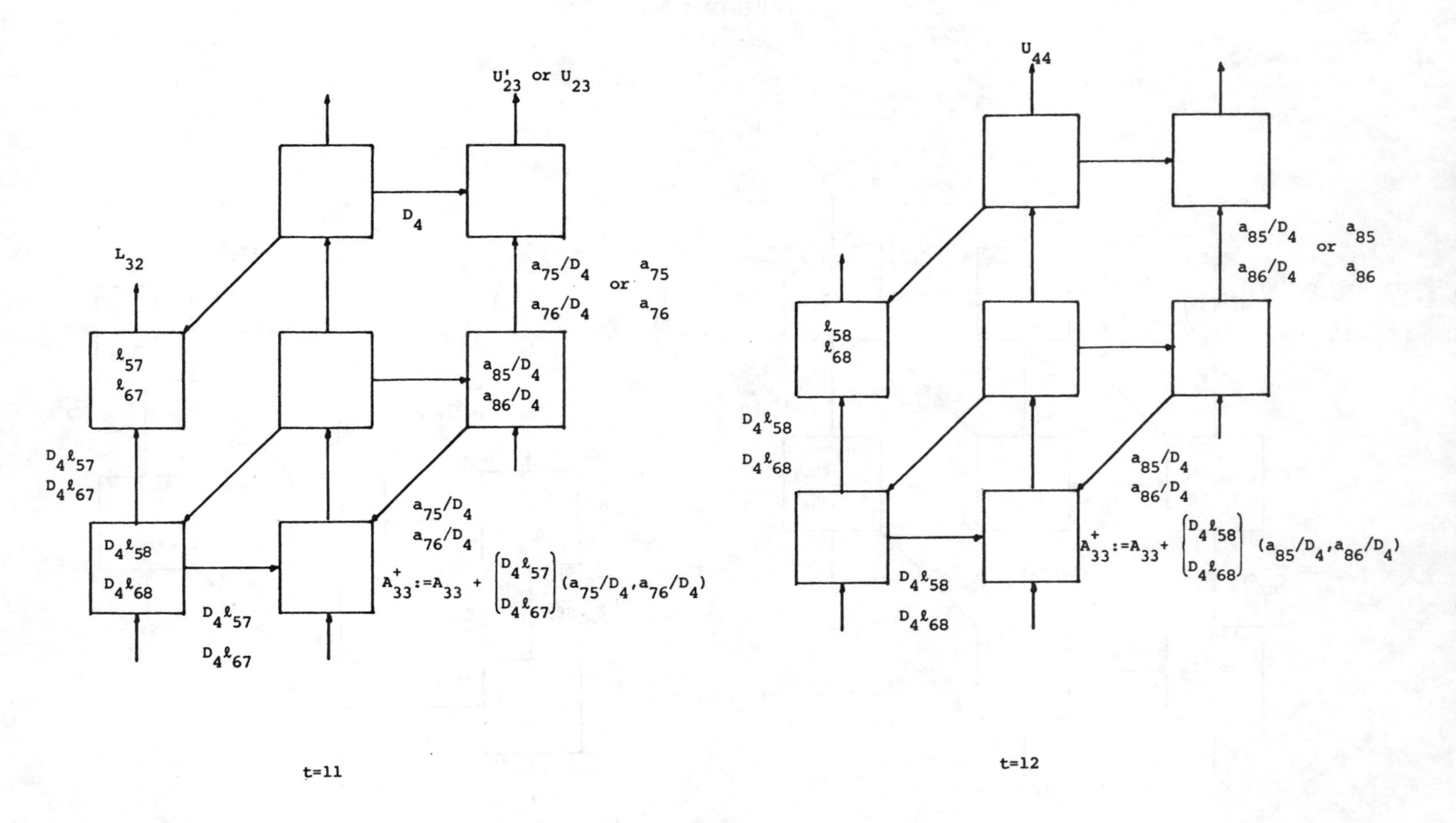

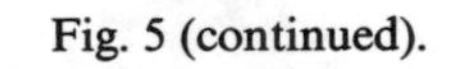

Fig. 5 (continued).

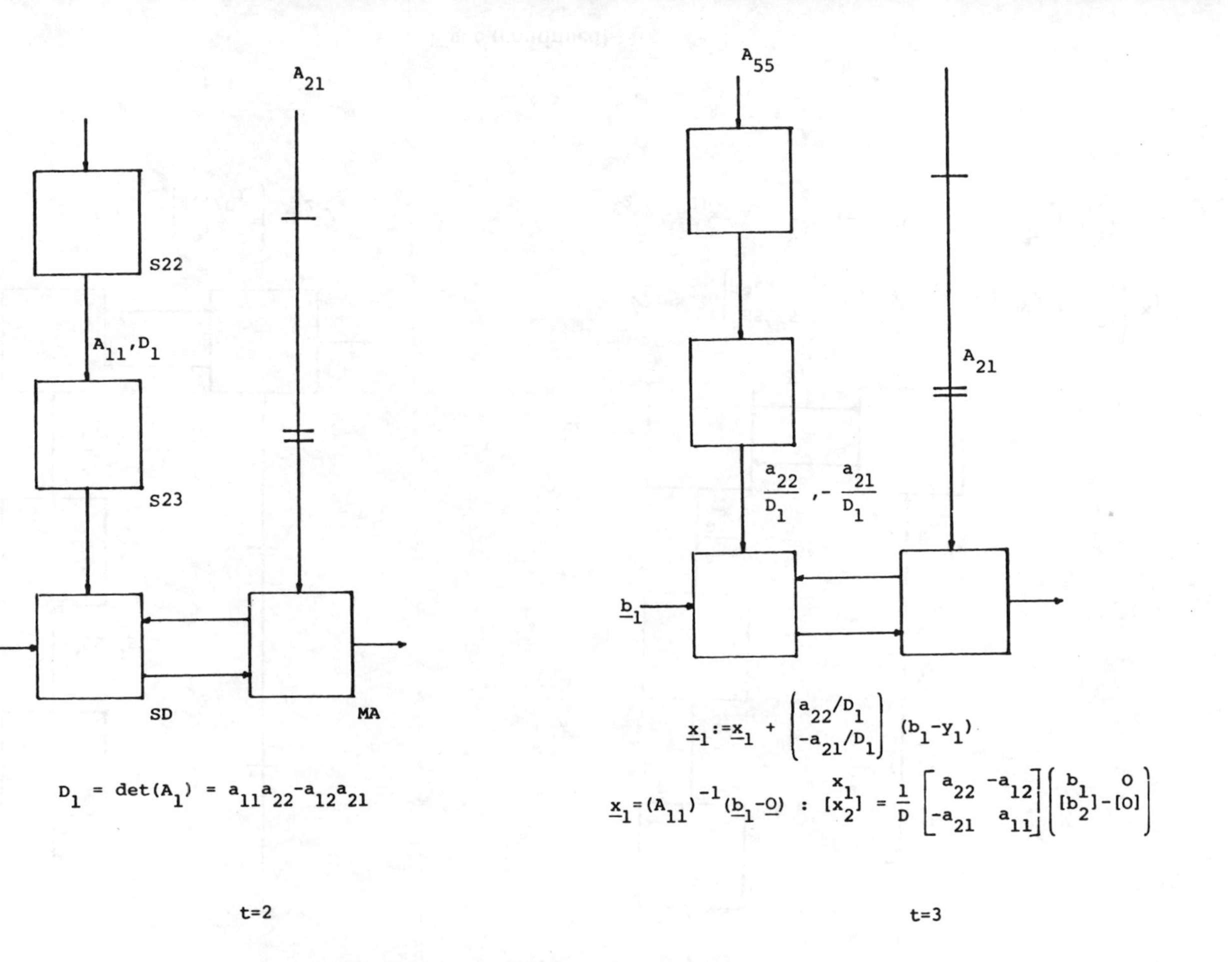

Fig. 6. Block (2×2) RF triangular system solution.

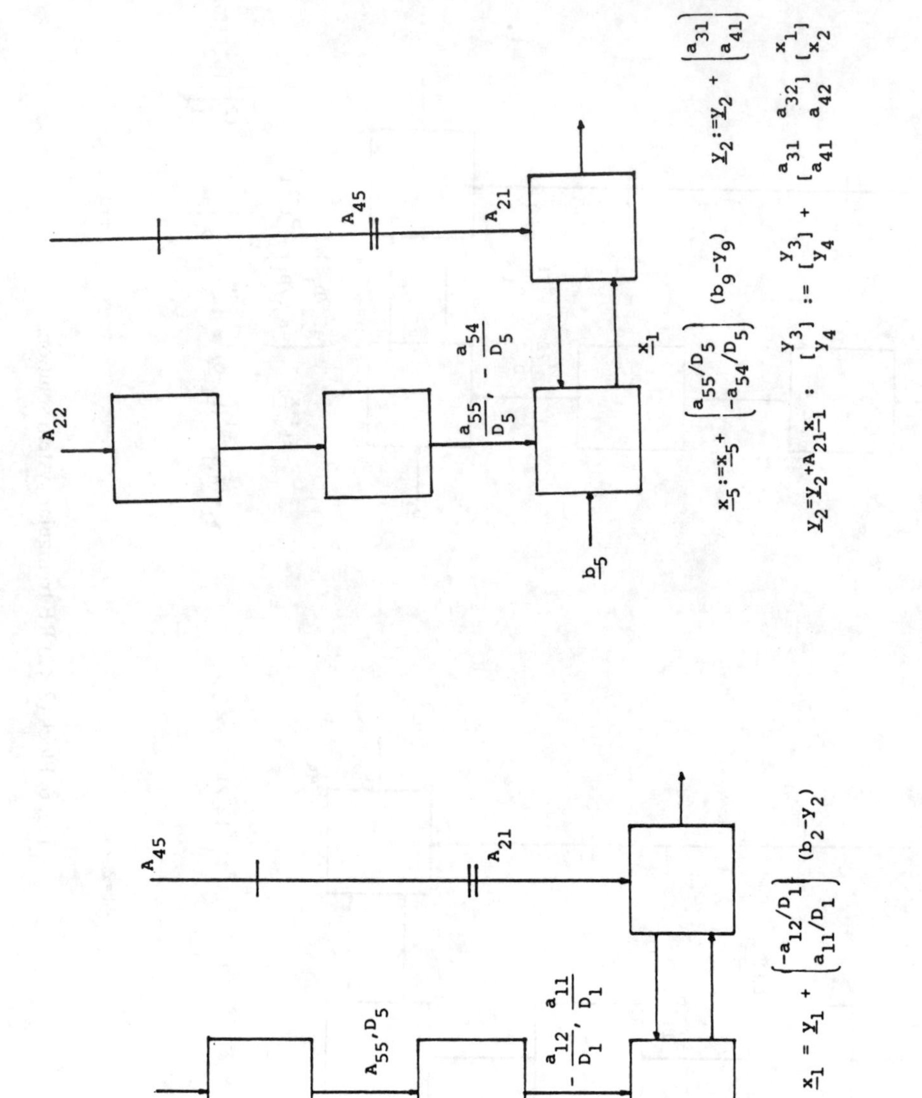

Fig. 6 (continued).

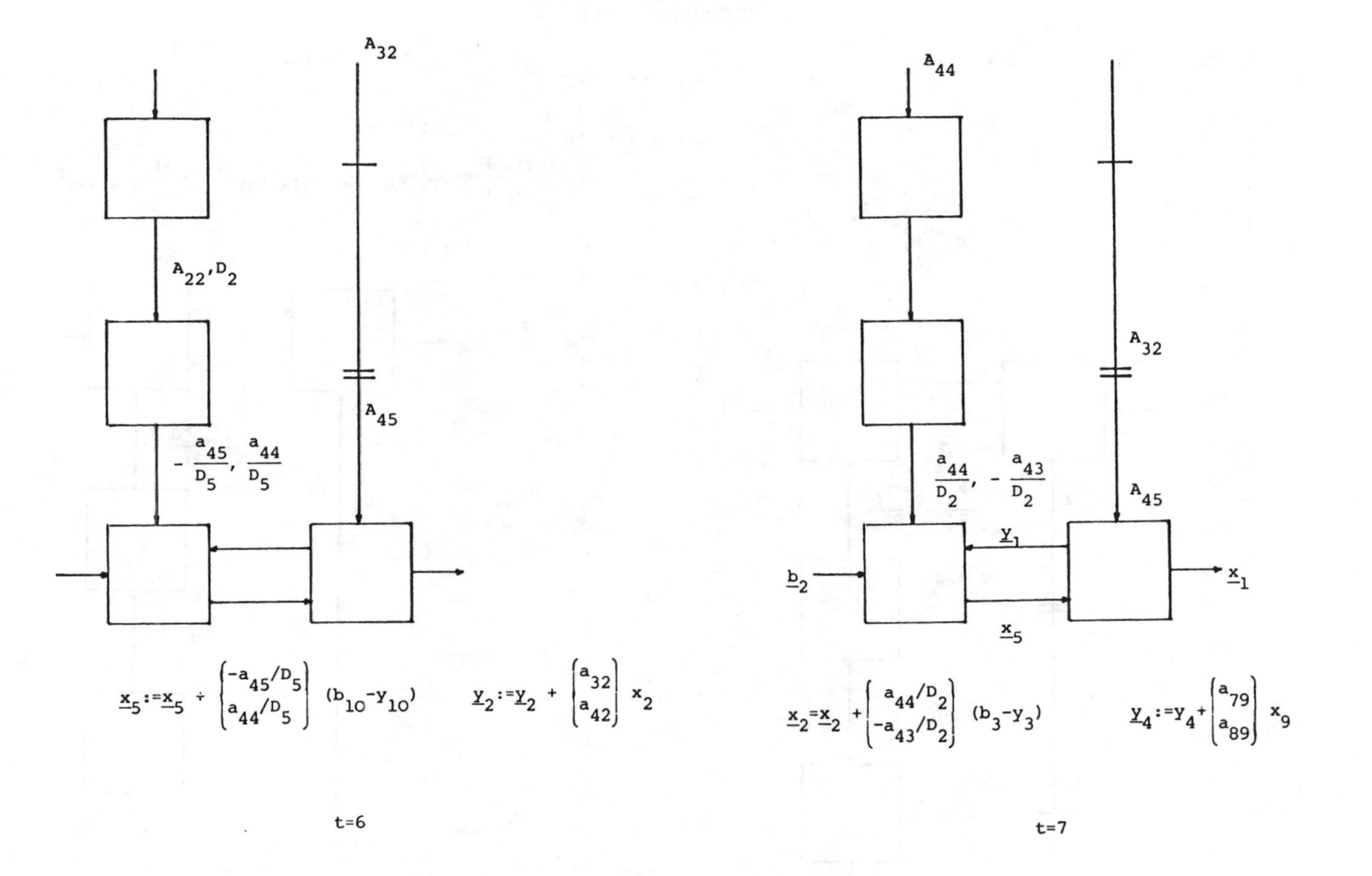

Fig. 6 (continued).

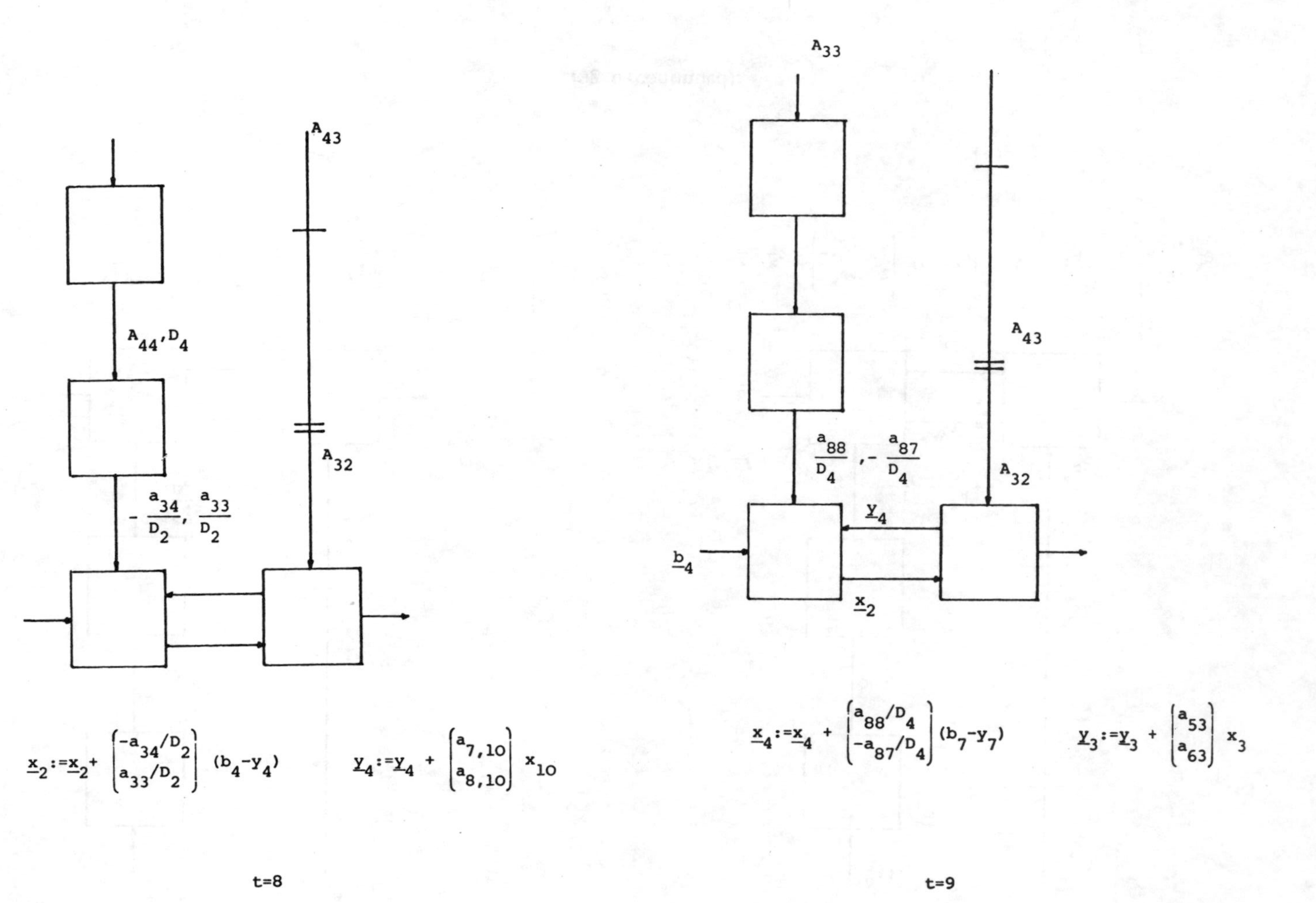

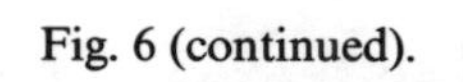
Fig. 6 (continued).

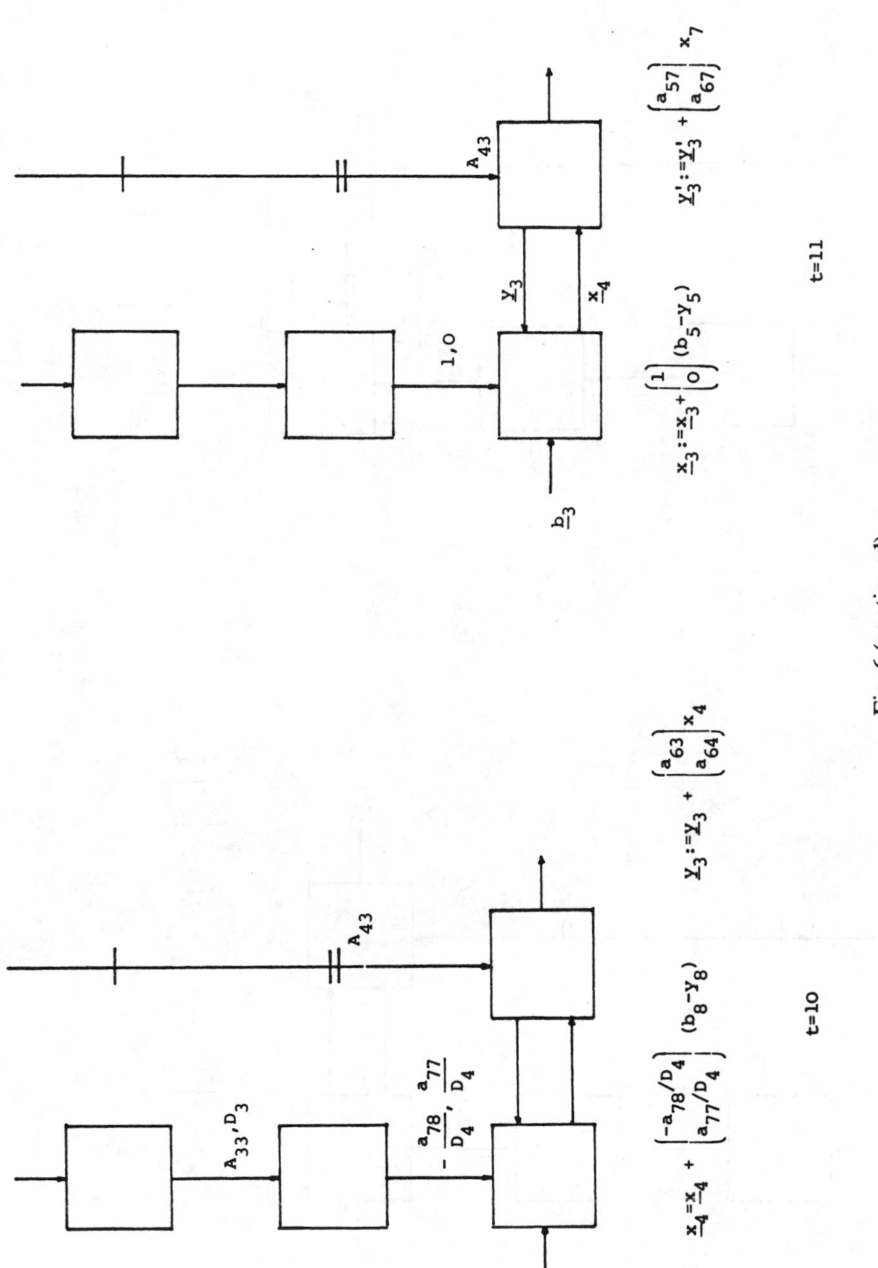

Fig. 6 (continued).

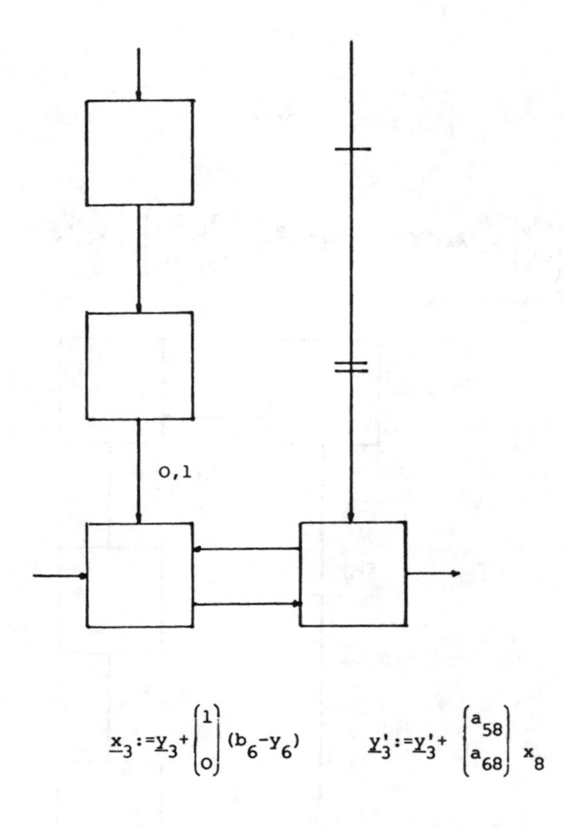

o,1
$\underline{x}_3:=\underline{y}_3+\begin{pmatrix}1\\o\end{pmatrix}(b_6-y_6)$
$\underline{y}_3':=\underline{y}_3'+\begin{pmatrix}a_{58}\\a_{68}\end{pmatrix}x_8$
t=12

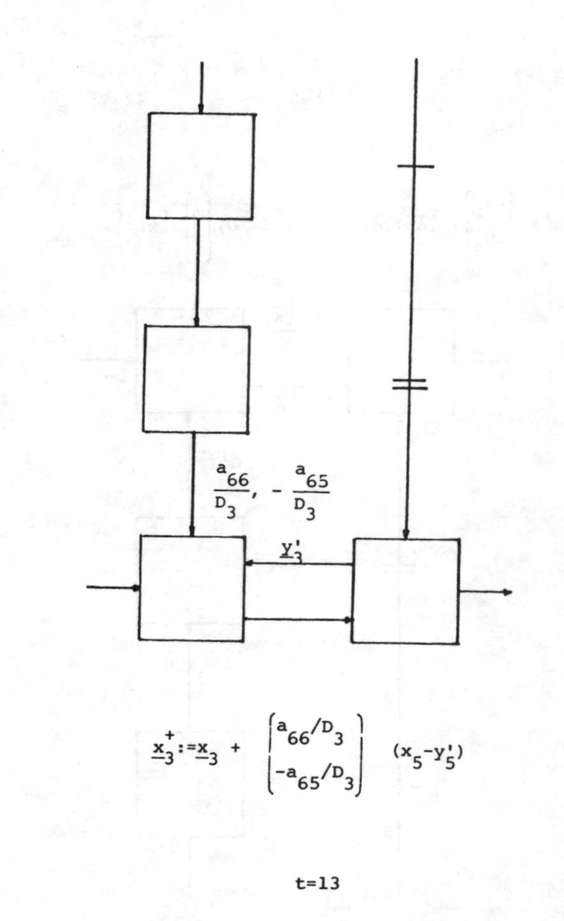

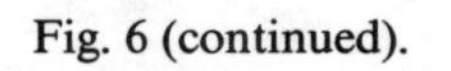
Fig. 6 (continued).

4. Conclusions

The block (2×2) RF method when extended to matrices with bandwidth greater than 7, i.e. block quindiagonal etc., imposes a more complicated resolution of the confrontation of the two streams, as explained in [6]. A measure of efficiency for the method can be the ratio r of the order of the matrix over its bandwidth: it is obvious that the method is efficient for $r \gg 1$, so that the solution of the central subsystems can be regarded as negligible.

The RF method can be applied in the LU method with neighbour pivoting in order to double the efficiency of the array; a similar increase in efficiency can be achieved for the triangular system solution. Furthermore, it would be interesting to investigate the application of the RF method on the QR decomposition arrays with or without pivoting.

References

[1] Abdel Kader, A.A., OCSAMO: A systolic array for matrix operations, in: *Proc. CONPAR 86* (Springer-Verlag, Berlin, 1986) 319–328.

[2] Ahmed, H.M., J. Delosme and M. Morf, Highly concurrent computing structures for matrix arithmetic and signal processing, *IEEE Comput.* **15** (1982) 65–82.

[3] Athale, R.A. and J.N. Lee, Optical processing using outer-product concepts, *Proc. IEEE* **72** (1984) 931–941.

[4] Avila, J.H. and P.J. Kuekes, A one gigaflop VLSI systolic processor, *Proc. SPIE, RTSP VI* **441** (1983) 159–165.

[5] Annaratone, M., et al., Extending the CMU warp machine with a boundary processor, *Proc. SPIE, RTSP VIII* **564** (1985) 56–65.

[6] Bekakos, M.P., A study of algorithms for parallel computers and VLSI systolic processor arrays, Ph.D. Thesis, Dept. of Computer Studies, LUT, 1986.

[7] Bocker, R.P., Algebraic operations performable with electrooptical engagement array processors, *Proc. SPIE*, Optical Information Processing, **388** (1983) 212–220.

[8] Casasent, D., Acoustooptic linear algebra processors; Architectures, algorithms and applications, *Proc. IEEE* **72** (1984) 831–849.

[9] Casasent, D., Ghosh, A., Optical linear algebra, *Proc. SPIE*, Optical Information Processing, **388** (1983) 182–189.

[10] Casasent, D., A. Ghosh and C.P. Neuman, Direct and indirect optical solutions to linear algebraic equations: Error source modeling", *Proc. SPIE, RTSP VI* **441** (1983) 201–208.

[11] Cosnard, M., Y. Robert and D. Trystram, Parallel solution of dense linear systems using diagonalization methods, *Int. J. Comput. Math.* (1988), submitted for publication.

[12] Evans, D.J. and M.P. Bekakos, On the implementation of acousto-optic cells for a 'Rotating' and 'Folding' algorithm for systolization, *Int. J. Comput. Math.* **20** (1986) 123–129.

[13] Evans, D.J., M.P. Bekakos and K.G. Margaritis, Optical 'Dequeues' for an 'R and F' systolic LU-factorization of tridiagonal systems, *Proc. 4th Int. Symp. Optic. Optoelectron. Applied Sci. Eng.*, Hague, (1987).

[14] Evans, D.J., K.G. Margaritis and M.P. Bekakos, On acousto-optic cell planes to map an RF algorithm using a 2-D Systolic geometry, *Optics Commun.* **63** (3) (1987) 147–152.

[15] Gentleman, W.M. and H.T. Kung, Matrix triangularization by systolic arrays, *Proc. SPIE, RTSP IV* **298** (1981) 19–26.

[16] Heller, D., Partitioning big matrices for small systolic arrays, in: T. Kailath et al., eds., *VLSI and Modern Signal Processing* (Prentice-Hall, Englewood Cliffs, NJ, 1985) pp. 185–199.

[17] Heller, D.E. and I.C.F. Ipsen, Systolic networks for orthogonal equivalence transformations and their applications, *Proc. Conf. Advanced Research in VLSI*, M.I.T., (1982) pp. 113–122.
[18] Huang, K.H. and J.A. Abraham, Algorithm based fault tolerance for matrix operations, *IEEE Trans. Comput.* **C-33** (1984) 518–528.
[19] Hwang, K. and F.A. Briggs, *Computer Architecture and Parallel Processing* (McGraw-Hill, New York, 1984).
[20] Hwang, K. and Y.H. Cheng, Partitioned matrix algorithms for VLSI arithmetic systems, *IEEE Trans. Comput.* **C-31** (1982) 1215–1224.
[21] Ipsen, I.C.F., Stable matrix computations in VLSI, Ph.D. Thesis, Pennsylvania State University, 1983.
[22] Jou, J.Y. and J.A. Abraham, Fault-tolerant matrix arithmetic and signal processing on highly concurrent computing structures, *Proc. IEEE* **74** (1986) 732–741.
[23] Kung, S.Y., On supercomputing with systolic/wavefront array processors, *Proc. IEEE* **72** (1984) 867–884.
[24] Kung, S.Y., et al., Wavefront array processor: Language, architecture and applications, *IEEE Trans. Comput.* **C-31** (1982) 1054–1066.
[25] Leiserson, C.E., Area efficient VLSI computation, Ph.D. Thesis, Dept. of Computer Science, CMU, 1981.
[26] Luk, F.T., Algorithm-based fault-tolerance for parallel matrix equation solvers, *Proc. SPIE, RTSP VIII* **564** (1985) 49–53.
[27] Luk, F.T. and H. Park, An analysis of algorithm-based fault-tolerance techniques, Technical Report, EE-CEG-86-11, School of Electrical Engineering, Cornell University, 1986.
[28] Margaritis, K.G., A study of systolic algorithms or VLSI processor arrays and optical computing, Ph.D. Thesis, Loughborough University of Technology, 1987.
[29] Mead, C.A. and L. Conway, *Introduction to VLSI Systems* (Addison Wesley, Reading, MA, 1980).
[30] Moldovan, D.I., On the analysis and synthesis of VLSI algorithms, *IEEE Trans. Comput.* **C-31** (1982) 1121–1126.
[31] Robert Y., Block LU Decomposition of a Band Matrix on a Systolic Array, *Int. J. Comput. Math.* **17** (1985) 295–315.
[32] Schreiber, R., On systolic array methods for band matrix factorizations, *BIT* **26** (1986) 303–316.
[33] Schreiber, R. and P.J. Kuekes, Systolic linear algebra machines in digital signal processing, in: T. Kailath et al., eds., *VLSI and Modern Signal Processing* (Prentice–Hall, Englewood Cliffs, NJ, 1985) pp. 389–405.
[34] Sorensen, D.C., Analysis of pairwise pivoting in Gaussian Elimination, *IEEE Trans. Comput.* **C-34** (1985) 274–278.
[35] Young, T.Y. and P.S. Liu, VLSI arrays for pattern recognition and image processing: I/O bandwidth considerations, in: K.S. Fu, ed., *VLSI for Pattern Recognition and Image Processing* (Springer-Verlag, Berlin, 1984) pp. 25–42.

Matrix Inversion by Systolic Rank Annihilation

D. J. EVANS and G. M. MEGSON
Department of Computer Studies, Loughborough University of Technology, Loughborough, Leicestershire, U.K.

The systolic principle is applied to the inversion of matrices by the methods of rank annihilation. The systolic arrays presented are particularly effective for computing the inverse of a matrix which differs only partially from a matrix with a known inverse. It is shown that the RANK-1 and RANK-2 annihilation schemes compete favourably with existing systolic schemes for arbitrary matrix inversion, by trading basic inner product cells for simple delay registers which consume less area. For Rank-1 annihilation we show that the computation time is $T=(3n-1)+4(n+2)r$, and for Rank-2, $T=(3n)+4(n+1)r$ where n is the order of the matrix and r the number of applications of the annihilation formula.

The technique generalises to arbitrary matrix inversion resulting in $O(n^2)$ computational schemes but unfortunately they do not compete with the systolic Gaussian Elimination methods which are of $O(n)$. However they provide a more general architecture applicable to non-linear problems where partial changes in matrices can easily be constructed with a relatively small amount of hardware.

KEY WORDS: Systolic control ring (SCR), systolic rank annihilation, OCCAM program simulation.

C.R. CATEGORIES:

1. INTRODUCTION

This paper considers the use of systolic array and wavefront processor techniques for the implementation of rank annihilation matrix techniques. In particular, it is shown that the most widely annihilation formulae for Rank-1 and Rank-2 modifications are best suited to the systolic approach of "on-the-fly" computation.

The rank annihilation scheme itself is important in a large number of application fields, some of which are:

1) STATISTICS: for updating correlation matrices

2) GRAPHICS: in spline approximation and refinement

3) NUMERICAL ANALYSIS: in the repeated solution of problems in which only minor modifications to boundary conditions are made. Also in Linear Algebra for the inversion of arbitrary matrices.

4) OPTIMISATION: for solving non-linear systems of equations, and updating the Jacobian matrix.

5) Any Linear Algebra problem where the number of matrix elements changed from one instance of the problem to another is small.

Recently, systolic arrays have been investigated [8], using hexagonal and orthogonal processor grids requiring $O(7n)$ time (where a time unit is a single ips) and $O(n^2)$ cells arbitrary matrix inversion. To date there appears to be no matrix inversion arrays which make use of the sparsity of a matrix or those which are sensitive to only small changes in the matrix, especially in an iterative type process. The rank-annihilation techniques are aimed at local changes in the coefficient matrix which can cause global changes in the resulting inverse. However, the annihilation concept seems ideally suited to the systolic approach.

In this paper wavefront processor algorithms are developed for Rank-1 and Rank-2 modifications which indicate the low processor efficiency involved thus forming a basis for the truly systolic approach. The systolic array algorithms are also for Rank-1 and Rank-2 modifications but use only a linear number of cells with the result that processor efficiency is dramatically improved for the device allowing high throughput rates.

2. RANK ANNIHILATION

The technique of rank annihilation employs the principle that given a matrix A (say), whose inverse is known then the inverse of a related matrix B whose elements are only partially different from A, can be found by a simple relationship between A^{-1} and B^{-1}.

As a simple introduction to the strategy suppose A and B are $n*n$ matrices, and u and v are n component vectors. If B differs from A only by changes in elements along a single row or a column, we can write B as

$$B = A + uv^T. \tag{1}$$

Then, it follows that,

$$B^{-1} = (A + uv^T)^{-1}, \tag{2}$$

which after some manipulation yields the Sherman–Morrison formula [4]

$$(A + uv^T)^{-1} = A^{-1} - \frac{(A^{-1}u)(v^T A^{-1})}{1 + v^T A^{-1} u}, \tag{3}$$

which produces the inverse of a matrix which differs from A by a rank of unity, or Rank-1.

The basic idea can be extended to produce a relationship between the inverses when B differs from A by changes in m rows or columns, leading to the Sherman–Morrison–Woodbury Formula or Rank-m method of the form,

$$(A + UV^T)^{-1} = A^{-1} - A^{-1}U(I + V^T A^{-1} U)^{-1} V^T A^{-1}. \tag{4}$$

and clearly when $m=1$ reduces to the Eq. (3). U and V are $n*m$ matrices producing vectors when $m=1$. As far as systolic implementation is concerned, these formulas have the advantage that they are relatively simple, as long as m is small. We shall actively pursue the $m=1$ and $m=2$ cases which are the most popular rank annihilation schemes of this kind. It should be noted that this controls the complexity of a systolic design as for $m=1$, $(I + V^T A^{-1} U)$ is a scalar and when $m=2$, it is a $2*2$ matrix; which can both be inverted easily.

Given the basic rank annihilation schemes the inverse of an arbitrary matrix can be found by setting $A = I$ and applying the

Rank-1 formula n times or the Rank-2 formula $\frac{1}{2}n$ times. Generally, the Rank-m formula can be applied n/m times, where it is assumed that n is divisible by m. For convenience in describing the systolic schemes, Eqs. (3) and (4) are partitioned in an obvious and appropriate way,

for Rank-1:

$$B^{-1}=A^{-1}-\frac{1}{(1+Z)}p, \tag{5a}$$

where

$$P=xy^T, \tag{5b}$$

$$A^{-1}u=x, \tag{5c}$$

$$v^TA^{-1}=y^T \tag{5d}$$

and

$$Z=v^Tx=\sum_{i=1}^{n} v_ix_i, \tag{5e}$$

where x, y are n component vectors, P is an $n*n$ matrix and z a scalar.

For Rank-2:

$$B^{-1}=A^{-1}-Py^T \quad \text{or} \quad B^{-1}=A^{-1}-xP^T, \tag{6a}$$

where

$$P=xC^{-1}, \quad \text{or } P^T=C^{-1}y^T, \tag{6b}$$

$$A^{-1}U=x, \tag{6c}$$

$$V^TA^{-1}=y^T, \tag{6d}$$

$$C^{-1}=(I+V^Tx)^{-1} \tag{6e}$$

with

$$C=\begin{bmatrix} a & b \\ c & d \end{bmatrix}$$

$$a=1+\sum_{i=1}^{n} v_{1i}x_{i1}, \tag{7a}$$

$$b=\sum_{i=1}^{n} c_{1i}x_{i2}, \tag{7b}$$

$$c=\sum_{i=1}^{n} v_{2i}x_{i1}, \tag{7c}$$

$$d=1+\sum_{i=1}^{n} v_{2i}x_{i2}. \tag{7d}$$

Hence,

$$C^{-1}=\frac{1}{ad-bc}\begin{bmatrix} d & -b \\ -c & a \end{bmatrix}. \tag{7e}$$

Finally notice that from (1) any matrix B can be constructed from a series of applications of the rank annihilation formulae which is the basis of the general inverse technique. More formally, we have,

$$B=A+\sum_{i=1}^{r} u_i v_i^T,$$

and when $r=n$ and $A=I$ we have the arbitrary inversion method. However if we impose a further constraint such that each row or column is modified only once, as would be the case in the arbitrary inversion it follows that all the $u_i, v_i, i=1(1)r$ can be precomputed. As an illustration consider the sequence of changes as i increases. These can be written as the recurrence,

$$A^{(0)}=A$$

$$A^{(i)}=A^{(i-1)}+u_i v_i^T$$

$$B=A^{(r)}.$$

If each modification affects a distinct row or column, the rows and columns not yet modified are identical to those of the original matrix A. Hence the u and v vectors transforming the row or column of A must be the same as the ones for the row changed from $A^{(i)}$.

Remark Notice that this strategy works only when either a sequence of row or column modifications, switching from row to column changes and vice versa which prohibits the precomputation of all u_i, v_i, $i=1(1)r$.

We shall see later that this property is useful in pipelining successive modification designs. The definitions of this section are straightforward and permit a top down design approach to the development of rank annihilation systolic arrays.

We now present two contrasting designs, a mesh connected scheme using wavefront concepts and a highly concurrent pipelined scheme.

3. MESH CONNECTED SCHEMES

In order to produce a wavefront model for the rank annihilation strategy we shall introduce a special kind of wavefront processor. The processor consists of an orthogonally connected square mesh of $(n+2)*(n+2)$ reduced instruction set (RISC) processors. An $n*n$ matrix of processors into which the known inverse A^{-1} can be loaded, is embedded inside a ring of processors comprising the first and last rows and columns of the mesh. This ring serves as a Systolic Control Ring (SCR) and the processors at grid positions $(1,1)$, $(1,n+2)$, $(n+2,n+2)$ and $(n+2,1)$ are controller units capable of generating a number of control signals in the horizontal and vertical directions. The remaining processors on the ring perform book-keeping and auxiliary operations to support the embedded mesh, while also relaying control signals from one controller to another.

Figure 1 shows the global connection pattern for the rank annihilation problem. The host interface consists of cells in the SCR along the first row and column. Controller $C1$ initiates the algorithm, after A^{-1} has been loaded into the wavefront mesh and the vectors u and v for a Rank-1 scheme have been loaded into their correct cells, u in the first row and v in the first column. The rest of the control ring

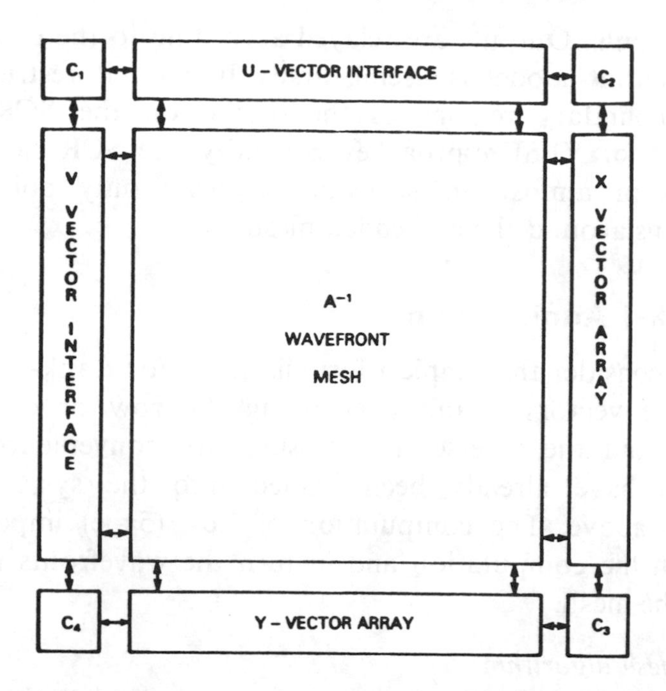

Figure 1(a) Global arrangement of rank annihilation systolic mesh.

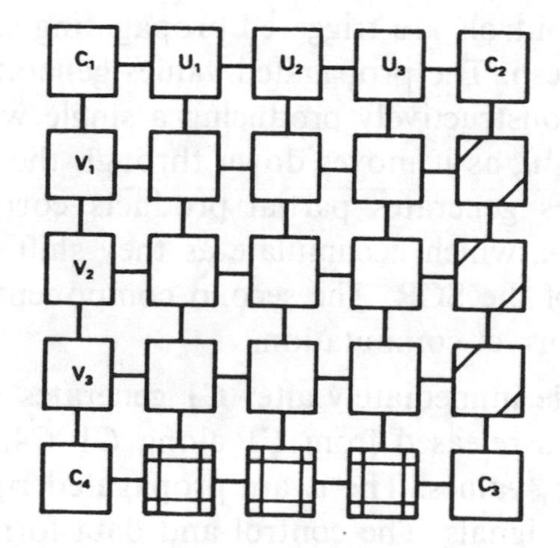

Figure 1(b) 3×3 example mesh connections

comprises cells on the last row and column providing auxiliary storage and computational power to collect the partial results of Eqs. (5a–e) and (6a–e). The mesh itself is an orthogonally connected square of identical processors, capable of receiving a wavefront control signal from all directions (north, south, east or west) and performing appropriate inner product operations using data on different inputs. Outputs are relayed according to the control values. This wavefront model is useful, firstly because it restricts different cells to boundary regions of the design on the SCR, which is convenient for VLSI approaches. Secondly, the SCR can generate a wavefront in almost any orientation using only point to point connections around the embedded mesh.

3.1 Rank-1 Annihilation

We shall consider the simplest formula first. For Rank-1 annihilation we require vectors u and v encoding the row or column to be modified, and the inverse A^{-1}. Assume for convenience that these quantities have already been loaded into the systolic mesh as described above. The computation of Eqs. (5a–e) imposes a strict regime on the computation and in turn the wavefronts necessary to control the mesh.

Rank-1 mesh algorithm
Step 1 a) SCR cell $C1$ generates a control moving right along $C1$ to $C2$. As the control passes through the cells containing the u_i values, a set of controls are triggered propagating signals and u_i data down into the mesh. The propagated values generate two wavefronts which interfere constructively producing a single wavefront $W1$. $W1$ computes $x=A^{-1}u$, as it moves down through the mesh. One of the component waves generates partial products corresponding to the values x_i, $i=1(1)n$, which accumulate as they shift right towards the $C2$–$C3$ portion of the SCR. The second component triggers the cells to perform the correct computation.

b) On the cycle immediately after $C1$ generates the $C1$–$C2$ signal, a similar signal is released from $C1$ along $C1$–$C4$. This triggers the cells containing v_i values. The v_i are propagated right into the mesh with the control signals. The control and data form two component waves which constructively interfere with each other producing a wavefront W_2. As control moves right accumulating partial product sums are shifted down producing the $y_i, i=1(1)n$ values.

W_1 and W_2 travel at the same speed with W_2 a cycle behind W_1 and so never interfere with each other. A side effect of the data flow is that the values x_i and v_i, $i=1(1)n$ are adjacent to each other with v_i one cycle behind x_i.

Step 2 This step begins when the $C1$–$C2$ control value on the SCR reaches $C2$. The $C1$–$C4$ reaches $C4$ a cycle later, and the first x_i, x_1 enters the x-vector store in the cell immediately below $C2$ in the SCR on the line $C2$–$C3$. The row of $C2$ simply passes the $C1$–$C2$ signal on to $C2$–$C3$. So x_1 enters the x-vector cell on the same cycle as the control signal, meanwhile on the $C4$–$C3$ section of the SCR nothing happens as y_1 is accumulating its last term and this is the leading term on W_2. x_1 is loaded into the cell, and the next n cycles sees the loading of all the remaining x_i, while the y_i are loaded into the y-vector store by the $C1$–$C4$ control which is relayed to $C4$–$C3$ by $C4$. Overlapped with this is the computation of the summation equation (5e). The close proximity of the v_i is brought into play. The x_i load signal is followed by another signal from $C2$, along with the starting value of the summation $(=1)$. This second signal travelling $C2$–$C3$, combines the v_i of W_2 with the loaded x_i pushing an accumulating total of $Z+1$ towards $C3$ on the SCR. Step 2 ends when this second signal reaches $C3$ loading $Z+1$ into $C3$. On this cycle the $C4$–$C3$ signal also reaches $C3$.

Step 3 $C3$ now takes control of the algorithm. The remaining Rank-1 steps require Eq. (5b) and Eq. (5a) to be computed. $C3$ generates two control signals, one along $C3$–$C2$, the other along $C3$–$C4$ (accompanied with the data result $Z+1$). As the signals move left and up, the y vector store modifies its stored element according to $y_i = 0 - y_i/(1+Z)$, $i = 1(1)n$, outputting the result into the mesh. The x-vector store simply pushes the x_i values left into the mesh. The y_i and x_i elements encroaching onto the mesh form two component waves which when they meet in a cell produces a wavefront W_3 moving from $C3$ to $C1$ diagonally through the mesh. $W3$ forms the A^{-1} modification $A_{ij}^{-1} - x_i y_j$, $i = 1(1)n$.

On reaching $C4$ and $C2$ the control values are passed on to traverse $C4$–$C1$, and $C2$–$C1$. On arriving at $C1$ the last modification of A^{-1} is performed and the mesh now contains B^{-1} and the Rank-1 modification is complete. The loading of new u and v vectors can be performed while the signals travel back to $C1$.

The main points to notice in this algorithm is that all mesh wavefronts are generated by the SCR signals, the mesh cells receiving their instructions for computation from the controls associated with the wavefront.

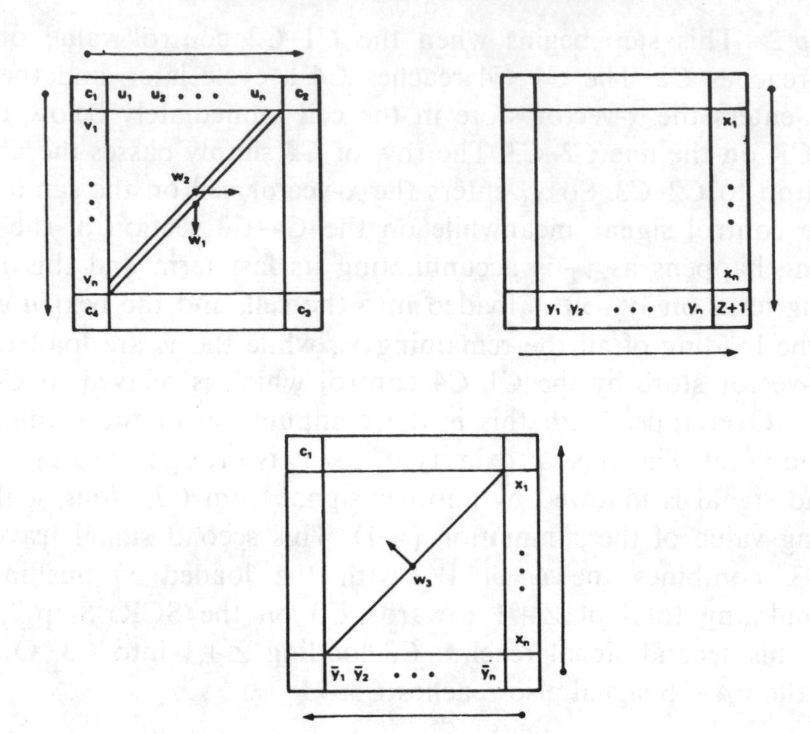

Figure 2(a) Rank 1 wavefronts.

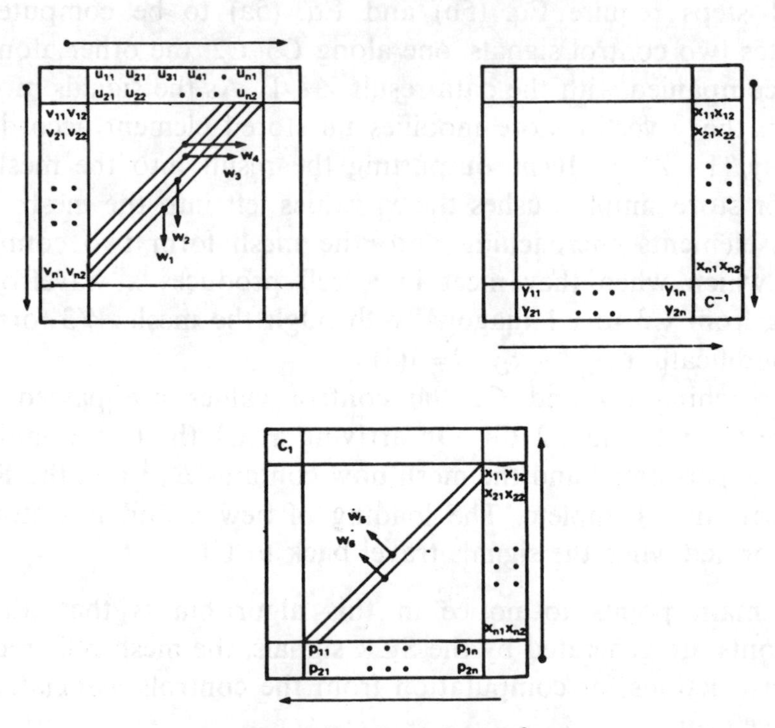

Figure 2(b) Rank 2 wavefronts.

Timing the Rank-1 scheme is simple and is evaluated by observing the wavefront diagrams in Figure (2a). Assuming one of the u, v vector interfaces contains n input lines we can load the starting matrix A^{-1} in $n+1$ cycles, this time is also the time for unloading the final matrix and also includes the time for setting up u and v. Thus, input and output requires $2(n+1)$ cycles. The rest of the algorithm is controlled by the time it takes for a signal to complete a circuit around the SCR. Trivially this is $4(n+1)$ when only a single signal is used. We also must account for the extra signals generated after the start signal at the intermediate controllers around the ring. $C2$ is the only controller to add a signal so we require only a single delay, and we derive the generalised timing

$$T=2(n+1)+r(4n+5),$$

where r is the number of Rank-1 modifications incorporated in a sequence of distinct row or column modifications.

When

$r=1, T=6n+7$ a single Rank-1 modification

$r=n, T=4n^2+7n+2$ arbitrary matrix inverse.

We see that we have an arbitrary matrix inverter which requires $O(n^2)$ time, which clearly does not compete with a systolic Gaussian Elimination (GE) technique requiring $0(7n)$ and $0(n^2)$ basic inner product cells. Although the array for $r \ll n$ has uses for modifying existing inverses for partially changed matrices.

Up to now we have been vague about the cells of the mesh and the cycle time. Considering the Rank-1 mesh algorithm above, it is clear that the cycle time of a cell is also a simple inner product. The SCR controllers $C1, C2, C3$ and $C4$ are trivial combinational logic, while the interface cells containing u and v are simple registers with additional logic to support the SCR. The x-vector store require a register to save the x_i, and an inner product circuit to add the partial product term as the $Z+1$ value passes through. Y vector cells similarly contain a register for the y_i, and the subtract/divide arrangement to modify the y_i using $Z+1$. The most complex cell is the embedded mesh cell, this contains a register to save the A_{ij}^{-1}

value and must perform

$$E = W + N * A \quad \text{with wavefront } W_1$$

$$S = N + E * A \quad \text{with wavefront } W_2$$

$$A = A + S * E \quad \text{with wavefront } W_3$$

where $A \equiv A_{ij}^{-1}$, and N, S, E, W are compass directions for input/output.

Thus in terms of basic inner product cells we require:

i) n^2 embedded cells

ii) $2n$ cells for the x and y-vector stores

TOTAL $= n(n+2)$ cells.

This is $2n$ more than the systolic GE algorithm for matrix inversion.

3.2 Rank-2 Annihilation

The mesh connected wavefront algorithm for Rank-1 annihilation can be easily extended to perform Rank-2 modifications. The global connection structure remains the same in the Rank-2 method as do the embedded mesh cells. The interface and vector stores on the SCR however need updating.

In this annihilation scheme we use the formulae Eqs. (6a–e) and Eqs. (7a–d). Equations (6c) and (6d) can be considered as two matrix vector and vector matrix operations respectively, given by,

$$A^{-1}U^{(1)} = x^{(1)}, \tag{8a}$$

$$A^{-1}U^{(2)} = x^{(2)}, \tag{8b}$$

$$V^{(1)T}A^{-1} = y^{(1)T}, \tag{8c}$$

$$V^{(2)T}A^{-1} = y^{(2)T} \tag{8d}$$

and

$$u = [u^{(1)}, u^{(2)}], \quad v^T = [v^{(1)T}, v^{(2)T}], \quad x = [x^{(1)}, x^{(2)}], \quad y^T = [y^{(1)T}, y^{(2)T}].$$

The wavefront algorithm is therefore modified to introduce more wavefronts. Again we assume that A^{-1} is loaded into the starting mesh and U and V contain $U^{(1)}, U^{(2)}$ and $V^{(1)}, V^{(2)}$.

Rank-2 mesh annihilation

Step 1 $C1$ starts the algorithm.

cycle 1: $C1$ generates a control travelling $C1$–$C2$ propagating $U^{(1)}$ forming a wavefront W_1 forcing partial products which accumulate left to right computing $X^{(1)}$ in Eq. (8a).

cycle 2: $C1$ generates a similar signal also on $C1$–$C2$ this time propagating $U^{(2)}$ into the mesh producing wavefront W_2, forcing accumulating partial products of Eq. (8b) and $X^{(2)}$ right.

cycle 3: $C1$ generates a signal downwards from $C1$–$C4$ triggering $V^{(1)}$ and forming a wavefront W_3, pushing partial products of $Y^{(1)}$ down $C4$–$C3$.

cycle 4: $C1$ generates a similar signal along $C1$–$C4$ using $V^{(2)}$ and forming $Y^{(2)}$ also moving down to $C4$–$C3$ forming W_4.

Step 1 ends when the first $C1$–$C2$ signal reaches $C2$, at this time the first $C1$–$C4$ signal is two cycles from reaching $C4$.

Step 2 At the start $C2$ simply relays incoming signals to the SCR along $C2$–$C4$. The X-vector cell immediately below $C2$ receives the $X_i^{(1)}$ and then $X_i^{(2)}$ loading them into registers controlled by the two controls relayed by $C2$. $C2$ follows the two relayed signals by two extra signals which synchronise with the $V_i^{(1)}$ and $V_i^{(2)}$ values as they reach the X-vector store. As the $V^{(1)}$ and $V^{(2)}$ values arrive they are used to compute the summations Eqs. (7a–d) resulting in the matrix C. These latter two controls emitted by $C2$ synchronise with the relaying of signals from $C1$–$C4$ to $C4$–$C3$ by $C4$, which are used to load the $Y_i^{(1)}, Y_i^{(2)}$ into the Y-vector cells. The step ends when the last $C4$–$C3$ signal penetrates through to $C3$, the X and Y matrices have been computed and C has been loaded into $C3$.

Step 3 a) $C3$ now takes control of the algorithm/mesh, and executes two cycles to compute C^{-1}. C is a $2*2$ matrix and the inverse is found by Cramer's rule.

b) $C3$ continues the SCR cycle next by generating two signals one on $C3$–$C2$ and the other along $C3$–$C4$ (accompanied by the first row of C^{-1}). Each Y-vector cell computes according to

$$\mathrm{OUT}_i^{(1)} = c_{11} * Y_i^{(1)} + c_{12} * Y_i^{(2)}.$$

$\mathrm{OUT}_i^{(1)}$ and $X_i^{(1)}$ are then inserted back into the mesh forming a constructive wavefront W_5 wherever they meet in a mesh cell.

c) $C3$ follows the signal in (b) above immediately with an identical signal, this time $C3$–$C4$ is accompanied by the 2nd row of C^{-1}. Again, the Y-vector cells compute

$$\mathrm{OUT}_i^{(2)} = c_{21} * Y_i^{(1)} + c_{22} * Y_i^{(2)}.$$

This result and $X_i^{(2)}$ are now output as component waves which form the wavefront W_6 by constructive interference.

d) As the two signals from (b) and (c) move toward $C4$ and $C2$ wavefronts W_5 and W_6 move diagonally from $C3$ to $C1$, the new inverse elements are generated by simple inner product type computations of the form,

$$A_{ij}^{-1} = A_{ij}^{-1} - X_i^{(1)} * \mathrm{OUT}_j^{(1)}.$$

$$B_{ij}^{-1} = A_{ij}^{-1} - X_i^{(2)} * \mathrm{OUT}_j^{(2)}.$$

e) The signals eventually reach $C4$ and $C2$ where they are relayed along $C2$–$C1$ and $C4$–$C1$, the next U and V can be loaded during the trip back to $C1$. On the cycle after the signals arrive at $C1$, the embedded mesh contains the new inverse.

The timing of the algorithm can be computed in a similar manner to the Rank-1 scheme. Notice that a cycle time is still a simple inner product, while the whole modification is related to the cost of a single trip around the SCR circuit and any added delays. Following a cycle for the extra signal sent by $C1$, and the two that $C2$ adds, together with the 2 cycles required by $C3$ to compute the $2*2$ inverse, and finally a cycle for the wavefront W_6, we get the generalised sequence modification time,

$$T = 2(n+2) + r(4(n+1)+6) = 2(n+2) + r(4n+10)$$

where loading and unloading require an extra cycle. Hence when

$r=1, \quad T=6n+14 \quad$ single Rank-2 modification.

$r=n/2, \quad T=2n^2+7n+4 \quad$ arbitrary matrix inverse.

The Rank-2 scheme appears superior to the Rank-1 scheme, especially for the arbitrary inversion problem requiring only half the time. This improvement over Rank-1 is due to the fact that we can affect the modification of two rows or two columns in a single sweep or cycle of the SCR. This improvement is achieved with only a modest increase in hardware.

The U and V interfaces now require two registers instead of one. The divider/subtract arrangement in the Y-vector store is changed to two multipliers and an adder, while the X-vector store replaces a single inner product cell by two. Both vector stores require two registers to store the X and Y values instead of the one register in the Rank-1 scheme. Finally, the controller $C3$ is no longer a simple combinational logic unit, Cramer's rule requires at most 6 ips to compute the $2*2$ inverse in parallel. Taking the Y-vector cells to be bounded by 2 ips cells in area our ips cell requirement is then increased to n^2+4n+6 cells.

3.3 Restrictions on the mesh rank annihilation schemes

We can now appreciate why our attention was restricted to the Rank-1 and Rank-2 schemes. Implementing Rank-3 would require $C3$ to compute the inverse of a $3*3$ matrix which would involve significant increases in hardware and more delays for the timing of a modification step. More general schemes would require $m*m$ inversion in $C3$. We would also have to expand the SCR to cope with the increased vector storage and computations. We have already used more hardware than the systolic GE inversion method with Rank-1 and its performance for general inversion was much better.

The computation of the $m*m$ inverse is more restrictive than it appears. Some improvement was obtained by going to Rank-2 at the expense of only a small increase in hardware. The Rank-2 improvement can be attributed to the increased number of wavefronts on the mesh at any one time. The only way to increase the wavefronts is to increase the modification size m. We conclude that the sparse wavefront patterns on the mesh provide low processor efficiency making this approach to rank annihilation schemes costly in area and in arbitrary inversion time also.

4. HIGHLY PARALLEL PIPELINED RANK ANNIHILATION

The starting point for the pipelined scheme is a review of the Rank-1 and Rank-2 schemes given by Eqs. (5a–e) and Eqs. (6a–e) noticing that the bulk of the calculations can be achieved by variations of matrix vector operations.

Consider the Rank-1 scheme, Eq. (5d) can be rewritten as the matrix vector problem,

$$(A^{-1})^T v = y. \tag{9}$$

Systolic arrays for the matrix vector problem are well known [1] and compute in a time of $T=2n+W$ where $W=$bandwidth of the matrix. The array requires W cells equivalent to an inner product step processor (multiply and add). For an arbitrary matrix W is bounded by $2n-1$, giving a matrix vector time of $T=4n-1$ using $2n-1$ ips cells.

4.1 Rank-1 pipeline

It appears that in order to compute x and y in Eq. (5c) and Eq. (5d) requires either two matrix vector arrays of $(2n-1)$ ips cells computing in parallel, or one array of $(2n-1)$ cells requiring two sequential applications of the algorithm in $T=(8n-2)$ cycles. However these two strategies contravene one of the basic principles of hard systolic algorithms. The matrix A^{-1} must be issued to more than one systolic device at the same instance in the parallel case, whilst in the sequential case, the matrix is issued to the same array twice. The key point is that the host memory issues the same data value more than once. Recalling the earlier work of the authors [2] on double pipes the execution time of an array acting on a single problem instance can be reduced by filling the dummy elements of the dataflow with real values. A similar argument can be applied to compute two problem instances in the same time on the same array. This latter method requires the dummy elements to be filled by interleaving two problems by staggering one in time relative to the other. Interleaving also has the effect of increasing processor efficiency. This interleaving property can be utilised to compute x and y on a single array in $T=(4n-1)$ cycles bringing A^{-1} from the host only once.

The essential point to note is that the matrix vector systolic arrays input matrix values along diagonals (see Figure 3a). All the elements input to a particular cell belong to the same sub(super)diagonal, and it follows that a matrix transpose is essentially a 180° rotation of the inputs about the central or diagonal cell.

Thus, for an arbitrary matrix with bounded bandwidth $W=2n-1$ the array is symmetrical hence the data rotation can be achieved implicitly by rotating the cells rather than the data. In real terms this means swapping the known and unknown input vectors and the basic cells reflected in the vertical axis. (See Figure 3b.)

By combining the matrix and its transpose matrix vector arrays allows the horizontal data streams to be interleaved by delaying the transposed data by a cycle, producing the array (Figure 4). The dummy elements in the matrix input are not filled, this requires some modifications to the basic ips cell. Essentially once the matrix input has entered the cell it must stay there for two cycles instead of one, which is achieved by a register. A control tag associated with the matrix input can be used to decide the state of the cell, setting data route switches to perform as an inner product cell for an array like Figure (3a) or Figure (3b) on alternate cycles. We conclude that the total time to compute x and y is $T=4n$ ips cycles using only $2n-1$ modified ips cells, with the matrix A brought only once from the host.

Remark We might attempt to improve the performance of the systolic array by compressing the elements of the A^{-1} array. However the horizontal data sequences have no spare dummy elements so no improvement is likely.

With x and y found, and having A^{-1} moving through the system systolically we have a good opportunity to complete the rest of the Rank-1 scheme using Eqs. (5a), (5b) and (5e). Equation (5e) is only a simple summation and can be evaluated by a single accumulating

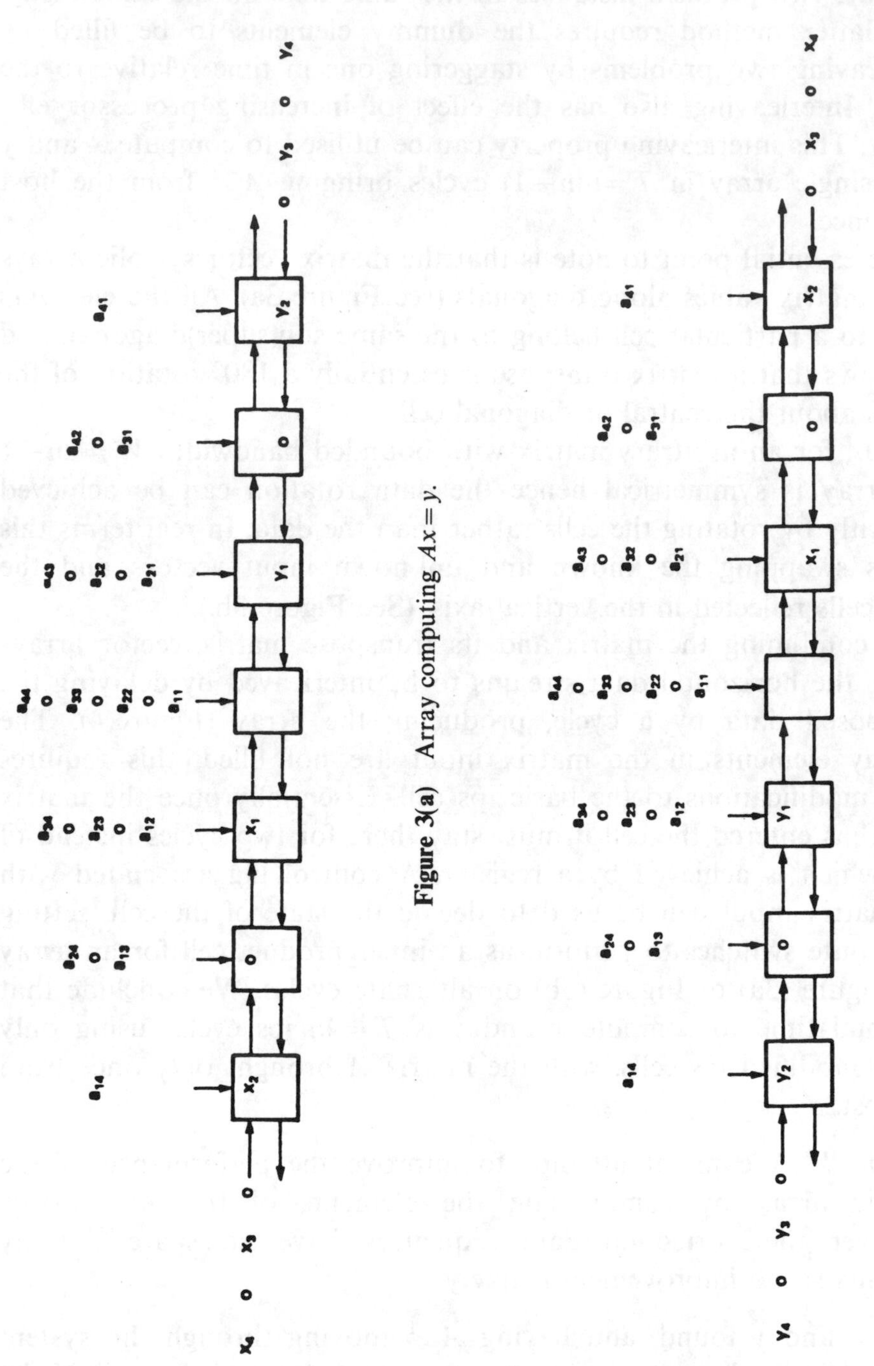

Figure 3(a) Array computing $Ax = y$.

Figure 3(b) Array computing $x^tA = y^t$.

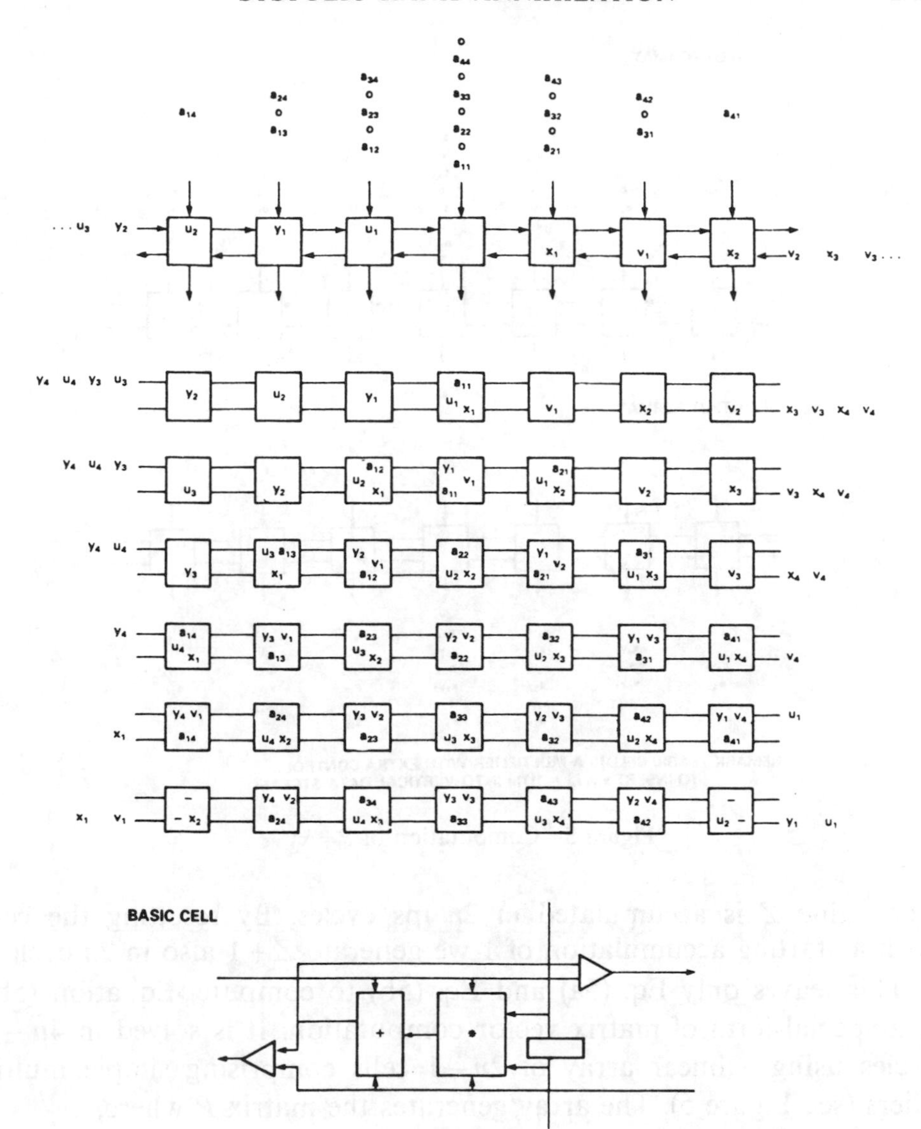

Figure 4 Systolic computation of $A^{-1}u=x$ and $v^tA^{-1}=y^t$ using interleaving.

inner product cell in n ips cycles. By considering Figure 4 we notice that a useful side effect of the interleaving of the matrix and transpose matrix vector problems places the x_i and v_i, $i=1(1)n$ in a suitable dataflow pattern to compute Eq. (5e) without any delay or synchronisation. The left side of the M–T–M array outputs the sequence

$$x_1v_1x_2v_2x_3v_3\ldots x_nv_n.$$

a) INPUT FORMAT

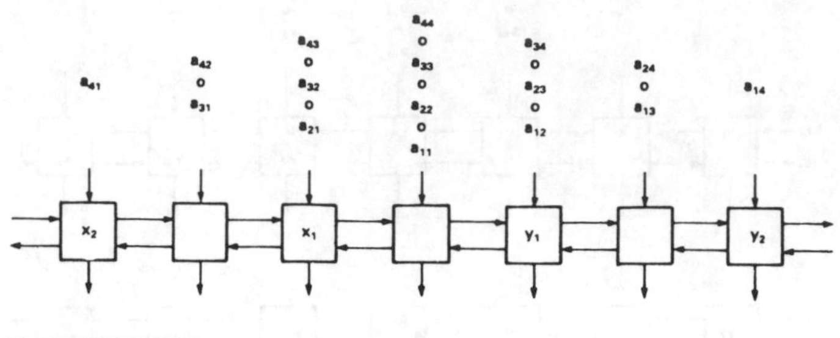

b) OUTPUT FORMAT

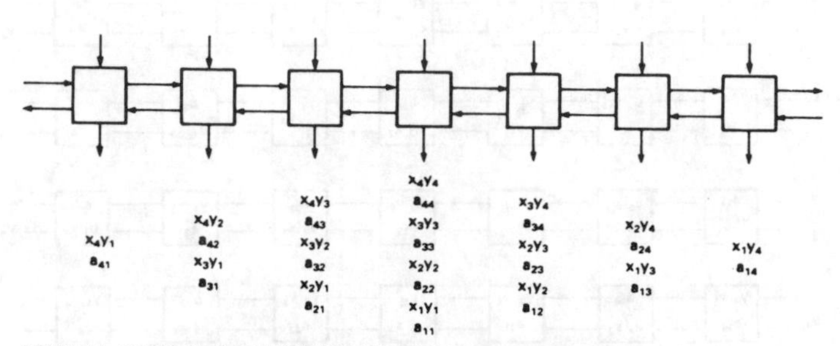

REMARK: BASIC CELL IS A MULTIPLIER WITH EXTRA CONTROL TO INSERT $x_i y_j$ $i,j = 1(1)n$ INTO VERTICAL DATA STREAM

Figure 5 Computation of $P = xy^t$.

The value Z is accumulated in $2n$ ips cycles. By resetting the cell with a starting accumulation of 1 we generate $Z+1$ also in $2n$ cycles.

This leaves only Eq. (5a) and Eq. (5b) to compute. Equation (5b) is a special form of matrix vector computation. It is solved in $4n-1$ cycles using a linear array of $2n-1$ cells comprising simple multipliers (see Figure 5). The array generates the matrix P where,

$$P = \begin{bmatrix} x_1y_1 & x_1y_2 & \cdots & x_1y_n \\ x_2y_1 & & & \vdots \\ \vdots & & & \vdots \\ x_ny_1 & x_ny_2 & \cdots & x_ny_n \end{bmatrix}.$$

The array introduces the matrix values of A^{-1} although we assume that $Z+1$ is not available in order to keep synchronization of the

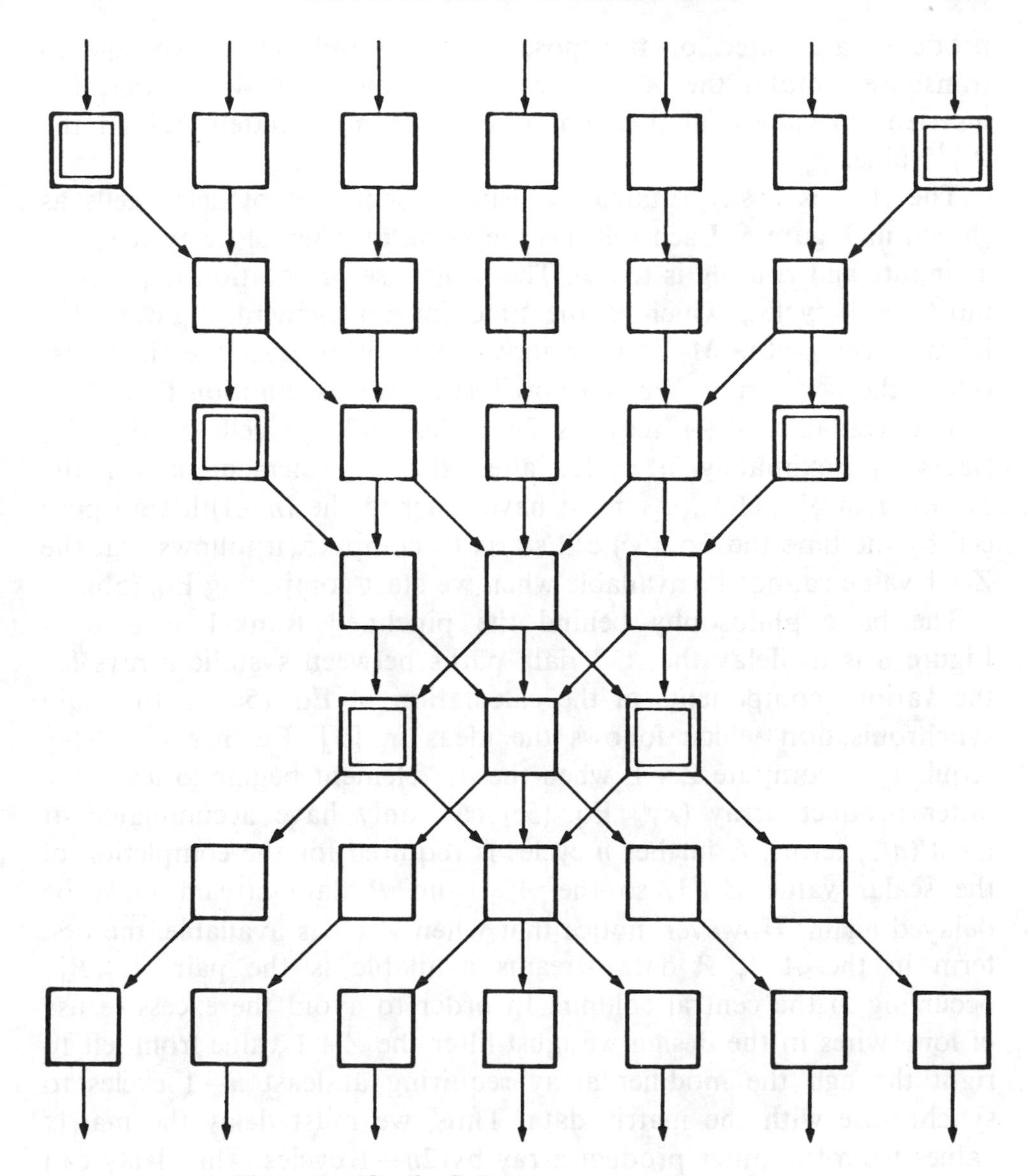

Figure 6 "On the fly" matrix transposition network.

different data streams. We will show later that the assumption on $Z+1$ is both realistic and practical. Recall that associated with the matrix values is a control tag. This can be used to trigger the array cells causing them to insert the P_{ij} elements into the dummy elements of the systolic data sequences of A^{-1}. The consideration of the localisation of data flow (see Figure 8) means that the systolic movement of x and y generated by the M–T–M array produces a rotated or transposed output, and it is necessary to transpose (rotate) the A^{-1} input–output by M–T–M so that the inserted P_{ij} values fill the dummy element immediately behind the A_{ij}^{-1} value.

The A^{-1} is easily rotated by using a sequence of delay cells as shown in Figure 6. Each cell has the same number of delay registers as inputs and routing is trivial. The transpose or rotation requires at most $2n$–1 cycles, which is the time for an element input to the leftmost cell of the M–T–M to move to position over the rightmost cell of the next array. We can now justify our assumption that $Z+1$ is not available. $Z+1$ requires $2n$ cycles and the cell for Eq. (5e) starts approximately n cycles after the a_{11} element leaves the M–T–M array. Thus, a_{11} must have reached the $(n-1)$th transpose cell by the time the Eq. (5e) cell starts to compute, it follows that the $Z+1$ value cannot be available when we start computing Eq. (5b).

The basic philosophy behind the pipelined Rank-1 scheme in Figure 8 is to delay the A^{-1} data paths between systolic arrays for the various components of the calculation of Eq. (5a–d), to retain synchronisation which follows the ideas in [3]. Further the delay required to compute $Z+1$, when the a_{11} element begins to leave the outer product array (xy^t) Eq. (5e) can only have accumulated at most $(n/2)$ terms. A further n cycles is required for the completion of the scalar value $Z+1$, so the A^{-1} and P data stream must be delayed again. However, notice that when $Z+1$ is available, the first term in the A^{-1}, P data streams available is the pair A_{11}, P_{11} occurring in the central column. In order to avoid the excessive use of long wires in the design we must filter the $Z+1$ value from left to right through the modifier array requiring at least $n-1$ cycles to synchronise with the matrix data. Thus, we must delay the matrix values from the outer product array by $2n-1$ cycles. This delay can be put to good use by rotating the values back to the normal form so that the final output will be identical to the input.

Figure 7 shows the modifier array, again consisting of $2n-1$ cells. Each cell is a simple divide subtract arrangement and can be considered as a simple ips cell for area considerations. The operation of the cell is in three stages:

i) The $Z+1$ value and a control tag moves into the cell from the left causing $Z+1$ to be loaded into a register.

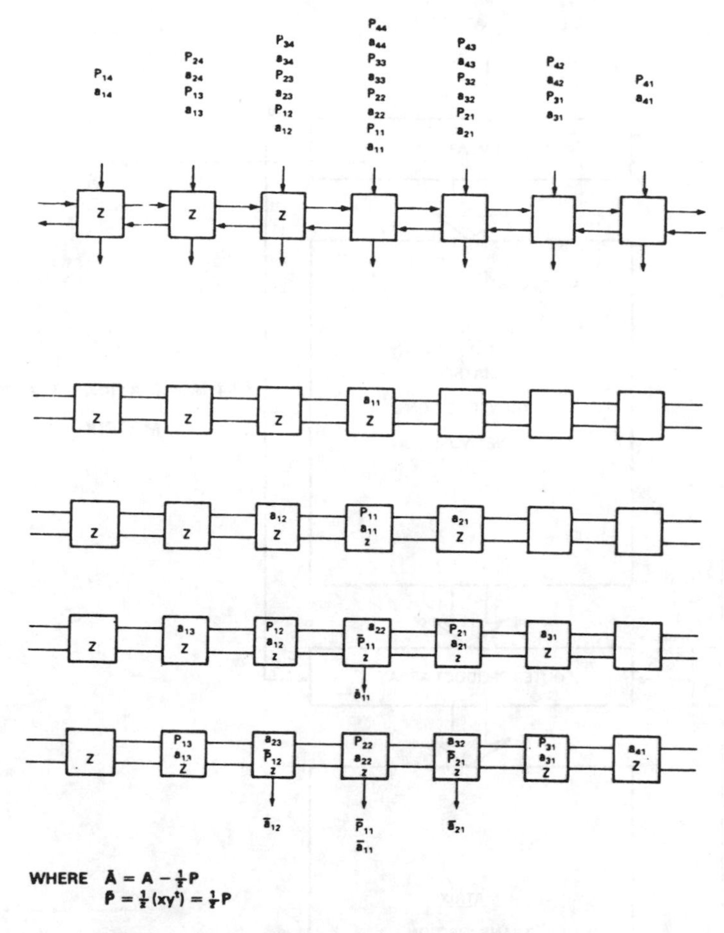

Figure 7 Systolic generation of new inverse.

ii) When the A_{ij} value reaches the cell its control tag is used to load it into a second register.

iii) The tag associated with the P_{ij}, formerly a dummy element executes a divide subtract operation, thus producing the modified A_{ij}^{-1} corresponding to B_{ij}^{-1}.

The delay of the result through the modifier is thus 2 ips cycles. This completes a single Rank-1 modification in a highly pipelined fashion. We have the vertical pipelining of the matrix data and also the horizontal piping of the u and v vectors and partial results.

The synchronisation of the various pipes and arrays can be observed from Figure 8, and can be used to derive the timing of the pipeline.

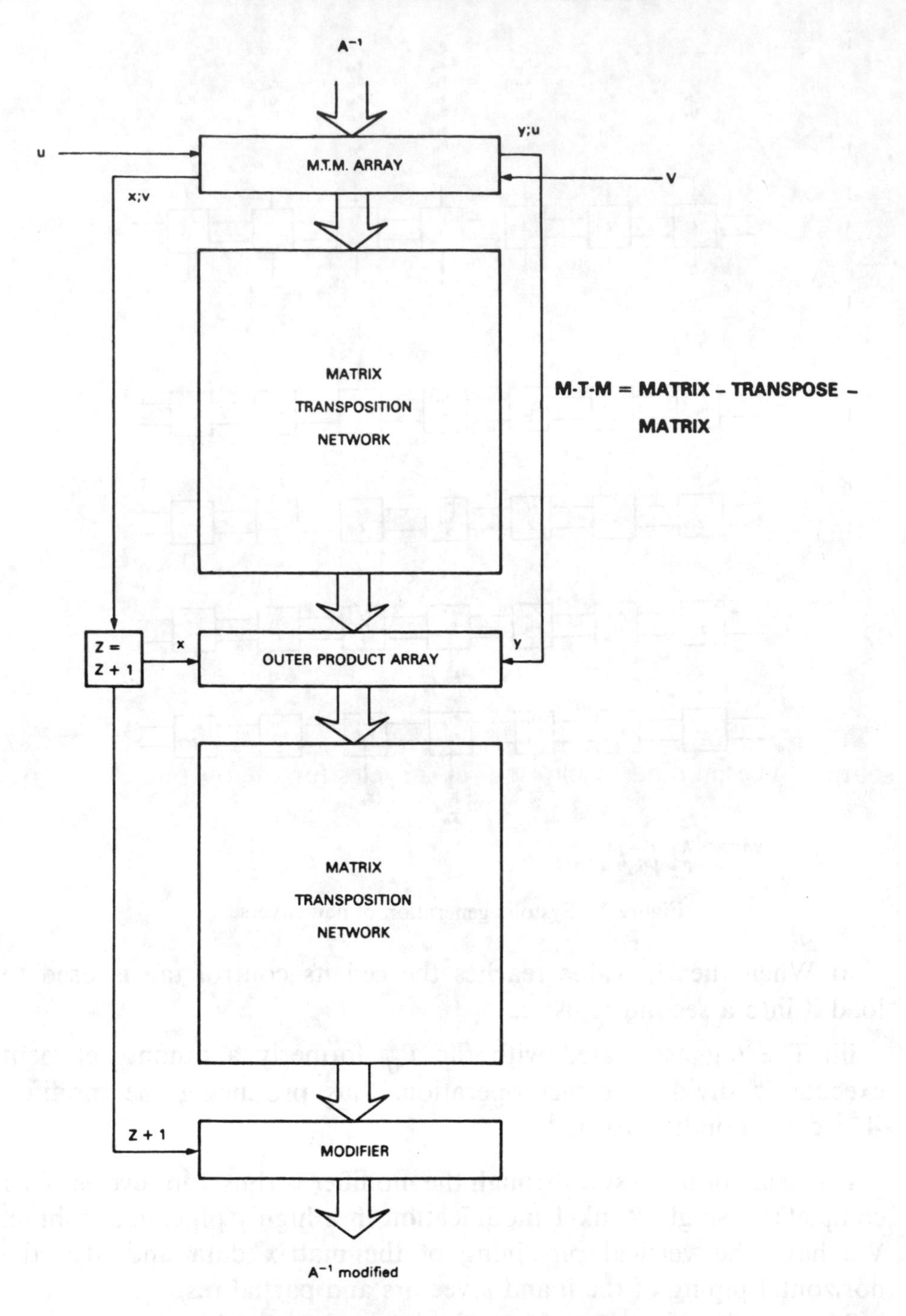

Figure 8 "On the fly" rank annihilation.

i) The M–T–M requires $n-1$ cycles to synchronise all data streams for starting the computation. After n cycles, the first x_i result is output, and a cycle later the first y_i. At this time A_{11} has penetrated $n-1$ cells of the transposition network.

ii) Each x_i, $i=1(1)n$ passes through the $Z+1$ cell, resynchronising it with y_i thus as x_1, y_1 enters the left most and right most cells of the outer product array, the matrix data enters the $n+1$ delay cell leaving $n-2$ cycles for the x_1 and y_1 to move the remaining $(n-2)$ cells and the matrix values to position themselves so they all meet in the central outer product cell.

iii) As explained previously, when the first term of the outer product array meet in the central cell, the $Z+1$ cell has only computed $\frac{1}{2}n$ terms, hence as the rotated A_{11}^{-1} and P_{11} values leave the outer product array, the $Z+1$ cell requires a further n cycles to complete the accumulation of the scalar and a further $(n-1)$ cycles to put the scalar through the $(n-1)$ cells to the left of the center cell in the modifier array. Inserting $(2n-1)$ cells to delay the leading matrix term the central diagonal of matrix inputs is synchronised with the $Z+1$ scalar. The transposition network therefore is sufficient to synchronise all matrix data while also rotating it for suitable output. The modifier itself requires 2 cycles for the output.

iv) There are $2n$ outputs thus the timing of the array is given by

$$T = \text{start up time} + \text{output time} + \text{pipeline latency}$$

$$= n-1 \quad + \quad 2n \quad + \quad 4n+2$$

$$= 3n-1 \quad + \quad 4n+2$$

Thus, a single Rank-1 modification can be computed in $(7n+1)$ which is again comparable to the time for an arbitrary matrix inversion using systolic Gaussian Elimination (GE). However, comparing the cell usage in the mesh scheme above we require $(n+2)*(n+2)$ ips cells and n^2 ips cells for the GE method. Here we require at most $3*(2n-1)+1=6n-3+1$ basic ips but $(8n^2-4n+1)$ delay registers in the transposition network. Thus we have traded basic inner product cells for delay registers which consume less area. As a result of the trade-off we have compacted the array. From Section 2 we can cascade the various Rank-1 modification steps under certain circumstances resulting in Figure (12a) giving a generalised timing

$$T = 3n - 1 + r(4n + 2)$$

with r, the number of applications of the formula. Notice however that the amount of hardware increases by a factor of r.

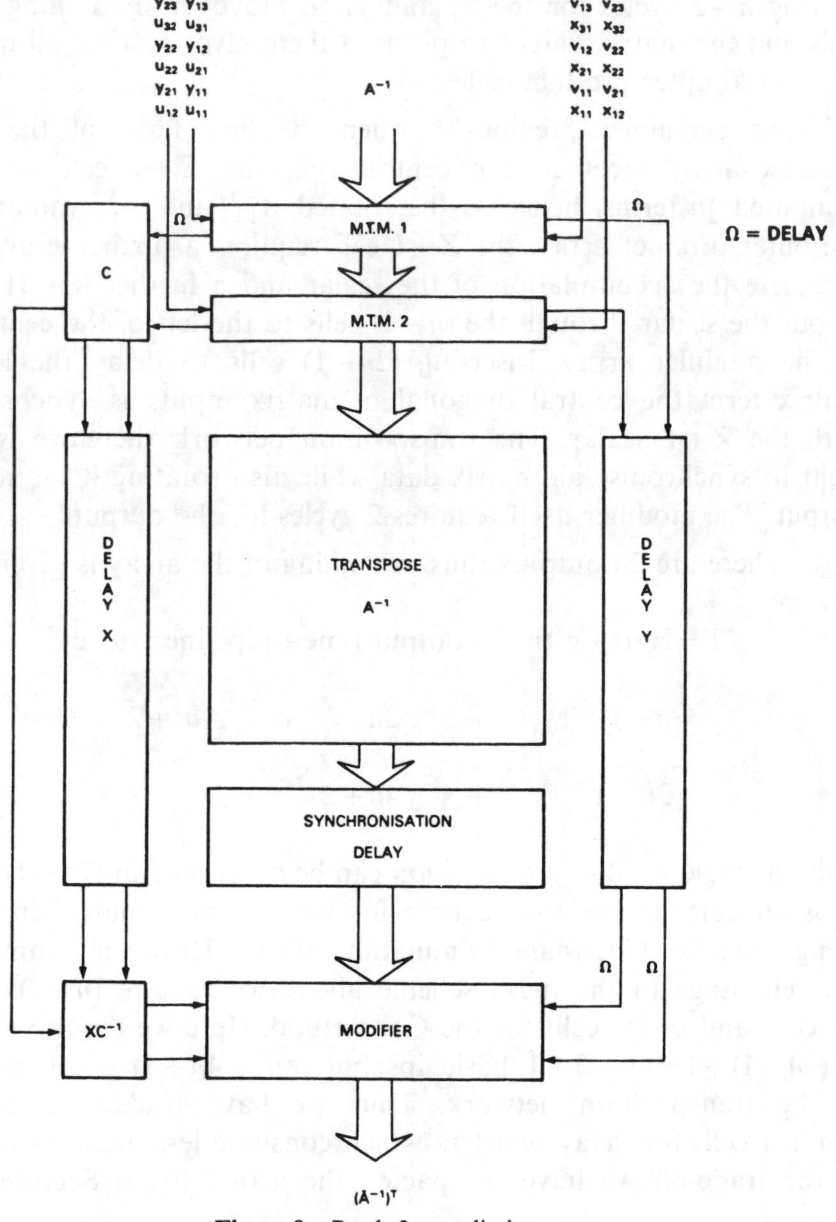

Figure 9 Rank 2 systolic inverter.

4.2 Rank-2 pipeline

Briefly we shall show that the rank annihilation pipeline can be extended to the Rank-2 method. The basic principle of the Rank-2 systolic inverter (Figure 9) is the same as the Rank-1 method except that the formulae equations (6a–e), equations (7a–e) are used. As for the mesh scheme, the problem can be reduced to a number of matrix vector operations sequenced on the pipeline in a similar manner to the ordering of the wavefronts.

We introduce two M–T–M arrays connected as shown in Figure 10, each array tier producing a column of the vectors x and y. The first row of cells producing the first column of X and Y, and the second row of cells the second and last column of X and Y. The output results of M–T–$M1$ must be delayed a single cycle so that they can synchronise with the results of M–T–$M2$. Thus elements of the first row of X enters the C cell on the same cycle. Again the useful side effect of the M–T–M arrays places the X and V values locally for easy computation of Eqs. (7a–d), using four inner product cells in a time $2n$. The scalars a and d are reset to starting values of 1, c and d set to zero. The C cell requires an extra two cycles to compute C^{-1} outputting the result to the XC^{-1} cell. The delay through the C cell for the incoming X values is a single ips cycle resynchronising them with the Y values, which are delayed in register queues long enough for C^{-1} to be produced.

The XC^{-1} array simply does what its label suggests, the computation is of the form,

$$\begin{bmatrix} x_{11} & x_{12} \\ x_{21} & x_{22} \\ \vdots & \vdots \\ x_{n1} & x_{n2} \end{bmatrix} \begin{bmatrix} c_{11} & c_{12} \\ c_{21} & c_{22} \end{bmatrix} = \begin{bmatrix} p_{11} & p_{12} \\ \vdots & \vdots \\ p_{n1} & p_{n2} \end{bmatrix} \quad \text{where } C^{-1} = \begin{bmatrix} c_{11} & c_{12} \\ c_{21} & c_{22} \end{bmatrix}.$$

Thus,, after C^{-1} is loaded into XC^{-1} four multipliers and two

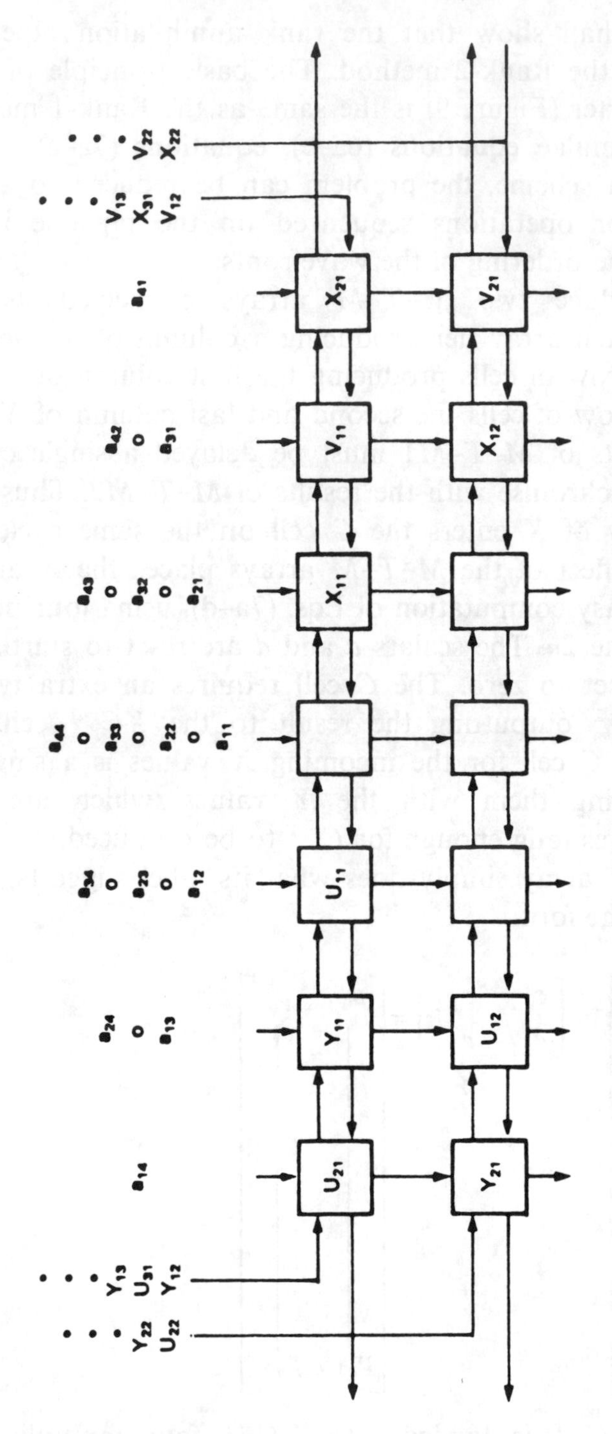

Figure 10 Rank 2 M–T–$M1$ and M–T–$M2$ arrays in starting positions.

adders are required to generate

$$P_{i1} = x_{i1}c_{11} + x_{i2}c_{21}$$

and

$$P_{i2} = x_{i1}c_{12} + x_{i2}c_{22}, \quad i = 1(1)n,$$

in a single ips cycle for each i. Clearly the columns of Y^T are now delayed by a single cycle to resynchronise with the modified X or rather P rows. Finally, it remains to compute the new inverse according to the formula,

$$B_{ij} = A_{ij}^{-1} = A_{ij}^{-1} - [P_{i1}y_{1j} + P_{i2}y_{2j}], \quad i, j = 1(1)n$$

which can be performed by two multipliers and two subtracters, requiring two ips cycles at the most to output the modified matrix elements, although data movement is controlled by a single ips cycle. The dummy element in the A^{-1} input covers the extra computation time. The extra values in the X and Y outputs emerging from the delay queues can be easily masked out and form no real problem. The modifier array is shown in Figure 11. It remains to assess the timing and cell usage, the data flow permitting us to calculate the number of registers required in the synchronisation delay section of Figure 9.

i) The time to set up the two M–T–M arrays is $n-1$ cycles. C receives its first results after $n+1$ cycles, and starts computing the matrix C. At the time x_{11}, x_{12} enter C the value A_{ii} from A^{-1} enters the nth cell in its column of the transposition network. This leaves $n-1$ delays left to compute C^{-1} which requires $2n+2$ cycles.

ii) It follows that we must add a total of $2n+1$ synchronisation delays for C^{-1} to complete and generate the first few rows of P required to push row 1 of P through the modifier to synchronise with the matrix values for final modification. This makes the number of delays between the modifier and the M–T–M array to be $4n$.

We obtain the timing for a single Rank-2 array as

$$T = 3n - 1 + (4n + 4) = 3n - 1 + 4(n + 1) = 7n + 4$$

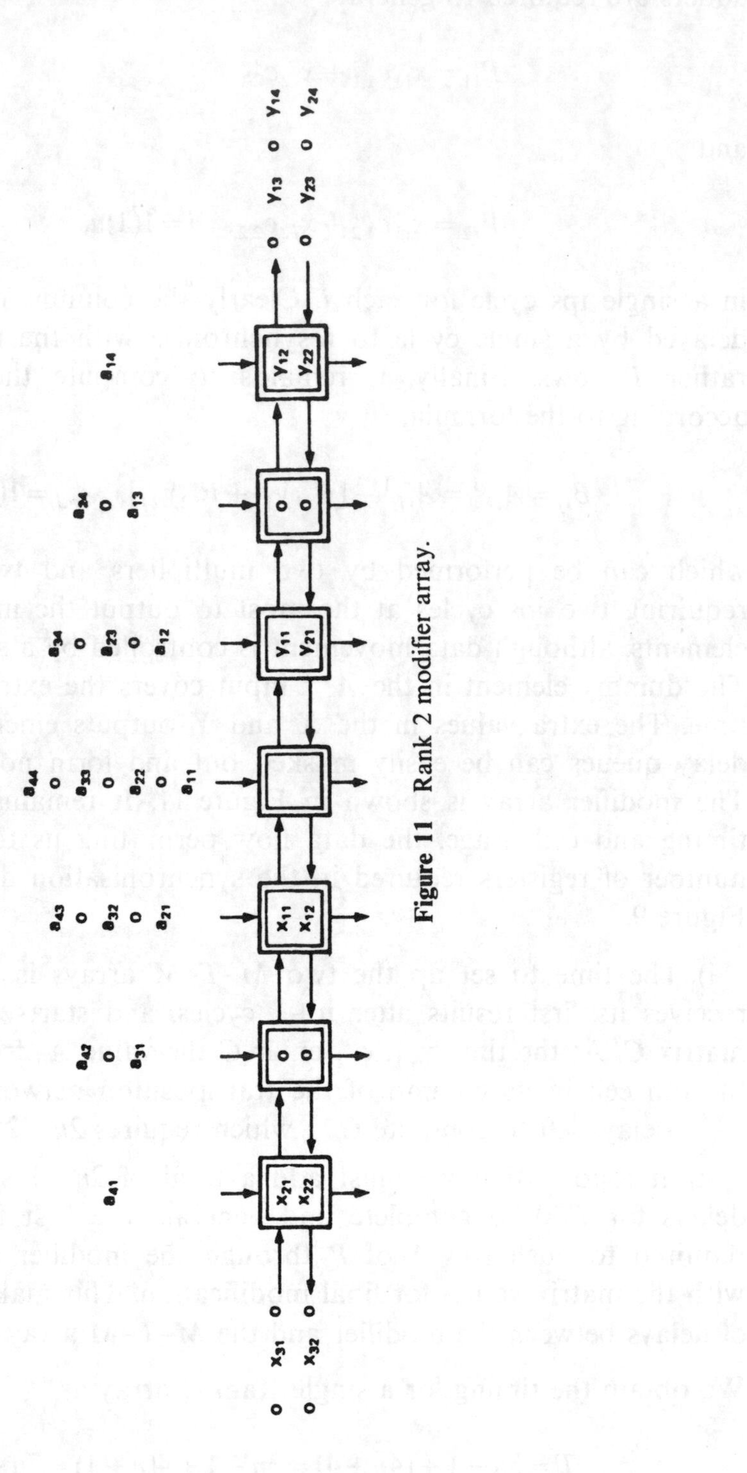

Figure 11 Rank 2 modifier array.

generalising to the cascading form in Figure (12b) the timing is

$$T = 3n - 1 + 4r(n+1)$$

with r, the number of applications of the modification formula. The cell requirement is also simply computed as

i)	M–T–M arrays	$(4n-2)$ ips
ii)	C cell (at most)	8 ips
iii)	XC^{-1}	4 ips
iv)	modifier 2 ips per cell	$(4n-2)$ ips
TOTAL		$(8n+8)$ ips

Also there are $4n$ delays on each matrix sequence between M–T–$M2$ and the modifier for an approximate bound, giving $8n^2$–$4n$ delay registers in all for the matrix and at most $2n$ cells for the delay queues in X and Y giving $8n^2$ delay registers in all. Again we have the property that the delay registers have replaced ips cells when compared with the mesh scheme and systolic GE methods. We have increased the hardware over the Rank-1 scheme by $2n+8$ ips and $8n-2$ registers for no improvement in computation time for a single application of the formulae. For general r values in the cascaded form, the Rank-2 scheme is faster than the Rank-1 and would require less hardware. For instance, the arbitrary matrix inversion problem is computed with,

$$r = n,\ T = 3n - 1 + n(4n+2) = 4n^2 + 5n - 1 \quad \text{Rank-1}$$

$$r = \tfrac{1}{2}n,\ T = 3n - 1 + 2n(n+1) = 2n^2 + 5n - 1$$

but we need to use only $\frac{1}{2}n$ cascaded arrays for Rank-2 rather than n for Rank-1. As we do not double the hardware in Rank-2 compared with Rank-1 we must save some hardware by using Rank-2 in preference to Rank-1.

Remark Cascading can only be applied when we perform a sequence of modifications to either rows or columns, and the modifications are made to distinct rows (columns) on each step. If

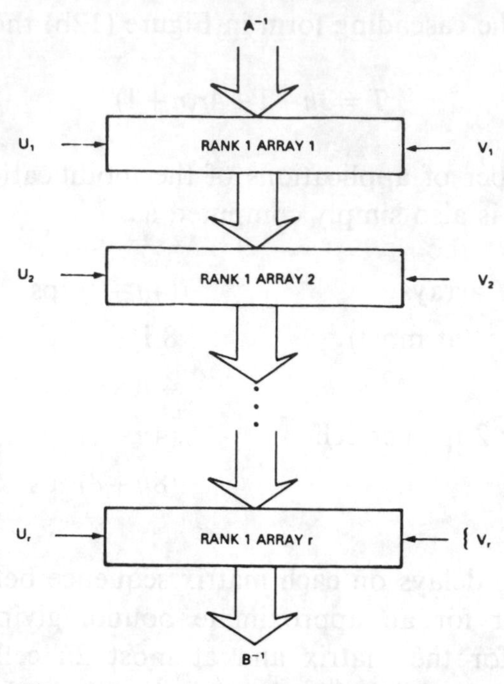

Figure 12(a) A cascaded Rank-1 scheme.

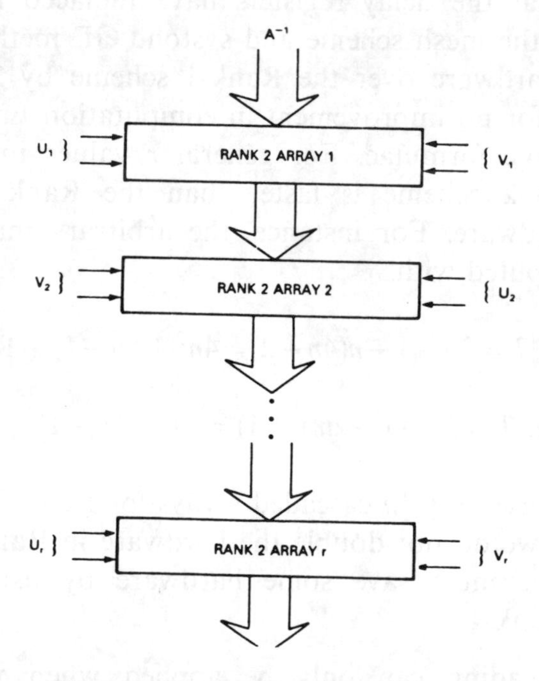

Figure 12(b) A cascaded Rank-2 scheme.

this is not the case, the U and V vectors on successive levels cannot be precomputed.

5. CONCLUSIONS

Two approaches to the Rank Annihilation method of matrix inversion were considered, a mesh connected wavefront model incorporating a Systolic Control Ring (SCR) and a highly parallel pipelined scheme.

Both methods were examined for Rank-1 and Rank-2 annihilation and the extension to more general methods Rank-m is shown to be too complex for current systolic concepts. The wavefront scheme required $O((n+2)^2)$ basic cells and computed a Rank-1 scheme in $O(6n)$, by extension the Rank-2 was solved in the same time with an increase of basic cells related to n, rather than n^2. Both wavefront methods suffered from low processor efficiency resulting from sparse wavefront projections. The Rank-2 increased the number of wavefronts and indicated that a Rank-m, $(m \geqq 3)$ method would dampen the sparsity as $m \rightarrow n$. The approach for $m \geqq 3$ was limited due to the complexity of computing an $m*m$ matrix inverse as an auxiliary problem.

Instead we switched our attention to a more systolic approach which reduced area and increased processor efficiency by trading some of the mesh processors for simple delay registers. Using the pipeline strategy we were able to reduce the number of basic cells to $O(n)$ while retaining $O(n^2)$ delay registers. Rank-1 pipelining resulted in a timing $O(7n)$ as did Rank-2 pipelining. In the cases where either a sequence of row modifications or column modifications were made to distinct rows or columns, a generalised cascaded array was constructed with timing $O(4rn)$ for Rank-1 and Rank-2; with Rank-2 clearly superior for a long sequence of modifications. The cascaded form of the annihilation pipelines allowed the techniques to be applied to arbitrary matrix inversion. A comparison with arbitrary matrix inversion using a systolic Gaussian Elimination (GE) approach requiring $O(n^2)$ cells and computing in a time $O(7n)$ showed that in hardware terms the mesh connected scheme was closely related. Although timings of $O(n^2)$ for pipelined and mesh schemes show the systolic GE superior for arbitrary inversion, in relation to time. The pipeline cascaded schemes cannot compete on area terms for arbitrary inversion where the generalised formula has $r=n$ (Rank-1) and $r=n/2$ (Rank-2), requiring r copies of a single modification pipeline.

We conclude that our pipelined arrays are suitable for minimising area when the inverses of matrices are required which differ from a matrix with known inverse, where systolic GE methods would have to re-compute the entire inverse.

Finally, we remark upon some features of the design which may concern traditional systolic array designers. Notice that designs of Figure 8 and Figure 9 introduce long wires, which in general are proportional to n in length. Normally this would introduce problems of clock skew and latch mistimings in the outer product array. However, notice that the transposition networks can be removed if we assume each array to input and output to the host machine. This has two effects, firstly the long wire is no longer a problem as the arrays are locally connected. Secondly, the transposition arrays contribute $O(n^2)$ delay cells, removing them reduces the design to only a linear number of processing elements with high efficiency. The removal of the transposition networks has another side-effect in that successive problems cannot be pipelined as rapidly, as the host must store the intermediate outputs of each sub-array. This memory problem can be reduced by partitioning the host memory into sections associated with each array, resulting in a reduction in the maximum problem size solvable. If the transposition networks are retained allowing the maximum throughput, the long wire is still of no importance due to the recent developments in optical interconnection technology [7]. Here fibre optics are used to distribute the clock signal around the system at the speed of light (no clock skew) while on chip silica wave guides could be used to achieve the same distribution on the chip. Hence we conclude that the Rank Annihilation systolic array is a soft-systolic design, in that the introduction of optical clock distribution will allow it to become hard systolic. On the other hand the removal of the transposition network would make the design hard systolic now at the expense of high throughput. We conclude that the arrays described here are in a migration state from soft to hard systolic under the influence of technology conditions changing over time as discussed previously [6], [2].

The systolic Rank-1 problem was simulated using an OCCAM program producing simple test results confirming the timings given above, see authors for program.

References

[1] C. E. Leiserson, Area efficient VLSI computation, Ph.D. Thesis, Oct. 1981.

[2] G. M. Megson and D. J. Evans, Soft-systolic pipelined matrix algorithms, pp. 171–180 in: *Parallel Computing '85*, Eds. M. Feilmeier, G. Joubert and U. Schendel, North-Holland, 1986.

[3] P. M. Dew, Systolic matrix iterative algorithms, in: *Parallel Computing '83*, Eds. M. Feilmeier, G. Joubert and U. Schendel, North-Holland, 1984, pp. 483–488.

[4] J. R. Westlake, *Numerical Matrix Inversion and Solution of Linear Equations*, Wiley & Sons Inc., U.S.A., 1968, pp. 32–34.

[5] A. Ralston and H. S. Wilf (eds.), *Mathematical Methods for Digital Computers*, Wiley & Sons Inc., U.S.A., 1960, pp. 73–77.

[6] G. M. Megson and D. J. Evans, Design and simulation of systolic arrays, Report 230, Comp. Stud., Loughborough University of Technology.

[7] Goodman, Leonberger, Kung and Athale, Optical interconnections for VLSI systems, *IEEE Proc.* **72**, No. 7.

[8] G. Rote, A systolic array algorithm for the algebraic path problem, *Computing* **34** (1985), 191–219.

Appendix 1

NUMERICAL EXAMPLES

i) Consider the 3∗3 *case*,

$$A=\tfrac{1}{9}\begin{bmatrix}-2 & 5 & -1\\ 4 & -1 & 2\\ -3 & 3 & 3\end{bmatrix},\quad A^{-1}=\begin{bmatrix}1 & 2 & -1\\ 2 & 1 & 0\\ -1 & 1 & 2\end{bmatrix}$$

now put,

$$B=\tfrac{1}{9}\begin{bmatrix}-2 & 5 & -1\\ 4 & -1 & 2\\ -3 & 3 & 3\end{bmatrix}+\begin{bmatrix}1\\0\\0\end{bmatrix}[2\ \ 1\ \ 3]$$

$$=\tfrac{1}{9}\begin{bmatrix}16 & 14 & 26\\ 4 & -1 & 2\\ -3 & 3 & 3\end{bmatrix}$$

using the systolic array with A^{-1} as input and $u=(1,0,0)^t$,

$v=(2,1,3)^t$. The output is shown in Tables 1 and 2. The large number of zeros indicates the delay for results to filter through the network. Timing is given as $T=7n+1$. For $n=3$, $T=22$ and an extra cycle is added to the data to shut down the array but plays no part in the algorithm.

The inverse is given as,

$$B^{-1}=\begin{bmatrix} 0.5 & -0.5 & -3.5 \\ 1 & -4.0 & -4.0, \\ -0.5 & 2.5 & 4.5 \end{bmatrix}$$

The simple check $BB^{-1}=I$ shows that the systolic array performs correctly. Finally, the generalised timing indicated that the first

Table I Output of modified inverse.

IOCCAM—Start run

0.000	0.000	0.000	0.000	0.000	0.000	0.000
0.000	0.000	0.000	0.000	0.000	0.000	0.000
0.000	0.000	0.000	0.000	0.000	0.000	0.000
0.000	0.000	0.000	0.000	0.000	0.000	0.000
0.000	0.000	0.000	0.000	0.000	0.000	0.000
0.000	0.000	0.000	0.000	0.000	0.000	0.000
0.000	0.000	0.000	0.000	0.000	0.000	0.000
0.000	0.000	0.000	0.000	0.000	0.000	0.000
0.000	0.000	0.000	0.000	0.000	0.000	0.000
0.000	0.000	0.000	0.000	0.000	0.000	0.000
0.000	0.000	0.000	0.000	0.000	0.000	0.000
0.000	0.000	0.000	0.000	0.000	0.000	0.000
0.000	0.000	0.000	0.000	0.000	0.000	0.000
0.000	0.000	0.000	0.000	0.000	0.000	0.000
0.000	0.000	0.000	0.000	0.000	0.000	0.000
0.000	0.000	0.000	0.000	0.000	0.000	0.000
0.000	0.000	0.000	0.500	0.000	0.000	0.000
0.000	0.000	−0.500	0.000	1.000	0.000	0.000
0.000	−3.500	0.000	−4.000	0.000	−0.500	0.000
0.000	0.000	−4.000	0.000	2.500	0.000	0.000
0.000	0.000	0.000	4.500	0.000	0.000	0.000
0.000	0.000	0.000	0.000	0.000	0.000	0.000
0.000	0.000	0.000	0.000	0.000	0.000	0.000

Time

OCCAM—Run finished

Table II Output of modified and original inverses interleaved.

OCCAM—Start run

0.000	0.000	0.000	0.000	0.000	0.000	0.000
0.000	0.000	0.000	0.000	0.000	0.000	0.000
0.000	0.000	0.000	0.000	0.000	0.000	0.000
0.000	0.000	0.000	0.000	0.000	0.000	0.000
0.000	0.000	0.000	0.000	0.000	0.000	0.000
0.000	0.000	0.000	0.000	0.000	0.000	0.000
0.000	0.000	0.000	0.000	0.000	0.000	0.000
0.000	0.000	0.000	0.000	0.000	0.000	0.000
0.000	0.000	0.000	0.000	0.000	0.000	0.000
0.000	0.000	0.000	0.000	0.000	0.000	0.000
0.000	0.000	0.000	0.000	0.000	0.000	0.000
0.000	0.000	0.000	0.000	0.000	0.000	0.000
0.000	0.000	0.000	0.000	0.000	0.000	0.000
0.000	0.000	0.000	0.000	0.000	0.000	0.000
0.000	0.000	0.000	0.000	0.000	0.000	0.000
0.000	0.000	0.000	1.000	0.000	0.000	0.000
0.000	0.000	2.000	0.500	2.000	0.000	0.000
0.000	−1.000	−0.500	1.000	1.000	−1.000	0.000
0.000	−3.500	1.000	−4.000	0.000	0.500	0.000
0.000	0.000	−4.000	2.000	2.500	0.000	0.000
0.000	0.000	0.000	4.500	0.000	0.000	0.000
0.000	0.000	0.000	0.000	0.000	0.000	0.000
0.000	0.000	0.000	0.000	0.000	0.000	0.000

Time

OCCAM—Run finished.

element emerges in the array after only $4n+2$ cycles. For $n=3$ this is 14 cycles, this too is correct, when we neglect the time for initial synchronisation of (2 cycles).

ii) *Throughput test*:

Next consider the problem of inputting separate problem instances in pipelined fashion. Choose three matrices (not necessarily of the same order) $A_1^{-1}, A_2^{-1}, A_3^{-1}$ and we require $B_1^{-1}, B_2^{-1}, B_3^{-1}$ using row and column modifying vectors u_i $i=1(1)3$, $v_j=j(1)3$. How close can we place separate instances $A_1^{-1}, A_2^{-1}, A_3^{-1}$ so they can be pipelined through the circuit. That is, what is the periodicity to input separate problems on the array. Table 3 gives us the answer. We used the same problem in (i) three times so correctness is easily verified. A separate problem instance can be input 2 cycles after the end of the

Table 3 Pipelining of separate problem instances.

OCCAM—Start run

0.000	0.000	0.000	0.000	0.000	0.000	0.000
0.000	0.000	0.000	0.000	0.000	0.000	0.000
0.000	0.000	0.000	0.000	0.000	0.000	0.000
0.000	0.000	0.000	0.000	0.000	0.000	0.000
0.000	0.000	0.000	0.000	0.000	0.000	0.000
0.000	0.000	0.000	0.000	0.000	0.000	0.000
0.000	0.000	0.000	0.000	0.000	0.000	0.000
0.000	0.000	0.000	0.000	0.000	0.000	0.000
0.000	0.000	0.000	0.000	0.000	0.000	0.000
0.000	0.000	0.000	0.000	0.000	0.000	0.000
0.000	0.000	0.000	0.000	0.000	0.000	0.000
0.000	0.000	0.000	0.000	0.000	0.000	0.000
0.000	0.000	0.000	0.000	0.000	0.000	0.000
0.000	0.000	0.000	0.000	0.000	0.000	0.000
0.000	0.000	0.000	0.000	0.000	0.000	0.000
0.000	0.000	0.000	1.000	0.000	0.000	0.000
0.000	0.000	2.000	0.500	2.000	0.000	0.000
0.000	−1.000	−0.500	1.000	1.000	1.000	0.000
0.000	−3.500	1.000	−4.000	0.000	0.500	0.000
0.000	0.000	−4.000	2.000	2.500	0.000	0.000
0.000	0.000	0.000	4.500	0.000	0.000	0.000
0.000	0.000	0.000	0.000	0.000	0.000	0.000
0.000	0.000	0.000	0.000	0.000	0.000	0.000
0.000	0.000	0.000	0.000	0.000	0.000	0.000
0.000	0.000	0.000	1.000	0.000	0.000	0.000
0.000	0.000	2.000	0.500	2.000	0.000	0.000
0.000	−1.000	−0.500	1.000	1.000	1.000	0.000
0.000	−3.500	1.000	−4.000	0.000	0.500	0.000
0.000	0.000	−4.000	2.000	2.500	0.000	0.000
0.000	0.000	0.000	4.500	0.000	0.000	0.000
0.000	0.000	0.000	0.000	0.000	0.000	0.000
0.000	0.000	0.000	0.000	0.000	0.000	0.000
0.000	0.000	0.000	0.000	0.000	0.000	0.000
0.000	0.000	0.000	1.000	0.000	0.000	0.000
0.000	0.000	2.000	0.500	2.000	0.000	0.000
0.000	−1.000	−0.500	1.000	1.000	1.000	0.000
0.000	−3.500	1.000	−4.000	0.000	0.500	0.000
0.000	0.000	−4.000	2.000	2.500	0.000	0.000
0.000	0.000	0.000	4.500	0.000	0.000	0.000
0.000	0.000	0.000	0.000	0.000	0.000	0.000
0.000	0.000	0.000	0.000	0.000	0.000	0.000

Time

OCCAM—Run finished.

previous problems input. The extra delay between problems is caused by the modifier delay. It follows that successive problems can be computed in a time

$$T=(3n-1)+k(2n)$$

where k = number of problems pipelined.

We conclude that the Rank Annihilation array is highly pipelined and achieves good throughput rates.

Systolic LU-Factorization "Dequeues" for Tridiagonal Systems

M. P. BEKAKOS and D. J. EVANS
Department of Computer Studies, Loughborough University of Technology, Loughborough, Leicestershire, UK

By the introduction of a new "rotate" and "fold" concept in the LU-decomposition strategy on a hexagonal systolic array of processors, the efficiency of the solution scheme for tridiagonal systems is shown to be increased.

The design of the pipelined systolic algorithm is then presented for both odd or even matrices.

KEY WORDS: Pipelined systolic algorithm, LU decomposition, "rotate and fold", hexagonal array processing.

C.R. CATEGORIES: C1.2, F1.1, F2.1, B7.1

1. INTRODUCTION

The factorization of a matrix $A=(a_{ij})$ into lower and upper triangular matrices L and U (i.e., LU-decomposition) has been proved possible on hexagonal systolic arrays (see [6, 7]), where it is assumed that the matrix A has the property that its LU-decomposition can be done without pivoting. This is true, for example, when A is a symmetric positive-definite, or an irreducible, diagonally dominant matrix. Usually this condition is not a problem, since most of the systems encountered in practice are diagonally dominant; on the other hand, there is currently no efficient way of incorporating a pivot strategy into existing vector or parallel algorithms.

Once the L and U factor matrices are computed, it is relatively easy to solve the resulting triangular linear systems.

The hex-connected systolic array of processors to implement the LU-decomposition, displayed in Figure 1, is constructed as follows. The processors below the upper boundaries are the standard IPSP's (Inner Product Step Processors) and are hex-connected in exactly the same manner as the matrix multiplication computing network. The processor at the top, denoted by a circle, is a special processor. It computes the reciprocal of its input and pumps the result southwest, and also pumps the same input northwards unchanged. The other processors on the upper boundaries are again IPSP's, but their orientation is changed: the processors on the upper left boundary are rotated 120 degrees clockwise; the processors on the upper right boundary are rotated 120 degrees counterclockwise. Again, the flow of data in the array is indicated by arrows.

The following theorem and lemma obtained from [7] illustrate the computational complexity of the scheme.

THEOREM 1.1 *Let $A=(a_{ij})$ be a $(n \times n)$-band matrix with bandwidth $w=p+q-1$, then a processor array having no more than pq hex-*

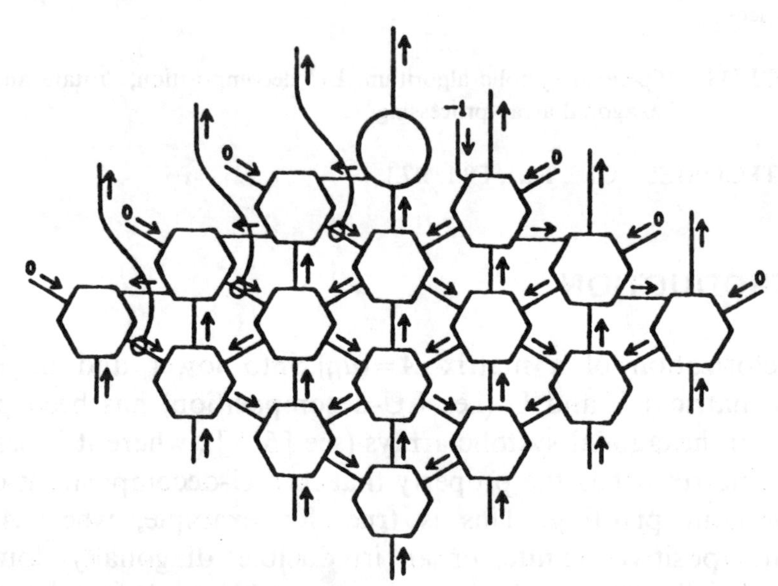

Figure 1 Hexagonal array of processors for pipelining the LU-decomposition of a $(n \times n)$-band matrix with bandwidth $w=7$.

connected processors can compute the LU-decomposition of A in $3n+min(p,q)$ *time-units.*

LEMMA 1.1 *Let* $A=(a_{ij})$ *be a* $(n\times n)$*-dense matrix, then* n^2 *hex-connected processors can compute the L and U matrices in* $4n$ *time-units.*

It should be noted that these complexities include I/O, control, and data movement.

The general recurrences for the pipelined systolic evaluation of the triangular matrices $L=(l_{ij})$ and $U=(u_{ij})$ for any $(n\times n)$-band matrix with bandwidth $w=p+q-1$ are the following:

$$a_{ij}^{(1)}=a_{ij}$$

$$a_{ij}^{(k+1)}=a_{ij}^{(k)}+l_{ik}(-u_{kj})$$

$$l_{ik}=\begin{cases} 0 & \text{if } i<k \\ 1 & \text{if } i=k \\ a_{ik}^{(k)}u_{kk}^{-1} & \text{if } i>k \end{cases} \tag{1.1}$$

$$u_{kj}=\begin{cases} 0 & \text{if } k>j \\ a_{kj}^{(k)} & \text{if } k\leq j. \end{cases}$$

2. THE "ROTATE" AND "FOLD" CONCEPT

When we apply the new rotate and fold concept, i.e., by proceeding from the top and bottom of the matrix simultaneously we obtain "two" LU-factorization streams functioning concurrently in opposite directions, one from the top downwards and the other vice versa. Certainly, the two factorization streams should confront each other in the center of the matrix and the degree of difficulty in handling the factorizing procedures in this part is directly dependent upon the size and the semi-bandwidths of the matrix. Herein, we shall exemplify the concept of the systolic LU-factorization dequeue for tridiagonal matrices.

THEOREM 2.1 *Let* $A=(a_{ij})$ *be a* $(n\times n)$*-band matrix with semi-bandwidths* $p=q=2$*, then by applying the "rotate" and "fold" tech-*

nique the factorization of A into the L and U matrices can be done in $\lceil 3n/2 \rceil + min(p, q)$ *time-units, using a hex-connected systolic network of pq processors. These will be described in Paradigm* (2.1a) *and* (2.1b) *for the cases when n is odd or even.*

Paradigm 2.1a We now discuss the case when n-odd and A is partitioned about the center element such that $A = A_1 + A_2$.

For the "rotate" and "fold" strategy the recurrences given in (1.1) will be modified to the following:

$$a_{ij}^{(1)} = a_{ij}$$

$$a_{ii}^{(2)} = a_{ii}^{(1)} + l_{i,i-1}(-u_{i-1,i})$$

$$l_{ir} = \begin{cases} 0 & \text{if } i < r \\ 1 & \text{if } i = r \\ a_{ir}^{(1)} u_{rr}^{-1} & \text{if } i > r \end{cases} \qquad \text{(1st stream)} \quad (2.1)$$

$$u_{rj} = \begin{cases} 0 & \text{if } r > j \\ a_{rj}^{(2)} & \text{if } r = j \neq 1 \\ a_{rj}^{(1)} & \text{if } r < j, r = j = 1 \end{cases}$$

and

$$a_{ij}^{(1)} = a_{ij},\ a_{ii}^{(3)} = a_{ii}^{(2)}$$

$$a_{ii}^{(2)} = a_{ii}^{(1)} + l_{i,i+1}(-u_{i+1,i})$$

$$l_{ir} = \begin{cases} 0 & \text{if } i > r \\ 1 & \text{if } i = r \\ a_{ir}^{(1)} u_{rr}^{-1} & \text{if } i < r \end{cases} \qquad \text{(2nd stream)} \quad (2.2)$$

$$u_{rj} = \begin{cases} 0 & \text{if } r < j \\ a_{rj}^{(2)} & \text{if } r = j \neq n \\ a_{rj}^{(1)} & \text{if } r > j, r = j = n. \end{cases}$$

Comment Since n is chosen to be odd, each stream in the destream procedure modifies the center element of the matrix middle row; hence, the $a_{(n+1)/2,(n+1)/2}$ element (only) is modified twice. This implies that after its first modification this element has to be collected "on-the-fly" from the output of the cell and brought back into the serial stream to re-enter that cell (in the same time-unit step) for the next modification.

The efficiency achieved in [6] from the single stream LU-decomposition scheme is: $E=1/3$, since in any row or column of the hex-connected systolic array, similar to the matrix multiplication case, only one out of every three consecutive processors is active at a given time.

The dequeue of data resulting when applying the "rotate" and "fold" concept to the two half-A matrices A_1, A_2, to take advantage of the process as "dormant" instances occur in the above implementation, together with the systolic array, are given in Figure 2.

The efficiency achieved has now been increased to: $E=1/2$, whereas the number of IPS cells is the same as for the usual LU-decomposition array.

For exemplary purposes of the "rotate" and "fold" concept, in Figure 3 are displayed all the computational steps required on this hex-connected systolic array of processors. For this specific paradigm, the LU-factorization is computed in $\lceil 3n/2 \rceil + \min(p,q) = 16$ time-units, instead of $3n + \min(q,q) = 29$ time-units that would be required normally.

Paradigm 2.1b We now discuss the case when n-even.

The recurrences that we shall use in this case are:

$$a_{ij}^{(1)} = a_{ij},\ a_{ii}^{(3)} = a_{ii}^{(2)}$$

$$a_{ii}^{(2)} = a_{ii}^{(1)} + l_{i,i-1}(-u_{i-1,i})$$

$$l_{ir} = \begin{cases} 0 & \text{if } i<r \\ 1 & \text{if } i=r \\ a_{ir}^{(1)} u_{rr}^{-1} & \text{if } i>r \end{cases} \quad \text{(1st stream)} \quad (2.3)$$

$$u_{rj}=\begin{cases} 0 & \text{if } r>j \\ a_{rj}^{(2)} & \text{if } r=j\neq 1 \\ a_{rj}^{(1)} & \text{if } r<j, r=j=1. \end{cases}$$

and

$$a_{ij}^{(1)}=a_{ij}$$

$$a_{ii}^{(2)}=a_{ii}^{(1)}+l_{i,i+1}(-u_{i+1,i})$$

$$l_{ir}=\begin{cases} 0 & \text{if } i>r \\ 1 & \text{if } i=r \\ a_{ir}^{(1)}u_{rr}^{-1} & \text{if } i<r \end{cases} \qquad \text{(2nd stream)} \qquad (2.4)$$

$$u_{rj}=\begin{cases} 0 & \text{if } r<j \\ a_{rj}^{(2)} & \text{if } r=j\neq n \\ a_{rj}^{(1)} & \text{if } r>j, r=j=n. \end{cases}$$

The dequeue of data resulting when applying the "rotate" and "fold" concept to the matrices A_1 and A_2, to benefit from the processors inactive instances occurring in the single LU-decomposition stream, together with the systolic array, are given in Figure 4.

The efficiency achieved has again been increased to: $E=1/2$, whereas the number of IPS cells is the same as for the usual LU-decomposition array.

Again for exemplary purposes of the "rotate" and "fold" concept for this case, in Figure 5 are displayed the computational steps required on this hex-connected systolic array of processors. For this specific paradigm, the LU-factorization is computed in $3n/2+\min(p,q)+1=9$ time-units, instead of $3n+\min(p,q)=14$ time-units that would be required normally.

Note that, the element $a_{((n/2)+1),((n/2)+1)}$ after its first modification has to be collected "on-the-fly" from the output of the cell and brought back into the serial stream to re-enter that cell, with a time-unit delay, for the next modification by the first wave.

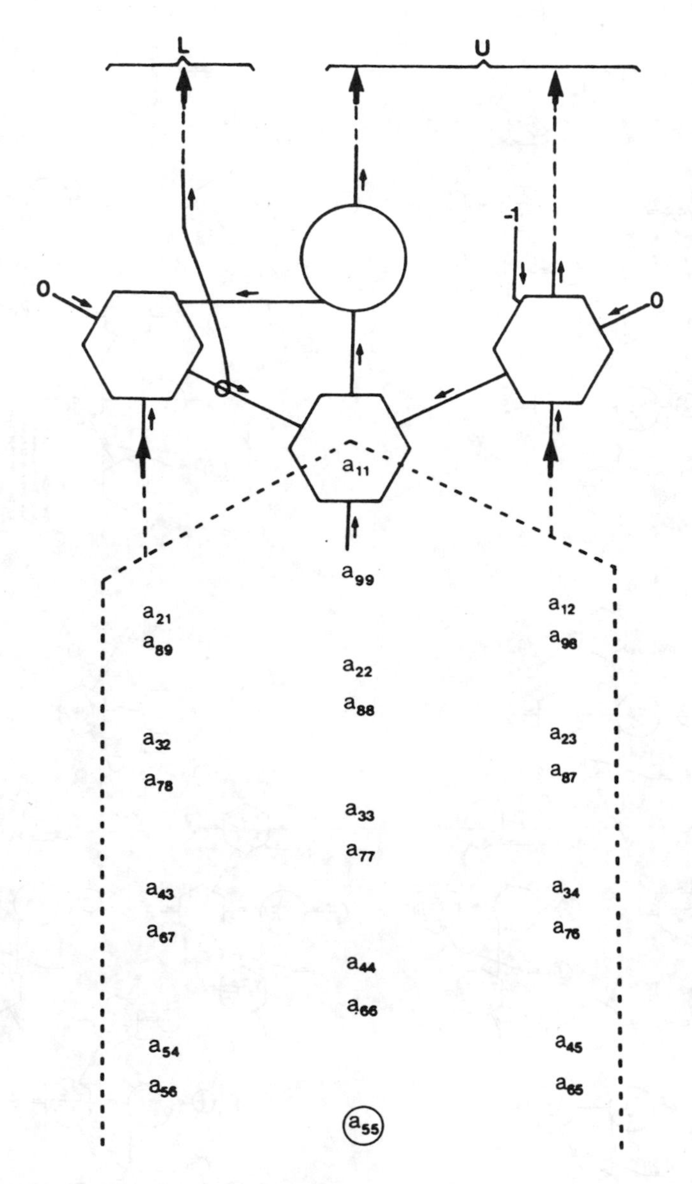

Figure 2 The dequeue of data for the LU-factorization on a hexagonal systolic array (for $p=q=2$ and $n=9$).

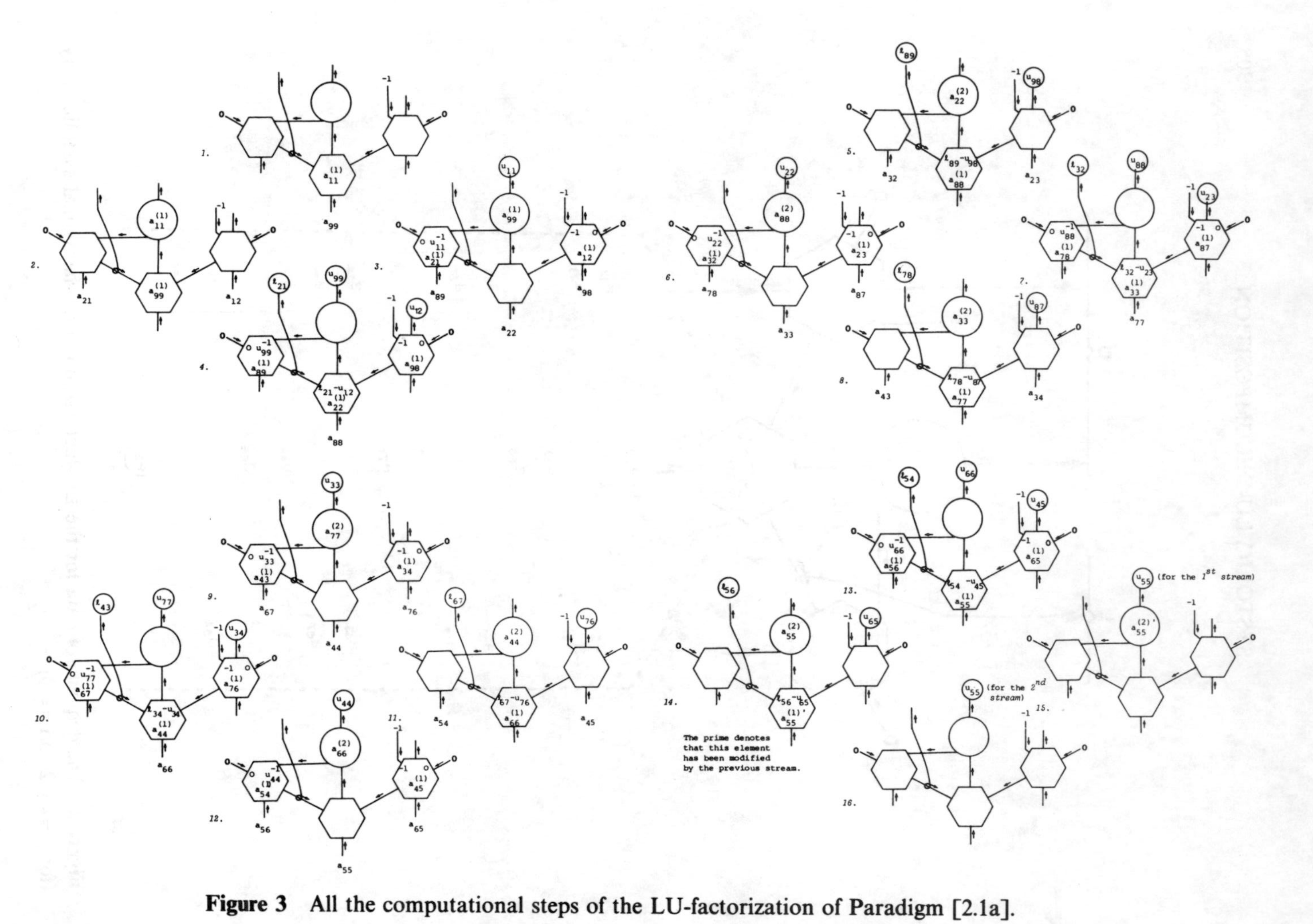

Figure 3 All the computational steps of the LU-factorization of Paradigm [2.1a].

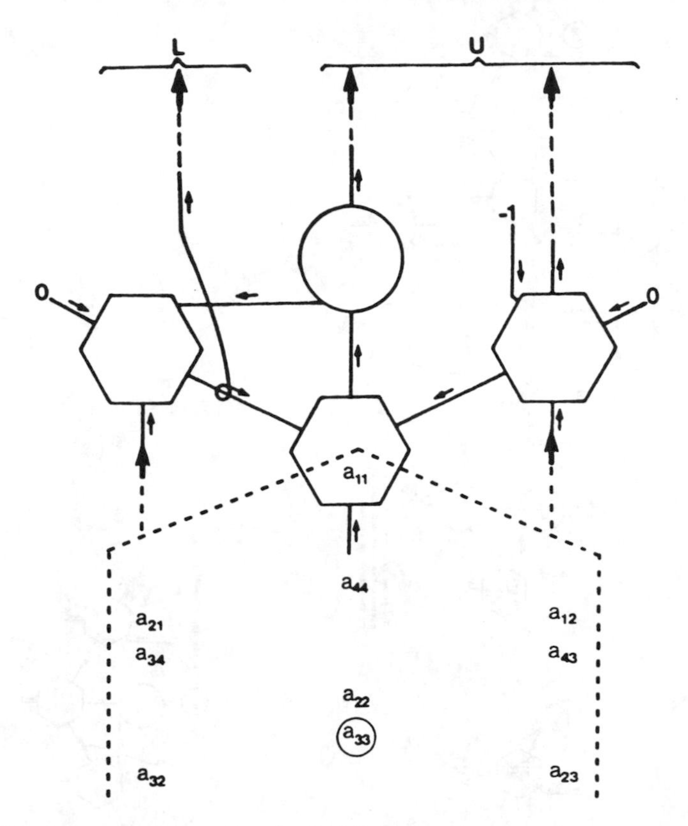

Figure 4 The dequeue of data for the LU-factorization on a hexagonal systolic array (for $p=q=2$ and $n=4$).

3. DEQUEUES FOR SOLVING TRIANGULAR LINEAR SYSTEMS

This section concerns itself with the solution of the corresponding triangular linear systems resulting from the factorization of the numerical examples of Paradigms [2.1, a and b] by applying the "rotate" and "fold" concept again, to obtain a similar increased efficiency as that for the factorization.

Suppose that we want to solve a linear system: $A\mathbf{x}=\mathbf{b}$. In fact, after the LU-decomposition process we have to solve two triangular linear systems:

$$\left.\begin{array}{l}\mathbf{L}\mathbf{y}=\mathbf{b}\\ \mathbf{U}\mathbf{x}=\mathbf{y}\end{array}\right\}. \tag{3.1}$$

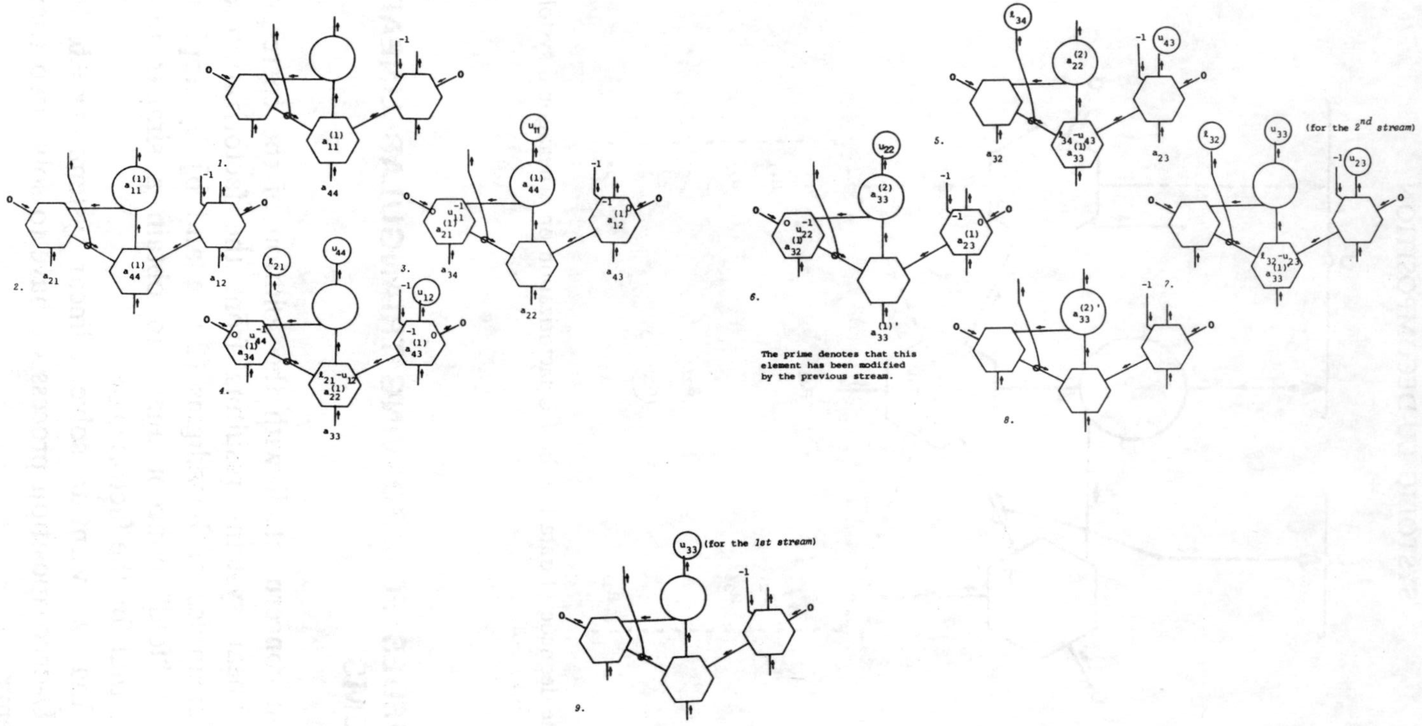
1.
2.
3.
4.
5.
6.
7.
8.
9.
The prime denotes that this element has been modified by the previous stream.
(for the 2nd stream)
(for the 1st stream)

Figure 5 All the computational steps of the LU-factorization of Paradigm [2.1b].

An upper triangular linear system can always be rewritten as a lower triangular linear system without any loss of generality.

Herein, we shall investigate both problems individually, making use of the same systolic network introduced in [6]; but, instead of a single data stream we shall form an appropriate dequeue for each of the problems above, to solve two triangular linear systems in the same amount of time required for the solution of one in the above implementation.

Further, note that, we shall tackle both these problems for the correspondingly resulting triangular linear systems of the above paradigms, in order to cover the instances of n being odd and even.

Case A (of Paradigm 2.1a)

i) Lower Triangular Linear Systems

Let $A=(a_{ij})$ be a non-singular $(n\times n)$-band lower triangular matrix. Suppose that A and vector $\mathbf{b}=(b_1,\ldots,b_n)^T$ are given. The problem is to compute $\mathbf{x}=(x_1,\ldots,x_n)^T$ such that: $A\mathbf{x}=\mathbf{b}$. The vector $\mathbf{x}$ can be computed by the following recurrences:

$$\begin{gathered} y_i^{(1)}=0 \\ y_i^{(2)}=y_i^{(1)}+a_{i,i-1}x_{i-1}=a_{i,i-1}x_{i-1} \\ x_i=(b_i-y_i^{(2)})/a_{ii},\ (\text{when } i=1, \text{ then } y_1^{(2)}=y_1^{(1)}). \end{gathered} \tag{3.2}$$

More specifically, let us consider the lower triangular linear system and compute the solution vector $\mathbf{x}$, using the above recurrences, i.e.,

$$\begin{array}{c} \uparrow \\ q \\ \downarrow \\ \\ \\ \\ \end{array} \underset{A}{\begin{bmatrix} a_{11} & & & \bigcirc \\ a_{21} & a_{22} & & \\ & \ddots & \ddots & \\ \bigcirc & & a_{n,n-1} & a_{n,n} \end{bmatrix}} \underset{\mathbf{x}}{\begin{bmatrix} x_1 \\ x_2 \\ \vdots \\ x_n \end{bmatrix}} = \underset{\mathbf{b}}{\begin{bmatrix} b_1 \\ b_2 \\ \vdots \\ b_n \end{bmatrix}}. \tag{3.3}$$

In this case (i.e., tridiagonal matrix), the bandwidth of the matrix is $w=q=2$. The above given recurrences, in (3.2), can be evaluated by the algorithm and network almost identical to those used for the band matrix-vector multiplication problem. The outline of this systolic network is illustrated in Figure 6.

On this network, the y_i, which are initially zero, move leftwards through the network, while the x_i, a_{ij}, and b_i are moving as indicated in Figure 6. The left-end processor is special in that it performs $x_i \leftarrow (b_i - y_i)/a_{ii}$. In fact, the special processor introduced in the LU-decomposition problem is a special case of this more general processor.

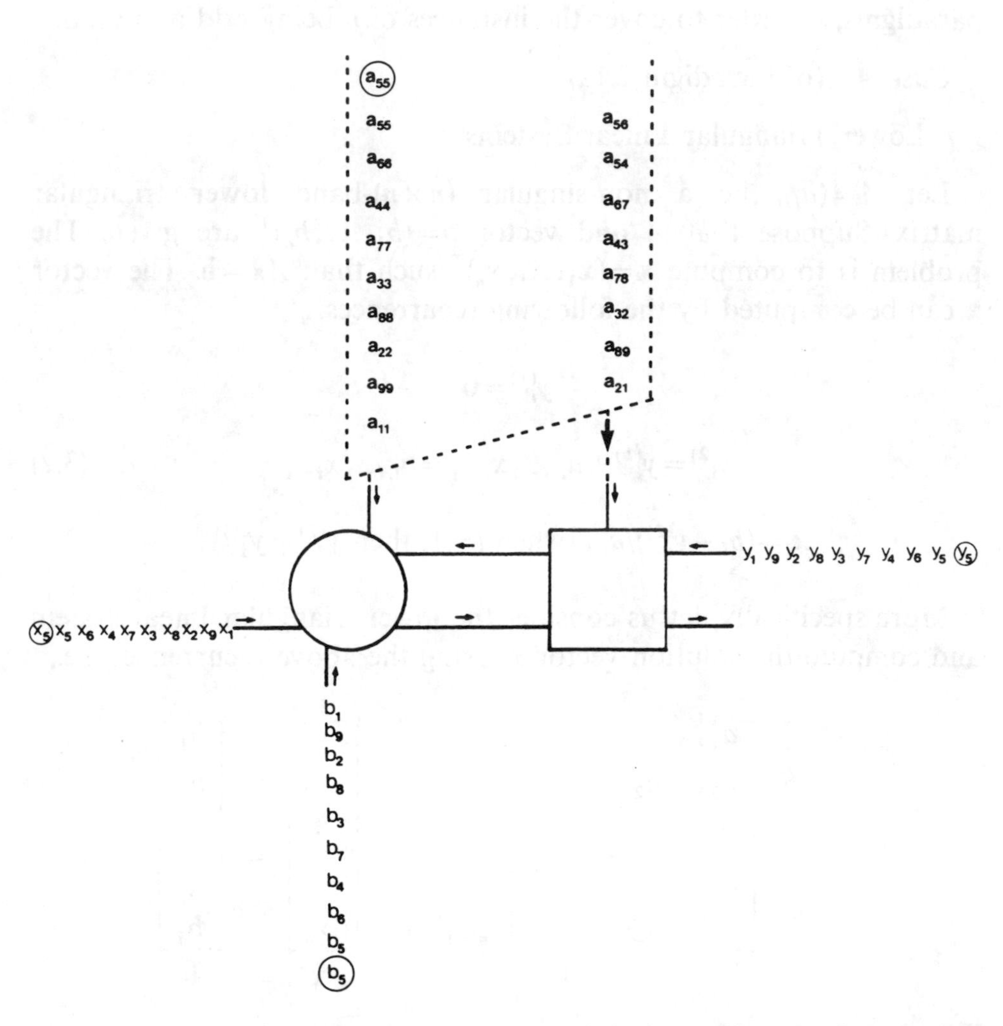

Figure 6 The dequeue of data for the solution of the lower triangular linear system of Paradigm [2.1a] on the linearly connected systolic array (for $w = q = 2$).

Each y_i accumulates an inner product term in the square processor as it moves to the left. At the time y_i reaches the left-end processor, it has, in general, the value $a_{i1}x_1+a_{i2}x_2+\cdots+a_{i,i-1}x_{i-1}$ and consequently the x_i computed by the formula above at the processor will have the correct value. From [6] we obtain the following Theorem 3.1.

THEOREM 3.1 *Let* $A=(a_{ij})$ *be a non-singular* $(n\times n)$*-band lower triangular matrix of bandwidth* $w=q$. *Suppose that A and a n-vector* $\mathbf{b}=(b_1,\ldots,b_n)^T$ *are given. Then, a n-vector* $\mathbf{x}=(x_1,\ldots,x_n)^T$ *such that:* $A\mathbf{x}=\mathbf{b}$, *can be computed in* $2n+q$ *time-units on a linearly connected systolic network of w IPSP's.*

Again, we shall take advantage from the fact that the number of processors required by the network can be reduced to $w/2$. In particular, we shall make use of the gaps (i.e., "idle" processors) by coalescing with the first stream of data, a second stream, thus halving the total solution time.

More specifically, we have to solve two lower triangular linear systems, i.e.,

$$\underbrace{\begin{bmatrix} a_{11} & & & \\ a_{21} & a_{22} & \bigcirc & \\ \ddots & \ddots & & \\ \bigcirc & & \ddots & \\ & & a_{k,k-1} & a_{k,k} \end{bmatrix}}_{A_1} \underbrace{\begin{bmatrix} x_1 \\ x_2 \\ \vdots \\ x_k \end{bmatrix}}_{\mathbf{x}_1} \quad \underbrace{\begin{bmatrix} b_1 \\ b_2 \\ \vdots \\ b_k \end{bmatrix}}_{\mathbf{b}_1}$$

and

$$\underbrace{\begin{bmatrix} a_{n,n} & & & \\ a_{n-1,n} & a_{n-1,n-1} & \bigcirc & \\ \ddots & \ddots & & \\ \bigcirc & & \ddots & \\ & & a_{k,k+1} & a_{k,k} \end{bmatrix}}_{A_2} \underbrace{\begin{bmatrix} x_n \\ x_{n-1} \\ \vdots \\ x_k \end{bmatrix}}_{\mathbf{x}_2} \quad \underbrace{\begin{bmatrix} b_n \\ b_{n-1} \\ \vdots \\ b_k \end{bmatrix}}_{\mathbf{b}_2}$$

where $k=(n+1)/2$.

The recurrence formulae for the system: $A_1\mathbf{x}_1=\mathbf{b}_1$ will be:

$$y_i^{(1)}=0$$

$$y_i^{(2)}=y_i^{(1)}+a_{i,i-1}x_{i-1}=a_{i,i-1}x_{i-1} \tag{3.4}$$

$$x_i=(b_i-y_i^{(2)})/a_{ii}, \text{ (when } i=1 \Rightarrow y_1^{(2)}=y_1^{(1)})$$

and for the system: $A_2\mathbf{x}_2=\mathbf{b}_2$:

$$y_i^{(1)}=0$$

$$y_i^{(2)}=y_i^{(1)}+a_{i,i+1}x_{i+1}=a_{i,i+1}x_{i+1} \tag{3.5}$$

$$x_i=(b_i-y_i^{(2)})/a_{ii}, \text{ (when } i=n \Rightarrow y_n^{(2)}=y_n^{(1)}).$$

Then we can state the following:

THEOREM 3.2 *An $(n\times n)$-band lower triangular linear system: $A\mathbf{x}=\mathbf{b}$, with* coefficient matrix bandwidth $w=q=2$, *can be solved in $n+2\bar{p}+q^*$ time-units applying the "rotate" and "fold" concept on a linearly connected systolic network of* w *IPPS's.*

Proof (See results of Figure 7).

The dequeue of data resulting when applying the above concept to the two half-parts A_1 and A_2 of the original matrix A, together with the systolic array are given in Figure 6.

Remark The efficiency achieved has been increased from: $E=1/2$ to 1, whereas the number of IPSP's is the same as in [6].

In Figure 7 are illustrated all the computational steps required on this linear systolic network of processors. Note from this figure that the common elements to both streams, denoted by circles in Figure 6, are subject to special handling. For this specific paradigm, the solution of the lower triangular linear system is computed in 13 time-units, since (n, p) are (odd, even)† instead of 20 time-units that would be required normally. It is obvious that the new technique, for n very large, pipelines the solution twice as fast.

---

*Note that, $\bar{p}$ is either $p-1$ or p (p is the seni-bandwidth of the original matrix) depending on the combination (odd/even) of (n, p).

†Hence $\bar{p}=p-1$.

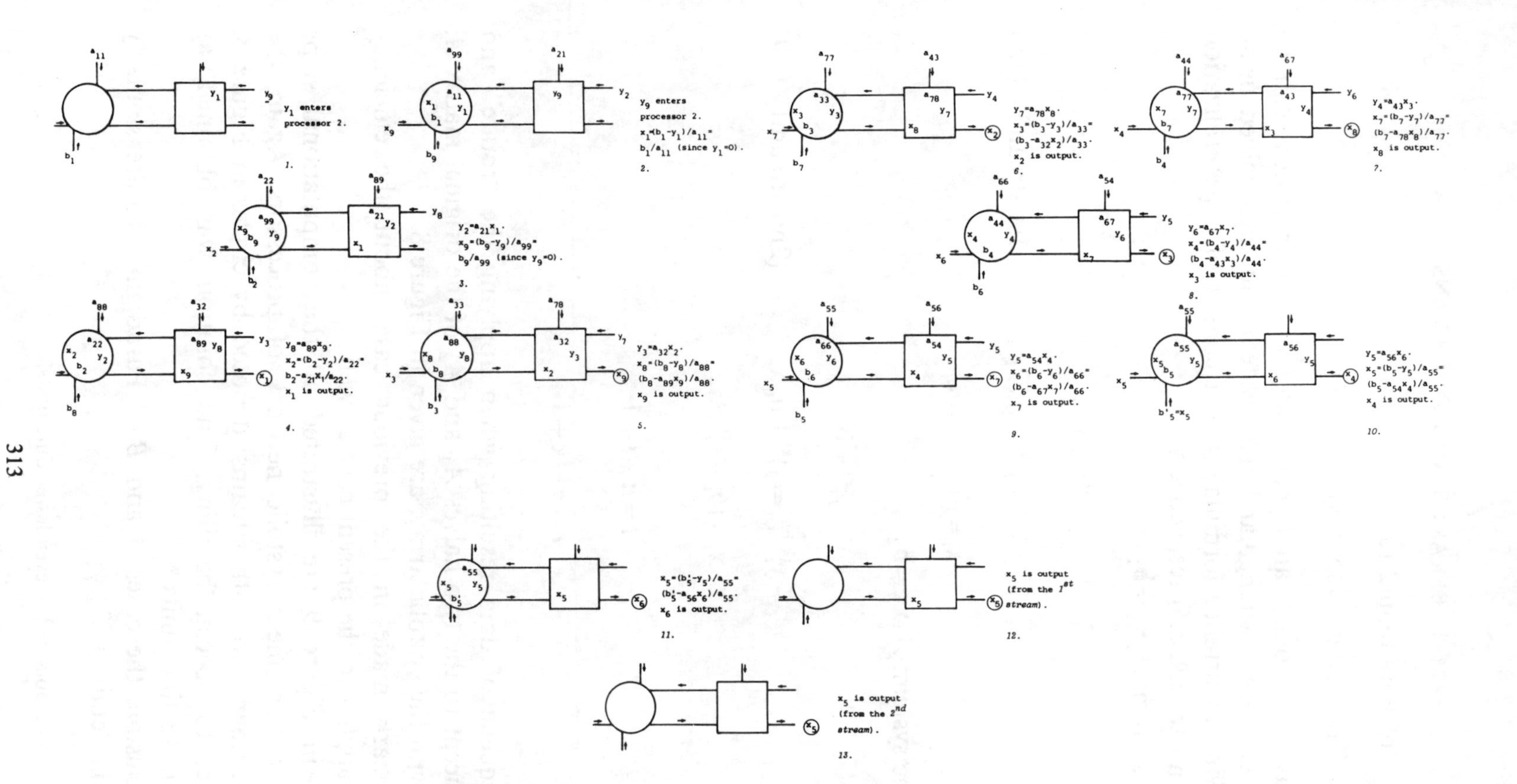

Figure 7 All the computational steps of the solution of the lower triangular linear system of Paradigm [2.1a].

Case B (of Paradigm 2.1a)

ii) Upper Triangular Linear Systems

As we have mentioned, an upper triangular linear system can be rewritten as a lower triangular linear system and then solved using the following recurrence formulae which consist of a generalization of the formulae (3.4, 3.5) respectively:

For the system: $A_1\mathbf{x}_1=\mathbf{b}_1$

$$\begin{aligned} y_i^{(k)}&=0\\ y_i^{(k+1)}&=y_i^{(k)}+a_{ik}x_k \qquad \text{(1st stream)} \qquad (3.6)\\ x_i&=(b_i-y_i^{(i)})/a_{ii} \end{aligned}$$

and for the system: $A_2\mathbf{x}_2=\mathbf{b}_2$

$$\begin{aligned} y_t^{(k)}&=0\\ y_t^{(k+1)}&=y_t^{(k)}+a_{tr}x_r \qquad \text{(2nd stream)} \qquad (3.7)\\ x_t&=(b_t-y_t^{(i)})/a_{tt}, \end{aligned}$$

where

$$\begin{aligned} t&=n-i+1\\ r&=n-k+1. \end{aligned}$$

The dequeue of data resulting when applying the "rotate" and "fold" concept to the two halves A_1 and A_2 of the original matrix A, together with the systolic array, are given in Figure 8.

The remark made in the previous case, about the efficiency achieved, applies to the present case as well.

Finally, in Figure 9 are illustrated all the computational steps required on this linear systolic network of processors. Again, the common elements to both streams, denoted by circles in Figure 8, are subject to special handling. The solution was obtained, as expected, in 13 time-units.*

In conclusion, the cases A and B of Paradigm 2.1b are similarly discussed in detail in [2].

---

*In fact it can be obtained in even fewer time-units.

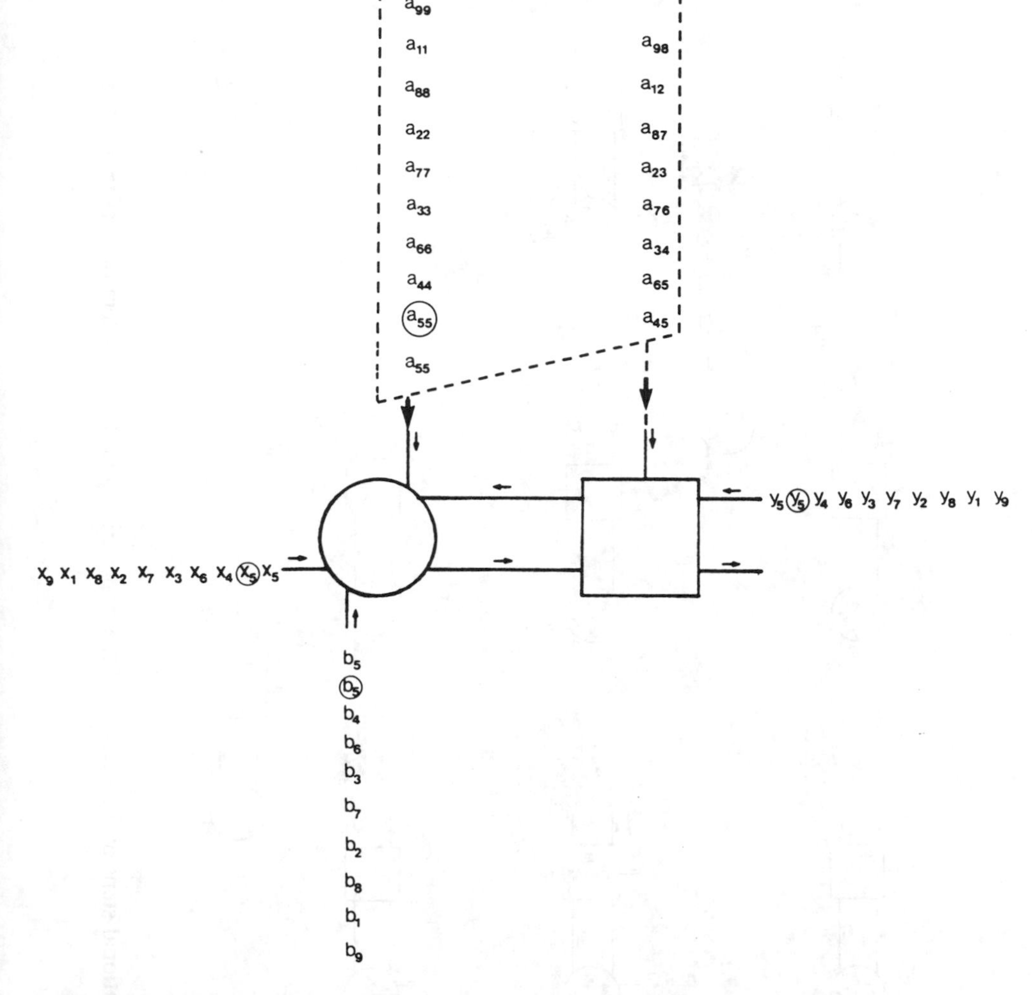

Figure 8 The dequeue of data for the solution of the upper triangular system of Paradigm [2.1a] on the linearly connected systolic array (for $w=q=2$).

4. GENERAL COMMENTS: THE PIVOTING PROBLEM, AND ORTHOGONAL FACTORIZATION

Research in interconnection networks and algorithms has been traditionally motivated by large scale parallel array computers, such as the ILLIAC IV. The technique presented herein was motivated by the advance in *VLSI*, albeit this is certainly applicable to any parallel computer complex. We have exemplified that many basic computations can be performed very efficiently by special-purpose

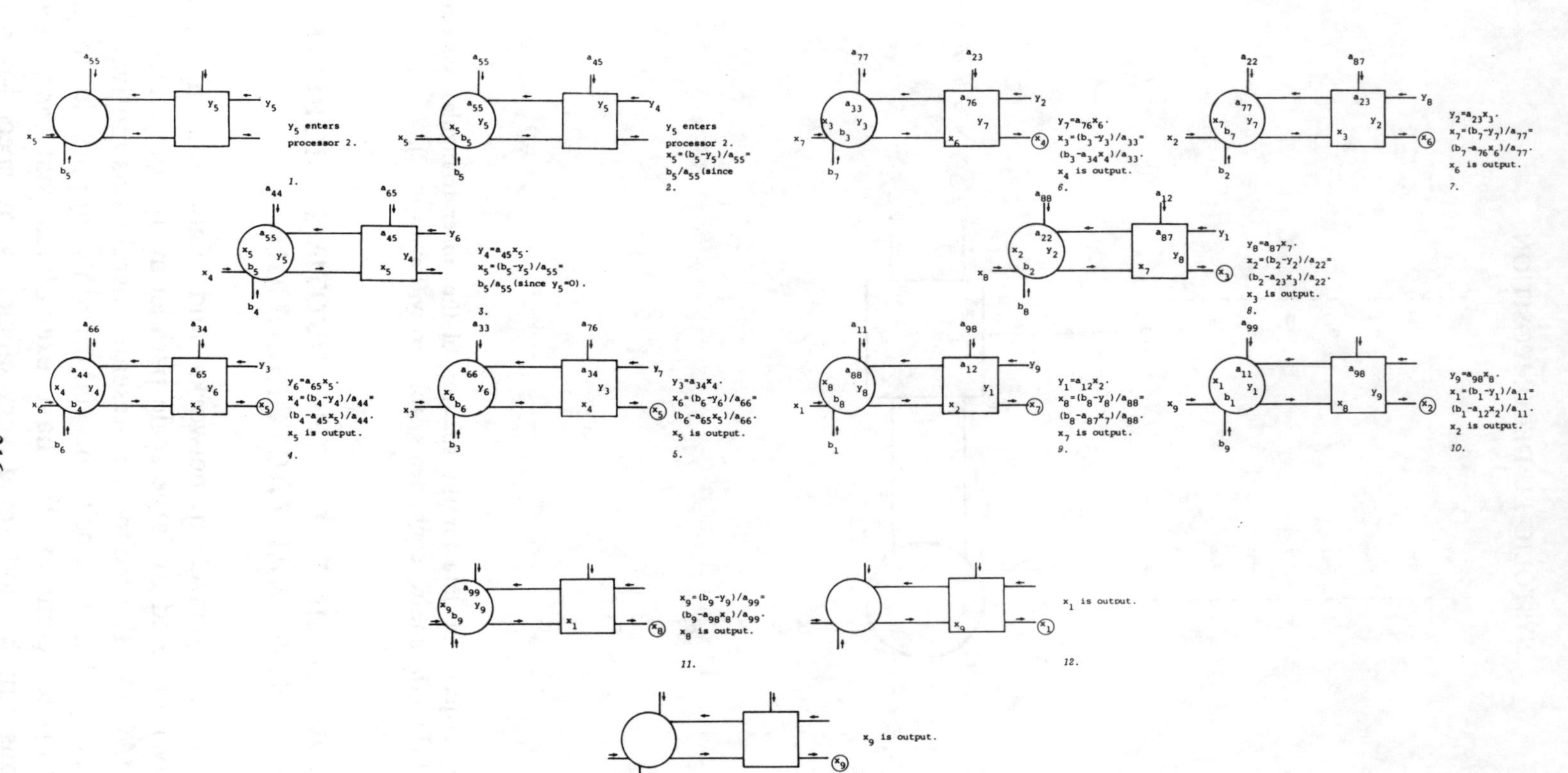

Figure 9 All the computational steps of the solution of the upper triangular linear system of Paradigm [2.1a].

multiple processor networks, which may be built very cheaply using the evolving *VLSI* technology. The important feature, common to all algorithms presented, is that their data flows are very simple and regular, and they are "pipelinable" algorithms.

In respect of the mathematical side, in everything that has been discussed previously, we have assumed that the matrices have the property, i.e., positive definite, that makes unnecessary the use of a pivoting strategy when the Gaussian elimination is applied to them. What, however, should one do if the matrices do not have this property? Note that, the Gaussian elimination becomes very inefficient on mesh-connected processors, if pivoting is necessary.

This question has motivated the consideration of Givens's transformation (see Hammarling [4]) for triangularizing a matrix, which is known to be a numerically stable method.

It turns out that, like Gaussian elimination without pivoting, the orthogonal factorization based on Givens's transformation can be implemented naturally on mesh-connected processors, although a pipelined implementation appears to be more complex. Sameh and Kuck, in [8], considered parallel linear system solvers based on Givens's transformation, but they did not give solutions to the processor communication problem considered here.

5. THE MATHEMATICAL EQUIVALENCE OF THE FOLDING ALGORITHM

Gaussian elimination can simply be described as a series of transformations that takes the matrix A into a triangular form given by,

$$N_{n-1}N_{n-2}\cdots N_1 A = U, \tag{5.1}$$

where N_i is the transformation that eliminates the elements a_{ji}, $i<j\leqq n$. We may introduce a series of similar transformations N_i' [5], which alternately with the N_i transformations eliminate the elements a_{ji}, $1\leqq j<i$. Thus, the folding algorithmic concept can be visualized as a Gaussian forward elimination process carried out from the "top" of the matrix and a Gaussian backward elimination process performed from the "bottom" of the matrix.

The transformation matrices N_i and N_i' are given by,

$$N_i=(n_{jk})=\begin{cases}-\dfrac{a_{jk}}{a_{kk}} & \text{for } k=i,\ k<j\leqq n\\ 1 & \text{for } j=k\\ 0 & \text{elsewhere}\end{cases}\tag{5.2}$$

and

$$N_i'=(n_{jk}')=\begin{cases}-\dfrac{a_{jk}}{a_{kk}} & \text{for } k=i,\ 1\leqq j<i\\ 1 & \text{for } j=k\\ 0 & \text{elsewhere.}\end{cases}\tag{5.3}$$

Let us suppose that n is odd (i.e., $n=2m+1$), then the elimination processes yield the following matrix identity:

$$N_{m+2}'N_m\cdots N_{n-1}'N_2N_n'N_1A=M,\tag{5.5}$$

where M has the form:

O

$\frac{n+1}{2}$

middle row

O

$\frac{n+1}{2}$

(5.6)

In a similar manner, if n is even (i.e., $n=2m$), the elimination processes when carried out produce the matrix identity:

$$N_{m+2}'N_{m-1}\cdots N_{n-1}'N_2N_n'N_1A=M',\tag{5.7}$$

where M' has the form:

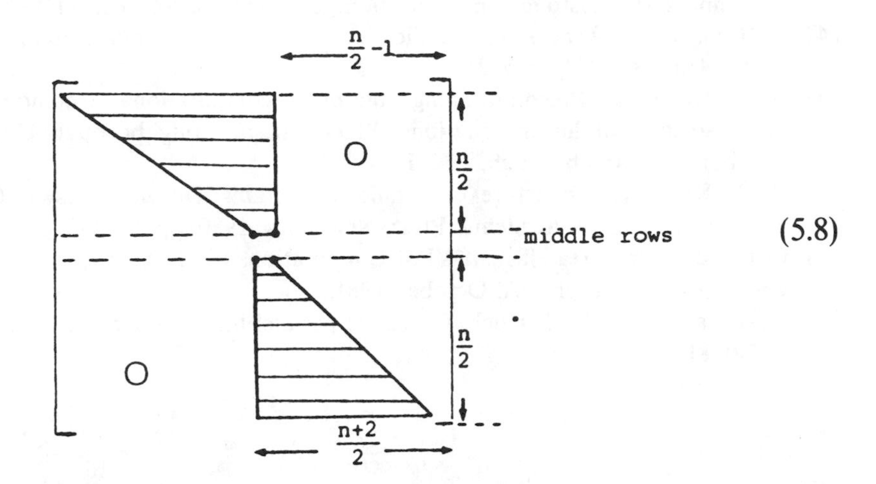

(5.8)

For the solution of the transformed system with a coefficient matrix as in (5.6) a backward and a forward substitution process commencing at the $(n+1)/2$ diagonal element is required; whereas, for the system with a coefficient matrix as in (5.8), we first have to solve the resulting (2×2) central subsystem, before carrying out the necessary backward and forward substitution processes from the $(n/2)+1$ and $n/2$ diagonal elements, respectively.

Finally, it is obvious that for a full general matrix the folding method requires more eliminations (i.e., more numerical operations) compared with the standard unidirectional Gaussian elimination. It is also interesting to note that the solution process for the folding method is effected from the center values of the matrix outwards, thus reducing the growth of the rounding error in the forward and backward substitution processes.

In conclusion, the folding method is very suitable for pipelining on *VLSI* systolic arrays of banded matrices.

References

[1] M. P. Bekakos and D. J. Evans, The exposure and exploitation of parallelism on fifth generation computer systems. In: *Proc. of Parallel Computing '85*, Berlin, FRG, September 1985; also *Parallel Computing 85*, M. Feilmeier *et al.* (eds.), North-Holland, 1986, pp. 425–442.

[2] M. P. Bekakos, A study of algorithms for parallel computers and VLSI systolic processor arrays, Ph.D. Thesis, Loughborough University of Technology, Loughborough, UK, 1986.

[3] M. P. Bekakos and D. J. Evans, A "rotating" and "folding" algorithm using a two-dimensional systolic communication geometry, *Par. Comp.* **4** (1987), 221–228.
[4] S. Hammarling, A note on modifications to the Givens' plane rotation, *J. Inst. Math. Appl.* **13** (1974), 215–218.
[5] M. Hatzopoulos, Preconditioning and other computational techniques for the direct solution of linear equations, Ph.D. Thesis, Loughborough University of Technology, Loughborough, UK, 1974.
[6] H. T. Kung, The structure of parallel algorithms. In: *Advances in Computers*, Yovits (ed.), Vol. 19, Academic Press, New York, 1980, pp. 65–112.
[7] C. E. Leiserson, Area-efficient VLSI computation, Ph.D. Thesis, Carnegie-Mellon University, Pittsburgh, PA, October, 1981.
[8] A. H. Sameh and D. J. Kuck, On stable parallel linear system solvers, *JACM* **25** (1978), 81–91.

Compact Systolic Arrays for Incomplete Matrix Factorisation Methods

D. J. EVANS and G. M. MEGSON

Department of Computer Studies, Loughborough University of Technology, Loughborough, Leicestershire, UK

The preconditioning strategy of the incomplete factorisation methods for the numerical solution of sparse linear systems is developed for VLSI architectures to yield compact systolic factorisation arrays.

KEY WORDS: Incomplete factorisation, fill-in, sparse linear systems.

C.R. CATEGORIES: F.1.1, F.2.1, B.7.1.

1. INTRODUCTION

In this paper the technique of Extended to the Limit (EL) sparse LU factorisation methods for the solution of large spare linear systems is applied to Systolic Architectures to produce compact area efficient designs with reduced computation time and a small number of input and output connections.

Incomplete factorisation strategies are becoming increasingly popular as methods for solving large linear systems because they allow the amount of storage required for the triangular matrix factors to be controlled, and also allow a large reduction in the arithmetic work required for a complete factorisation. When solving a large sparse system of linear equations of the form,

$$A\mathbf{u}=\mathbf{d}, \tag{1.0}$$

the use of direct methods like factorisation admit the possibility of fill-in during the production of the factors, often destroying the sparse nature of the matrix or of bands within a matrix. From a numerical point of view it is useful to be able to control the fill-in factor so that the number of non-zeroes created is relatively small in relation to the number of non-zeroes of A. Further, the control of the fill-ins must be performed in such a way that the available memory can be used efficiently, and that the algorithm is both easily specified and efficient. A number of possibilities for omitting fill-in elements are available some of which involve:

a) restricting specified locations of the coefficient matrix to be filled in or preserved.
b) distributing storage equally among rows of the matrix.
c) neglect only small elements—on the basis that they will not affect the result significantly.
d) fill the storage then neglect any fill-ins that cannot be accommodated.

We shall concentrate on the technique of Evans and Lipitakis [2] known as extending to the limit (EL) factorisation which can be classed as a method of type (a) as it restricts fill-ins to certain diagonals of the matrix. In particular, the way that the control of fill-in elements translates to area usage of a systolic array is explored with respect to the EL strategy. We can translate fixed storage to fixed area and consider the implementation of an incomplete factorisation process on a systolic array which is smaller or more compact than would normally be required. Throughout the report a working knowledge of arrays from [3] and [11] is assumed.

1.1 Sample problems

In order to assess the success of the compact or incomplete systolic arrays, produced in this paper the following problems derived from 2-D and 3-D parabolic and elliptic equations are presented. These have certain advantages for besides being large and sparse systems they have also been studied in great detail in [1] with respect to the EL factorisation, and give a good basis from which to assess new systolic EL type algorithms.

Problem I:

The parabolic equation,

$$\frac{\partial^2 u}{\partial x^2}+\frac{\partial^2 u}{\partial y^2}=0, \quad (x,y)\in R\equiv(0,1)\times(0,1) \tag{1.1}$$

with boundary conditions $u(x,y)=0$, produces a system based on the five-point finite difference formula,

$$4u_{ij}-u_{i+1,j}-u_{i-1,j}-u_{i,j+1}-u_{i,j-1}=0 \tag{1.2}$$

when a mesh with spacing Δx in the x-direction and Δy in the y-direction is superimposed on R to give gridpoints $u(i\Delta x, j\Delta y)=u_{ij}$. If there are n^2 internal grid points a column-wise ordering of the points results in a real quindiagonal matrix of order n^2, with the form,

$$A=\begin{bmatrix} A_1 & -I & & & 0 \\ -I & \ddots & \ddots & & \\ & \ddots & \ddots & \ddots & \\ & & \ddots & \ddots & -I \\ 0 & & & -I & A_n \end{bmatrix},$$

$$A_i=\begin{bmatrix} 4 & -1 & & & 0 \\ -1 & \ddots & \ddots & & \\ & \ddots & \ddots & \ddots & \\ & & \ddots & \ddots & -1 \\ 0 & & & -1 & 4 \end{bmatrix} \quad 1\leqq i\leqq n, \tag{1.3}$$

with $\mathbf{d}$ and $\mathbf{u}$ being $n^2 * 1$ known and unknown vectors respectively, of (1.0) and $\mathbf{d}$ is made up from the boundary conditions of the problem.

Problem II:

The 3-D problem of the form,

$$\frac{\partial^2 u}{\partial x^2}+\frac{\partial^2 u}{\partial y^2}+\frac{\partial^2 u}{\partial z^2}=0 \tag{1.4}$$

$$(x, y, z) \in R \equiv (0,1)\times(0,1)\times(0,1),$$

and boundary conditions,

$$u(x, y, z)=0$$

which produces a cube dissected by a three dimensional grid with spacing Δx, Δy, Δz and when $\Delta x=\Delta y=\Delta z$ produces the seven-point finite difference formula,

$$\begin{aligned}6u_{i,j,k}-u_{i+1,j,k}-u_{i-1,j,k}-u_{i,j+1,k}-u_{i,j-1,k}\\ -u_{i,j,k+1}-u_{i,j,k+1}\end{aligned} \tag{1.5}$$

specified at the n^3 gridpoints in R with $u(i\Delta x, j\Delta y, k\Delta z)=u_{i,j,k}$. Ordering these points produces a linear system similar to the problem I with A an n^3 matrix with seven non-zero diagonals and of form,

$$A=\begin{bmatrix} A_1 & -I & & 0 \\ -I & \ddots & \ddots & \\ & \ddots & \ddots & -I \\ 0 & & -I & A_n \end{bmatrix}_{n^3 * n^3} \tag{1.6}$$

with

$$A_i=\begin{bmatrix} B_{j+1} & -J & & 0 \\ -J & \ddots & \ddots & \\ & \ddots & \ddots & -J \\ 0 & & -J & B_{j+n} \end{bmatrix}_{n^2 * n^2}$$

and

$$B_j = \begin{bmatrix} 6 & -1 & & & 0 \\ -1 & \ddots & \ddots & & \\ & \ddots & \ddots & \ddots & \\ & & \ddots & \ddots & -1 \\ 0 & & & -1 & 6 \end{bmatrix}_{n*n} \quad i=1,\ldots,m$$

$$j=(i-1)m, \quad i=1,\ldots,m$$

with **I** and **J** the $n^2 * n^2$ and $n*n$ identify matrices respectively, and **u** and **d** as n^3*1 vectors.

Clearly when n is large these matrices fulfil the requirement of large sparse and banded matrices where even the band is sparse in structure.

1.2 The Extension to the limit (EL) sparse factorisation method

The EL factorisation procedure was introduced by Evans and Lipitakis in [2], and produces various approximate and exact factorisations of the matrix A in (1.0). Essentially, an approximate factorisation method is obtained if A is replaced by $(A+\bar{R})$ such that,

$$A+\bar{R}=L_sU_s, \tag{1.7}$$

with L_s and U_s as sparse lower and upper triangular factors respectively, whose product approximates A, with $\bar{R}$ the error of approximation. The concept of a limit is produced by generating a sequence of algorithms which compute various L_s, U_s and $\bar{R}$ matrices,

$$\begin{array}{lcl} \text{COMPUTATION} & & \text{ALGORITHM} \\ A+\bar{R}_1=L_{s_1}U_{s_1} & \Rightarrow & \text{FACTOR(1)} \\ A+\bar{R}_2=L_{s_2}U_{s_2} & \Rightarrow & \text{FACTOR(2)} \\ \vdots \quad \vdots & & \vdots \\ A+\bar{R}_k=L_{s_k}U_{s_k} & \Rightarrow & \text{FACTOR}(k) \\ \vdots \quad \vdots & & \vdots \\ A+\bar{R}_z=L_{s_z}U_{s_z} & \Rightarrow & \text{FACTOR}(z) \end{array} \tag{1.8}$$

which produce a corresponding sequence of decreasing residuals, such that,

$$\lim_{k \to z} (A - L_{s_k} U_{s_k})\mathbf{u} = \mathbf{0}. \tag{1.9}$$

When $k = z$, the residual is zero and Factor(z) is the optimal or complete factorisation. The problem is to define an algorithm which allows free movement up and down the sequence of Factor(k) algorithms permitting arbitrary approximations based on storage requirements or computational complexity. Clearly there are many such sequences of matrices $\bar{R}_k$, which produce such corresponding algorithms. What Evans and Lipitakis [2] achieved was the discovery of a Global Algorithm which embodied all the Factor(k) algorithms of a sequence thus implicitly defining the $\bar{R}_k$ sequence. Further this factorisation algorithm (ALUBOT) permitted movement up and down the sequence by the specification of a small number of parameters. In fact the parameter guides were related to the number of zero diagonals in A which were permitted the fill-in during the factorisation process. For the sake of completeness we briefly review the method and details can be found in [1] and [2]. Consider solving (1.0) when the coefficient matrix A is derived from Problem I, a complete or exact factorisation (neglecting rounding error) is given by

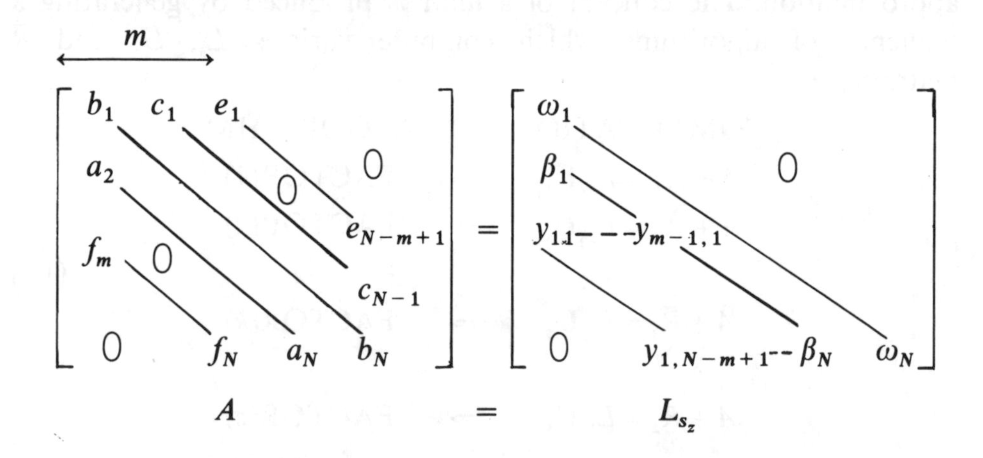

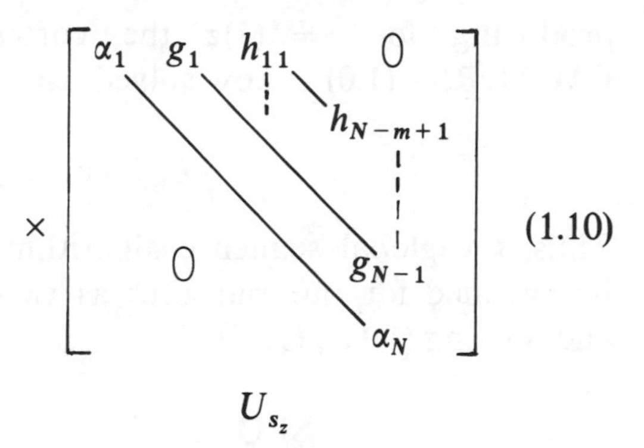

(1.10)

with $N=n^2$ and $m=n+1$. For the incomplete factorisation we specify a value r which dictates the number of non-zero diagonals to be retained for fill-in, moving left from the outermost diagonals e_i and f_i in A. This results in an approximate factorisation of the form,

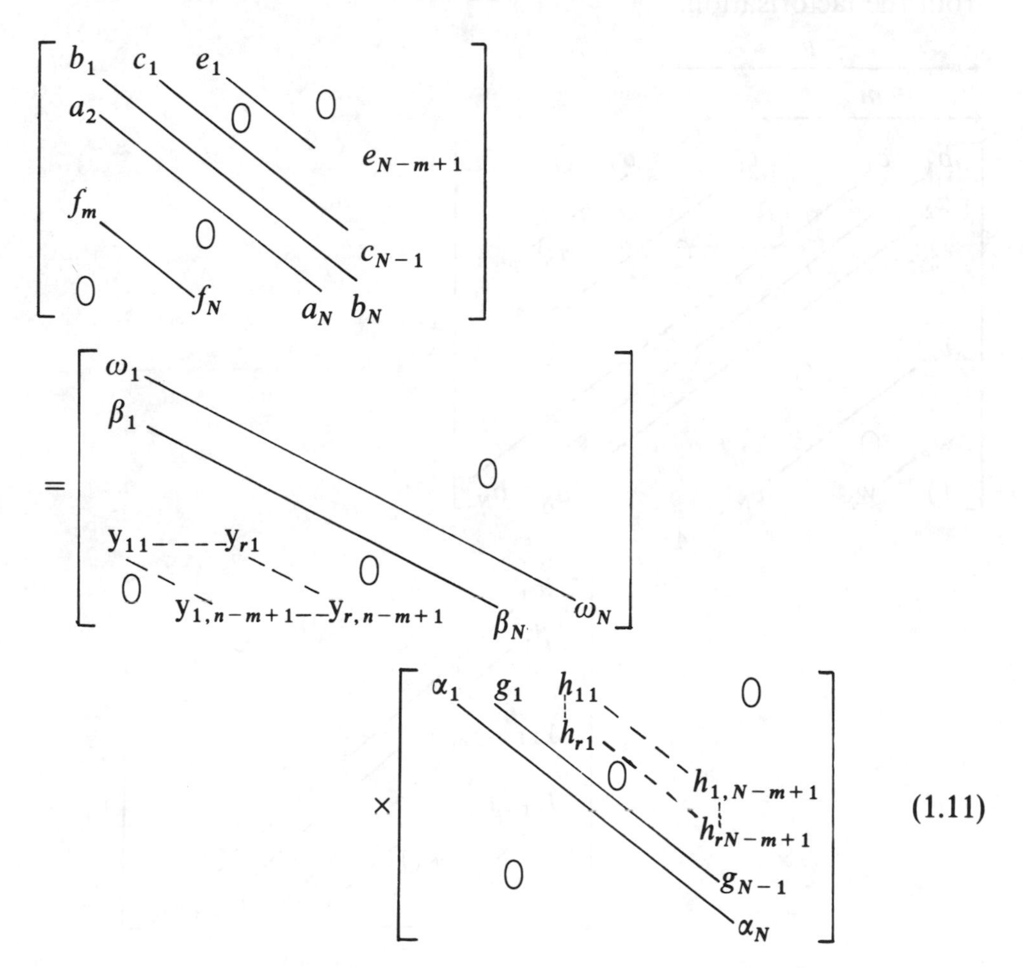

(1.11)

producing for $r=1(1)z$ the corresponding sequence algorithm FACTOR(r). (1.0) is now solved using the two coupled systems,

$$L_{s_r}\mathbf{y}=\mathbf{d},\ U_{s_r}\mathbf{u}=\mathbf{d}. \tag{1.12}$$

Thus, the global sequence algorithm (the ALUBOT algorithm) can be encoded for our purposes as two procedures called sequentially and solving (1.11), (1.12),

SEQ
ALUBOT(N, m, r, A) (1.13)
FBSUBS($N, m, r, L_{s_r}, U_{s_r}, \mathbf{d}$)

Similarly, a sequence controller for solving problem II is derived from the factorisation,

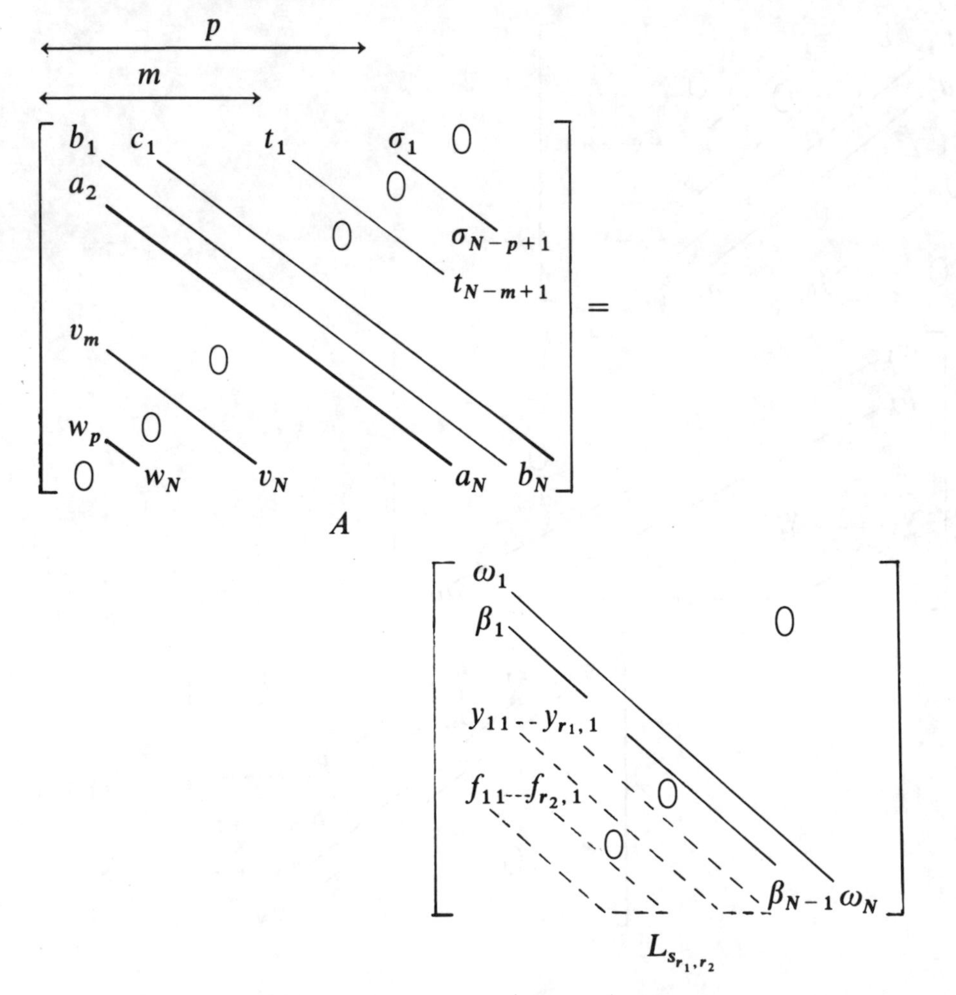

$$\times \begin{bmatrix} \alpha_1 & g_1 & & h_{1,1} & & t_{1,1} & \\ & \ddots & \ddots & \vdots\;h_{r_1,1} & \ddots & \vdots\;t_{r_2,1} & \ddots \\ & & \ddots & \ddots & 0 & \ddots & \\ & 0 & & \ddots & \ddots & g_{N-1} & \\ & & & & & \alpha_N & \end{bmatrix} \quad (1.14)$$

$$U_{s_{r_1,r_2}}$$

producing the sequence for the solution of

SEQ
  ALUBOT-2(N,m,p,r_1,r_2,A) (1.15)
  FBSUBS-2$(N,m,p,r_1,r_2,L_{s_{r_1,r_2}},U_{s_{r_1,r_2}},\mathbf{d})$

whilst the generalisation to more bands is trivial. Clearly, the method saves storage over the complete factorisation, and computation time too as the fill-ins outside the bands are simply neglected (i.e. not computed). The above algorithms indicate that we use the approximate factors as a direct method for solving (1.0), implicitly accepting the error of approximation in the solution. Alternatively the above system can be solved directly and then iterated using iterative refinement to converge to the answer using residuals. Such iteration schemes have compared favourably with standard methods for solving these problems. From [1] it is known that the number of retained diagonals r_i is small, for instance $0<r_i\leqq 4$ will give a good initial approximation for the iterative scheme. Notice that the retained diagonals are grouped or locally placed for each outer diagonal which is a promising systolic property. We are interested in mapping the concept of an approximation sequence of algorithms like the one embodied above into a systolic array design structure.

2. SYSTOLIC 'EXTENSION TO THE LIMIT' (EL) ARCHITECTURES

In considering the hardware cost of solving systems like (1.10) and (1.11), the LU factorisation of a matrix of a given bandwidth $W=$

$p+q-1$ (say) can be computed on a hexagonally-connected systolic array of pq inner product cells in a time $T=3N+\min(p,q)$, where p and q are the number of super- and sub-diagonals in the band respectively [3], [4]. Thus, for problem I, with $N=n^2$, and $W=2n+1$, we require $(n+1)^2$ cells and $T=[3n^2+(n+1)]$ inner product steps to factorise the matrix A. Similarly for problem II where $N=n^3$ and $W=2n^2+1$, requires $(n^2+1)^2$ cells and time $T=3n^3+(n^2+1)$. Clearly for large grid sizes in the finite difference approximation, the resulting systems are very large and the number of cells in the designs is enormous. Even though VLSI technology has developed at a rapid pace it is still unlikely that $0(n^4)$ cells is practical in the case of problem II.

The problem arises from the fact that although the bands in (1.11) and (1.14) are extremely sparse, the LU factorisation creates fill-in into the zero elements of the array. Consideration of the hexagonal scheme of [3] indicates that wherever a fill-in occurs a cell must be retained in the array design. Thus for a complete factorisation, the whole of the zero region between the central band and the outer diagonals fills in producing a high cell count. It should now be self evident that an incomplete factorisation like those of the previous section admit the possibility of large cell savings. Consequently we define an approximation sequence of systolic architectures which scales the accuracy of factorisation with increases in the cell count and hence area. This view of systolic array derivation is supported by Thomson and Tucker [5], in which a semi-formal model of algorithm designs suggests a sequence of draft designs, formalised by algorithm transformations as a result of design decisions. In the present context we view draft algorithms as different systolic arrays, and algorithm transformations as movements up and down the approximation sequence (1.8), with design decisions based on area and accuracy requirements controlled by the parameters r_i, $i=1(1)z$.

2.1 First level compaction

Consider now the algorithms (1.13) and (1.15) in conjunction with Figure 1. Let W_s be the semi-bandwidth of A, and W_i be the interior bandwidth of the central band. From a close observation of the standard hex of [3], two features are evident.

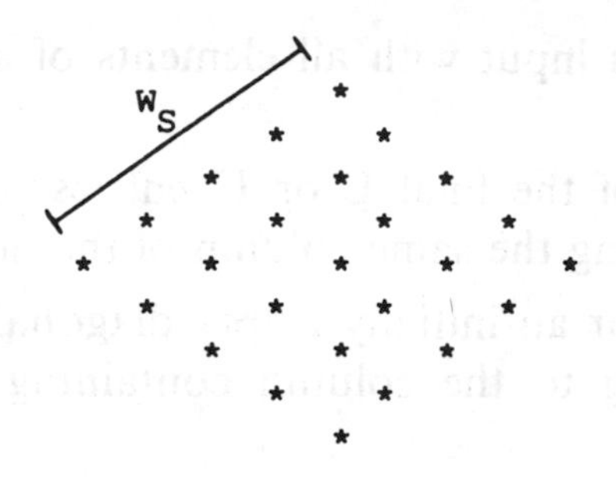

a) A complete hex

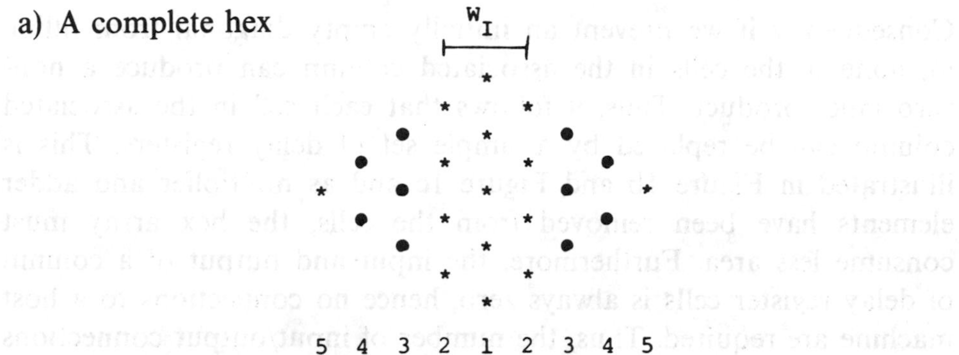

b) Compacted hex for 2-D problem I ($m=5, r=1$)

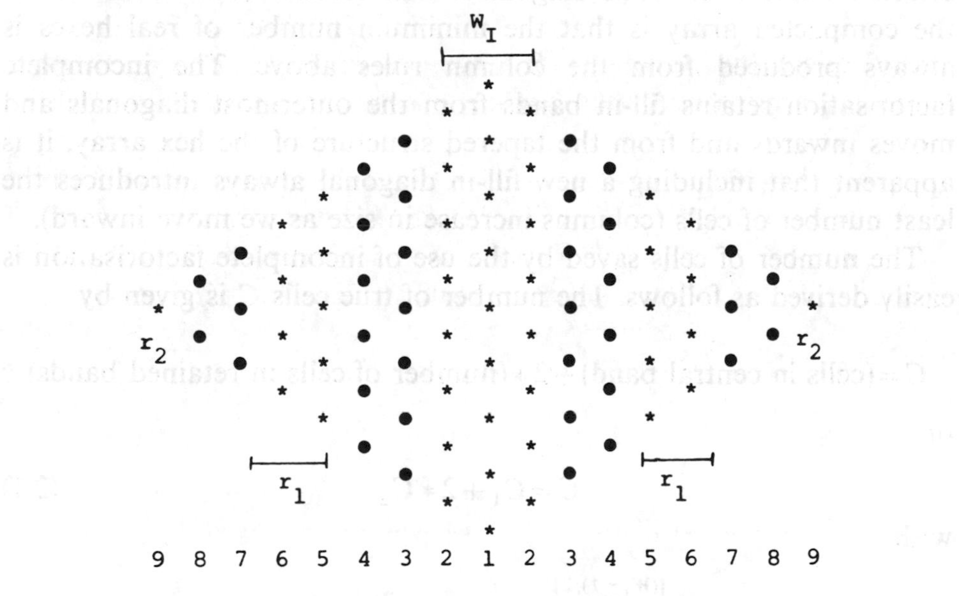

c) Larger compacted hex for 3-D problem II ($m=6$, $p=9$, $r_1=2$, $r_2=1$) ⋆ = true hex cell, ● = delay (dummy) hex cell

Figure 1 Cell replacement and compaction for EL algorithm.

a) The matrix A is input with all elements of a diagonal entering the same cell.
b) Accumulation of the final L or U entries of a diagonal occur only in cells along the same column of the hex.
c) A fill-in entry for an initially empty diagonal can only occur in a cell belonging to the column containing the input cell for that diagonal.

Consequently if we prevent an initially empty diagonal from filling in, none of the cells in the associated column can produce a non-zero inner product. Thus, it follows that each cell in the associated column can be replaced by a simple set of delay registers. This is illustrated in Figure 1b and Figure 1c and as multiplier and adder elements have been removed from the cells, the hex array must consume less area. Furthermore, the input and output of a column of delay register cells is always zero, hence no connections to a host machine are required. Thus, the number of input/output connections is reduced to the number of non-zero diagonals of A plus the selected number of fill-in diagonals. One further useful attribute of the compacted array is that the minimum number of real hexes is always produced from the column rules above. The incomplete factorisation retains fill-in bands from the outermost diagonals and moves inwards and from the tapered structure of the hex array, it is apparent that including a new fill-in diagonal always introduces the least number of cells (columns increase in size as we move inward).

The number of cells saved by the use of incomplete factorisation is easily derived as follows. The number of true cells C is given by

$$C = (\text{cells in central band}) + 2 * (\text{number of cells in retained bands})$$

or

$$C = C_1 + 2 * C_2 \tag{2.1}$$

with

$$C_1 = W_s + 2 \sum_{i=1}^{\lfloor (W_I - 1)/2 \rfloor} (W_s - i)$$

$$= W_s + 2W_s \left\lfloor \frac{W_I - 1}{2} \right\rfloor - \left\lfloor \frac{W_I - 1}{2} \right\rfloor \left(\left\lfloor \frac{W_I - 1}{2} \right\rfloor + 1 \right) \tag{2.2}$$

and

$$C_2 = \sum_{i=1}^{r} \{(W_S - t + 1) - i + 1\} = (W_s - t + 2)r - \frac{r}{2}(r+1) \qquad (2.3)$$

For Figure 1b, when r = number of retained bands, and t is the location of the first retained band (in the case shown $t=5$).

The saving in cells which is also proportional to the number of delay registers is given by,

$$S = W_s^2 - C, \qquad (2.4)$$

for Figure 1c the calculation is generalised resulting in

$$C = C_1 + 2 * (\text{sum of cells in individual bands}) = C_1 + 2 * C_3$$

and with r_i, $i=1(1)k$ denoting the number of retained diagonals in k individual bands, and t_i, $i=1(1)k$ the column index numbers of the first retained diagonal from the central column or leading diagonal of each band. The contribution of a single band Δr_i gives C_3 the simple form,

$$C_3 = \sum_{j=1}^{k} \Delta r_j, \qquad (2.5a)$$

and

$$\Delta r_j = \sum_{i=1}^{r_j} \{(W_s - t_j + 1) - i + 1\}$$
$$= (W_s - t_j + 2)r_j - \frac{r_j}{2}(r_j + 1) \qquad (2.5b)$$

Hence,

$$C_3 = \sum_{j=1}^{k} (W_s - t_j + 2)r_j - \frac{1}{2}\sum_{j=1}^{k} r_j(r_j+1) \qquad (2.5c)$$

and the saving is again produced by (2.4). Thus for problem I, $W_s = n+1$, $W_I = 3$, $t = n-2$, $r = 4$,

$$C_1 = (n+1) + 2(n+1-1) = 3n+1 \qquad (2.6a)$$

$$C_2=((n+1)-(n-2)+2)4-\tfrac{4}{2}(5)=5*4-2*5=10 \quad (2.6b)$$

$$C=C_1+2*C_2=3n+1+20=3n+21. \quad (2.6c)$$

producing a saving,

$$S=(n+1)^2-3n-21=n^2-n-20, \quad (2.6d)$$

For Problem II, $W_s=n^2+1$, $W_I=3$, $t_1=n-2$, $r_1=4$, $t_2=n^2-2$, $r_2=4$,

$$C_1=(n^2+1)+2(n^2)=3n^2+1 \quad (2.7a)$$

$$C_3=\Delta r_1+\Delta r_2 \quad (2.7b)$$

$$\Delta r_1=((n^2+1)-(n-2)+2)4-2(5)=(n^2-n+5)4-10 \quad (2.7c)$$

$$\Delta r_2=((n^2+1)-(n^2-2)+2)4-10=10 \quad (2.7d)$$

$$C=11n^2-8n+21, \quad (2.8a)$$

giving a saving,

$$S=n^4-9n^2+8n-20. \quad (2.8b)$$

The same principle can be easily extended to matrices in which the band structure is not symmetric (as opposed to symmetric elements). Figure 2 illustrates the situation where $W=p+q-1$ (instead of $W=2W_s-1$) as previously. The number of cells C is now

$$C=(\text{cells in bands from left of array})+(\text{cells in central band})$$

$$+(\text{cells in bands from right of array}) \quad (2.9)$$

$$C=C_1+C_2+C_3$$

If we put $W_s=\min(p,q)$ which in Figure 2a is q and compute as follows,

$$C_2=2W_s+\sum_{i=1}^{[(W_I-1)/2]}(W_s-i)$$

$$=2W_s+W_s\left\lfloor\frac{W_I-1}{2}\right\rfloor-\frac{1}{2}\left\lfloor\frac{W_I-1}{2}\right\rfloor\left(\left\lfloor\frac{W_I-1}{2}\right\rfloor+1\right) \tag{2.10}$$

$$C_1=\sum_{j=1}^{k_0}\Delta r_j=\sum_{j=1}^{k_0}(W_s-t_j+2)r_j-\frac{1}{2}\sum_{j=1}^{k_0}r_j(r_j+1). \tag{2.11}$$

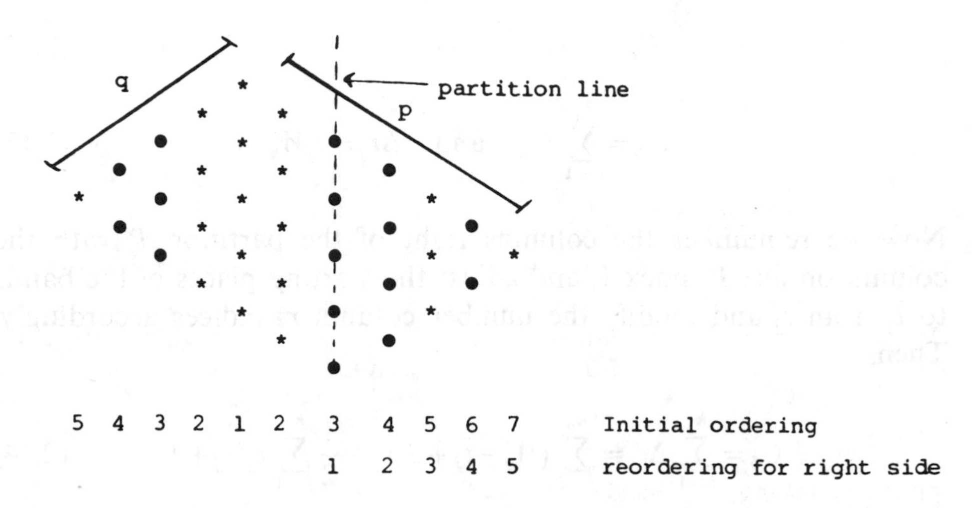

a) Hex skewed right

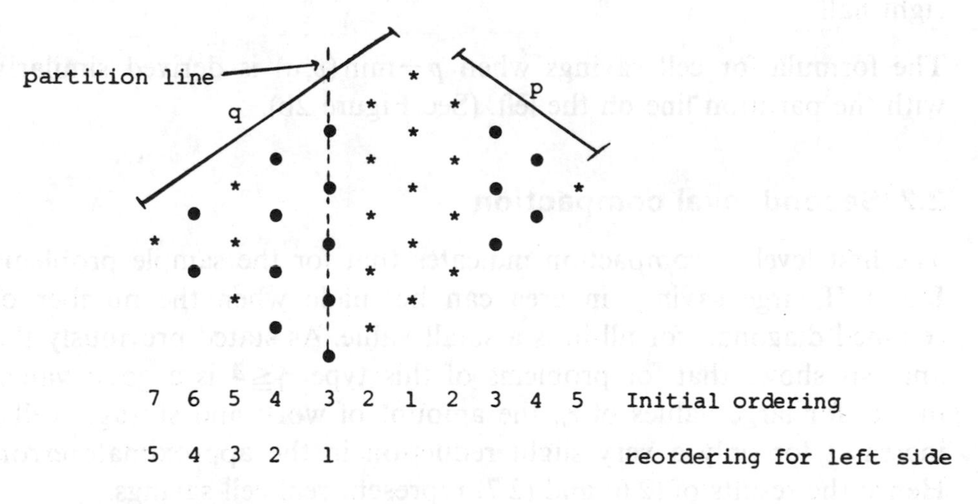

b) Hex skewed left

Figure 2 Non-symmetric array compaction

C_3 is now more complex and requires the addition of a partition line P. The partition is placed so that columns to the right of the line are strictly decreasing in cell count, while columns to the left and right of the partition and the centre column of the hex contain W_s cells each. Now

$$C_3 = (\text{total cells left of partition}) + (\text{total cells right of partition})$$

$$= \qquad C_4 \qquad + \qquad C_5 \tag{2.12}$$

$$C_4 = \sum_{i=1}^{k_1} \Delta r_i \quad \text{and} \quad \Delta r_j = r_j W_s \tag{2.13}$$

Now we re-number the columns right of the partition P with the column on line P index 1, and adjust the starting places of the bands to $\bar{t}_j$ from t_j and modify the number column r_j indices accordingly. Then,

$$C_5 = \sum_{i=1}^{k_2} \Delta \bar{r}_i = \sum_{j=1}^{k_2} (W_s - \bar{t}_j + 2)\bar{r}_j - \frac{1}{2}\sum_{j=1}^{k_2} \bar{r}_j(\bar{r}_j + 1) \tag{2.14}$$

with k_0 = number of bands in left half, $k_1 + k_2$ = number of bands in right half.

The formula for cell savings when $p = \min(p, q)$ is derived similarly with the partition line on the left. (See Figure 2b).

2.2 Second level compaction

The first level of compaction indicates that for the sample problems I and II large savings in area can be made when the number of retained diagonals for fill-in is a small value. As stated previously the analysis shows that for problems of this type $r_i \leq 4$ is a good value; in fact for large values of r_i, the amount of work and storage (cells) increases for only a very slight reduction in the approximate error. Hence the results of (2.6) and (2.7) represent real cell savings.

Now returning to Figure 1, it is evident that the introduction of an incomplete factorisation, in which some diagonals were exempted from fill-in allowed some columns of cells into which those diagonals were input to be traded for delay registers. We shall call these primary neutral cells. Observation of the dataflow of first level compaction with primary neutral cells indicates that further secondary neutral cells are introduced into the array. A second level compaction trades these secondary cells for even more delay registers. In order to identify these secondary cells, we proceed as follows. We trace out the path of an element input to a column of primary neutral cells (delays). In general, this path consists of two stages:

Stage 1 A vertical movement to the upper hex boundary; by the first level compaction it is clear that only zeroes are input and no modifications to the zero values are made on this journey.

Stage 2 Once in the cell on the boundary, the hex cell computes a multiplier for cells on the left, and negates values on the right. As one of these values is zero (we started in a primary column) the south-west output or south-east output of cells on the right and left is zero. Hence, any cells on the south-west or south-east path from a boundary cell belonging to a primary column never modify values moving vertically.

Figure 3 shows the compaction of the systolic array. Steps (a)–(b) consist of the first level compaction, while Step (c) illustrates paths for identifying secondary neutral cells. As no modification occurs in secondary neutral cells they are effectively delay cells and the adder multiplier arrangement can be removed creating even more savings in area. Figure 3c indicates the structure of the array after the second level compaction (the cuts will be explained shortly), from which it is trivial to deduce that the number of true hexes is now proportional to the number of non-zero diagonals of A plus the number of diagonals retained for fill-in.

Although the compacted arrays have drastically reduced the area of the hex required for an incomplete factorisation, the computation time has remained unchanged. Only area has been traded while retaining synchronisation of the original systolic array by delay cells. The original hex itself will be faster than the sequential factorisation by a sequential incomplete factorisation strategy will save computation time over a complete factorisation. Yet the compacted arrays appear not to have inherited this property. The reason is that with synchronisation of the original hex retained by delay registers

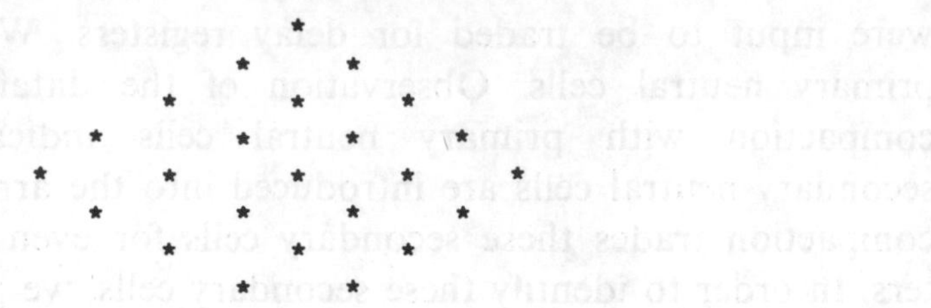

a) Standard hex array

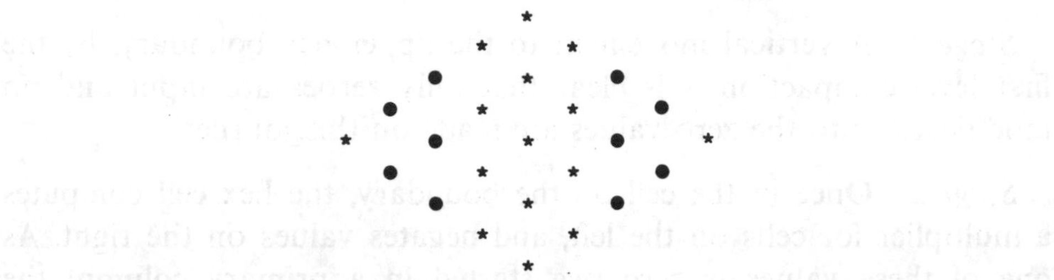

b) Non-retained diagonal cells replaced (by delay)

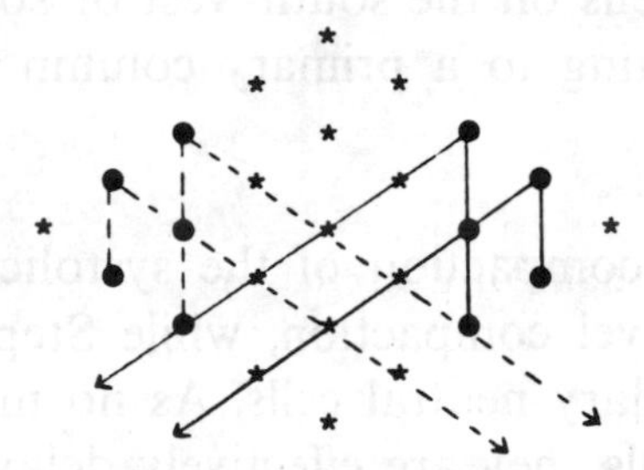

c) Identify secondary neutral cells (replace with delays)

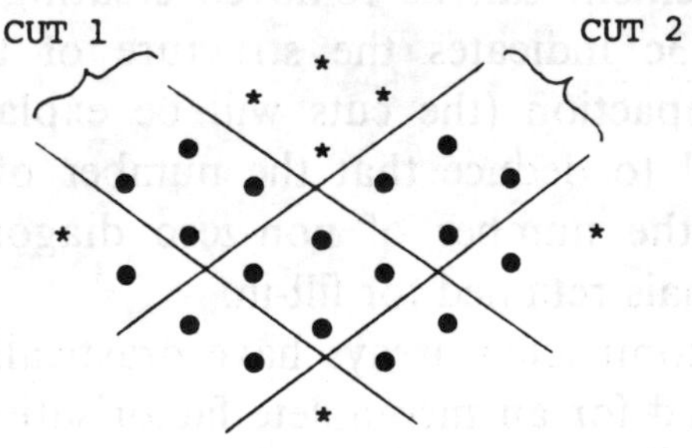

d) Define a synchronisation cuts (and remove delays)

e) Compact hex array for approximate factorisation

Figure 3 Construction of incomplete factorisation array

computations neglected by the sequential algorithms are only partially neglected in the array. By partially neglected we mean that although no computation takes place the time consumed by the computation is still present. Our gains in area result only from the observations that waiting consumes less area than calculation.

3. A SYSTOLIC INCOMPLETE FACTORISATION ARRAY

The only way to reduce the time of the hexagonal array is to change the length of the input data stream or remove part of the array which contributes a timing delay. The former scheme can be applied only in special cases such as tridiagonal matrices or by the use of block matrix calculations and are beyond the scope of this report, but are detailed elsewhere [1]. The latter scheme of reducing the size of the hex by removing cells will be considered here.

First and second level compactions of the normal hex have introduced many delay registers, whose only function is synchronisation. Effectively the second level compaction reveals how many true hex cells are required for the computation, and the delay registers can be regarded as simply consuming area.

3.1 Cutting compacted arrays

In Figure 3d, the concept of a cut is introduced. Basically, a cut removes delay cells from a compacted array which retains the hexagonal pattern, making it easy to resynchronise input elements. Figure 3e illustrates the effect of cutting the compacted array, while Figure 4 illustrates the second level compaction and cutting of a more complex array which would occur for problem II. Notice that the total number of cells after cutting is w^2 where w is the number of non-zero diagonals of A plus retained fill-in diagonals, similar to the true number of hexes from the second level compaction. However now there are no delay cells (for uniformity, the delays in the cut hex are replaced by true hexes).

In fact after a cut we can generally expect a saving of over 75% of the original number of hex cells which is a dramatic reduction for large n in problems I and II.

However making the cuts, destroys the factorisation process described in Section (1.2) by essentially destroying synchronisation

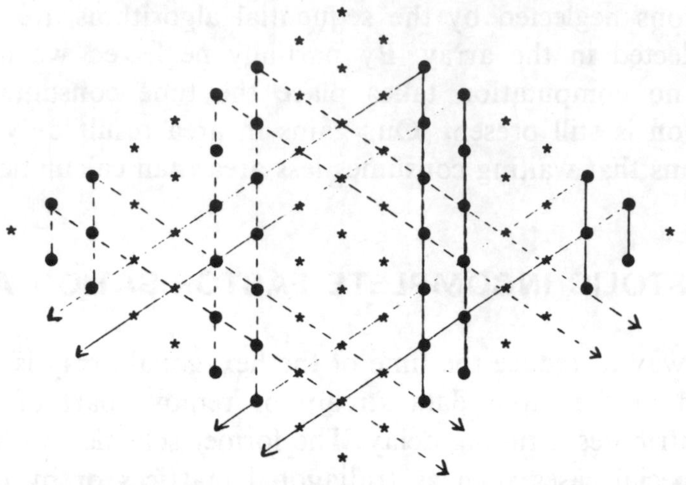

a) Identify secondary neutral cells

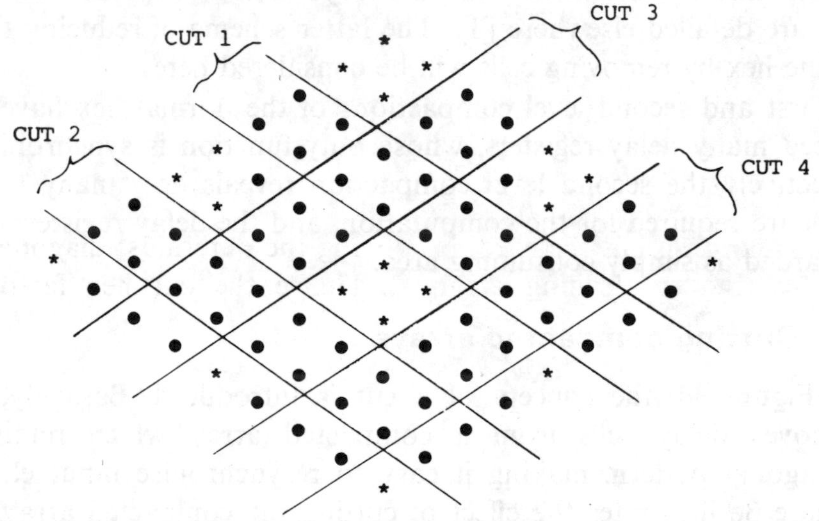

b) Make synchronisation cuts

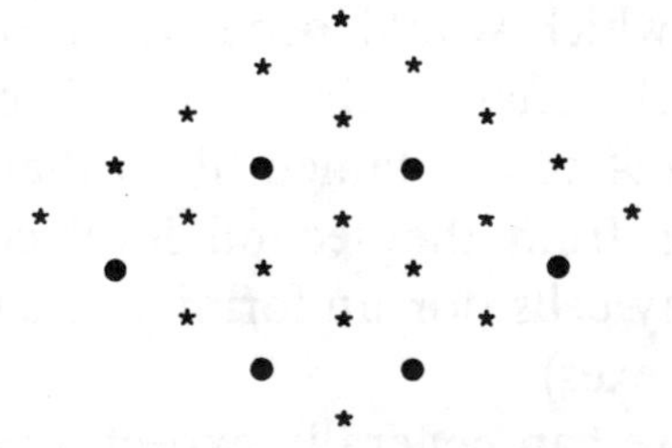

c) Compacted array

Remark The original hex required $9*9=81$ hex cells, Compacted array $5*5=25$ hex cells a saving of 56 cells which is almost 75% area reduction.

Figure 4 Incomplete array construction for Figure 1c.

between the central band and the retained outer diagonals of A. In order to take advantage of the cut array we must answer two questions:

1. What kind of factorisation does the cut hex compute?
2. Is this factorisation a good incomplete factorisation for the original problem?

Later we show that you can get a reasonably good incomplete factorisation from the cut hex.

3.2 The systolic incomplete factorisation (SIF) algorithm

The first question can easily be answered by observing the hex computation. The cuts in the compact array remove columns containing primary neutral cells, while secondary primary cells are removed to create edges which allow the dissected parts of the array to be fitted together in a smaller hex. This reduces the computation time by reducing the time an element takes to pass through the array, but also has the effect of moving the outermost diagonals of A in towards the leading diagonal. Hence the cut hex factorises a matrix with the same order as A but with a narrower bandwidth. Consequently for the model problems A gets transformed by a cut as follows:

$$\overset{\longleftarrow\; 2+r \;\longrightarrow}{} \qquad \bar{A} = \begin{bmatrix} b_1 & c_1 & & e_1 & & & \\ a_2 & \ddots & \ddots & & \ddots & e_{N-m+1} & 0 \\ & \ddots & \ddots & 0 & & \ddots & \\ f_m & & \ddots & \ddots & & & \ddots \\ & \ddots & 0 & & \ddots & \ddots & \\ 0 & & f_N & & & \ddots & c_{N-1} \\ & & & \ddots & a_N & & b_N \end{bmatrix}_{N*N} \quad \text{for } (1.10) \qquad (3.1)$$

$$\overbrace{\qquad\qquad\qquad\qquad}^{2+r_1+r_2} \qquad \overbrace{\qquad\qquad}^{2+r_1}$$

$$\bar{A} = \begin{bmatrix} b_1 & c_1 & t_1 & & \sigma_1 & & 0 \\ a_2 & \ddots & \ddots & \ddots & & 0 & \sigma_{N-p+1} \\ v_m & & \ddots & \ddots & t_{N-m+1} & & \vdots \\ & \ddots & 0 & \ddots & 0 & \ddots & \\ w_p & 0 & v_N & & \ddots & \ddots & \\ & w_N & & \ddots & & & c_{N-1} \\ 0 & & \ddots & & & a_N & b_N \end{bmatrix}_{N*N} \text{ for } (1.14) \qquad (3.2)$$

Now performing an exact or complete factorisation on the above matrices produces the factors generated by the cut hex, and the zero diagonals between the central band and the outer diagonals of the L_s and U_s becomes dense, with fill-ins. The problem now is that we have factorised a matrix for a totally different problem to the one used in the EL factorisations specified earlier. Therefore we would expect to get poor approximate solutions by solving the coupled system (1.12). A further problem is encountered when the outer diagonals are moved inwards, unknown entries are created at the end of diagonals indicated above by dashed lines. The choice of these elements may be crucial to a good approximate (incomplete) factorisation. Careful consideration of the solution process for the linear system reveals that we can compensate for this seemingly bad factorisation by performing the solution of the coupled systems (1.12) with matrices L_s and U_s output from the cut hex, but with the fill-in and outer diagonals shifted back to the original positions. This has two advantages, firstly we solve a system which has similar structure to the original problem, and secondly the fill-ins associated with elements created to fill unknown entries on the shifted diagonals are shifted out of the matrix hopefully reducing their effect on the final solution. In short the cut hex derives a new incomplete factorisation which is performed as follows:

1) Form a new matrix $\bar{A}$, by removing diagonals which will not fill-in within the band, moving the outer diagonals and the

retained fill-in diagonals towards the leading diagonal. Factorise this matrix into two complete factors L_s and U_s.

2) Modify L_s and U_s to form new sparse matrices $\bar{L}_s$ and $\bar{U}_s$ by shifting the retained fill-in diagonals back to their original positions, neglecting any terms that fall outside the matrix. $\bar{L}_s$ and $\bar{U}_s$ have the same form as the RHS of (1.11) and (1.14). Solve the coupled system corresponding to (1.12) using the above factors.

The method itself can be expressed in the notation of (1.13) as

$$\begin{array}{l}\text{SEQ}\\ \quad \text{ALUBOT}(N, r+2, r, \bar{A})\\ \quad \text{FBSUBS}(N, r+2, r, L_{s_r}, \bar{U}_{s_r}, \mathbf{d})\end{array} \tag{3.3}$$

The method was tested for various values of m, with N fixed, and with m fixed and variable N, with the number of retained diagonals r plotted against the Euclidean Error vector norm. Tests were repeated once for problem I with $d=(1,1,\ldots,1)$ and again with $u=(1,1,\ldots,1)$ when d_i was simply the sum of the coefficients for row i of A. The results are shown in graphs A1–A6, Appendix 2, where EXT-1, and EXT0 indicate that new elements created by shifting the outer diagonals in were set to -1 or zero respectively.

Remark In the graphs for variable N and m fixed the normalised Euclidean error vector norm was used.

The results indicated that the cut hex with the shifted solution process produced an incomplete factorisation process which retained the desirable property of rapid error decay with the retention of the first few fill-in diagonals. The results were not as good as for the EL algorithm, and we should consider retaining approximately $r=7$ diagonals rather than $r=4$. The cases for $M=40$, $N=50$ indicated that the approximation error may grow as the bandwidth is increased. Further tests for large N with fixed m were carried out, showing that if the error did diverge with more diagonals then they did so at a much slower rate than the initial error decrease for retaining a small number of bands. In fact the more diagonals included the slower the divergence rate became. For instance with $N=200$, $M=150$ approximately 10 extra diagonals had to be included to alter the error norm by 0.1. Further experimentation indicated that this divergence occurred when $m>N/2$, as the problems considered in this paper all have $m<N/2$ no problems should occur. Possible explanations for this behaviour are related to the distance that the outer diagonals are moved in order to accommodate them on the cut hex, and the interference of the factorisation process with the narrower bandwidth affecting more central band elements.

3.3 An alternative factorisation

In contrast to the incomplete factorisation above, an alternative method was devised. In this second method, the cut hex was modified by adding delay registers between the cells in columns adjacent to each other but in a separate fill-in band and re-timing the outer diagonal inputs so that they are still synchronised with the data sent from the reciprocal cell of the hex array. This had two effects, i.e.,

i) The factorisation of the central band proceeded in parallel with the production of fill-in entries.

ii) The fill-in entries themselves could not affect the factorisation of the central band.

Consequently the matrix being factorised had the same structure as A from the outset, but the central elements of the matrix performed as a tridiagonal factorisation, the fill-in diagonals being computed from multipliers calculated from the tridiagonal form. The basis for this idea was that the multiplier of problems I and II, were all of the same order of magnitude and might produce good approximate answers. Unfortunately this proved to be a bad method with the results indicated in Appendix A7–A9. We see that no rapid drop in the error norm occurs no matter how many diagonals are retained. Initially, the results looked promising as the error is approximately the same as that gained by the first method with an even smaller number of diagonals retained, and no divergence occurred. The problem is however that if we get a large error initially there is no chance of decreasing it by adding more diagonals, as there is with the first scheme.

Finally, the effect on the methods by the size of co-diagonal and outer diagonals was considered, for the cases when the:

a) Co-diagonals were strong and outer diagonals weak

b) Both co-diagonals and outer diagonals were of same order and

c) Co-diagonals were weak and the outer diagonals strong.

These results are shown in A10–A12. Observe that the values when the co-diagonals and outer diagonals are strong or weak relative to each other. For EXT -1 an error reduction is always achieved; while for EXT0, divergence occurs albeit slowly. The alternative scheme the delayed cut hex algorithm retains attributes of small error decreases no matter how many diagonals are allowed to fill in.

The results indicate that it is better to adopt the SIF algorithm with the cut hex when there is a likelihood of weak or strong co- and outer diagonals relative to each other, and the newly created elements which occur due to shifting the outer diagonals are set to -1. With the standard case when co- and outer diagonals are approximately equal, EXT0 seems best (see A1–A4). For a more general form of the SIF, with different problems in which the elements on the same diagonal are different, the created elements must be chosen from an averaging, min, or max type criterion.

Applying the SIF algorithm to optimise area reduction gives higher approximation errors when compared with the EL algorithm, but as the table below clearly shows the reduction in cell count in the systolic arrays more than compensates. Indeed it would be impractical to build large hex arrays, but the small circuits produced by the SIF algorithm may be feasible.

Also noticed from the table, the property of fixed hardware for increasing problem sizes. This is due to the fact that the cut hex for SIF has a size dependent upon the number of non-zero diagonals of A and the retained fill-in diagonals. Whereas the complete LU factorisation array is variable, with bandwidth related to problem size n. Consequently, the large grids (smaller step-sizes) applied to the sample problems tends to give the cut hex almost 100% area savings. Further, if we construct a cut hex with 256 hexes, we could solve both problems I and II without padding the input for the smaller problem, but allowing $r > 7$, for I and $r_1 = r_2 = 7$ for II. Thus, a fixed sized array can be constructed giving variable accuracy for a range of problems.

3.4 Sparse forward and backward substitution arrays

Now we concentrate on solving the coupled system (1.12). The cut hex produces a compact output of the factors with the form similar to the RHS of (1.10) and with $m=r+2$ for problem I and $m=r_1+2$, $p=r_1+r_2+2$ for problem II. After shifting to compensate for the factorisation error the matrices resemble (1.11) and (1.14) respectively. These L and U factors are sparse in form, and it follows that specially reduced forward and backward substitution arrays can be constructed in a similar manner to compaction for the hex. Essentially we remove inner product cells corresponding to zero diagonals and insert delay registers. Such substitution networks are described in [6], with the number of cells proportional to the number of non-zero diagonals in the matrix. Computation time however is unaffected and is given by $T=2N+q$ with q the total bandwidth of the input matrix including zero and non-zero diagonals in the main band.

As stated above, the speed of the systolic array can only be affected by reducing the number of cells hence latency of the design, or by reducing the length of the input sequence. In the former scheme the delays associated with zero diagonals would again be cut out of the array. Unfortunately this provides an array which produces the exact solution to (3.1) or (3.2) with a resulting high approximation error. The idea of shifting the factors output by the hex array was introduced to avoid this.

The latter scheme of reducing input length is possible. Robert [9], and Robert and Tchuente [10] introducing a special bi-linear array for solving a triangular system with a null first (sub- or super-diagonal for lower and upper triangular matrices respectively). The array required a time $T=N+\lceil q/2\rceil+1$ and an extra adder cell. This network can be utilised if we shift co-diagonals as well as the fill-in diagonals to give matrices of the form (4.1), computing the solution in time $T=N+\lceil m/2\rceil+1$ for each factor, which saves N cycles per factor. The problem now is that the algorithm has changed, the effect of using the cut hex, and this substitution array for the diagonal extensions EXT-1 EXT0 and the delayed algorithm are shown in B1–B9 Appendix 3.

From which we observe that EXT-1 has its error reduced by a small amount, while EXT0 is increased, and the delayed array is about the same. We can safely conclude that using the substitution array of [10] does not affect the approximation significantly. In fact we compensate when EXT-1 is used for the initial shift in of the outer diagonals for the cut hex factorisation.

$$
\bar{L}_s=\begin{bmatrix}
\omega_1 & & & & & \\
0 & \ddots & & & 0 & \\
\beta_1 & \ddots & \ddots & & & \\
y_{11}\text{—}y_{r1} & & \ddots & \ddots & & \\
\ddots & 0 & & \ddots & \ddots & \\
y_{1,n-m+1}\text{—}y_{r,n-m+1} & & & \beta_{N-2} & 0 & \omega_N
\end{bmatrix},
$$

$$
\bar{U}_s=\begin{bmatrix}
\alpha_1 & 0 & g_1 & & h_{11} & \\
 & \ddots & \ddots & \ddots & \vdots & \ddots \\
 & & \ddots & \ddots & h_{r1} & \ddots \\
 & & & \ddots & 0 & h_{r,N-m+1} \\
 & 0 & & & \ddots & g_{N-1} \\
 & & & & & 0 \\
 & & & & & \alpha_N
\end{bmatrix}
\tag{4.1}
$$

We have again chosen the optimal systolic array and analysed the effect of the modified computation on the approximation to the exact solution. The effect on timing is given by,

$$
\begin{aligned}
T &= \text{factorisation cost} + 2 * \text{substitution cost} \\
&= 9N + W + m,
\end{aligned}
\tag{4.2}
$$

with the ordinary substitution arrays of [3] producing

$$
T=\begin{cases} 9n^2+n+r+3 & \text{for I} \\ 9n^3+n^2+r_1+r_2+3 & \text{for II} \end{cases}
\tag{4.3}
$$

and with the substitution arrays of [10]

$$
T=(3N+W)+2\left(N+\left\lceil \frac{m}{2} \right\rceil+1\right) \leqq 5N+W+m+4
\tag{4.4}
$$

$$
T=\begin{cases} 5n^2+n+r+7 & \text{for I} \\ 5n^3+n^2+r_1+r_2+6 & \text{for II} \end{cases}
\tag{4.5}
$$

compared with the sequential ALUBOT algorithms of [1] which gives a total operation count of

$$T=\begin{cases}\frac{1}{2}n^2(n^2+5n+5) & \text{for I}\\ \frac{1}{2}n^3(n^4+5n^2+2n+8) & \text{for II}\end{cases} \tag{4.6}$$

N.B. introduce $\frac{1}{2}$ as an array cycle is 2 operations (add, mult).

Table I Cell requirements for cut hex (SIF) and complete factorisation hex arrays.

			SIF problem I		*SIF problem II*		*Complete LU*	
n	m	p	$r=7$	$r=4$	$r_1=r_2=7$	$r_1=r_2=4$	*I*	*II*
10	11	101	81	36	256	100	121	10 201
20	21	401	81	36	256	100	441	160 801
30	31	901	81	36	256	100	961	811 801
40	41	1 601	81	36	256	100	1 681	2.5×10^6
50	51	2 501	81	36	256	100	2 601	6.2×10^6
100	101	10 001	81	36	256	100	10 201	1.0×10^8

4. CONCLUSIONS

The concept of incomplete and in particular EL LU factorisation, has been applied to hexagonal arrays for LU factorisation. The idea of a sequence of arrays with varying hardware providing an exact and range of approximate factorisations gave rise to a method of array compaction. First primary neutral cells corresponding to unretained bands in the incomplete factorisation were identified and replaced by simple delay registers. Once primary cells were introduced secondary neutral cells could be identified permitting further cell replacement. As an hexagonal inner product was replaced by three delay registers, or in real terms, the removal of an adder and multiplier for each cell replaced, the area of the array was reduced. In general, a saving of over 75% of the cells is achieved while leaving the computation time unchanged.

Identification of primary and secondary cells was called first and second level compaction, as it reduced area while preserving the EL incomplete algorithm by using the delay cells as synchronising elements. As a result the computation time of the array remained unchanged. In order to remove delay cells a new systolic incomplete algorithm was presented. The method by which the factorisation was constructed is unique. As far as the authors are aware it is the first time an optimal systolic array has been constructed, and the corresponding numerical algorithm derived from it. The systolic incomplete factorisation (SIF) algorithm produced was examined in detail and found to possess the features of rapid error reduction for small numbers of retained fill-in diagonals. Although the results were not as good as the EL algorithm, we trade-off the large saving in array area with increased computation due to the error of approximation in solving the coupled system (1.12) iteratively to converge onto the exact answer.

The SIF algorithm was tested for sample problems and produced designs with cell requirements $0(W_1^2)$ and $0(W_2^2)$ for the 2-D and 3-D problems respectively, where $W_1=3+r_1$, $W_2=4+r_1+r_2$ with r_i the number of diagonals retained in the factorisation and the constant terms, the semi-bandwidth of non-zero diagonals in the original matrix. This is contrasted with $0(n^2)$ and $0(n^4)$ cells with the ordinary hex array for a complete factorisation. Perhaps the most interesting feature of the compaction technique and the SIF algorithm is that the structure of the systolic array is identical to the traditional architecture of [3]. However, the reduction in cells makes the SIF more practical to construct in VLSI technology than a complete hex for the sample problems.

Consequently the SIF method provides a technique for building a finite sized systolic array which can solve both the 2-D and 3-D problems, but producing varying approximations due to the constrained values or r_i implicit in the fixed array size. Hence we have illustrated a systolic array design sequence of draft algorithms based on algorithm transformations derived from design decisions of approximation accuracy, area usage and computation time. The result was the SIF algorithm, an area efficient incomplete factorisation systolic array. The consequences for future systolic applications is clear, we can define a sequence of designs which trade accuracy against circuit area producing fast, economic, and practical parallel devices, for quick approximations to problems. Such low cost devices will be useful add-ons for existing architectures.

References

[1] E. A. Lipitakis, Computational and algorithmic techniques for the solution of elliptic and parabolic differential equations in two and three space dimensions, Ph.D. Thesis, 1978, L.U.T.

[2] E. A. Lipitakis and D. J. Evans, Solving non-linear elliptic difference equations by extendable sparse factorisation procedures, *Computing* **24** (1980), 325–339.

[3] C. E. Leiserson, Area-efficient VLSI computation, Ph.D. Thesis, 1981, CMU.

[4] C. Mead and L. Conway, *Introduction to VLSI*, Ch. 8, Addison-Wesley, 1980.

[5] B. C. Thompson and J. V. Tucker, *Theoretical Considerations in Algorithm Design*, Proceedings of NATO Advanced Study Institute "Fundamental Algorithms for Computer Graphics", 1985.

[6] M. Berzins, T. F. Buckley and P. M. Dew, Systolic matrix iterative algorithms. In: *Parallel Computing 83*, M. Feilmeier, G. Joubert, and U. Schendel (eds.), Elsevier Science Publishers (North-Holland), 1984.

[7] D. J. Evans and E. A. Lipitakis, On sparse LU factorisation procedures for the solution of parabolic differential equations in three space dimensions, *Intern. J. Comp. Math.* **7**, Sec. B (1979), 315–338.

[8] D. J. Evans (ed.), *Preconditioning Methods Theory and Applications*, Gordon and Breach Science Publishers, 1983.

[9] Y. Robert, Block LU decomposition of band matrix on a systolic array, *Intern. J. Comp. Math.* **17** (1985), 295–316.

[10] Y. Robert and M. Tchuente, *Parallel Solution of Band Triangular Systems on VLSI Arrays with Limited Fan-out*, International Workshop on Modelling and Performance Evaluation of Parallel Systems, M. Becher (ed.).

[11] H. T. Kung and M. S. Lam, Wafer-scale integration and two-level pipelined implementations of systolic arrays, *Journal of Parallel and Distributed Computing* **1** (1984), 32–63.

Appendix 1

MAIN NUMERICAL ROUTINES

a) ALUBOT $(N, m, r, A)\{$

/* Tri-diagonal Factorisation */

$w_1 = b_1;\ d_1 = a_2;\ g_1 = c_1/w_1$

FOR $i = 2$ TO $m-2$

$\{w_i = b_i - d_{i-1} * g_{i-1};\ d_i = a_{i+1};\ g_i = c_i/w_i\}$

$w_{m-1} = b_{m-1} - d_{m-2} * g_{m-2}$

/* Rest of Factorisation */

FOR $j = 1$ TO $N - m + 1$

$\{e_{1,j} = v_{j+m-1};\ h_{1,j} = u_j/w_j$

$g_{m+j-2} = c_{m+j-2}/w_{m+j-2};\ d_{m+j-2} = a_{m+j-1}$

IF $r - j + 1 >= 2$ THEN

$\{$FOR $i = 2, r - j + 1$

```
            {e_ij = -g_{i+j-2} * e_{i-1,j};
             h_ij = -d_{i+j-2} * h_{i-1,j}/w_{i+j-1}
            }
        }
    ELSE IF NOT ((j=1) OR (r=1)) THEN
        {IF j>r THEN IP=2 ELSE IP=r-j+2;
        r1=r+1
        FOR i=IP TO r
        {z=0;
            FOR k=1 TO i-1 {z=z+e_kj * h_{k-i+r_1, I+J-r_1}};
            E_ij = -g_{i+j-2}*e_{i-1,j-z};
            z=0;
            FOR k=1 TO i-1 {z=z+e_{k-i+r_1, i+j-r_1} * h_{k,j}}
            h_ij=(-D_{i+j-2} * h_{i-1,j} - z)/w_{i+j-1}
            };
        };
    i=r; z=0
    FOR k=1 TO i {z=z+e_kj * h_{k,j}}
    w_{m+j-1} = b_{m+j-1} - d_{i+j-1} * h_ij - g_{i+j-1} * e_ij
              - d_{i+j-1} * g_{i+j-1} - z
}
```

Remark $v=f$, $u=e$, $e_{ij}=y_{ij}$ in (1.10) taken from [1]

b) Delayed cut hex algorithm

```
DCHEX (N, m, r, A)
/* Tri-diagonal Factorisation */
w_1 = b_1; d_1 = a_2; g_1 = c_1/w_1;
FOR i=2 TO N-1
    {w_i = b_i - d_{i-1} * g_{i-1}; d_i = a_{i+1}; g_i = c_i/w_i}
    w_N = b_N - d_{N-1} * g_{N-1}
FOR j=1 TO N-m+1
    {e_1j = v_{j+m-1}; h_{1,j} = u_j/w_j;
    FOR k=2 TO r+1
    {h_{k,j} = -h_{k-1,j} * d_{k+j-2}/w_{k+j-1};
     e_{k,j} = -e_{k-1,j} * g_{k+j-2}
    }
}
```

The dataflow for the cut hex—and addition of registers is easily derived from this algorithm.

Appendix 2

SYSTOLIC INCOMPLETE FACTORISATION (SIF) ANALYSIS

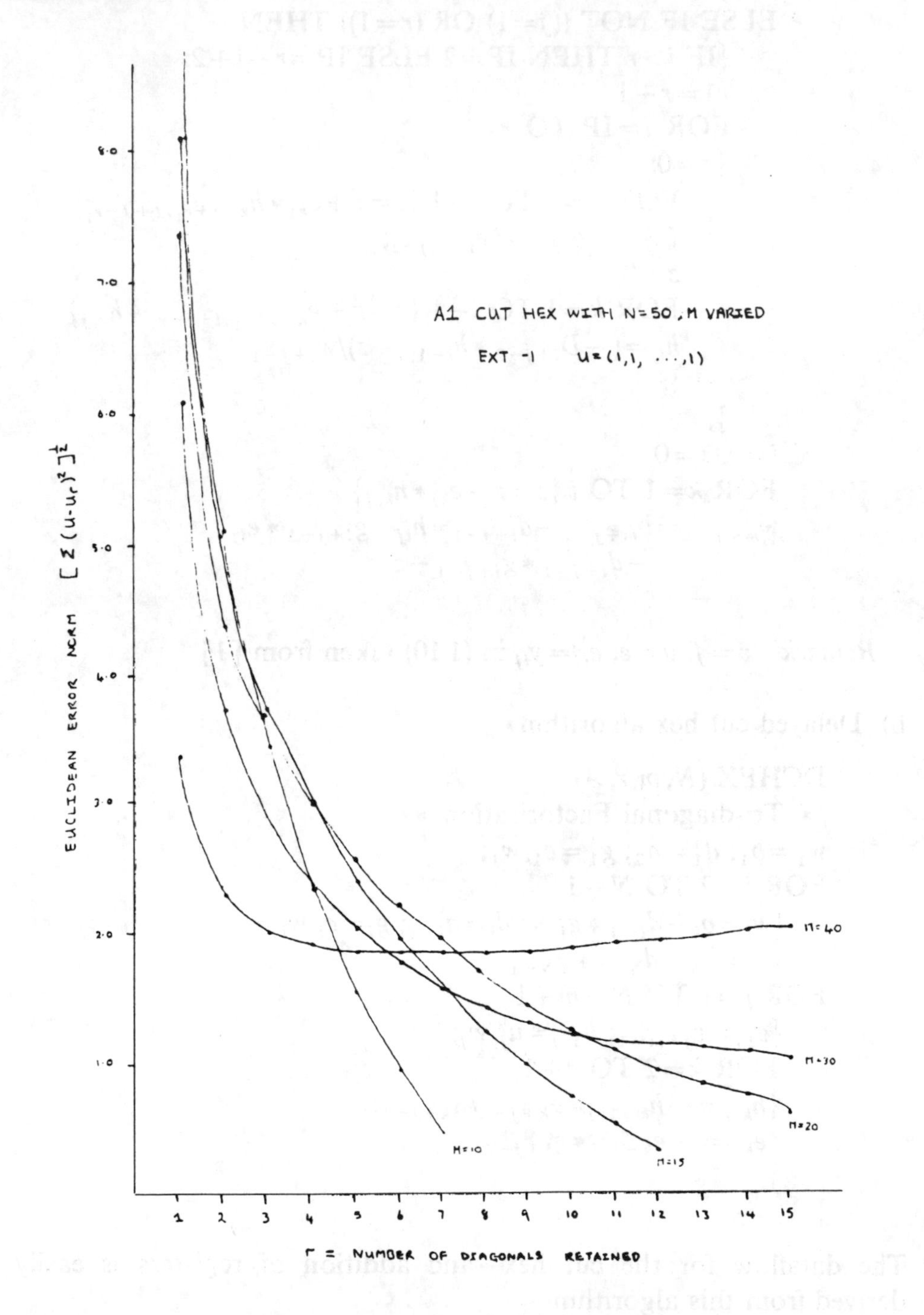

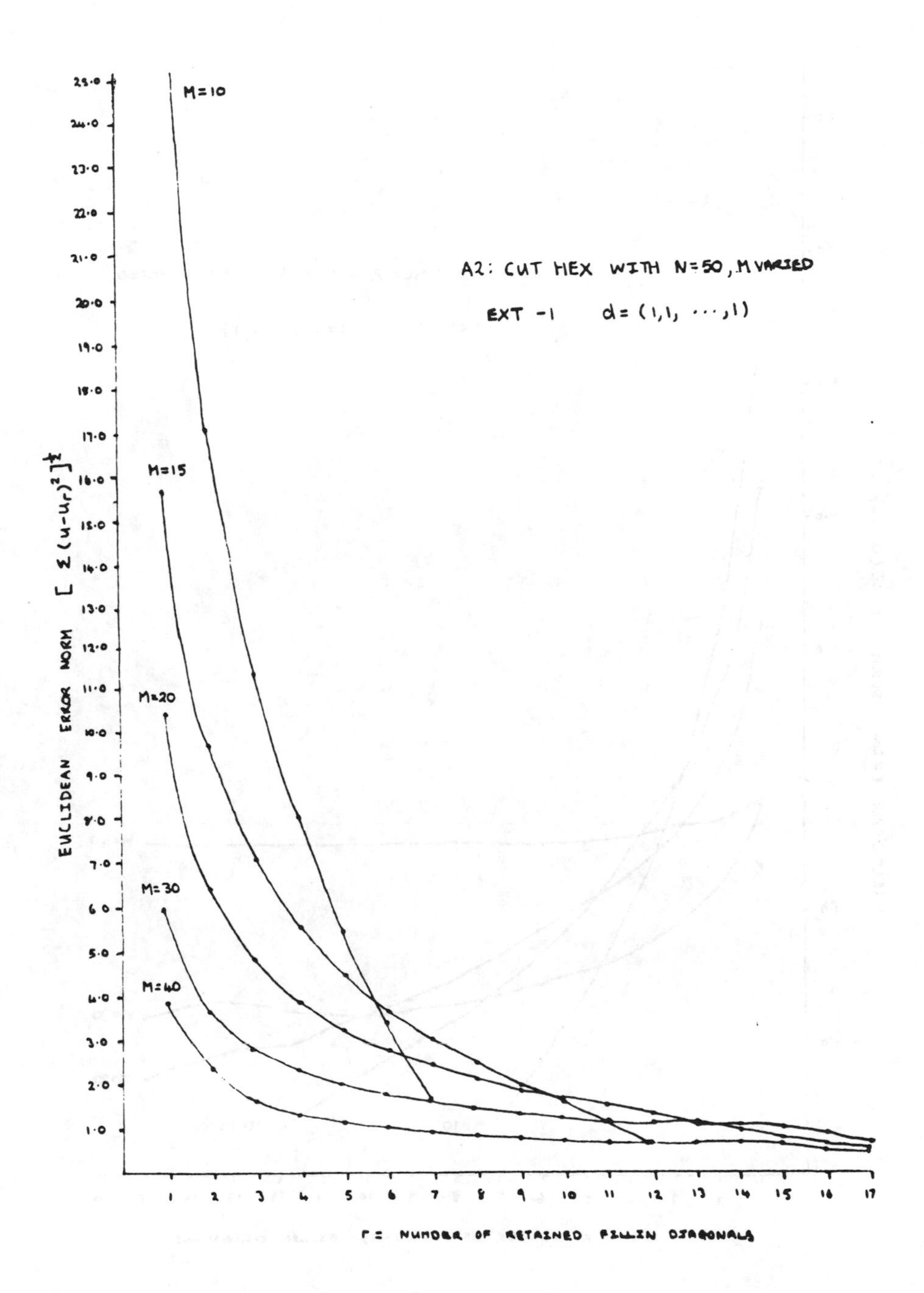
A2: CUT HEX WITH N=50, M VARIED
EXT -1 d = (1,1, ···,1)
M=10
M=15
M=20
M=30
M=40
EUCLIDEAN ERROR NORM [Σ (u-u_r)^2]^½
r = NUMBER OF RETAINED FILLIN DIAGONALS
25·0
24·0
23·0
22·0
21·0
20·0
19·0
18·0
17·0
16·0
15·0
14·0
13·0
12·0
11·0
10·0
9·0
8·0
7·0
6·0
5·0
4·0
3·0
2·0
1·0
1
2
3
4
5
6
7
8
9
10
11
12
13
14
15
16
17

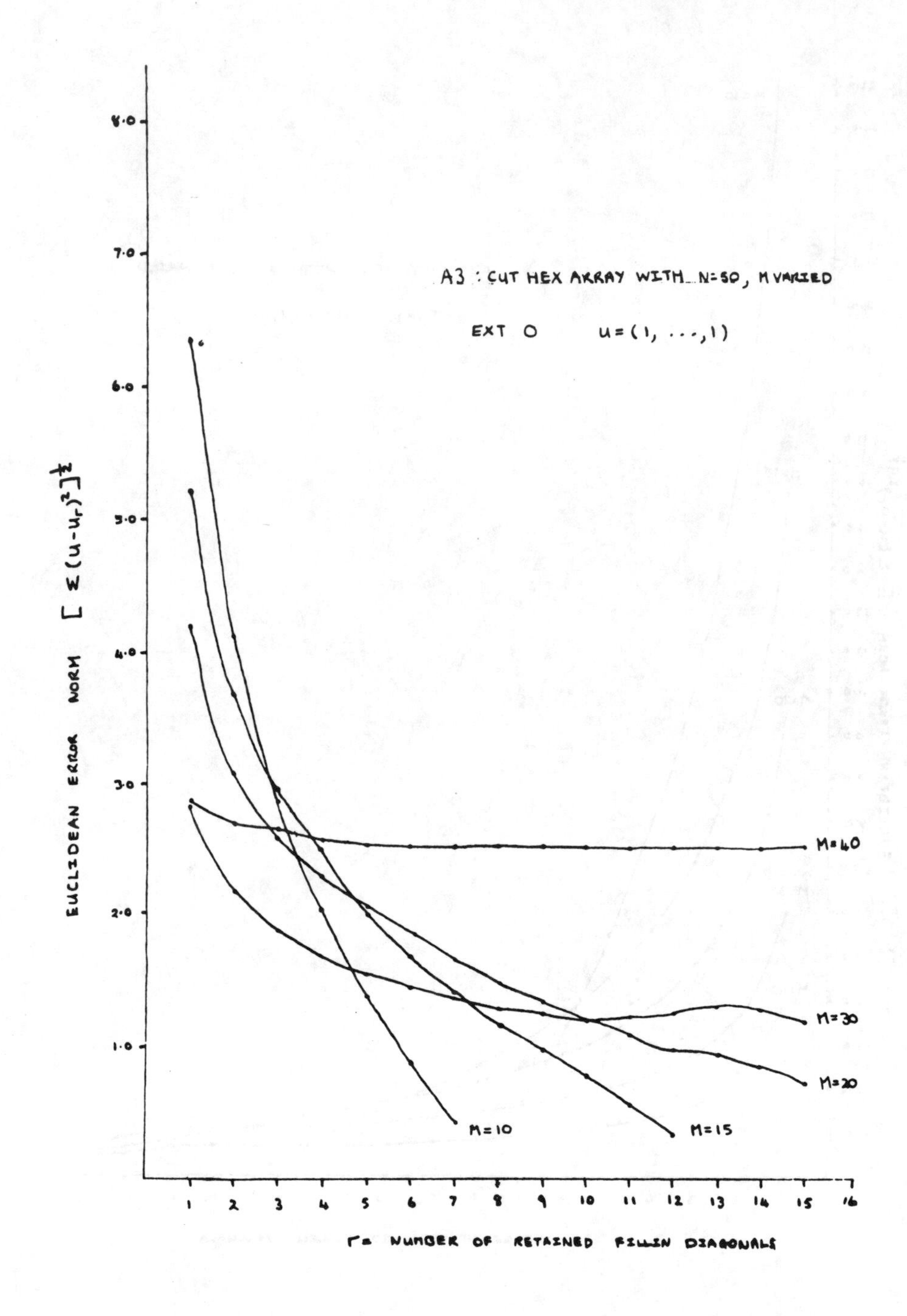
A3 : CUT HEX ARRAY WITH N=50, M VARIED
EXT O u=(1, ..., 1)
EUCLIDEAN ERROR NORM [Σ(u-u_r)^2]^½
r = NUMBER OF RETAINED FILLIN DIAGONALS
M=40
M=30
M=20
M=10
M=15

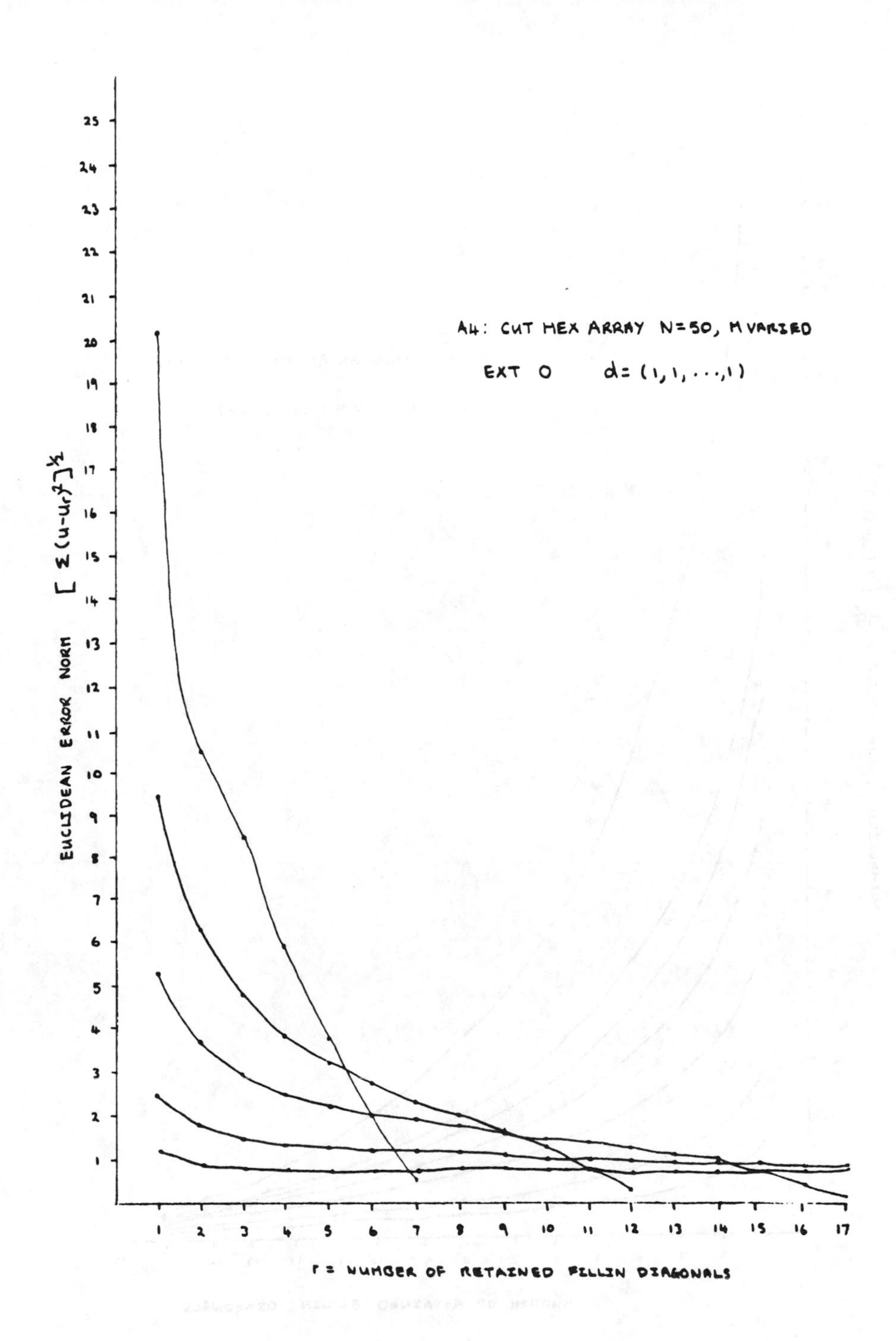
A4: CUT HEX ARRAY N=50, M VARIED
EXT 0 d=(1,1,...,1)
EUCLIDEAN ERROR NORM [Σ(u-ur)²]½
r = NUMBER OF RETAINED FILLIN DIAGONALS
1
2
3
4
5
6
7
8
9
10
11
12
13
14
15
16
17
18
19
20
21
22
23
24
25

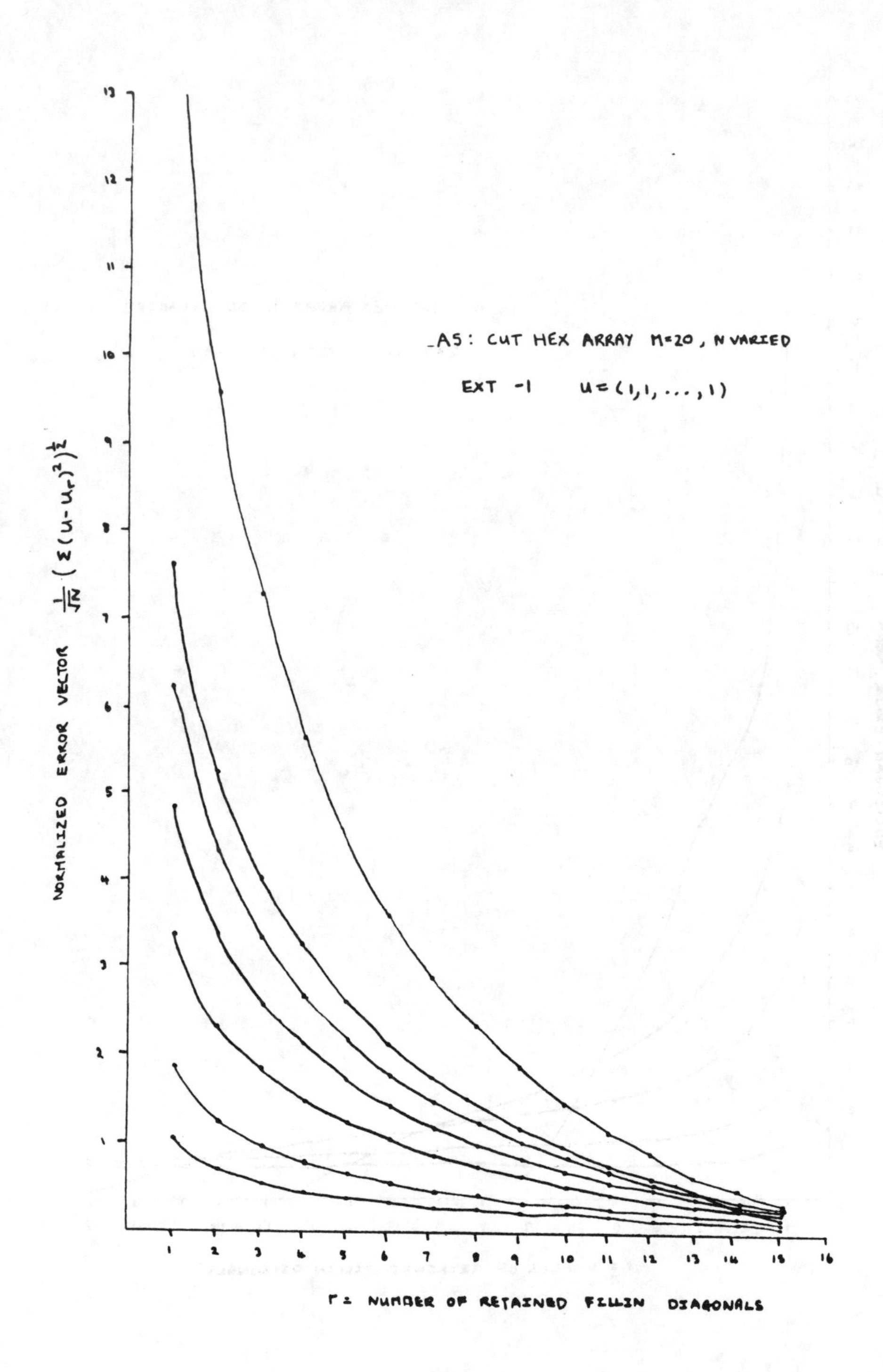
AS: CUT HEX ARRAY M=20, N VARIED
EXT -1 U=(1,1,...,1)
NORMALIZED ERROR VECTOR
r = NUMBER OF RETAINED FILLIN DIAGONALS

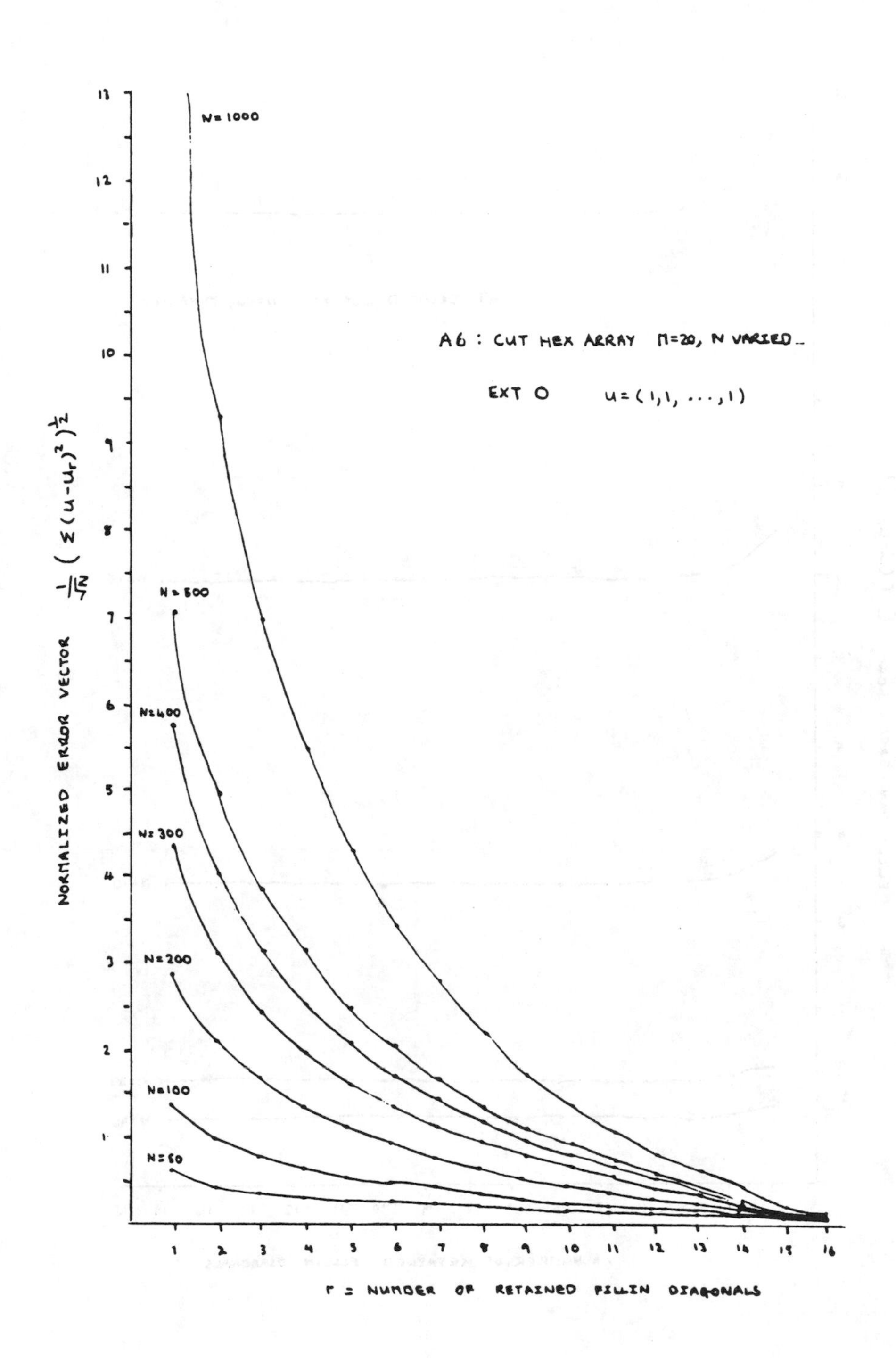
A6 : CUT HEX ARRAY M=20, N VARIED
EXT O u = (1, 1, ..., 1)
N = 1000
N = 500
N = 400
N = 300
N = 200
N = 100
N = 50
NORMALIZED ERROR VECTOR $\frac{1}{\sqrt{N}} \left(\Sigma (u - u_r)^2 \right)^{\frac{1}{2}}$
r = NUMBER OF RETAINED FILLIN DIAGONALS
1 2 3 4 5 6 7 8 9 10 11 12 13 14 15 16
1 2 3 4 5 6 7 8 9 10 11 12 13

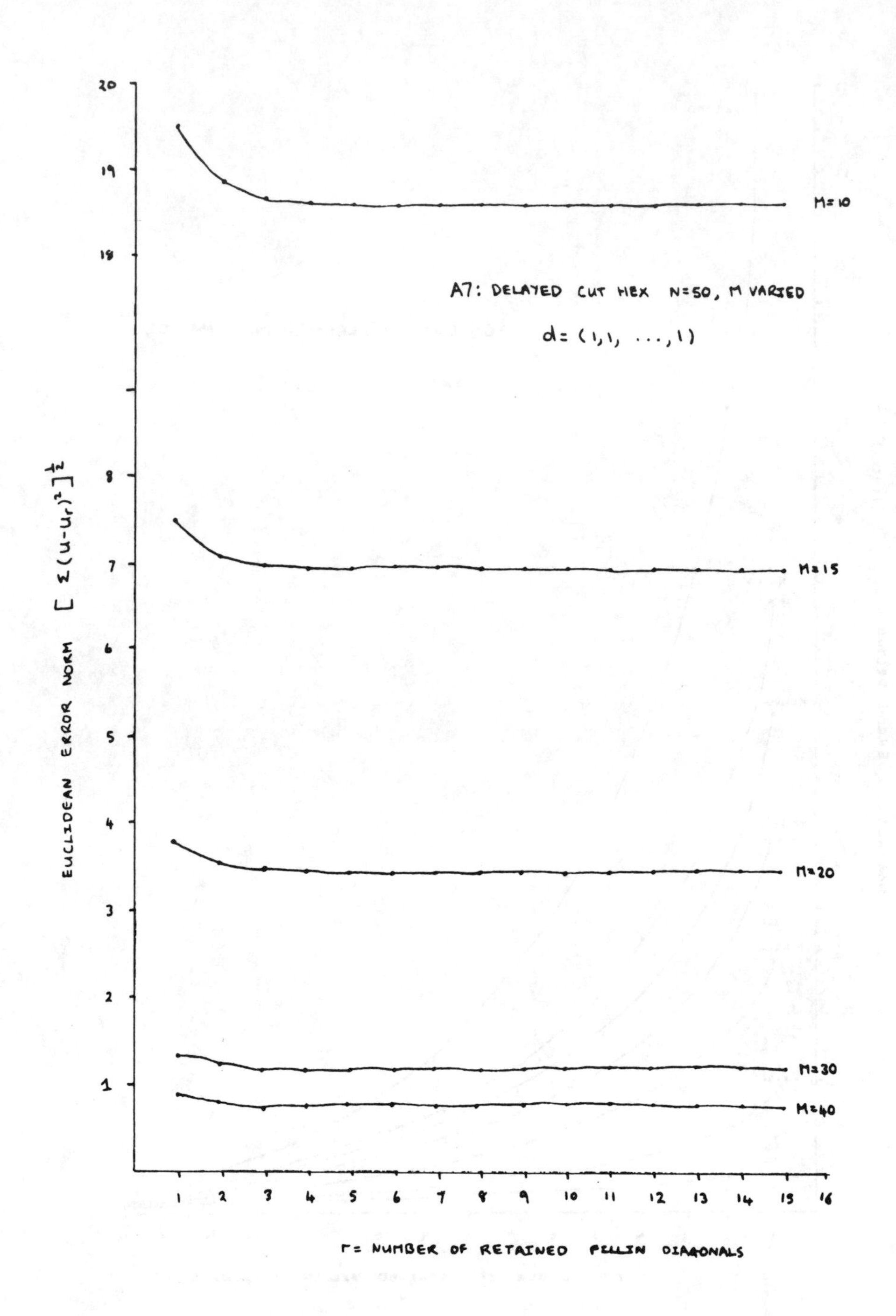

A7: DELAYED CUT HEX N=50, M VARIED
d= (1,1, ...,1)
M=10
M=15
M=20
M=30
M=40
EUCLIDEAN ERROR NORM [Σ(U-U_r)^2]^1/2
r= NUMBER OF RETAINED FILLIN DIAGONALS
20
19
18
8
7
6
5
4
3
2
1
1 2 3 4 5 6 7 8 9 10 11 12 13 14 15 16

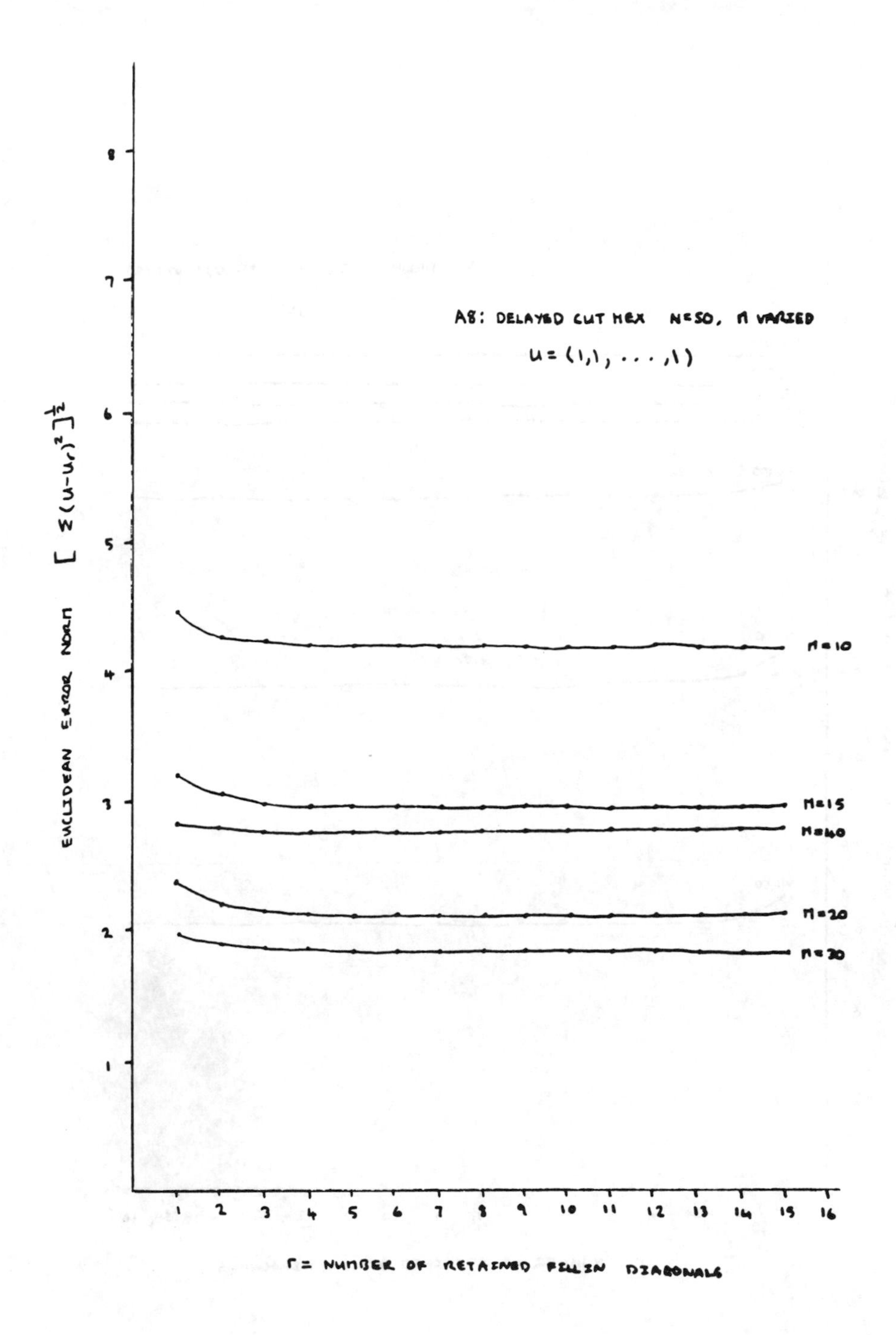
A8: DELAYED CUT HEX N=50, M VARIED
U = (1,1, . . . ,1)
EUCLIDEAN ERROR NORM [Σ(U-U_r)²]^½
M=10
M=15
M=40
M=20
M=30
1
2
3
4
5
6
7
8
1 2 3 4 5 6 7 8 9 10 11 12 13 14 15 16
r = NUMBER OF RETAINED FILLIN DIAGONALS

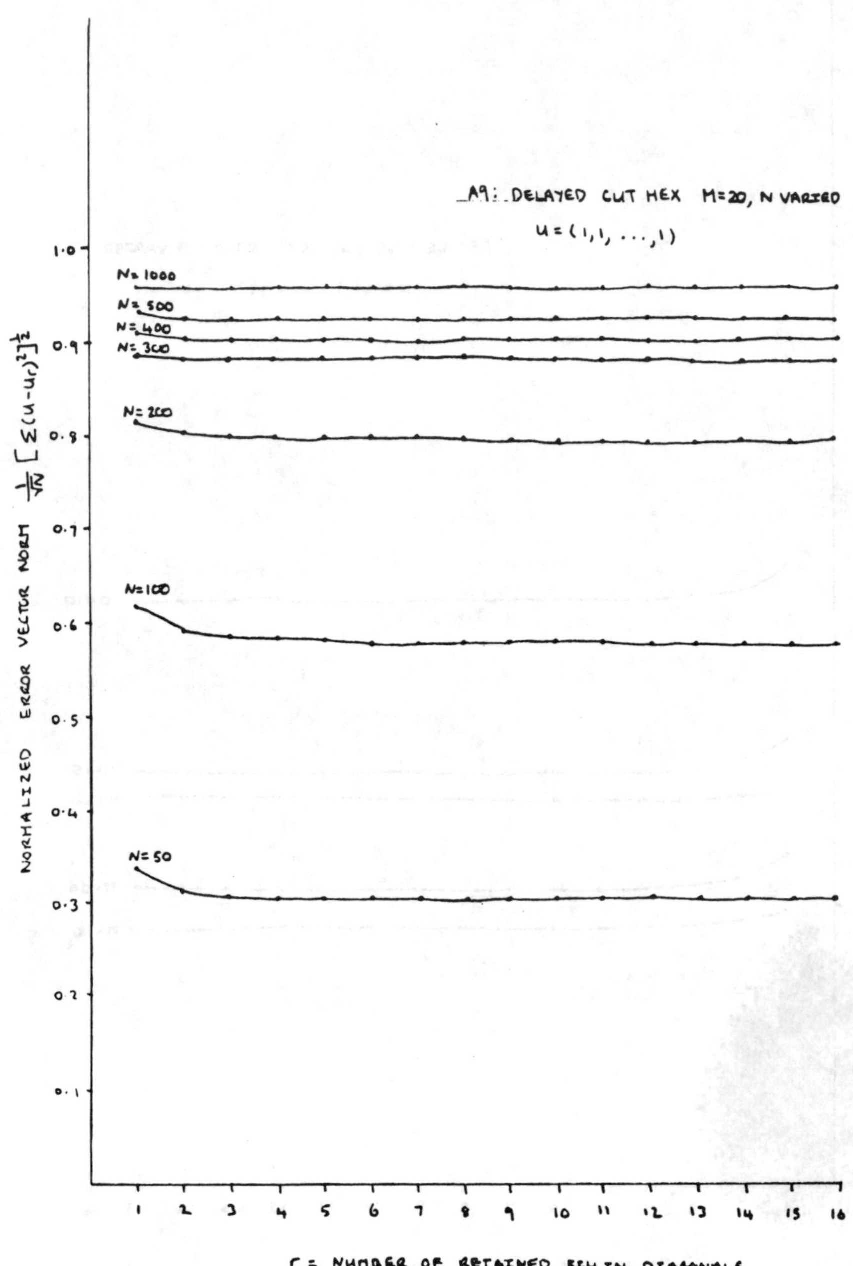
A9: DELAYED CUT HEX M=20, N VARIED
U = (1, 1, ..., 1)
N=1000
N=500
N=400
N=300
N=200
N=100
N=50
1·0
0·9
0·8
0·7
0·6
0·5
0·4
0·3
0·2
0·1
1 2 3 4 5 6 7 8 9 10 11 12 13 14 15 16
NORMALIZED ERROR VECTOR NORM $\frac{1}{\sqrt{N}}\left[\Sigma(u-u_r)^2\right]^{\frac{1}{2}}$
r = NUMBER OF RETAINED FILLIN DIAGONALS

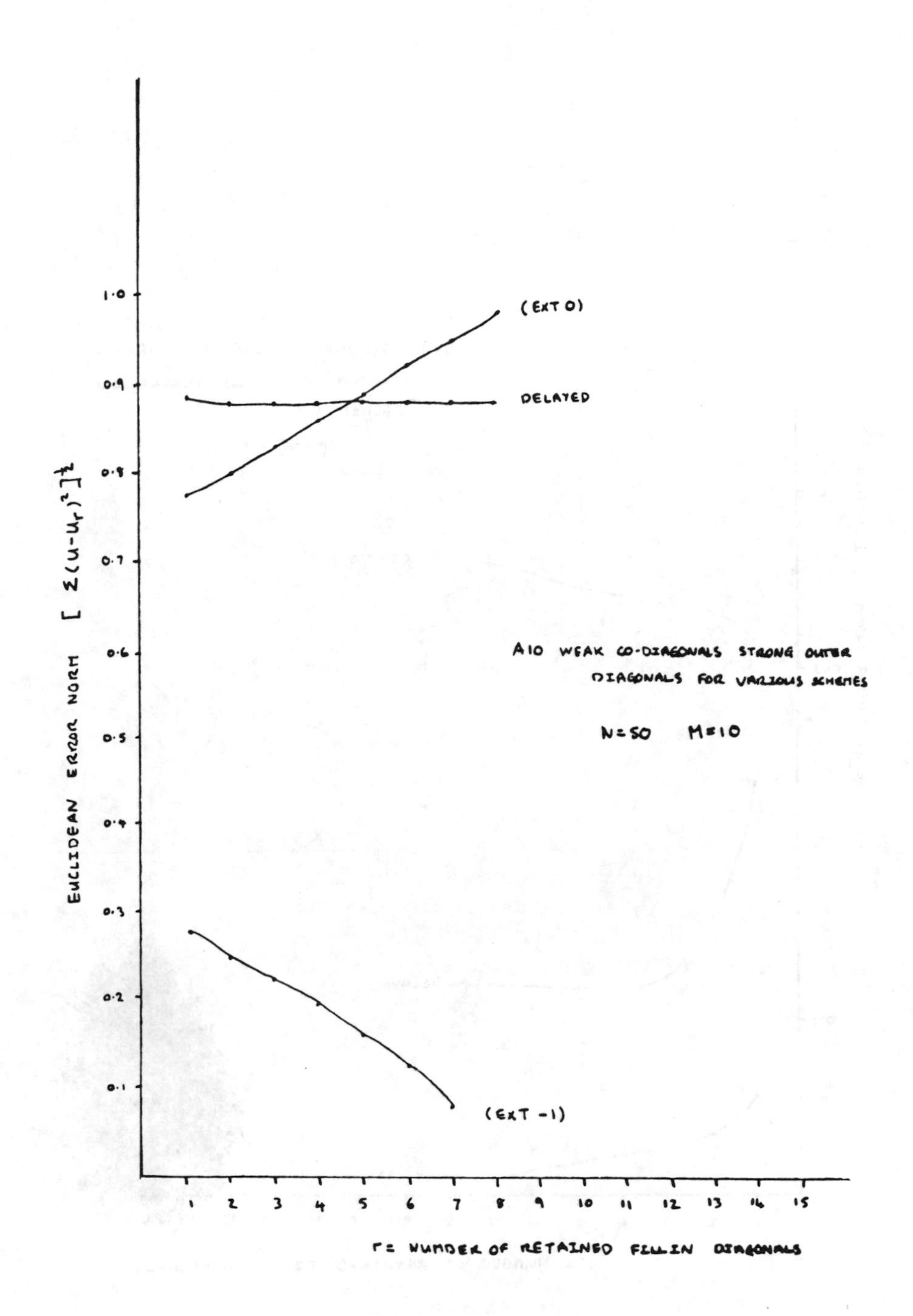

(EXT 0)
DELAYED
(EXT -1)
A10 WEAK CO-DIAGONALS STRONG OUTER DIAGONALS FOR VARIOUS SCHEMES
N=50 M=10
EUCLIDEAN ERROR NORM [Σ(U-U_r)^2]^½
r= NUMBER OF RETAINED FILL IN DIAGONALS
1·0
0·9
0·8
0·7
0·6
0·5
0·4
0·3
0·2
0·1
1 2 3 4 5 6 7 8 9 10 11 12 13 14 15

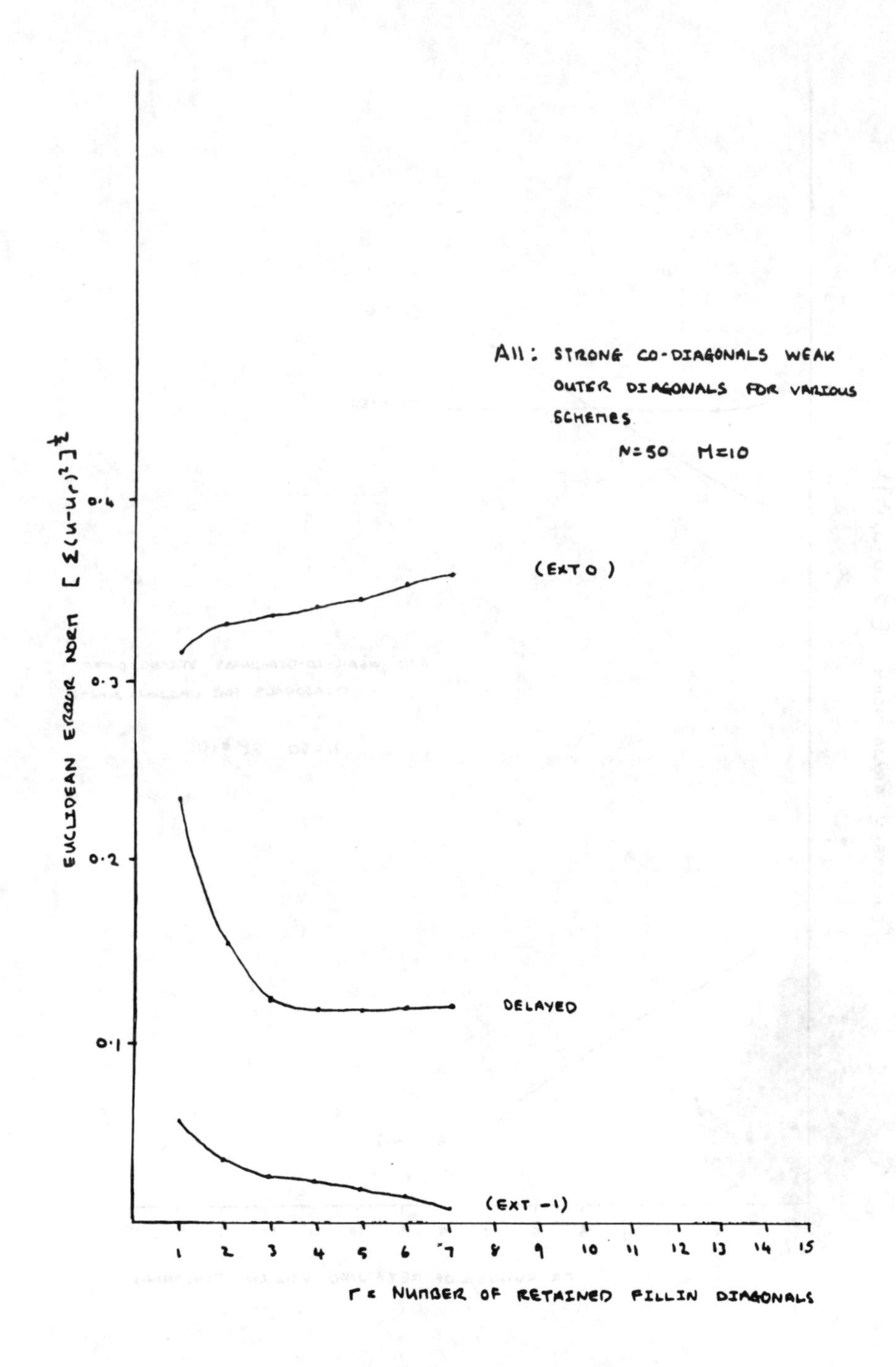

A11: STRONG CO-DIAGONALS WEAK OUTER DIAGONALS FOR VARIOUS SCHEMES
N=50 M=10
(EXT 0)
DELAYED
(EXT -1)
EUCLIDEAN ERROR NORM
0·1
0·2
0·3
0·4
1 2 3 4 5 6 7 8 9 10 11 12 13 14 15
r = NUMBER OF RETAINED FILLIN DIAGONALS

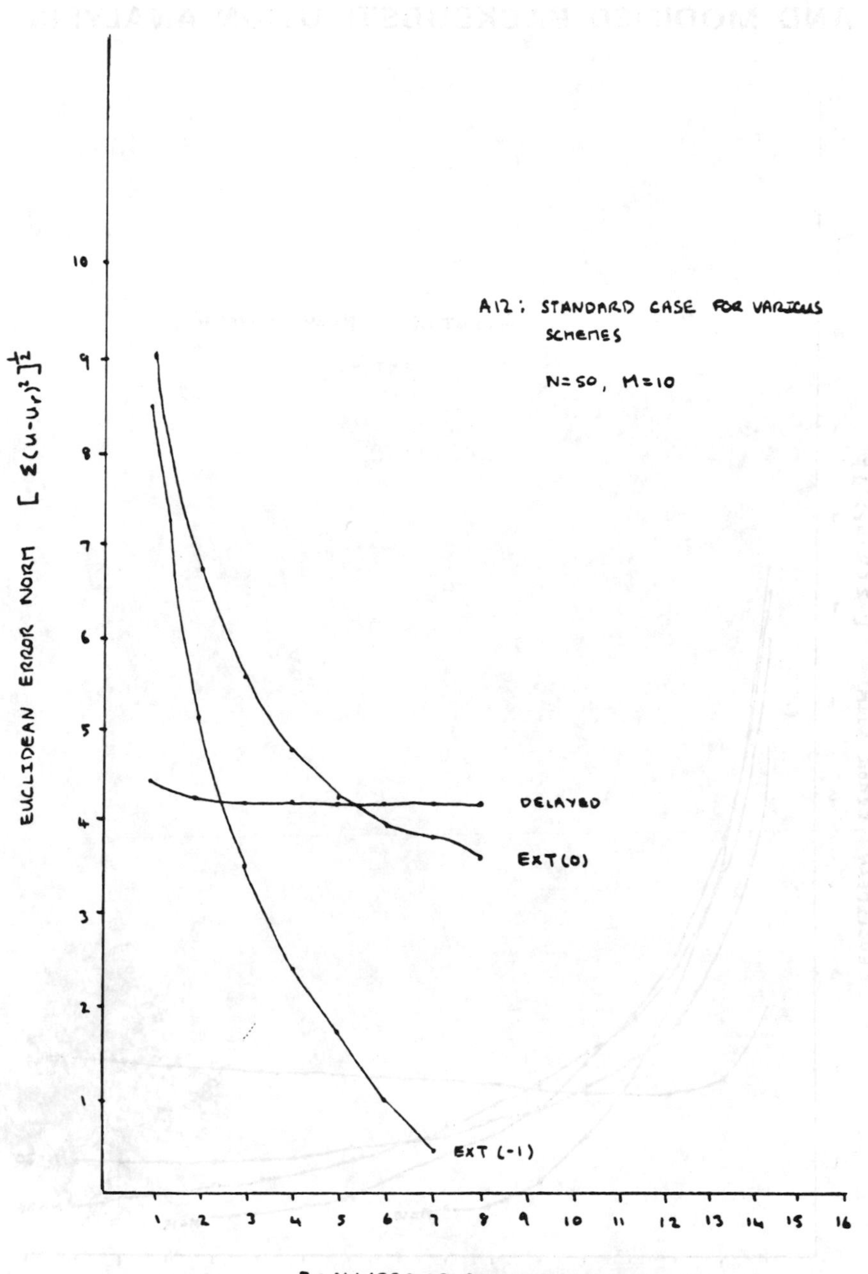
A12: STANDARD CASE FOR VARIOUS SCHEMES
N=50, M=10
EUCLIDEAN ERROR NORM
DELAYED
EXT(0)
EXT (-1)
r = NUMBER OF RETAINED FILL IN DIAGONALS

Appendix 3

SIF AND MODIFIED BACKSUBSTITUTION ANALYSIS

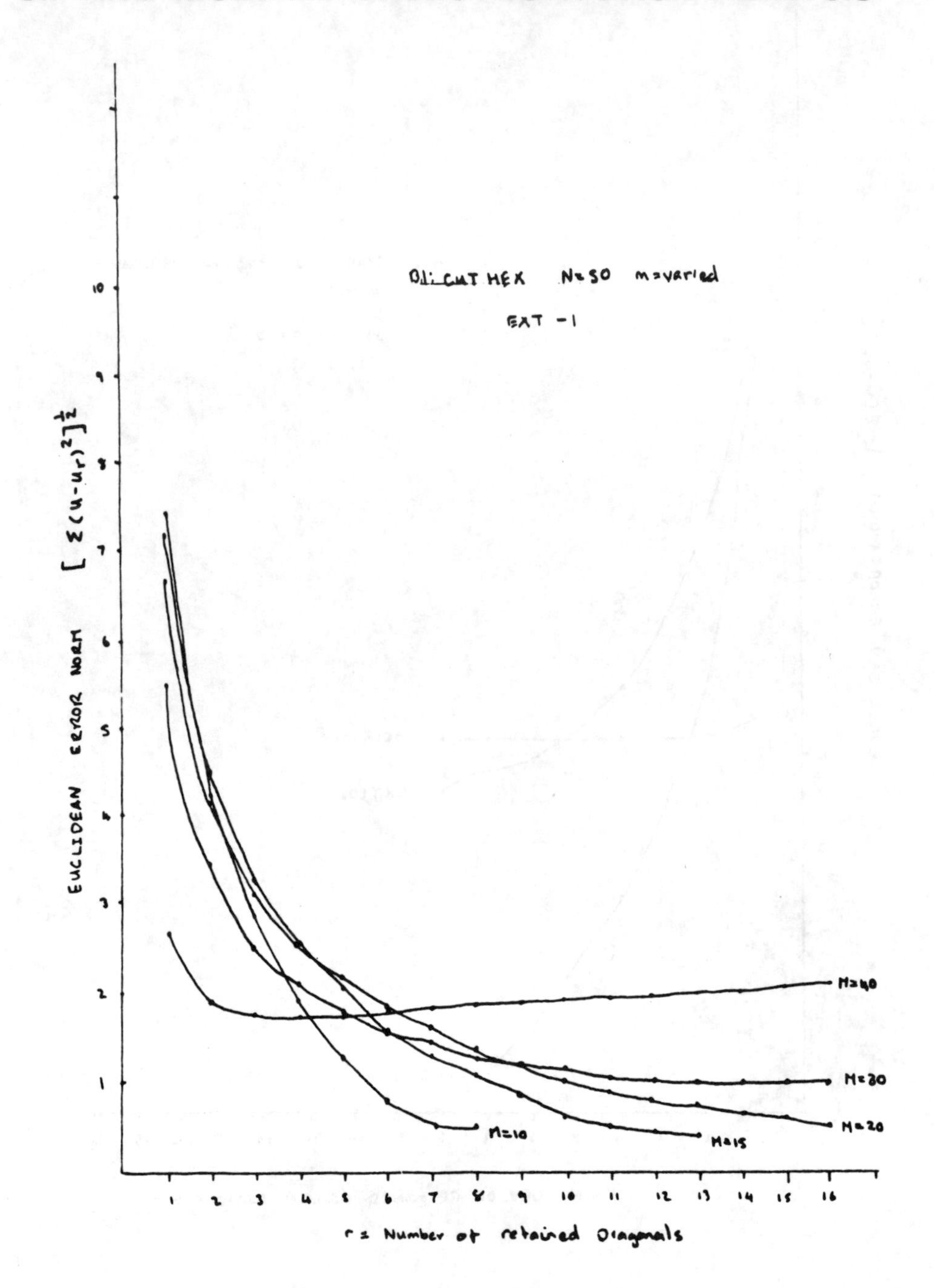

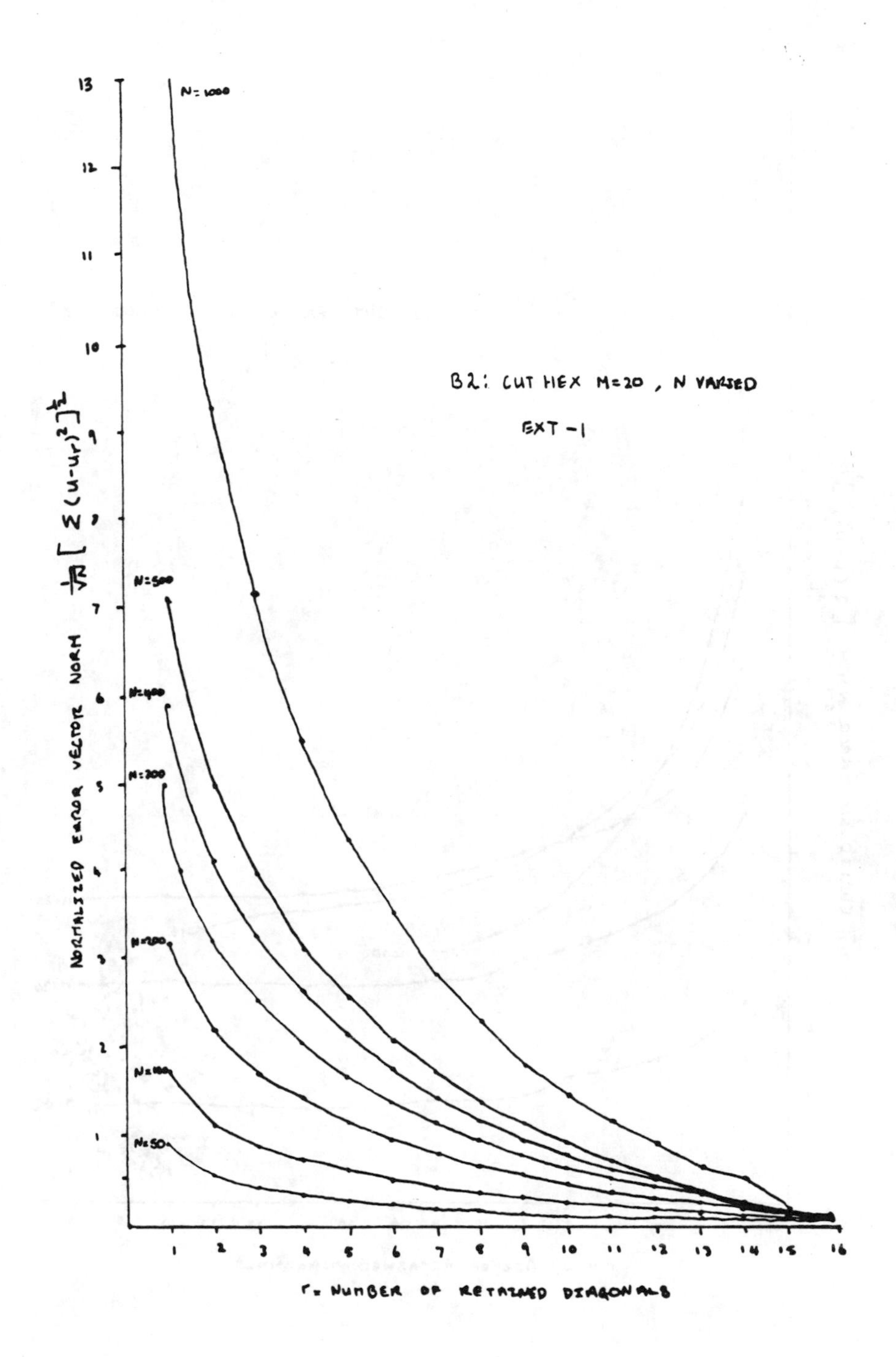
B2: CUT HEX M=20, N VARIED
EXT-1
N=1000
N=500
N=400
N=300
N=200
N=100
N=50
NORMALIZED ERROR VECTOR NORM
r = NUMBER OF RETAINED DIAGONALS

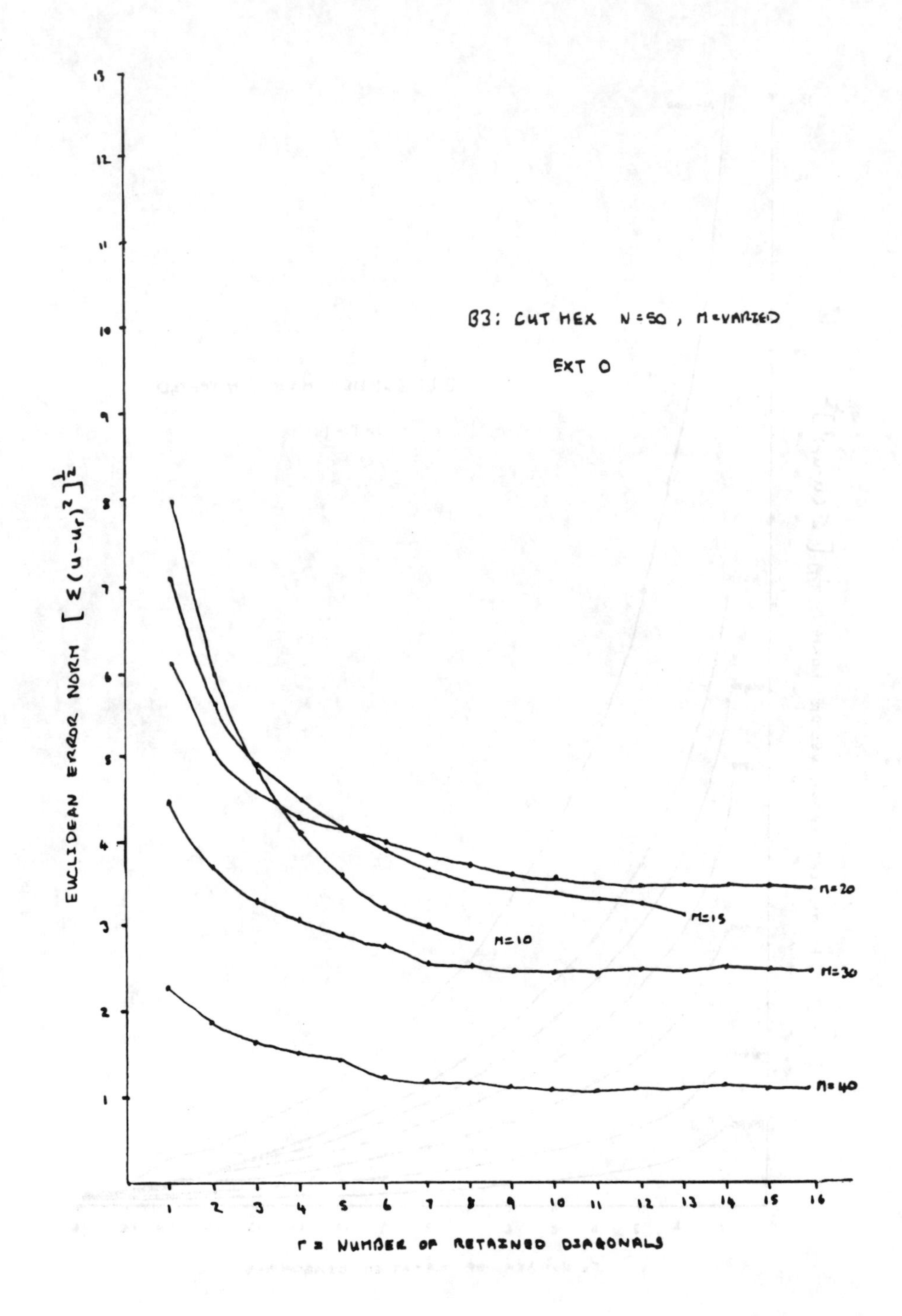
B3: CUT HEX N=50, M=VARIED
EXT 0
EUCLIDEAN ERROR NORM $[\Sigma(u-u_r)^2]^{\frac{1}{2}}$
r = NUMBER OF RETAINED DIAGONALS
M=10
M=15
M=20
M=30
M=40

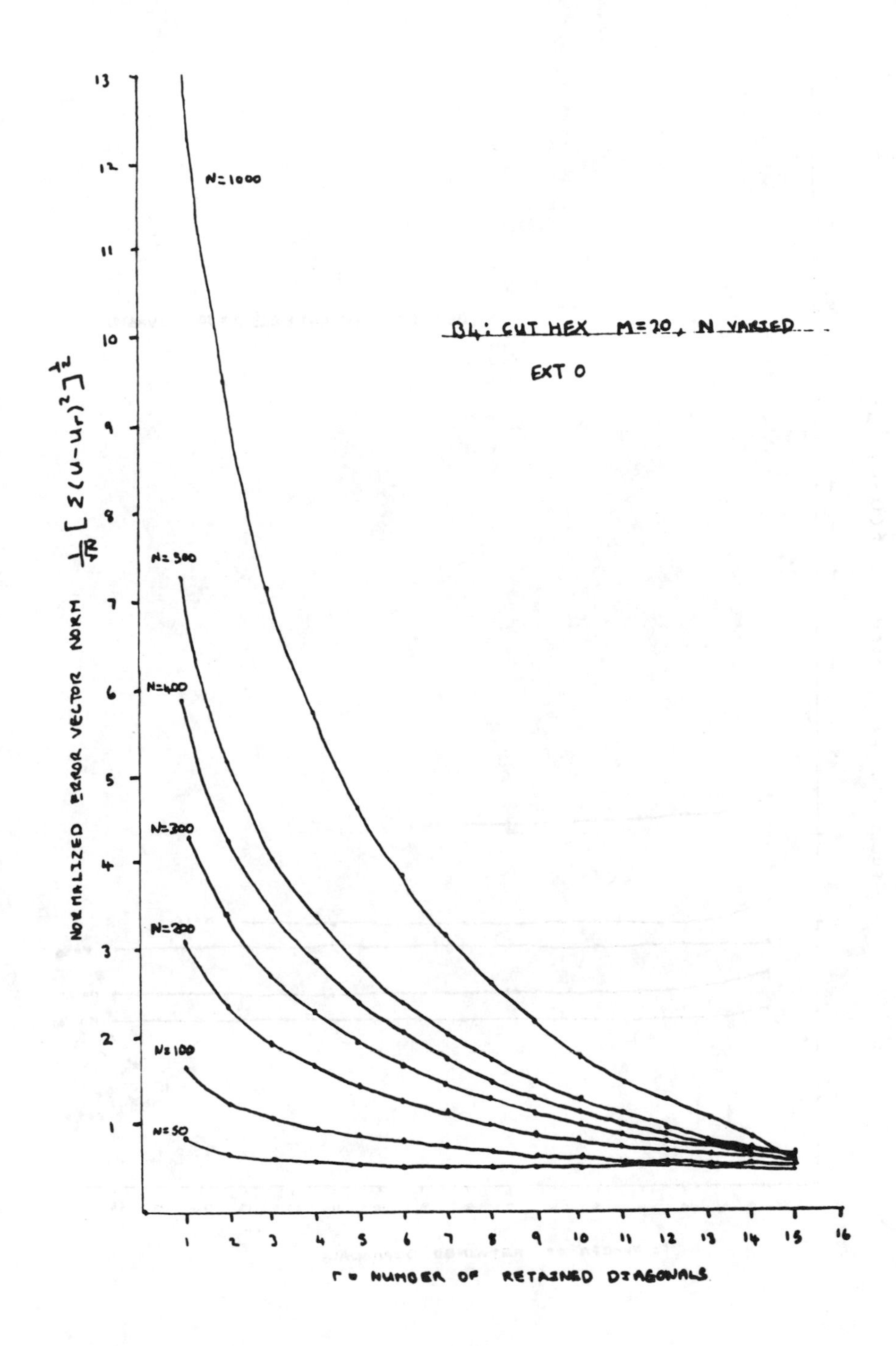
B4: CUT HEX M=20, N VARIED
EXT 0
N=1000
N=500
N=400
N=300
N=200
N=100
N=50
NORMALIZED ERROR VECTOR NORM
r = NUMBER OF RETAINED DIAGONALS

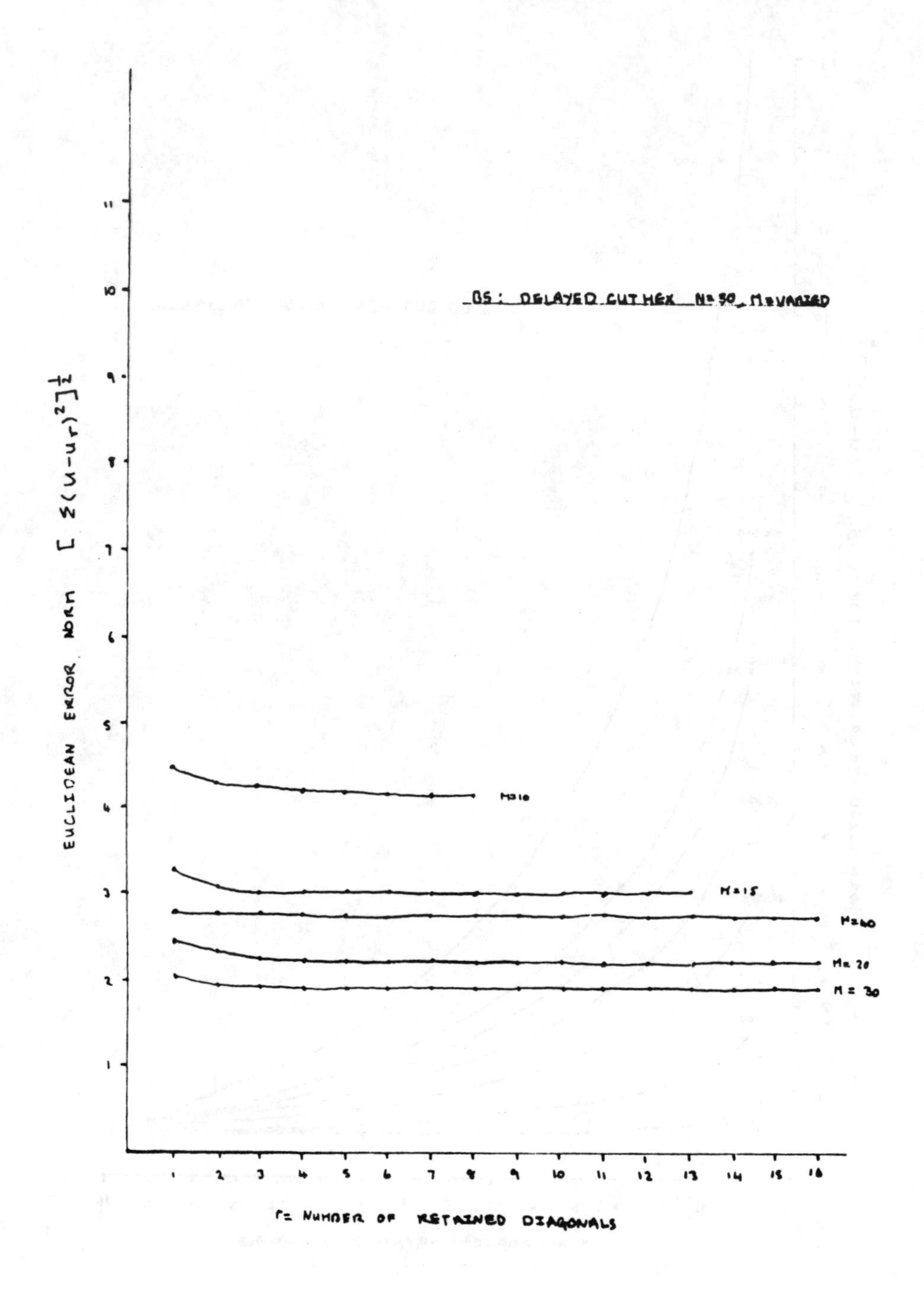
BS: DELAYED CUTHEX N=50, M=VARIED
EUCLIDEAN ERROR NORM $[\Sigma(U-U_r)^2]^{\frac{1}{2}}$
M=10
M=15
M=20
M=30
r= NUMBER OF RETAINED DIAGONALS
1 2 3 4 5 6 7 8 9 10 11 12 13 14 15 16

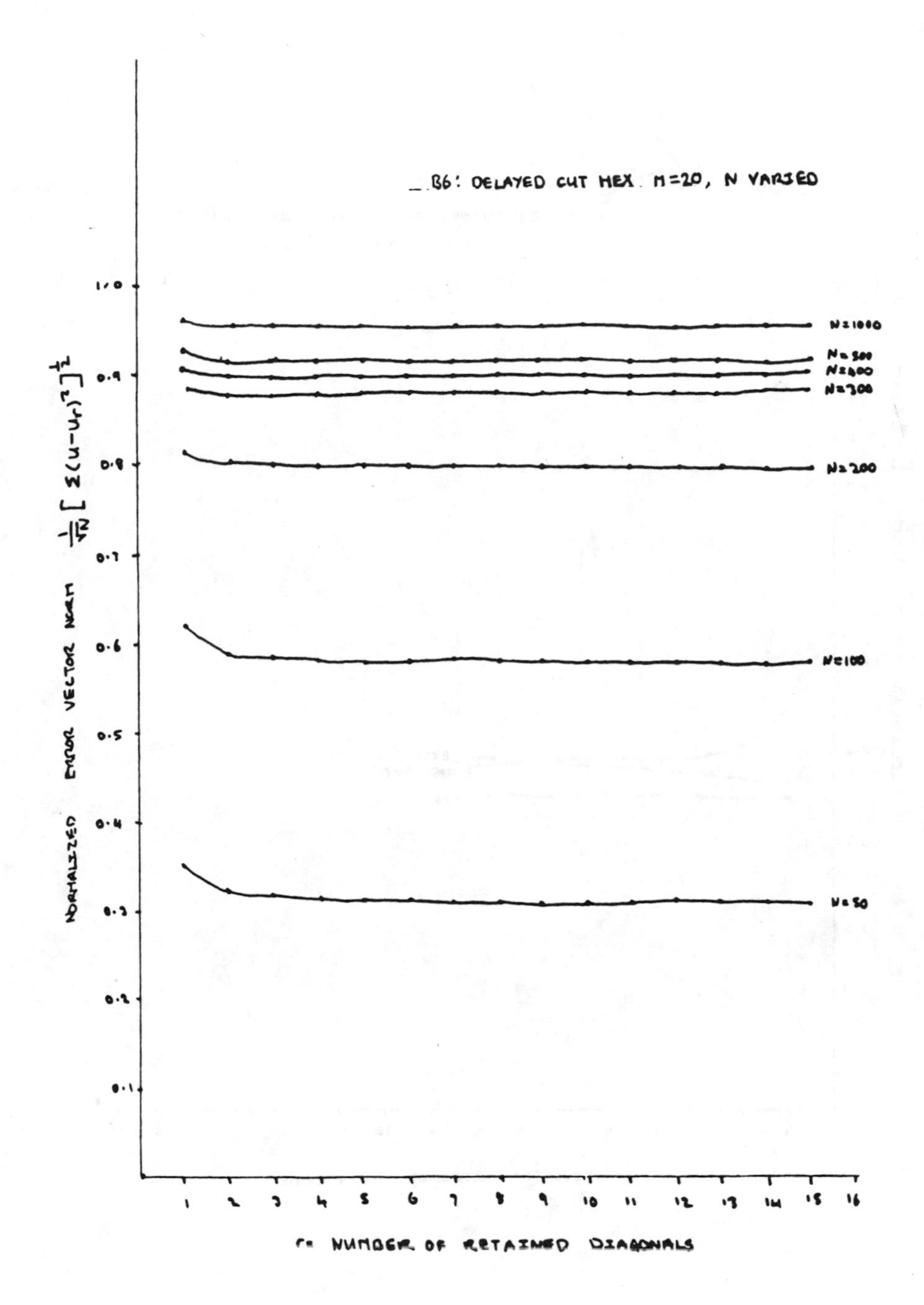
B6: DELAYED CUT HEX. M=20, N VARIED
NORMALIZED ERROR VECTOR NORM
1/0
0·9
0·8
0·7
0·6
0·5
0·4
0·3
0·2
0·1
N=1000
N=500
N=400
N=300
N=200
N=100
N=50
1
2
3
4
5
6
7
8
9
10
11
12
13
14
15
16
r= NUMBER OF RETAINED DIAGONALS

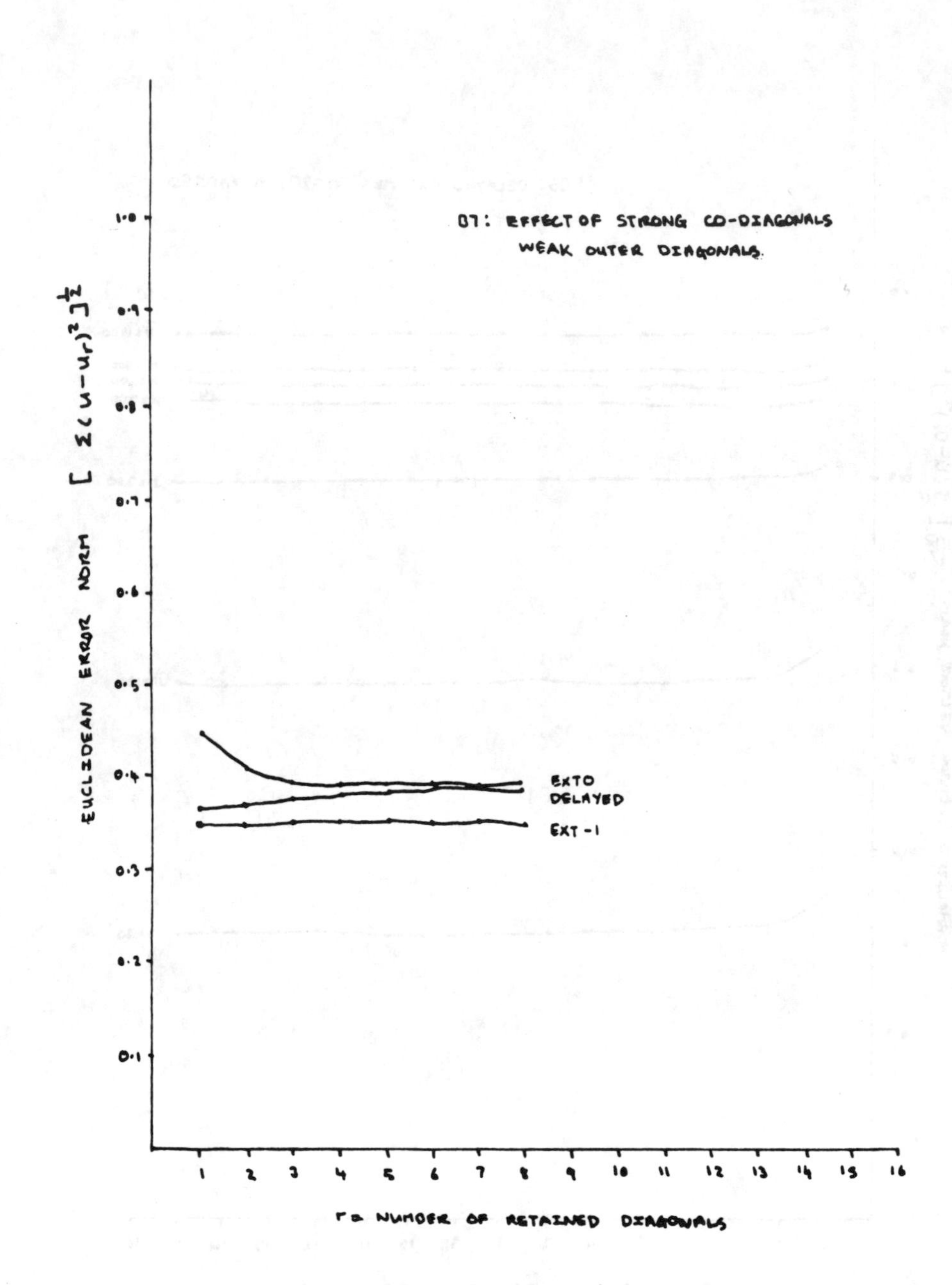
D7: EFFECT OF STRONG CO-DIAGONALS
WEAK OUTER DIAGONALS.
EUCLIDEAN ERROR NORM [Σ(u − u_r)²]^½
1·0
0·9
0·8
0·7
0·6
0·5
0·4
0·3
0·2
0·1
EXTO
DELAYED
EXT - 1
1 2 3 4 5 6 7 8 9 10 11 12 13 14 15 16
r = NUMBER OF RETAINED DIAGONALS

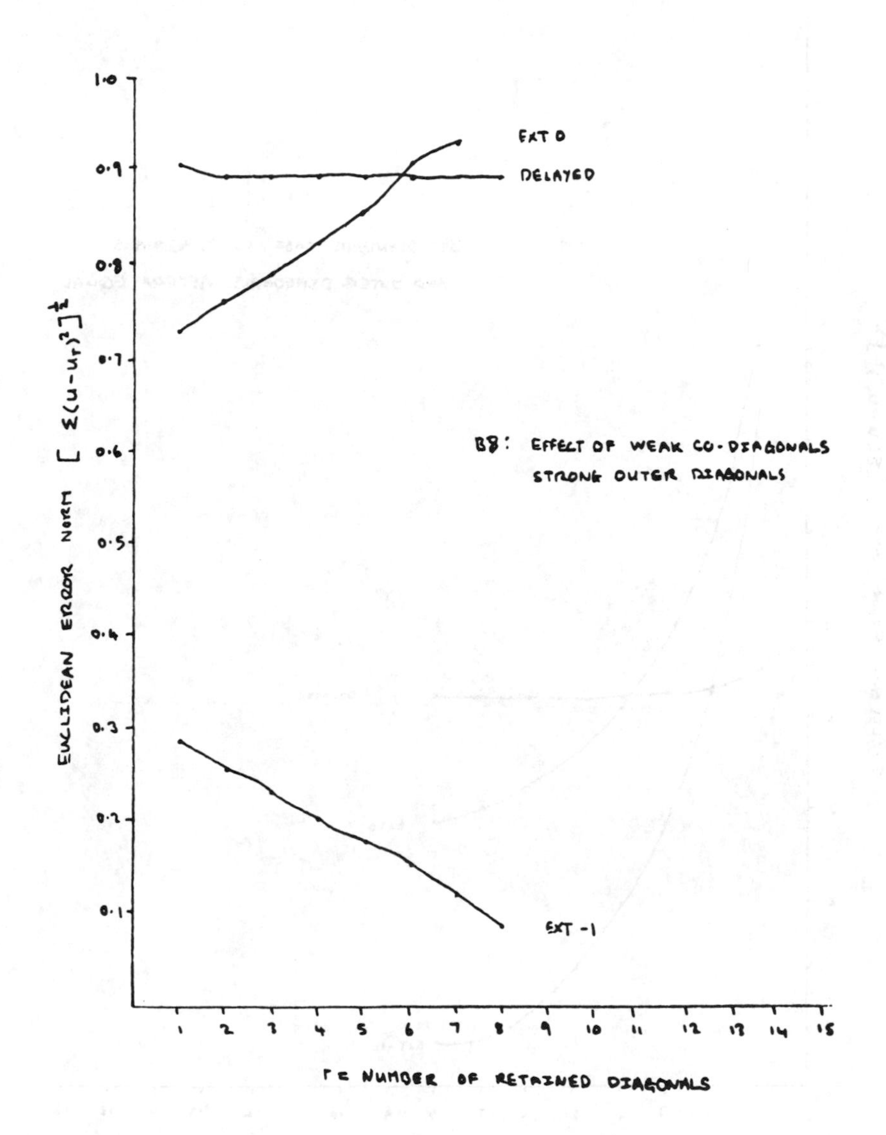

1.0
0.9
0.8
0.7
0.6
0.5
0.4
0.3
0.2
0.1
EXT 0
DELAYED
B8: EFFECT OF WEAK CO-DIAGONALS
STRONG OUTER DIAGONALS
EXT -1
EUCLIDEAN ERROR NORM [Σ(u-u_r)²]^½
1 2 3 4 5 6 7 8 9 10 11 12 13 14 15
r = NUMBER OF RETAINED DIAGONALS

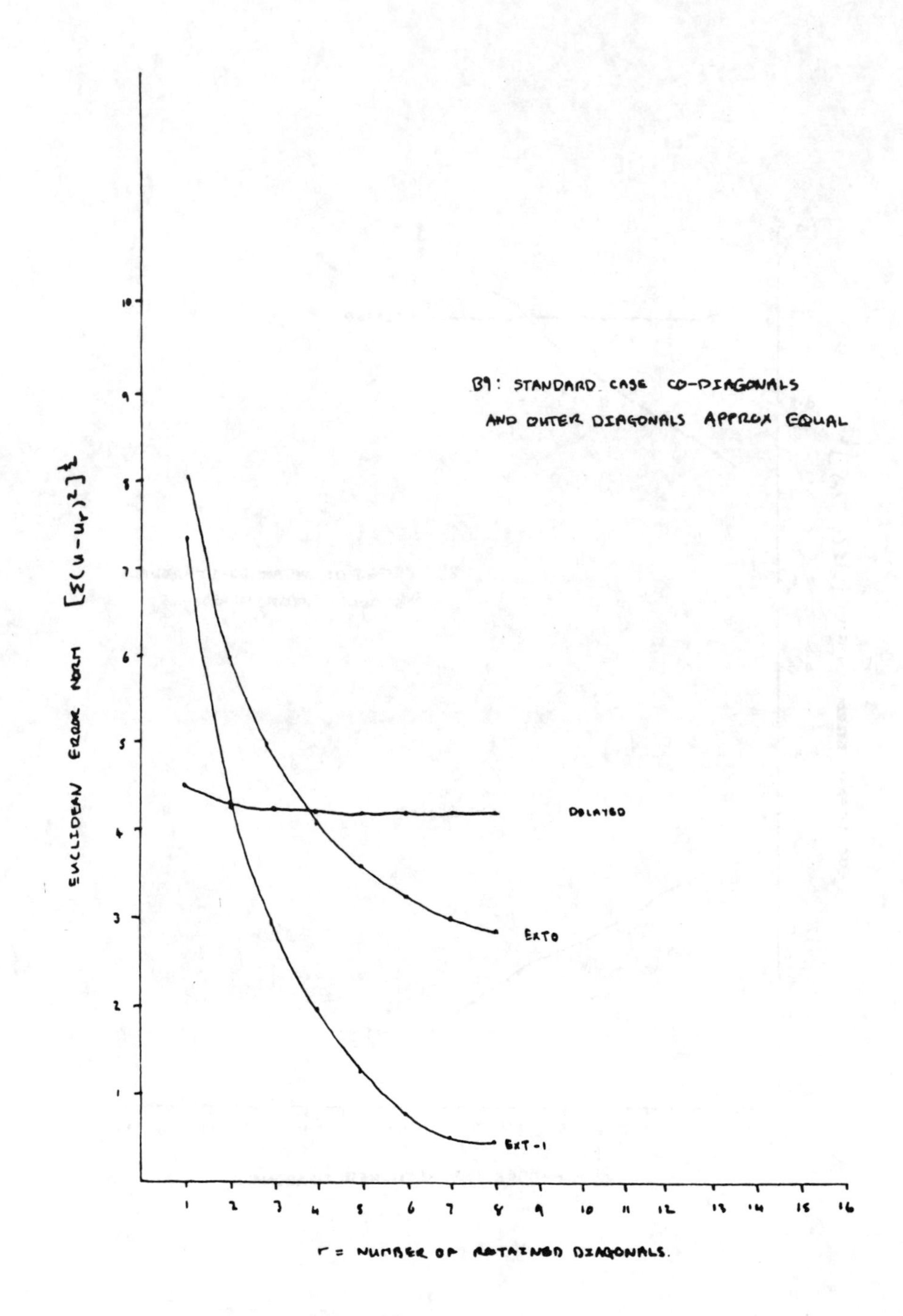
B9: STANDARD CASE CO-DIAGONALS
AND OUTER DIAGONALS APPROX EQUAL
EUCLIDEAN ERROR NORM [Σ(u−u_r)²]½
DELAYED
EXT0
EXT-1
r = NUMBER OF RETAINED DIAGONALS.

EIGENVALUE-EIGENVECTOR COMPUTATIONS

SYSTOLIC DESIGNS FOR THE CALCULATION OF THE EIGENVALUES AND EIGENVECTORS OF A SYMMETRIC TRIDIAGONAL MATRIX

D. J. EVANS and K. MARGARITIS

Department of Computer Studies, Loughborough University of Technology, Loughborough, Leicestershire, U.K.

This paper presents systolic algorithms for the calculation of the eigenvalues and eigenvectors of a symmetric tridiagonal matrix using the methods of bisection and inverse iteration respectively. A single array design is considered, where the use of only one array of linearly connected systolic processors solves the problem at the expense of more complex cell definition and control mechanisms.

KEY WORDS: Systolic algorithm, symmetric tridiagonal matrix, eigenvalues and eigenvectors, bisection, inverse iteration.

C.R. CATEGORIES: G.1.3, F.1.1, F.2.1.

1. THE CALCULATION OF THE EIGENVALUES OF A SYMMETRIC TRIDIAGONAL MATRIX

Given a symmetric tridiagonal matrix of order n, with diagonal elements $a_1, a_2, \ldots, a_n$ and off-diagonal elements $b_2, b_3, \ldots, b_n$ with $b_1 = 0$. Then for any number x let the Sturm sequence defined as:

$$p_0(x)=1, \quad p_1(x)=a_1-x,$$
$$p_i(x)=(a_i-x)p_{i-1}(x)-b_i^2 p_{i-2}(x), \qquad i=2,3,\ldots,n, \tag{1}$$

then, $p_n(x)$ is the characteristic polynomial of the matrix. Furthermore the number $S(x)$ of disagreements in sign of the sequence $p_1(x), p_2(x), \ldots, p_n(x)$ indicates the number of roots of the polynomial, i.e. the eigenvalues of the matrix, that are smaller than x [1] (see Figure 1). Bisection is used to isolate the roots of $p_n(x)$; the termination criterion is

$$(\beta-\alpha) \leqq \varepsilon, \tag{2}$$

where ε is a predefined small positive quantity and $[\alpha, \beta]$ is the current bisection interval. If the termination criterion is not satisfied the new bisection interval is determined as follows: Let R be the order of the root being sought, i.e. the smallest root has order 0, the next 1, etc.; then

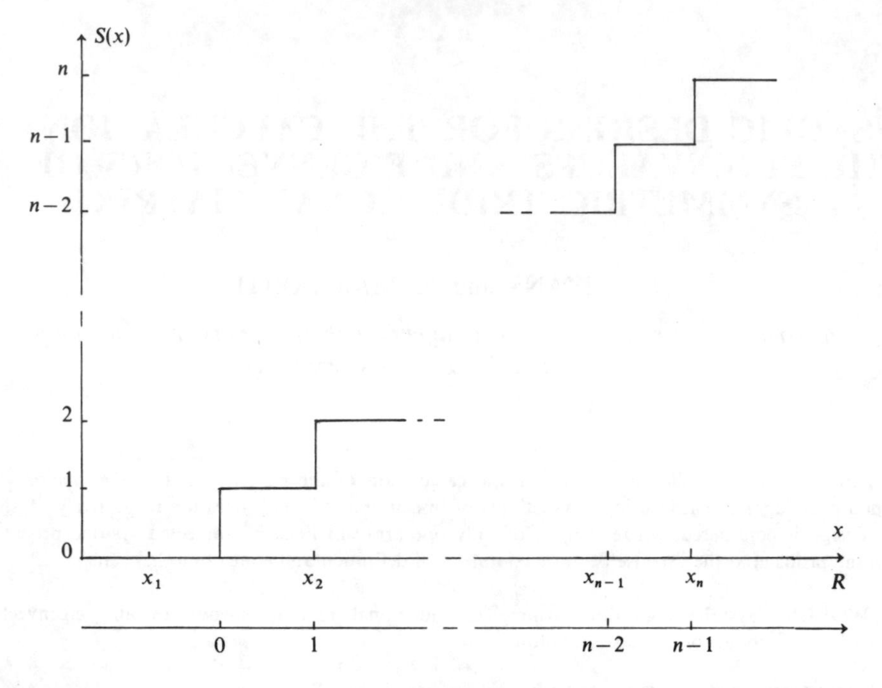

Figure 1 Roots for Sturm sequence polynomial.

if

$S(x) > R$

$x := (\alpha + x)/2$; new interval is $[\alpha, x]$ (3)

true

$x := (x + \beta)/2$; new interval is $[x, \beta]$.

An eigenvalue with multiplicity $m > 1$ can also be detected as $S(x)$ produces a "jump" of m at x. Therefore the bisection process returns to the multiple root for m times before the next eigenvalue is located.

In [2] a modification of the Sturm sequence was proposed because it was noticed that it is quite common for $p_i(x)$ to produce overflow. Thus, instead of calculating the sequence (2), an alternative sequence is computed, defined as,

$$q_i(x) = p_i(x)/p_{i-1}(x), \qquad i = 1, 2, \ldots, n, \tag{4}$$

and calculating using the recurrence,

$$\begin{aligned} q_1(x) &= a_1 - x, \\ q_i(x) &= (a_i - x) - b_i^2/q_{i-1}(x), \qquad i = 2, 3, \ldots, n. \end{aligned} \tag{5}$$

Some advantages of the modified sequence are the following: the q-sequence does not suffer from overflow problems; less computation is required since two multiplications are replaced by one division; the calculation of $q_i(x)$ requires only $q_{i-1}(x)$, while $p_i(x)$ requires both $p_{i-1}(x)$ and $p_{i-2}(x)$. On the other hand, recurrence (5) is unstable, while (1) is extremely stable, for it is possible for $q_{i-1}(x)$ to become zero for some i; however, in such cases the zero can be replaced by a suitably small quantity. Another advantage of the modified sequence is the fact that $S(x)$ is now given by the number of negative q's.

Using the Sturm sequence properties all n eigenvalues of a symmetric tri-diagonal matrix can be found in parallel [3, 4]. Bisection can be applied on each root independently from the other roots since the only criterion used is $S(x)$, as explained in (3) and in Figure 1. The initial interval can be common to all eigenvalues. Alternatively a different bisection interval can be supplied for each eigenvalue, based on good approximations in the cases where the matrix is frequently updated and the eigenvalue computation process is repeated.

A pipeline for the calculation of the Sturm sequence of polynomials is described herein, based on the recurrence (5) and it is compared to the pipeline proposed in [4] using relations (1). Then, the pipeline is incorporated in a systolic eigenvalue solver implementing the bisection process; several extensions are also briefly discussed.

2. STURM SEQUENCE PIPELINE

The main computational effort of the algorithm described is concentrated in the calculation of the Sturm sequence for a given x and the determination of the corresponding $S(x)$. A diagram of the pipeline calculating the Sturm sequence is shown in Figure 2. Each pipeline block is analysed in three parts: the first part is a simple delay element that allows the synchronous movement of the bisection distance d with the other data in the pipeline; the operation of d is analysed later on. A second usage for this simple communication channel is the preloading of the matrix diagonals, i.e. the polynomial coefficients, into the pipeline blocks during the setup phase.

The Q processor of the pipeline block calculates $q$out for given $q$in a, b and x; its functional specification can be:

```
if
   setup phase
      load a, b through x, d channels
if
   qin = 0
      q := sufficiently small quantity
   true
      q := qin
xout := xin
qout := (a − x) − (b/q).
```

The processor is quite complex as it involves two subtractions, one division and a

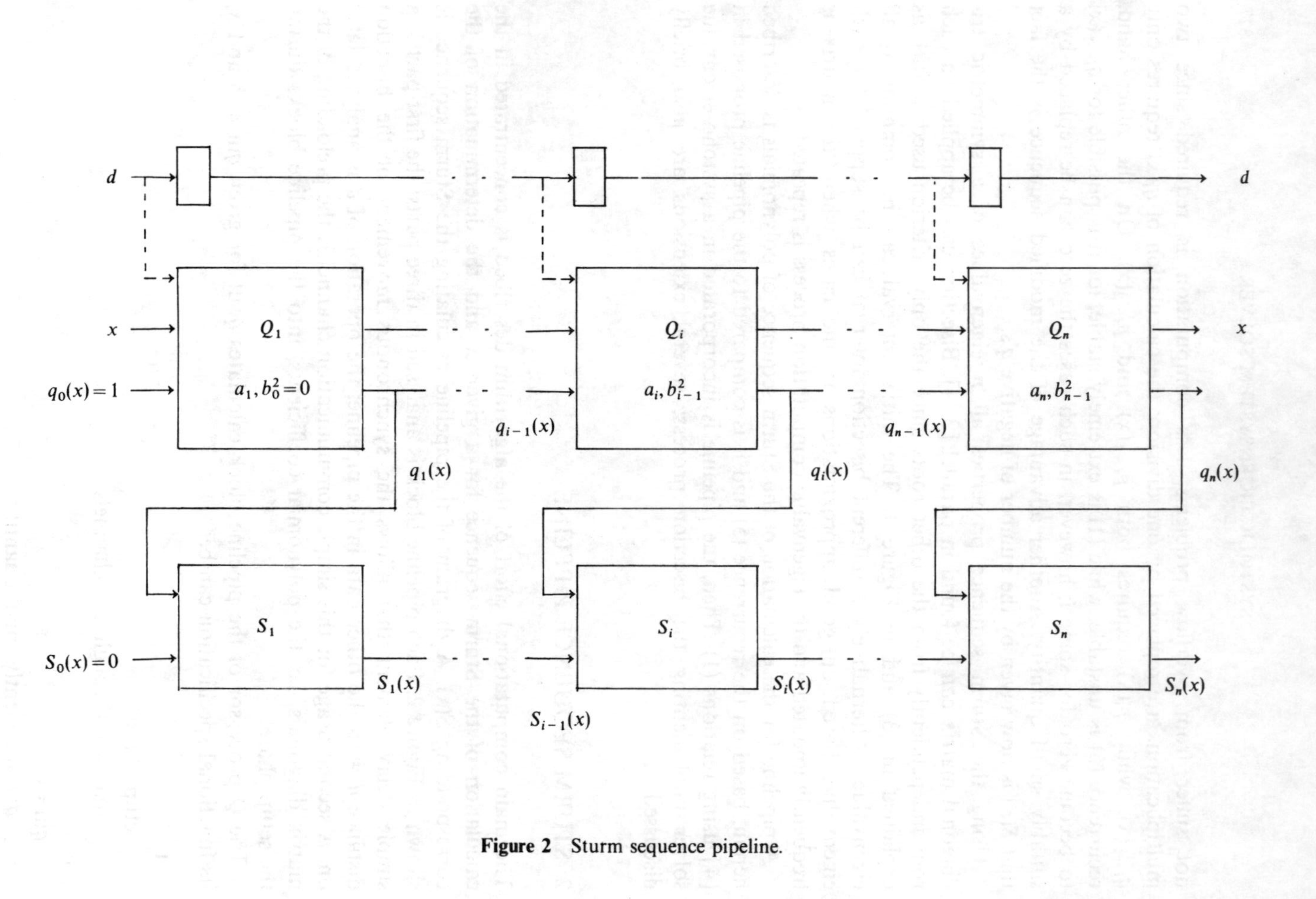

Figure 2 Sturm sequence pipeline.

comparison with zero; if the recurrence (1) is used the processor will require one additional input–output channel and it will involve two subtractions and two multiplications.

The S processor of the pipeline block computes Sout for given Sin and $q$in, i.e.,

if
  $q\text{in} < 0$
    $s\text{out} := s\text{in} + 1$
  true
    $s\text{out} := s\text{in}$.

Notice that the comparison with zero can be combined with the corresponding calculation in processor Q; observe also that this computation is considerably simpler than the sign-change detection required if the original Sturm sequence is used. From Figure 2 it is obvious that the computation of processor S_i is performed after that of Q_i, or, in other words, it is overlapped with the computation of Q_{i+1}. Therefore $S(x)$ is produced one cycle after $q_i(x)$. Another implication is that since the S_i processor is much simpler than Q_i, the S pipeline remains "idle" for some part of each clock cycle waiting for the Q pipeline to finish its computation. An alternative approach is to define as time unit for the pipeline the sequence of the computations in Q_i and S_i, and have no delay at the end of the pipeline. Finally, two unequal clock cycles may be considered or a data-driven (wavefront) computation.

3. SYSTOLIC EIGENVALUE SOLVER

The systolic eigenvalue solver is shown in Figure 3. The main part of the system is the n-processor pipeline calculating the Sturm sequence. The pipeline can be folded to form a systolic ring reflecting the iterative nature of the process, together with two controlling processors, the Bisection Test processor and the I/O controller.

The initial values are supplied to the first cell of the pipeline, while the nth cell provides $S(x)$ to the Bisection Test processor. The bisection interval corresponding to certain x should be present in the Bisection Test processor simultaneously with $S(x)$. This can be achieved in two ways: a local store can be placed in the system, keeping all bisection intervals and making them available in the proper cycle for updating; or a distributed memory, i.e. simple registers can allow the bisection interval to travel through the pipeline together with x and $S(x)$. The first approach introduces the need of large local storage and control mechanisms, while the second approach involves additional communication channels throughout the pipeline. The second technique is used here since it fits better with the pipeline scheme and it is also used for the preloading of the pipeline with the matrix diagonal elements.

In order to minimise computation and communication overheads a bisection interval $[\alpha, \beta]$ is expressed in the form of a bisection distance $d = (\beta - \alpha)/4$, and a bisection point $x = (\alpha + \beta)/2$. Then, (3) and (4) can be expressed as

$$d \leqq \varepsilon, \tag{6}$$

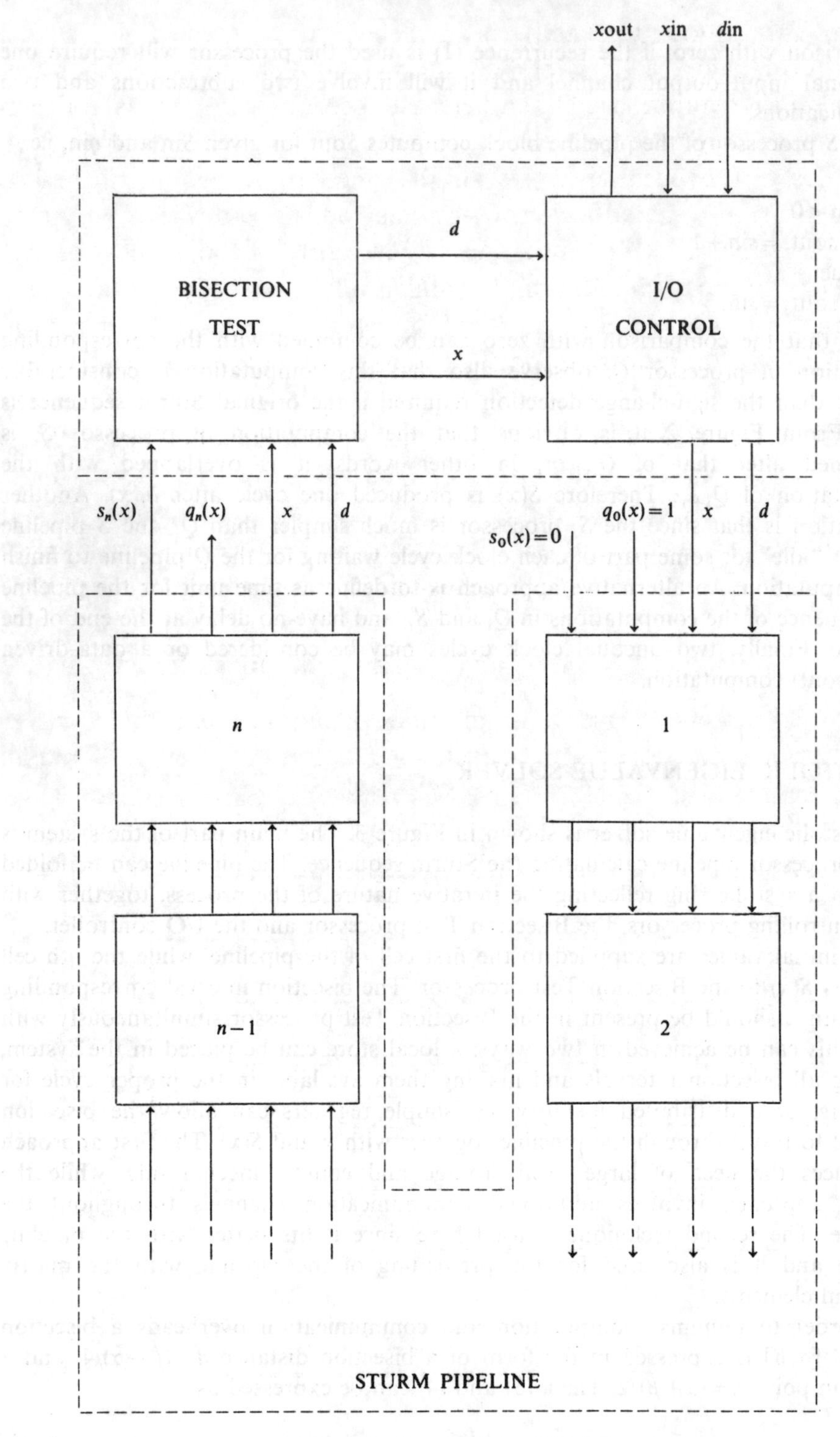

Figure 3 Systolic system overview.

and

$$
\begin{aligned}
&\text{if} \\
&\quad S(x) > R \\
&\qquad x := x - d \\
&\quad \text{true} \\
&\qquad x := x + d \\
&\quad d := d/2.
\end{aligned} \tag{7}
$$

The computation of (7) is performed in the Bisection Test processor, where the "order" R of the root is kept by means of a counter which is initialised during the setup phase to 0. Thus $S(x_1)$ is compared with 0, $S(x_2)$ with 1 and finally $S(x_n)$ with $n-1$; then the counter is reset to 0 for the next bisection cycle. The computation of the Bisection Test process is simple enough to be completed within a single pipeline time unit, and therefore it introduces only one cycle delay.

The I/O controller allows for the initial loading of the quantities $a_1, a_2, \ldots, a_n$ and $0, b_2, b_3, \ldots, b_n$ followed by $x_1, x_2, \ldots, x_n$ and $d_1, d_2, \ldots, d_n$. After that the ring is "closed" and the normal operation of the system is resumed as the output of the Bisection Test processor is routed back to the pipeline, as shown in Figure 3. The iterates of each bisection cycle are also output to the host where a convergence test similar to that of (6) is performed. The I/O controller imposes no additional delay in the computation since it comprises of simple multiplexers.

The systolic system proposed performs a single iteration for the n eigenvalues in $n+2$ steps, each step having the complexity of 2 subtractions and 1 division, needed for the calculations in a Q processor.

The eigenvalue solver can be extended to perform a convergence test for the iterates, as in (6), and then produces its own Reset signal, when all eigenvalues are found. Thus the host is only informed when the computation is finished and a new one can start. Furthermore, only certain eigenvalues can be found, either of specific order or in specific range: in the first case the order of the eigenvalue would be associated to x and be compared with $S(x)$; in the second case an additional convergence check is necessary, i.e.

$$q_n(x) \leqq 0, \tag{8}$$

since the eigenvalues outside the specified range will "converge" to the bounds of the interval but will produce a non-zero $q_n(x)$.

4. THE CALCULATION OF THE EIGENVECTORS OF A SYMMETRIC TRIDIAGONAL MATRIX

Consider a $(n \times n)$ symmetrical tridiagonal matrix A with diagonal elements $a_1, a_2, \ldots, a_n$, off-diagonal elements $b_1, b_2, \ldots, b_{n-1}$ and an eigenvalue of the matrix, λ. Then the eigenvector $\mathbf{x}$ of matrix A that corresponds to the eigenvalue can be calculated as the solution of the linear system of equations

$$(A - \lambda I)\mathbf{x} = \mathbf{d}, \tag{9}$$

where $\mathbf{d}$ is a suitably chosen vector. The Inverse Iteration method, described in [5] is as follows: if we apply the LU decomposition on $(A-\lambda I)$, then (9) can be solved by means of a forward and a backward substitution. If the eigenvalue is accurate then two iterations on (9) are more than adequate, provided that $\mathbf{d}$ is not completely deficient in the eigenvector to be computed, (see [1,6]).

The LU factors are determined by Gaussian elimination with partial pivoting applied to matrix $(A-\lambda I)$. There are $n-1$ major steps to the process, rows $i+1$, $i+2,\ldots,n$ being as yet unmodified at the beginning of the ith major step. The configuraton at the beginning and the end of a step is shown in Figure 4. Matrix U has now three diagonals in general, to allow for any interchanges that may occur, while L has only one subdiagonal stored as a vector, together with a record of the permutations occurred.

The ith step is as follows (see Figure 4):

if
  $|b_i|>|u_i|$
    $c_{i+1}=1$, i.e. interchange rows i and $i+1$
    $p_i-b_i, q_i=a_{i+1}-\lambda, r_i=b_{i+1}, x_{i+1}=u_i, y_{i+1}=v_i, z_{i+1}=0$
  true
    $c_{i+1}=0$, i.e. no interchange takes place
    $p_i=u_i, q_i=v_i, r_i=0, x_{i+1}=b_i, y_{i+1}=a_{i+1}-\lambda, z_{i+1}=b_{i+1}$
  $m_{i+1}=x_{i+1}/p_i$
  $z_{i+1}=0, u_{i+1}=y_{i+1}-m_{i+1}q_i, v_{i+1}=z_{i+1}-m_{i+1}r_i,$ (10)

with $b_n=0$. Now, in the special case, where,

$$\mathbf{d}=L\mathbf{e}, \quad \text{with} \quad \mathbf{e}^T=[1,1,\ldots,1], \tag{11}$$

then, from (9)

$$LU\mathbf{x}=L\mathbf{e}, \quad \text{or} \quad U\mathbf{x}=\mathbf{e}. \tag{12}$$

With this choice of $\mathbf{d}$, the first iterate of the eigenvector $\mathbf{x}, \mathbf{x}_1$ is determined by a back substitution only. Then, having obtained $\mathbf{x}_1, \mathbf{x}_2$ can be found by forward and backward substitution. The forward substitution can be performed either separately, using the multipliers and permutation information saved during the Gaussian elimination process; or, alternatively the Gaussian elimination with partial pivoting can be extended to the right-hand-side vector also, as shown in Figure 4, for an initial vector $\mathbf{x}$, and final vector $\mathbf{y}$:

if
  $c_{i+1}=1$
    $y_i=x_{i+1}, w_{i+1}=z_i$
  true
    $y_i=z_i, w_{i+1}=x_{i+1}$
  $z_{i+1}=w_{i+1}-m_{i+1}y_i.$ (13)

This approach is followed herein, since it allows for a more compact and general

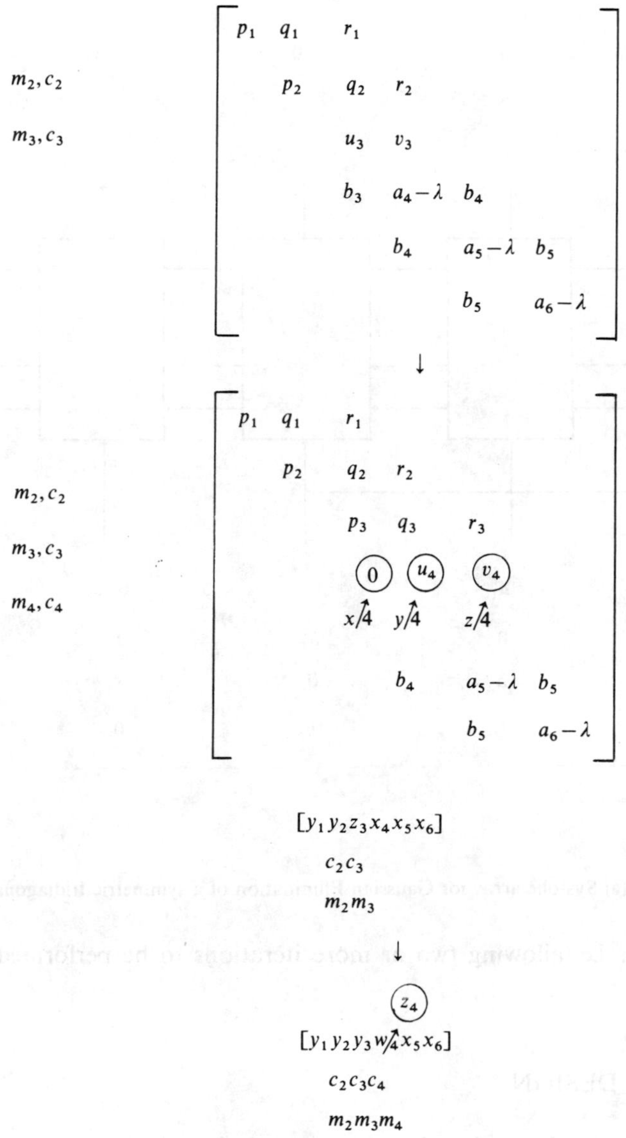

$$
\begin{matrix} \\ m_2, c_2 \\ m_3, c_3 \\ \\ \\ \\ \end{matrix}
\qquad
\begin{bmatrix}
p_1 & q_1 & r_1 & & & \\
 & p_2 & q_2 & r_2 & & \\
 & & u_3 & v_3 & & \\
 & & b_3 & a_4-\lambda & b_4 & \\
 & & & b_4 & a_5-\lambda & b_5 \\
 & & & & b_5 & a_6-\lambda
\end{bmatrix}
$$

$$\downarrow$$

$$
\begin{matrix} \\ m_2, c_2 \\ m_3, c_3 \\ m_4, c_4 \\ \\ \\ \end{matrix}
\qquad
\begin{bmatrix}
p_1 & q_1 & r_1 & & & \\
 & p_2 & q_2 & r_2 & & \\
 & & p_3 & q_3 & r_3 & \\
 & & \underset{x_4}{\textcircled{0}} & \underset{y_4}{\textcircled{u_4}} & \underset{z_4}{\textcircled{v_4}} & \\
 & & & b_4 & a_5-\lambda & b_5 \\
 & & & & b_5 & a_6-\lambda
\end{bmatrix}
$$

$$[y_1 y_2 z_3 x_4 x_5 x_6]$$

$$c_2 c_3$$

$$m_2 m_3$$

$$\downarrow$$

$$\textcircled{z_4}$$

$$[y_1 y_2 y_3 w_4 x_5 x_6]$$

$$c_2 c_3 c_4$$

$$m_2 m_3 m_4$$

Figure 4 Step 4 of Gaussian Elimination with partial pivoting.

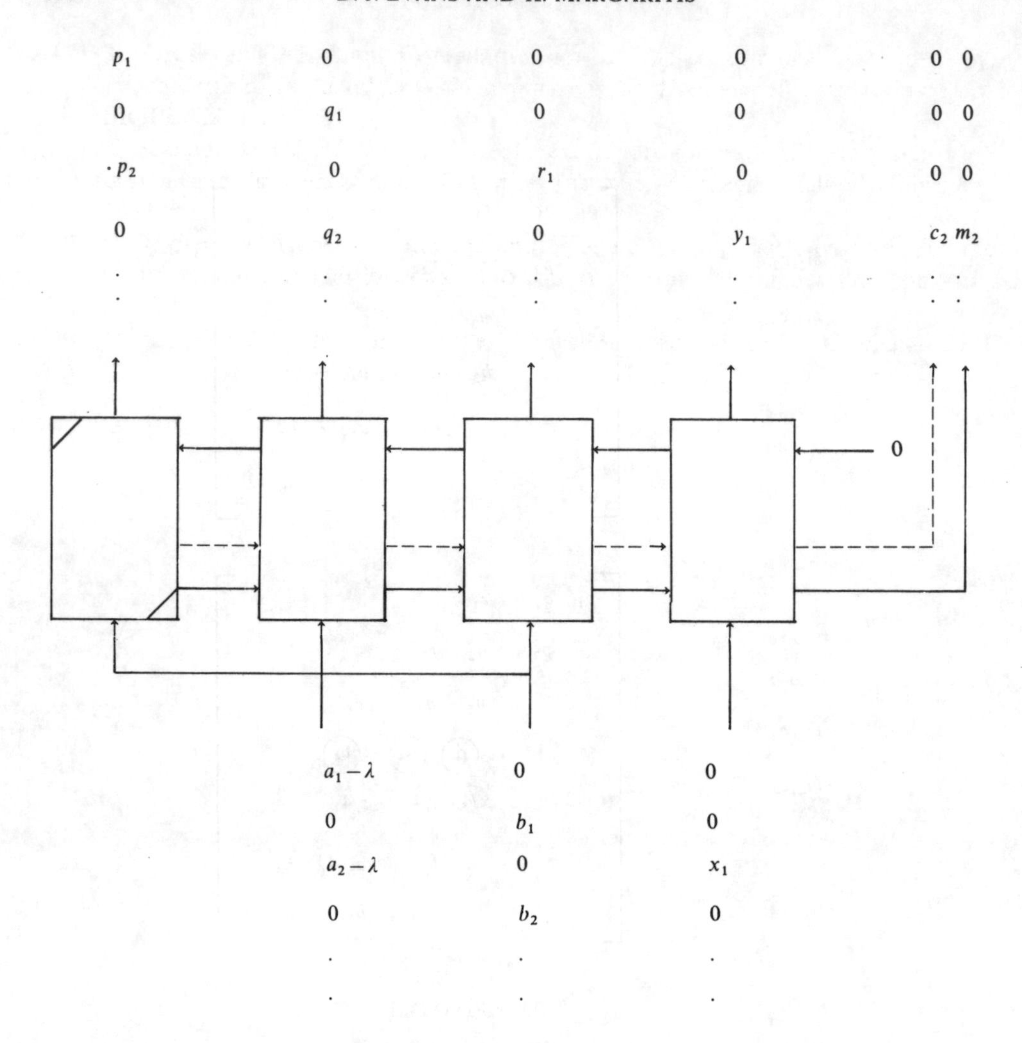

Figure 5 (a) Systolic array for Gaussian Elimination of a symmetric tridiagonal system.

purpose design, i.e. allowing two or more iterations to be performed on the same systolic array.

5. SYSTOLIC DESIGN

A systolic array implementing the computation described is shown in Figure 5(a) together with the I/O data sequences. The array accepts as input the diagonals of the matrix and the right-hand-side vector and produces as output matrix U, and three vectors corresponding to matrix L, permutation information and the modified right-hand-side vector. Three inputs are adequate since the matrix is symmetric. The array consists of four cells, a boundary cell to the left and three

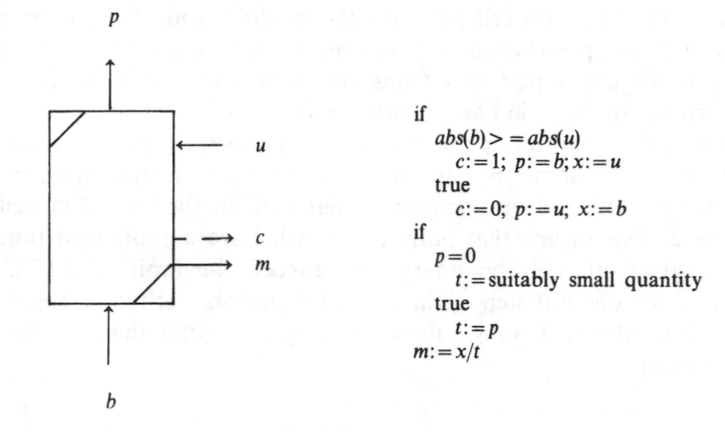
p
u
c
m
b
if
abs(b) > = abs(u)
c: = 1; p: = b; x: = u
true
c: = 0; p: = u; x: = b
if
p = 0
t: = suitably small quantity
true
t: = p
m: = x/t

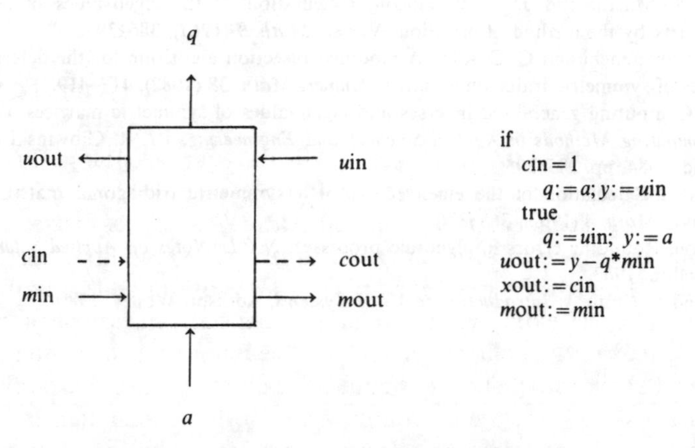
q
uout
uin
cin
cout
min
mout
a
if
cin = 1
q: = a; y: = uin
true
q: = uin; y: = a
uout: = y − q*min
xout: = cin
mout: = min

Figure 5 (b) Cell definitions.

IPS cells to the right augmented with a row interchange facility. The specification of the two types of cells is given in Figure 5(b). For the boundary cell, if before the division p is zero then it is substituted with a suitably small quantity. The computational complexity of the boundary cell is greater than one IPS and therefore the time unit of the computation of the array should be adjusted accordingly.

The boundary cell accepts u_i and b_i (see Figure 4) and decides on the next pivot row setting the flag c_{i+1} to indicate an interchange; it also computes the multiplier m_{i+1} for updating the adjacent non-pivot row. The two first IPS cells collect two adjacent row elements belonging in the same column: one from the east and one from the south. Thus they perform interchanges and modifications based on the values of c_{i+1} and m_{i+1} passed to them by the boundary cell. The IPS cells are active on alternate cycles and the boundary cell also computes c, m once every two cycles. The third IPS cell performs the modification of the right-hand-side vector, which for this purpose can be seen as an extra diagonal of the matrix. Thus the array of Figure 5 performs Gaussian elimination with partial pivoting and the forward substitution in $(2n+3)$ time units.

The back substitution process can be performed by the systolic array in [7]; however, since some p_i can be zero for some i, the boundary cell must be augmented with a device similar to that used in the boundary cell of the array of Figure 2. This means that both arrays will have a prolonged time cycle since the computation of their boundary cells exceed the limit of 1 IPS. The total time required for one full step of the Inverse Iteration method is, therefore $(4n+2)$ time units, since the back substitution can only start after the completion of the Gauss elimination.

References

[1] J. H. Wilkinson, *The Algebraic Eigenvalue Problem*, Clarendon Press, Oxford, 1965.

[2] B. Barth, R. S. Martin and J. H. Wilkinson, Calculation of the eigenvalues of a symmetric tridiagonal matrix by the method of bisection, *Numer. Math.* **9** (1967), 386–393.

[3] D. J. Evans, J. Shanehchi and C. C. Rick, A modified bisection algorithm for the determination of the eigenvalues of symmetric tridiagonal matrix, *Numer. Math.* **38** (1982), 417–419.

[4] R. Schreiber, Computing generalised inverses and eigenvalues of symmetric matrices using systolic arrays, in: *Computing Methods in Applied Sciences and Engineering, VI*, R. Glowinski *et al.* (eds.), North-Holland, 1984, pp. 285–295.

[5] J. H. Wilkinson, Calculation of the eigenvectors of a symmetric tridiagonal matrix by inverse iteration, *Numer. Math.* **4** (1962), 368–376.

[6] J. H. Wilkinson, Rounding errors in algebraic processes, *N.P.L. Notes on Applied Science*, No. 32, H.M.S.O., London, 1963.

[7] C. A. Mead and L. Conway, *Introduction to VLSI Systems*, Addison Wesley, 1980.

Systolic designs for eigenvalue-eigenvector computations using matrix powers

D.J. EVANS and K. MARGARITIS

Department of Computer Studies, Loughborough University of Technology, Loughborough, Leicestershire, UK LE11 3TU

Abstract. This paper describes systolic designs for the Power Method, Matrix Squaring Methods and Raised Matrix Power (RMP) Method, which is a combination of the two first methods, for the calculation of the dominant eigenvalue and the corresponding eigenvector of a matrix. The methods described are appropriate for non-sparse matrices with dominant eigenvalue. The designs were simulated and tested soft systolically in OCCAM.

Keywords. Vector computation, power method, matrix squaring method, soft-systolic simulation.

1. Introduction

Let the eigenvalues of a $(n \times n)$ matrix A be real and ordered according to absolute value so that,

$$|\lambda_1| > |\lambda_2| \geqslant |\lambda_3| \geqslant \cdots \geqslant |\lambda_n|. \tag{1}$$

Now let A operate repeatedly on a vector $\boldsymbol{u}$, which is a linear combination of the eigenvectors of A, [2],

$$\boldsymbol{u} = c_1 \boldsymbol{x}_1 + c_2 \boldsymbol{x}_2 + \cdots + c_n \boldsymbol{x}_n. \tag{2}$$

Then, through a series of successive matrix-vector multiplications, we have

$$A^k \boldsymbol{u} = \lambda_1^k \left\{ c_1 \boldsymbol{x}_1 + c_2 \left(\frac{\lambda_2}{\lambda_1} \right)^k \boldsymbol{x}_2 + \cdots + c_n \left(\frac{\lambda_n}{\lambda_1} \right)^k \boldsymbol{x}_n \right\}. \tag{3}$$

For k sufficiently large, we have

$$A^k \boldsymbol{u} \cong \lambda_1^k c_1 \boldsymbol{x}_1. \tag{4}$$

Thus, the dominant eigenvalue λ_1 is calculated as

$$\lambda_1 \cong (A^k \boldsymbol{u})_i / (A^{k-1} \boldsymbol{u})_i, \quad i = 1, 2, \ldots, n, \tag{5}$$

while $A^k \boldsymbol{u}$ is proportional to the corresponding eigenvector $\boldsymbol{x}_1$, providing no overflow occurs. The convergence rate is determined by

$$|\lambda_2 / \lambda_1|^k < \epsilon, \tag{6}$$

where ϵ is a specified tolerance.

Reproduced with permission of Elsevier Science Publishers B.V. (North Holland)
Parallel Computing 14(77 – 87)
1989

The Power Method algorithm with a normalising strategy included can be formulated as follows [1]:

given y_0, an arbitrary vector, and let the sequences z_k, y_k be defined by the equations

$$z_{j+1} = Ay_j, \quad y_{j+1} = z_{j+1}/\mu(z_{j+1}), \qquad j = 0, 1, \ldots, k-1, \tag{7}$$

with

$$\mu(x) = \max |x_i|, \quad i = 1, 2, \ldots, n. \tag{8}$$

For k sufficiently large,

$$y_k \cong x_1/\mu(x_1), \qquad \mu(z_k) \cong \lambda_1. \tag{9}$$

A modified version of the method is proposed in [3], where the normalisation factor $\mu(x)$ is defined as

$$\mu(x) = \text{the first non-zero element of } x. \tag{10}$$

Thus, the dominant eigenvalue λ_1 is given by the first non-zero component of z_k, and y_k corresponds to x_1 divided by its first non-zero component.

The scaling operation performed by $\mu(z_j)$ aims to prevent overflow or underflow, due to repeated matrix-vector multiplications. However, if (10) is used, the scaling may be cancelled, if the first non-zero element of z_j is either too big or too small. On the other hand, (8) imposes a considerable delay in the computations, since z_{j+1} cannot be computed before all z_j are known.

The convergence rate in (6) can be improved, if, instead of forming the sequence $A^k y_k$, a sequence of matrix powers is constructed directly [11]. The Matrix Squaring Method has the advantage that when A^j is computed, A^{2j} is obtained in one matrix-matrix multiplication. Hence the sequence A, A^2, $A^4, \ldots, A^{2^s}$ is built up.

For sufficiently large s [1],

$$A^{2^s} \cong \lambda_1^{2^s} c_1 x_1 x_1^{\mathrm{T}}. \tag{11}$$

Each column of A^{2^s} is proportional to x_1, each row is proportional to x_1^{T}. To obtain the eigenvalue λ_1 any column of A^{2^s} can be multiplied by A. Then (5) can be applied between the original column of A^{2^s} and that of AA^{2^s}.

Generally the successive squares will either increase or decrease rapidly in size and overflow or underflow may occur. This can be avoided by scaling each matrix by a power of two, chosen so that the element of maximum modulus has index zero. This introduces no additional rounding errors.

The convergence rate is determined by

$$|\lambda_2/\lambda_1|^{2^s} < \epsilon. \tag{12}$$

If A is not sparse, there is about n times as much work in one step of the Matrix Squaring Method as in one step of the Power Method. Thus, Matrix Squaring is more efficient for $|\lambda_2/\lambda_1|$ close to 1, $n \leqslant 2000$ and if A is not banded, since its sparseness is destroyed by Matrix Squaring [11].

A combination of the two methods can be considered as the application of a limited number of matrix squaring steps, followed by simple power method iterations. This combination, i.e. the Raised Matrix Power (RMP) method seems to be the most economical strategy for finding the dominant eigenvalue, and the corresponding eigenvector. The number r of matrix squaring steps depends on the ratio λ_2/λ_1, but a simple estimation can be derived:

$$r = \lfloor \log_2(\sqrt{n}) \rfloor. \tag{13}$$

Therefore, the first phase of the Raised Matrix Power (RMP) Method involves the calculation of $B = A^{2^r}$. Then, after k simple iterations

$$B^k \boldsymbol{u} = (\lambda_1')^k c_1 \boldsymbol{x}_1, \tag{14}$$

where

$$\lambda_1' = \lambda_1^{2^r} \tag{15}$$

as is shown in [3,11]. Thus, the dominant eigenvalue is obtained by a series of r square root extractions; $\boldsymbol{x}_1$ is collected as before.

The rate of convergence is now determined by

$$\left(|\lambda_2/\lambda_1|^{2^r} \right)^k < \epsilon. \tag{16}$$

For the first phase of the method the same limitations apply as in Matrix Squaring; however the overflow-underflow effect is now alleviated and similarly for r small the sparseness is not totally destroyed. Furthermore, the normalisation operation that is applied in the second phase of the method can keep the size of the numbers involved under control.

2. Systolic implementation

The main computational effort in the Power Method is the calculation of a series of matrix-vector multiplications, of the form

$$\boldsymbol{x}_{k+1} = A\boldsymbol{x}_k. \tag{17}$$

The Matrix Squaring Method is based on successive matrix-matrix multiplications

$$A_{k+1} = A_k A_k. \tag{18}$$

The Raised Matrix Power (RMP) Method combines both the computations (18) and (17).

Both computations are iterative, i.e. the output of the kth step forms the input of step $k+1$.

Following the definitions introduced in [4] an iterative process can be expanded either in time, yielding an iterative array, or in area, producing a pipeline. Figure 1 outlines these concepts for the case of computations (17) and (18). In general, there is an area-time tradeoff between the two approaches: the area expansion requires $\cong k$ times the area of time expansion, while the time expansion requires $\cong k$ times the computation time of the area expansion, where k is the number of iterations of the process. Both these approaches are discussed subsequently.

Basic components of the systolic implementation of recurrences (17) and (18) are matrix-vector and matrix-matrix multiplication arrays. The well-known designs in [5] are introduced mainly for banded matrix operations, and have been modified in [6,8], to yield unidirectional dataflow, compact I/O sequences, improved computation time and processor utilisation. These modified designs (see Figs. 2 and 3) allow for direct interconnection, and for extensive pipelining; therefore they are good candidates for the area expansion scheme. The same systolic arrays can also be used for full matrix computations, but with increased area requirements.

Systolic designs for full matrix computations have been proposed in [10]. A modification of these designs, that allow for successive full matrix-matrix multiplications is introduced in [9] (see Fig. 4). This array can be used for the time expansion scheme, since its output are fed back to the input. For matrix-vector multiplications, a similar array can be utilised, as shown in Fig. 5: it is a single-column version of the array in Fig. 4, with the addition of memory modules of size n words for each of the processors; for a similar configuration see [5,8].

Obviously, the arrays of Figs. 2 and 3 can be used for the time expansion scheme, if feedback lines with adequate FIFO memory modules are introduced. Furthermore other

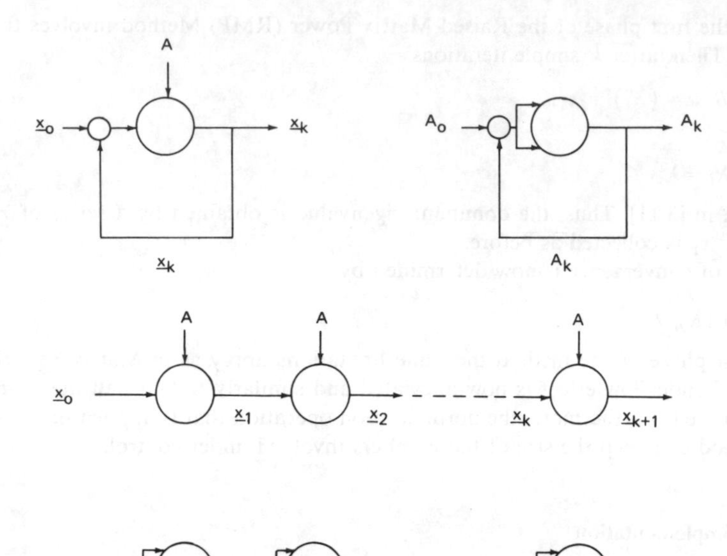

Fig. 1. (a) Time-expansion; (b) Area-expansion.

systolic networks have been proposed for the same operations, each one aiming to optimise certain specific aspects of the computation in question. However only the above mentioned seem to be suitable for the systolic implementation of iterative algorithms. Some of the possible systolic iterative designs are described in more detail in [7].

An additional complexity for the iterative algorithms discussed herein, is the normalisation or scaling that is necessary between two successive iteration steps.

For the Power Method, the technique in (8) requires a full search of the vector to be normalised; this imposes a delay of $\cong n$ cycles and therefore cancels the potential of pipelined

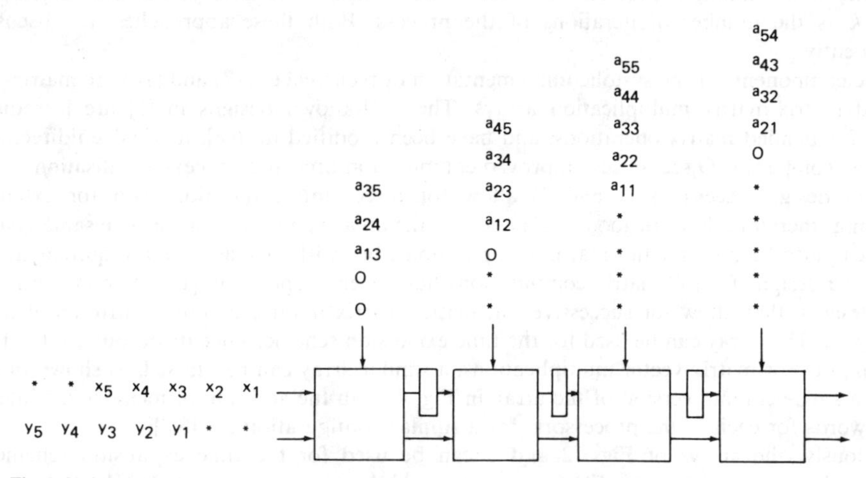

Fig. 2. Unidirectional matrix-vector multiplication systolic array ($Ax = y$, $w_A = p_A + q_A - 1$, $p_A = 3$, $q_A = 2$).

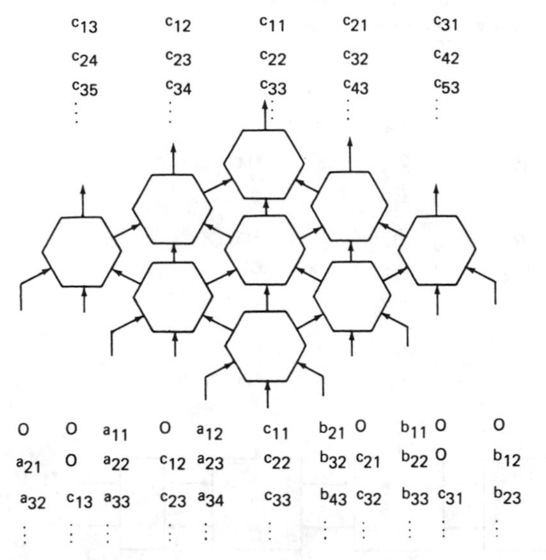

Fig. 3. Unidirectional matrix-matrix multiplication systolic array ($AB = C$, $w_A = w_B = 2p - 1 = 3$, $w_C = 2w - 1$).

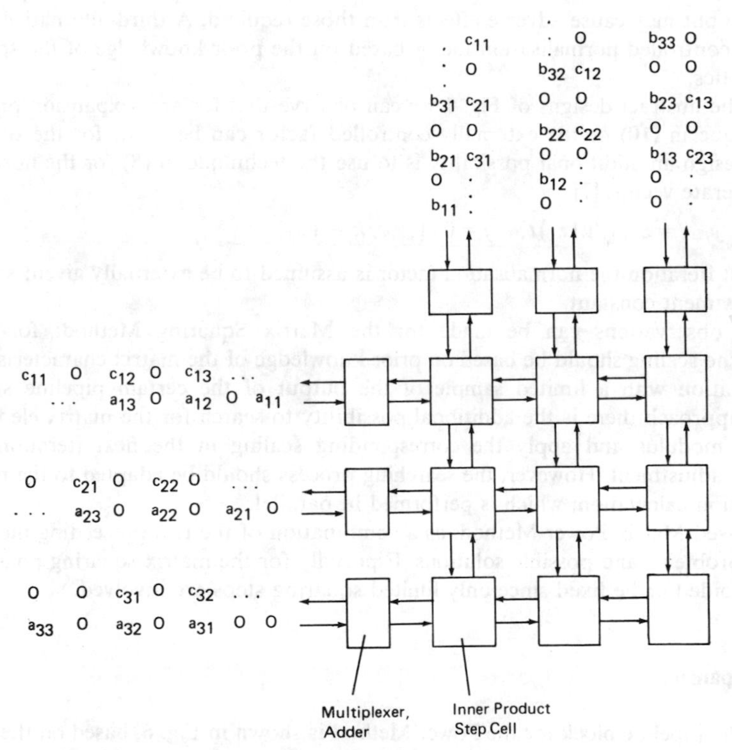

Fig. 4. Reusable matrix-multiplication array ($n = 3$, $AB = C$).

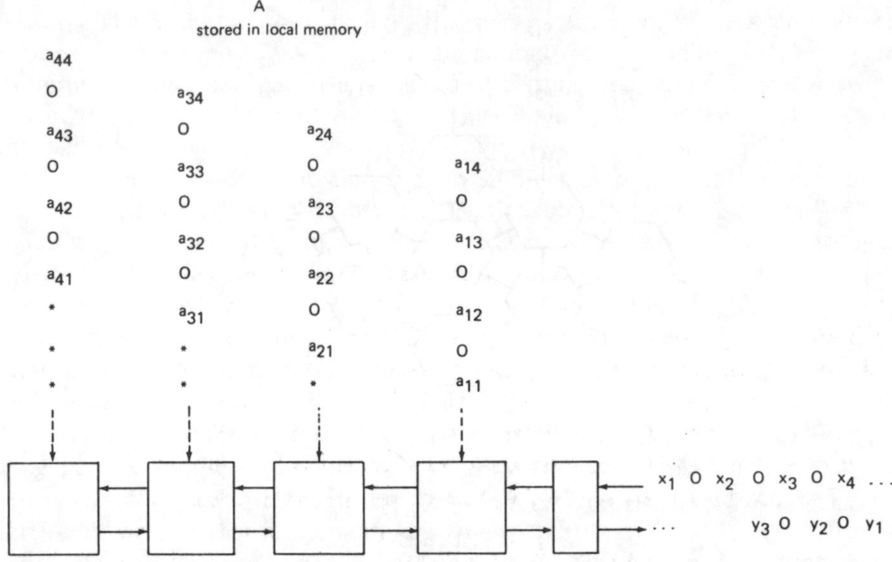

Fig. 5. Reusable matrix-vector multiplication array ($n = 4$, $Ax = y$).

implementation of the algorithm. As an alternative the technique in (10), requires only $\cong 1$ cycle delay but may cause adverse effects than those required. A third alternative would be an externally controlled normalisation factor based on the prior knowledge of the specific matrix characteristics.

Using the abstract designs of Fig. 1 we can observe that for area-expansion pipeline design the technique in (10) or the externally controlled factor can be used; for the time-expansion iterative design an additional possibility is to use the technique in (8) for the normalisation of the next iterate vector, [1],

$$y_{j+1} = z_{j+1}/\mu(z_j)t, \quad j = 0, 1, \ldots, k-1; \tag{19}$$

for the first iteration the normalisation factor is assumed to be externally given; similarly t is a given adjustment constant.

Similar observations can be made for the Matrix Squâring Method: for the pipeline approach the scaling should be based on prior knowledge of the matrix characteristics, possibly in combination with a limited sample of the output of the certain pipeline stage; for the feedback approach there is the additional possibility to search for the matrix element with the maximum modulus and apply the corresponding scaling in the next iteration, with some predefined adjustment. However, the searching process should be adapted to the matrix-matrix multiplication calculation, which is performed in parallel.

The Raised Matrix Power Method, as a combination of the two preceeding methods, shares the same problems and possible solutions. Especially for the matrix squaring phase the scaling may be avoided or be fixed since only limited squaring steps are involved.

3. Area expansion

A systolic pipeline block for the Power Method is shown in Fig. 6, based on the interconnection of unidirectional matrix-vector multiplication arrays as illustrated in Fig. 2.

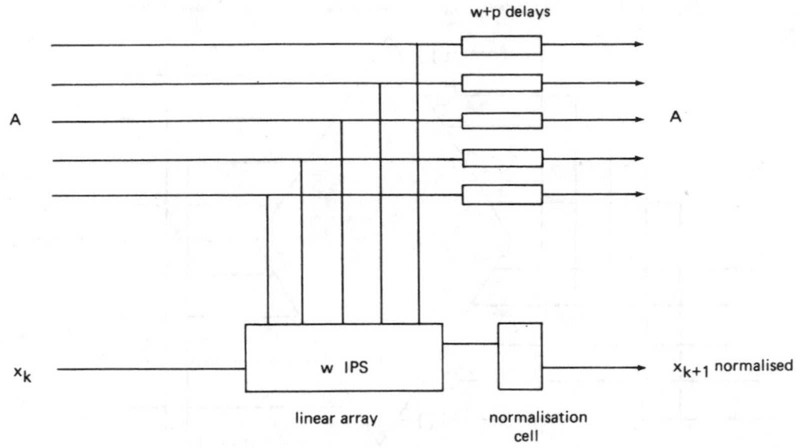

Fig. 6. Power method pipeline block for a banded matrix A with bandwidth $w = 2p - 1 = 5$. Area $\cong w + 1$, Delay $= w + p$.

The normalisation cell that is placed between two successive pipeline stages can operate using (10), or a systolically propagated normalisation factor, or, possibly, a combination of two, so that the dynamic range of the system is effectively used. Since the computational complexity of the normalisation cell is not greater than a comparison followed by a division (or shift-right operation), it can be assumed that the area-time complexity of this cell is not greater than 1 Inner Product Step (IPS).

Therefore, the Power Method for a $(n \times n)$ banded matrix can be performed on a pipeline as in Fig. 6, in

$$\text{time} = n + k(w + p) \text{ IPS cycles,} \tag{20}$$

where $w = p + q - 1$ is the bandwidth of the matrix (see [5,8]), k is the number of iterations. The area required is

$$\text{area} = k(w + 1) \text{ IPS cells.} \tag{21}$$

A soft-systolic simulation program for the pipeline of Fig. 6 is given in [7].

A systolic pipeline block for the Matrix Squaring Method is shown in Fig. 7, based on the interconnection of unidirectional matrix-matrix multiplication arrays, as illustrated in Fig. 3.

The major characteristic of this pipeline is the greater increase of the area requirements, since for each pipeline stage,

$$w_{out} = 2w_{in} - 1, \quad w_{in}, w_{out} \ll n, \tag{22}$$

where w_{in}, w_{out} are the bandwidths of input and output matrices. This fact imposes limits on the bandwidth w of matrix A, as well as the number s of pipeline blocks, i.e. the matrix squaring steps allowed. Thus, if the maximum area for a pipeline block is w_{max}^2, then

$$2w^{s-1} - (2^{s-1} - 1) \leqslant w_{max}. \tag{23}$$

The scaling post-processor is an extension of the normalisation cell used for the Power Method pipeline.

The simplest post-processor array can be one where a scaling factor is preloaded to w_{out} scaling cells performing a simple division (or shift-right).

In order for the dynamic range of the system to be effectively used the scaling should be adapted according to the matrix in question; however, this is very time consuming, if the whole

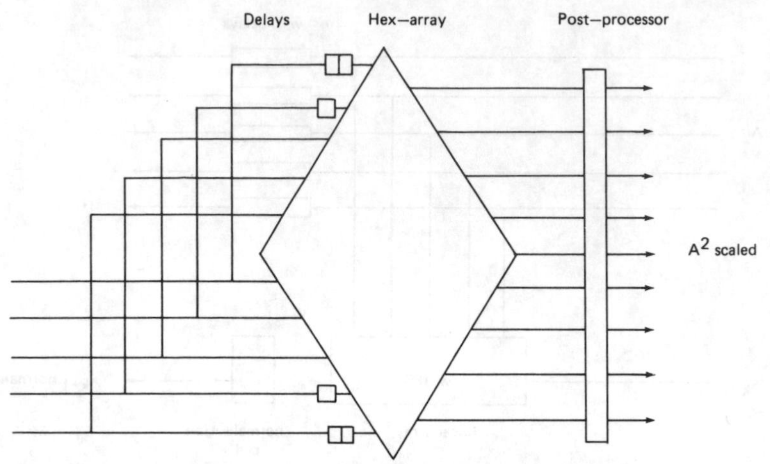

Fig. 7. Matrix Squaring Method pipeline block for a banded matrix A with bandwidth $w = 2p - 1 = 5$. Area $= w^2 + 2w - 1$, Delay $= w + 1$.

matrix is to be searched for its maximum element. Instead the first row and column can be searched and the scaling factor is determined using this data. For w_{out} small, the search can be performed serially, otherwise a systolic search tree might be used.

In the case of the simple scaling post-processor, the area-time requirements are

$$\text{area} = w_{out} \text{ IPS cells} \tag{24}$$

and

$$\text{delay} = 1 \text{ IPS cycle.} \tag{25}$$

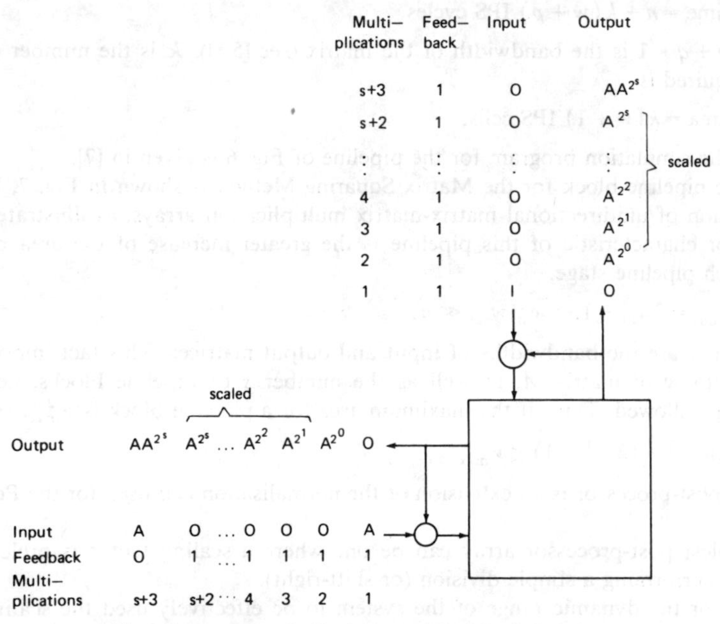

Fig. 8. Input, output, feedback sequence for Matrix Squaring.

Therefore, the Matrix Squaring Method for a $(n \times n)$ banded matrix can be performed on a pipeline as in Fig. 8 in

$$\text{time} = n + \sum_{i=0}^{s-1} \left\{ \left(2^i w - (2^i - 1)\right) + 1 \right\} \text{ IPS cycles} \tag{26}$$

and with area requirements

$$\text{area} = \sum_{i=0}^{s-1} \left\{ \left(2^i w - (2^i - 1)\right)^2 + \left(2^i w - (2^i - 1)\right) \right\} \text{ IPS cells.} \tag{27}$$

A soft-systolic simulation program for the pipeline of Fig. 7 with simple scaling post-processor is given in [7].

The Raised Matrix Power Method can be implemented by simply interconnecting the Matrix Squaring and the Power Method pipelines; the only interface required is a set of reformatting registers, imposing no additional delay.

Since only a limited number of matrix squaring steps will be used, the scaling processes in the matrix squaring pipeline can be relaxed. Another reason that dictates towards a small number of matrix squaring steps is the fact that the area of the Power Method pipeline blocks depends on the bandwidth of the output matrix of the Matrix Squaring pipeline.

4. Time expansion

The iterative systolic array for the Power Method is shown in Fig. 5, with the difference that the multiplexer cell now incorporates the normalisation process.

The normalisation process is similar to that used in the area expansion case. As an alternative, the normalisation cell can be modified to operate as it is suggested in (19). In both cases no additional delay is introduced, since the multiplexer complexity is kept within the limits of 1 IPS.

Therefore, the Power Method for a $(n \times n)$ full matrix can be performed on an iterative array as in Fig. 5 in

$$\text{time} = 2n(k+1) + 1 \text{ IPS cycles,} \tag{28}$$

the area required is

$$\text{area} = n + 1 \text{ IPS cells;} \tag{29}$$

each of the n cells of the main array has a memory module of n words.

A soft-systolic simulation program for the array of Fig. 5 is given in [7]. In the last cycle of the computation the unnormalised vector is also obtained so that the eigenvalue is determined.

The iterative systolic array for the Matrix Squaring Method is shown in Fig. 4, with the difference that the multiplexers include the scaling generation. The input-output-feedback sequence is given in Fig. 8.

The scaling process has similar complexities, as in the area expansion case for Matrix Squaring. Firstly, an externally determined scaling factor can be used; in order to improve the efficiency of the scaling process a sample of the matrix must be searched. However, the searching of the first column and row of the matrix imposes a delay of more than n cycles. Extending the idea in (19), the scaling factor calculated during iteration i can be used to determine the matrix of iteration $i+1$.

The final matrix-matrix multiplication, i.e. AA^{2^s}, can also be computed in the same array, provided that A is stored by the systolic array interface and produced in the final multiplication.

Therefore the Matrix Squaring Method for a full $(n \times n)$ matrix can be performed on an iterative array as in Fig. 4 in

$$\text{time} = 2n(k+2) + 1 \text{ IPS cycles;} \tag{30}$$

the area required is

$$\text{area} = (n+1)^2 \text{ IPS cells.} \tag{31}$$

A soft-systolic simulation program for the Matrix Squaring Method is given in [7]. After the last matrix squaring, one more matrix-matrix multiplication between A and A^{2^s} can be performed, to calculate λ_1.

The Raised Matrix Power Method can be performed on either the array of Fig. 4, or in a combination of the arrays in Figs. 4 and 5.

If the Matrix Squaring array is used, then the s squaring steps are followed by k degenerate matrix multiplications between A^{2^s} and $[x_i \mid 0]$, $i = 0, 1, \ldots, k$. Thus only $n+1$ of the $(n+1)^2$ processors are active. The remaining processors act as simple memory cells.

For full processor utilization, A^{2^s} can be passed to the Power Method array so that the Matrix Squaring array is free for another computation. This implies a delay of $2n$ cycles for the loading of A^{2^s} to the memory modules of the array in Fig. 5.

5. Conclusion

This paper discussed the systolic implementation of some iterative methods for the computation of the dominant eigenvalue and the corresponding eigenvector of a matrix. More specifically the Power Method, the Matrix Squaring Method and a combination of these two methods, called Raised Matrix Power Method are discussed.

Two alternative implementation approaches are investigated, i.e. area and time expansion. The Area Expansion technique produces systolic pipelines mostly suitable for banded matrices with $w \ll n$, where w is the bandwidth and n is the order of the matrix. The Time Expansion technique produces iterative arrays suitable for full matrices.

The pipeline designs allow for maximum throughput and computation time equal to $n + t(k, w)$, where k is the number of iterations; however the pipeline can perform only a pre-specified number of iterations and the area required increases with k. Especially for the Matrix Squaring pipeline this increase is very sharp and imposes an upper limit on kw.

The iterative design performs the iterations 'sequentially' and the computation time is approximately $2kn$; however the array can perform an unlimited number of iterations with no additional hardware. An upper limit on n is imposed by the fact that the area required is of order $(n+1)^2$ for the Matrix Squaring Method.

References

[1] H.J. Caulfield, J.H. Gruninger and W.K. Cheng, Using optical processors for linear algebra, in: *Proc. SPIE, Optical Information Processing* (1983).

[2] C.-E. Froberg, *Introduction to Numerical Analysis* (Addison-Wesley, Reading, MA (1965).

[3] W. Jennings, *First Course in Numerical Methods* (MacMillan, New York, 1964).

[4] L. Johnson and D. Cohen, A mathematical approach to modelling the flow of data and control in computational networks, in: H.T. Kung, R. Sproull and G. Steele, eds., *VLSI Systems and Computations* (Computer Science Press, Rockvill, MD, 1981).

[5] H.T. Kung and C.E. Leiserson, Algorithms for VLSI processor arrays, in: C. Mead and L. Conway, eds., *Introduction to VLSI Systems* (Addison-Wesley, Reading, MA, 1980) Sect. 8.3.

[6] G.-J Li and B.W. Wah, The design of optimal systolic arrays, *IEEE Trans. Comput.* **34** (1) (1985).
[7] K.G. Margaritis, A study of systolic algorithms for VLSI processor arrays and optical computing, Ph.D. Thesis, Loughborough University, 1987.
[8] K. Margaritis and D.J. Evans, Improved systolic matrix-vector multiplication, Technical Report CS-320, Loughborough University, 1986.
[9] P. Quinton, B. Jannault and P. Gachet, A new matrix multiplication systolic array, in: M. Cosnard et al., eds., *Parallel Algorithms and Architectures* (North-Holland, Amsterdam, 1986).
[10] J.M. Speiser and H.J. Whitehouse, Parallel processing algorithms and architectures for real time signal processing, in: *Proc. SPIE* **298**, Real Time Signal Processing IV (1981).
[11] J.H. Wilkinson, *The Algebraic Eigenvalue Problem* (Clarendon Press, Oxford, 1965).

LINEAR AND DYNAMIC PROGRAMMING

Path Planning on the Warp Computer: Using a Linear Systolic Array in Dynamic Programming*

F. BITZ and H. T. KUNG

Department of Computer Science, Carnegie Mellon University, Pittsburgh, Pennsylvania 15213, USA

Given a map in which each position is associated with a traversability cost, the path planning problem is to find a minimum-cost path from a source position to every other position in the map. The paper proposes a dynamic programming algorithm to solve the problem, and analyzes the exact number of operations that the algorithm takes. The algorithm accesses the map in a highly regular way, so it is suitable for parallel implementation. The paper describes two general methods of mapping the dynamic programming algorithm onto the linear systolic array in the Warp machine developed by Carnegie Mellon. Both methods have led to efficient implementations on Warp. It is concluded that a linear systolic array of powerful cells like the one in Warp is effective in implementing the dynamic programming algorithm for solving the path planning problem.

KEY WORDS: Path planning, parallel processing, dynamic programming, systolic array, computer systems organization, parallel processors, analysis of algorithms, image processing.

C.R. CATEGORIES: C.1.2, F.2.2, I.4.0.

*The research was supported in part by Defense Advanced Research Projects Agency (DOD) monitored by the Space and Naval Warfare Systems Command under Contract N00039-85-C-0134, and in part by the Office of Naval Research under Contracts N00014-87-K-0385 and N00014-87-K-0533.

1. INTRODUCTION

During 1984–87, Carnegie Mellon has developed a programmable systolic array machine called Warp [2, 3]. Currently produced by GE, the machine has a linear systolic array of 10 cells, each capable of delivering 10 million floating-point operations per second (10 MFLOPS). One of the first applications of the machine is navigation for robot vehicles using vision and laser range-finder [4, 6, 8, 9].

Related to robot navigation is path planning. Given a map on which each position is associated with a traversability cost, the path planning problem is to find a minimum-cost path from a source position to every other position in the map. We use a dynamic programming algorithm to solve the problem. Two general methods of mapping the dynamic programming algorithm onto the linear systolic array in Warp have been considered. Both methods have led to efficient implementations of the dynamic algorithm on Warp.

In the next section we define the path planning problem in mathematical terms. Section 3 describes the dynamic programming algorithm for solving the problem, and analyzes the number of times that the algorithm requires to scan the map. The linear systolic array in the Warp machine is briefly described in Section 4. Section 5 describes two general approaches of mapping the dynamic programming algorithm onto Warp and compares their performance. Some concluding remarks are given in the last section.

2. PATH PLANNING PROBLEM

A *map* is an $n \times n$ grid of positions, for some positive integer n. Each position x in the map is associated with a non-negative real number $tc(x)$ corresponding to the traversability cost of the position. For simplicity, we assume that the function tc is directional independent, i.e., it is as expensive to cross a position in one direction as in another. (For applications where this assumption is not true, tc should be defined as a vector function consisting of scalar functions corresponding to various directions.) Figure 1(a) gives traversability costs of positions in a map with $n=6$. In applications, n can be 512, 1024, or larger. Figure 2 gives traversability costs in a 400×380 map corresponding to an area near Denver. In the figure, a darker position has a higher traversability cost.

A position in the map has up to eight neighbors as shown in Figure 1(b). An edge (u, v), denoted by an arrow in Figure 3, is a directed line segment connecting two neighboring positions u and v in the map. The cost of edge (u, v) is

$$\frac{tc(u)+tc(v)}{2}, \quad \text{if the edge is horizontal or vertical,}$$

or

$$\left(\frac{tc(u)+tc(v)}{2}\right)\sqrt{2}, \quad \text{if the edge is diagonal,}$$

where $tc(u)$ and $tc(v)$ are traversability costs of u and v, respectively. The $\sqrt{2}$ multiplier reflects the added traveling distance due to the diagonal connection.

(a)

6	1	2	2	2	2
2	2	1	3	2	5
3	4	6	2	7	6
4	1	1	4	1	4
9	2	5	8	1	5
6	7	4	3	1	1

(b)

$n1$	$n2$	$n3$
$n8$	x	$n4$
$n7$	$n6$	$n5$

Figure 1 (a) Traversability costs in a 6×6 map, and (b) eight neighbors, $n1 \ldots, n8$, of position x.

A *path* from position A to position B is a sequence of edges $(s_1, s_2), (s_2, s_3), \ldots, (s_{p-1}, s_p)$ for some positive integer p, where $s_1 = A$ and $s_p = B$. The cost of a path is the sum of costs of all its edges. For example, the cost of the path shown in Figure 3 is

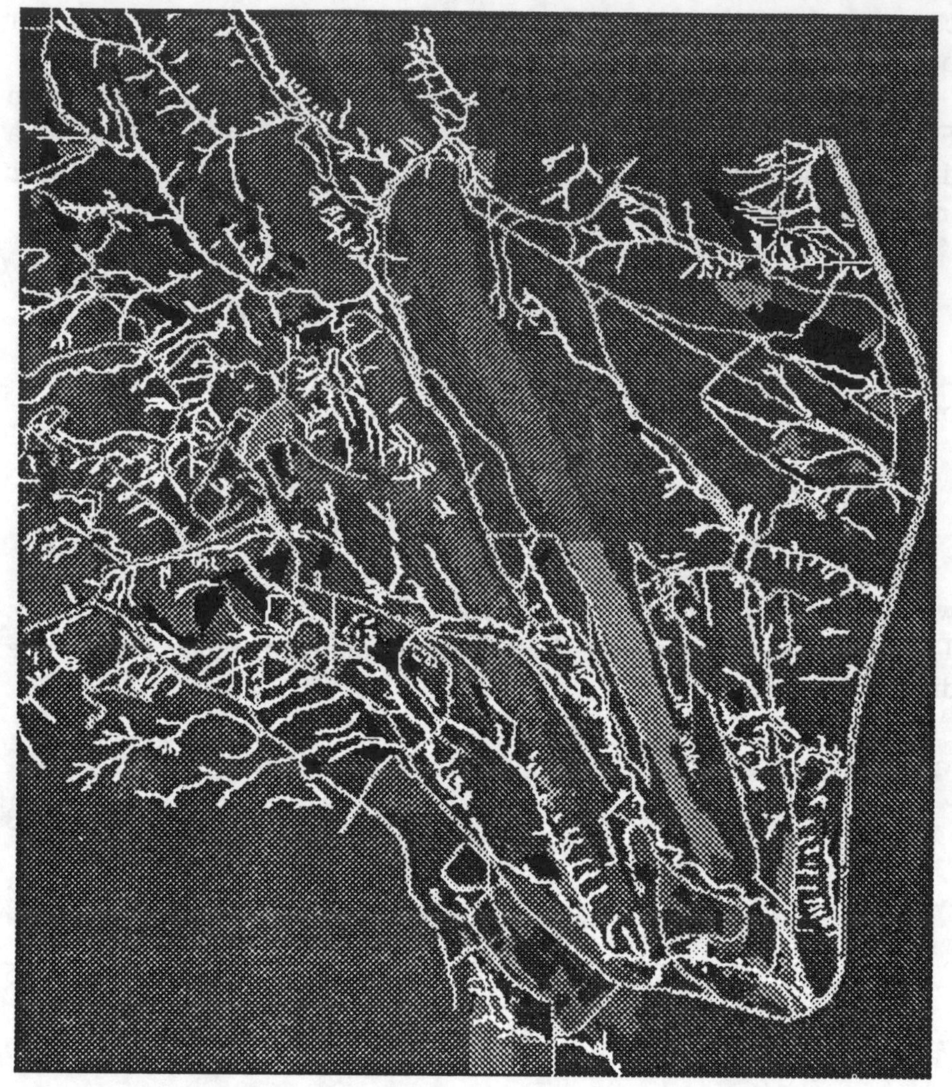

Figure 2 Traversability costs in a 400 × 380 map.

$$\left(\frac{4+1}{2}\right)\sqrt{2}+\left(\frac{1+2}{2}\right)\sqrt{2}+\left(\frac{2+1}{2}\right)\sqrt{2}+\left(\frac{1+1}{2}\right)+\left(\frac{1+3}{2}\right)\sqrt{2}.$$

The color labels of the edges in Figure 3 will be explained in Section 3.3

A *shortest path* (or minimum-cost path) from position A to position B has the minimum cost over all paths from A to B. The *path planning problem* considered in this paper is that given a

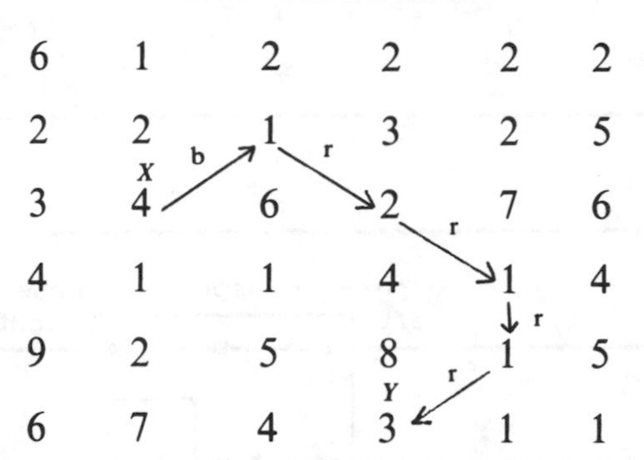

Figure 3 Path from X to Y, where edges labelled by r and b are red and blue edges, respectively.

position, called the *source*, we want to find a shortest path from it to *every* other position in the map. This is in contrast to another possible definition for the problem that calls for finding a shortest path from the source to only one other position.

3. DYNAMIC PROGRAMMING ALGORITHM

It is easy to see that every subpath of a shortest path is itself a shortest path. Therefore we can used the principle of dynamic programming [1]. That is, to find a shortest path from the source to a position we need only consider shortest paths from the source to the neighbors of the position. This motivates the dynamic programming algorithm described below.

Before the algorithm starts, the best known cost for every position in the map and its surroundings is assigned a value, 0 at the source and infinity at all other positions. The algorithm performs the so-called *red sweep* and *blue sweep*, described below. In each sweep the value of each position in the map is updated. At any time during the algorithm execution, the value of a position represents the cost of a current "shortest path", which the algorithm has recognized up to this moment, from the source to the position.

3.1 Red sweep

The red sweep scans in the row-major ordering as shown in Figure 4(a). During the sweep, a new value (subscripted by *new* in the equations below) is computed for every position in the map in the

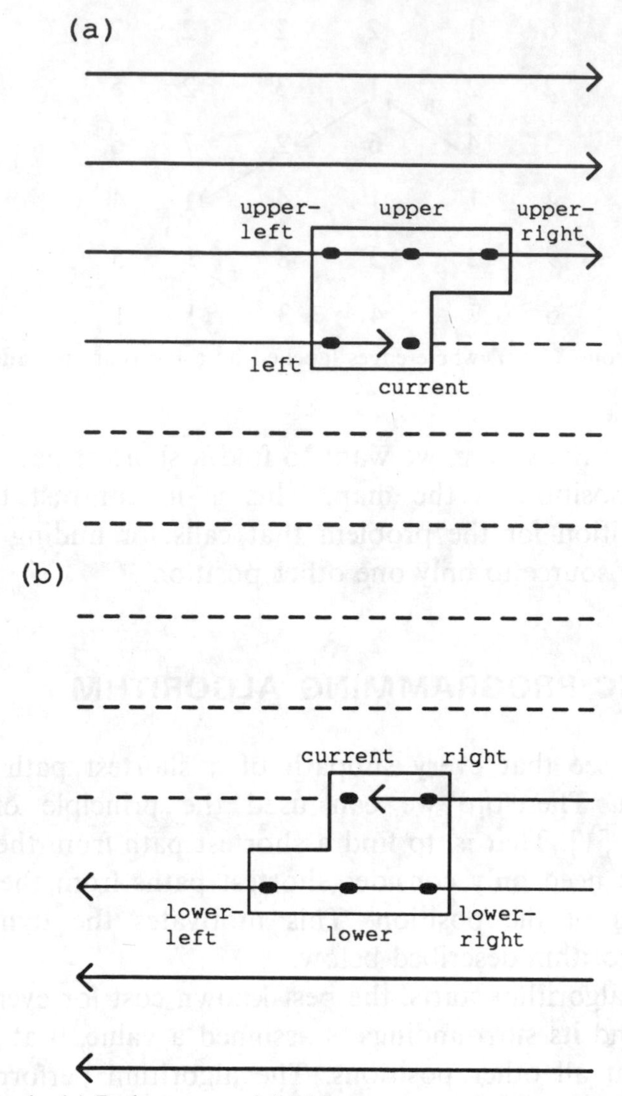

Figure 4 (a) Red sweep and (b) blue sweep, and the associated masks.

order of the sweep, using the old values (subscripted by *old*). That is, the new value of the current position of the sweeping, $current_{new}$, is computed by:

$$current_{new}$$

$$= \min\left(current_{old}, upper\text{-}left_{new} + \frac{tc(upper\text{-}left) + tc(current)}{2}\sqrt{2}, \right.$$

$$upper_{new} + \frac{tc(upper) + tc(current)}{2},$$

$$upper\text{-}right_{new} + \frac{tc(upper\text{-}right) + tc(current)}{2}\sqrt{2},$$

$$\left. left_{new} + \frac{tc(left) + tc(current)}{2} \right), \tag{1}$$

where quantities in the right hand side are associated with positions in the mask shown in Figure 4(a).

3.2 Blue sweep

The blue sweep scans the map in the reversed row-major ordering as shown in Figure 4(b). The new value of the current position of the sweeping, $current_{new}$, is computed by:

$current_{new}$

$$= \min\left(current_{old}, lower\text{-}left_{new} + \frac{tc(lower\text{-}left) + tc(current)}{2}\sqrt{2}, \right.$$

$$lower_{new} + \frac{tc(lower) + tc(current)}{2},$$

$$lower\text{-}right_{new} + \frac{tc(lower\text{-}right) + tc(current)}{2}\sqrt{2},$$

$$\left. right_{new} + \frac{tc(right) + tc(current)}{2} \right), \tag{2}$$

where quantities in the right hand side are associated with positions in the mask shown in Figure 4(b).

Note that the combination of the two masks associated with the red and blue sweeps allows the value of the current position to be updated using the values of all the eight neighboring positions.

Therefore if a sufficient number of red and blue sweeps are performed, the final value of every positions, which is the cost of a shortest path from the source to the position, will be obtained.

3.3 Number of required sweeps

The red and blue sweeps are performed alternatively until no values are changed in one sweep. The following analysis shows how the number of required sweeps depends on the given map and the source position.

Consider a shortest path from the source to any other position. We color its edges in red or blue according to their directions. That is, edges pointing to left, upper-left, upper, and upper-right directions are colored blue, whereas edges pointing to right, lower-right, lower, and lower-left directions are colored red. The colors of the edges shown in Figure 3 follow this coloring scheme.

Consider a subpath whose edges all have the same color, say, the red color. Suppose that when a red sweep is about to start, the head of the first edge in the subpath has already received its final value. Then it is easy to check that after the red sweep *all* positions on the subpath will receive their final values. This implies the following result:

THEOREM *Suppose that a shortest path from the source to any other position changes colors c times. Then if the red and blue sweeps are performed alternately, the final values of all positions on the shortest path will be obtained exactly after c or $c+1$ sweeps.*

Therefore the number of required sweeps is $C+1$ or $C+2$, where C is maximum number of color changes in a shortest path from the source to any other position. The total number of operations that the dynamic programming algorithm will take is $O(Cn^2)$, since each sweep performs the computation corresponding to Eq. (1) or (2) n^2 times. In the worse case C can be as large as $O(n^2)$. However, for real terrain maps C is expected to be much smaller than n. For the 400×380 map in Figure 2, C is about 40.

3.4 Obtaining a shortest path

When the algorithm is terminated, all positions in the map have received their final values. Using these values, it is straightforward to trace a shortest path to the source from any other position. That is, the next position on the path is a neighboring position whose value leads to the value of the current position using Eq. (1) or (2). This procedure continues until the source is reached. The total number of operations is $O(n)$.

One possible way to speed up the procedure is to maintain a back pointer in each position during red and blue sweeps. Every time when the value of the current position is updated, we also update the back pointer to point to the neighboring position whose value leads to the new value for the current position. Following these back pointers, a shortest path to the source from any other position can be easily obtained. However, using back pointers may have two drawbacks. First, storing the pointers consumes memory space, which may be a critical resource for some computers. Second, a back pointer in a position may be updated several times by different sweeps, and therefore extra computations may be performed. This may not increase the computation time significantly when a sweep is carried out by a number of processors in parallel, as to be described in the rest of the paper.

4. LINEAR SYSTOLIC ARRAY IN WARP

Systolic arrays have high communication bandwidth between neighboring cells, which are defined by a simple topology. As shown in Figure 5, in the case of the Warp processor array, this is the simplest possible topology, namely a linear array. The linear configuration was chosen for several reasons. First, a linear array is easier to program than other interconnection topologies; it is relatively simple for a programmer to envision a linear array and map his or her computation onto it.

Second, the linear interconnection is easy for hardware implementation. Because of its simplicity, the Warp array has a very clean backplane, capable of providing 40 Mbytes per second data bandwidth between every pair of neighboring cells. This high inter-cell communication bandwidth makes it possible to transfer large volumes of intermediate data between neighboring cells and thus supports fine-grain parallel processing. The linear interconnection also makes it easy to extend the number of cells in the array.

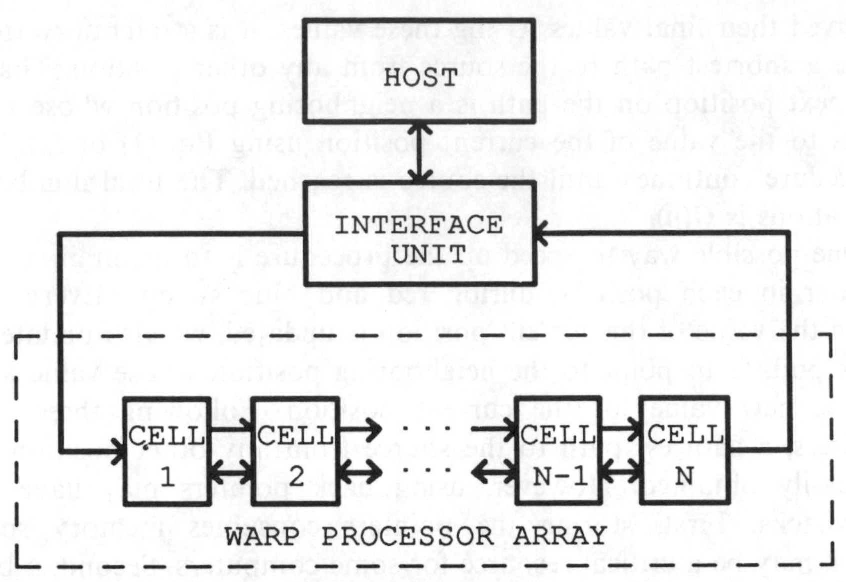

Figure 5 Warp system.

Third, a linear array has modest I/O requirements since only the two end-cells communicate with the outside world. Therefore there is no need to have an exotic host to provide very high bandwidth I/O with the array.

In contrast with some parallel arrays such as MPP [5] and the Connection Machine [10] that have 2-dimensional arrays of simple processors, each Warp cell in the Warp array is a powerful processor with the following features:

P1. Pipelined floating-point adder and multiplier, giving a total peak computation rate of 10 MFLOPS

P2. High data I/O bandwidth of 80 Mbytes per second

P3. 32K-word data memory

P4. Crossbar interconnection between data memory, input queues, register files, etc.

P5. Horizontally microcoded (>250 bits)

P6. Sequencer and 8K-word program memory

P7. Hardware flow control for inter-cell communication.

These features plus the optimizing Warp compiler [7] help to increase the Warp cell's performance and programmability.

5. TWO MAPPING METHODS

We consider two general methods for mapping the dynamic programming algorithm in Section 3 onto a linear systolic array such as the one in Warp. Suppose that the linear array has k cells, with $k \leq n$.

5.1 Vertical partitioning method

Recall that a red or blue sweep scans the map in the row-major or reversed row-major ordering. The vertical paritioning method assigns an equal number of consecutive columns of the map to each cell. This is depicted in Figure 6.

Consider first the red sweep. Cell 1, the left-most cell, starts computing on the first portion of the first scan line. Immediately after the value of the last position in this portion is computed, cell 1

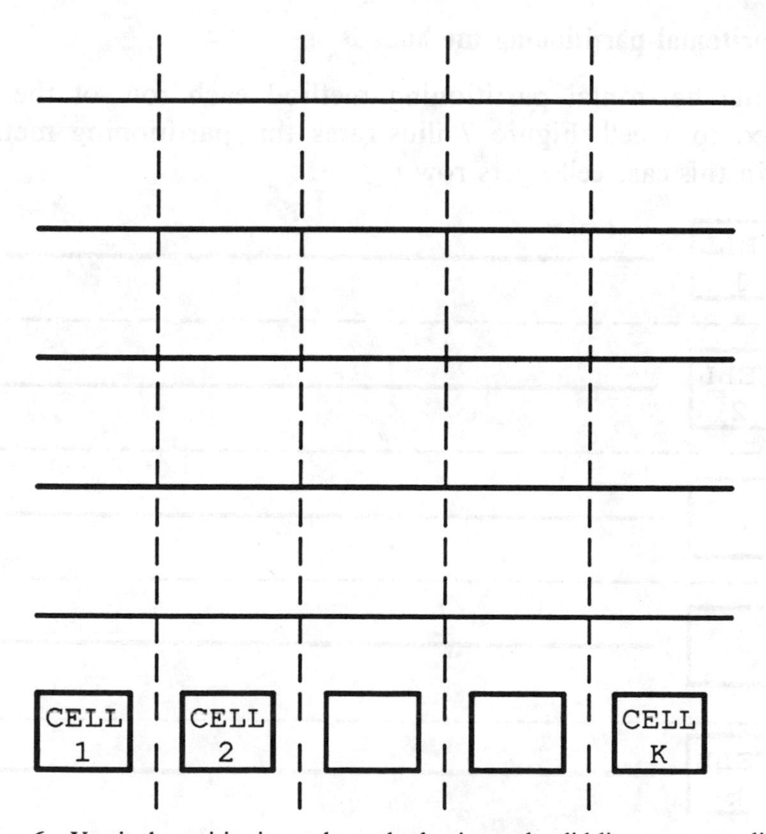

Figure 6 Vertical partitioning, where the horizontal solid lines are scan lines.

passes the value to cell 2. Cell 2 can then start computing on the second portion of the first scan line, while cell 1 starts computing on the first portion of the second scan line. The computed value of the first position in the second portion of the first scan line needs to be passed from cell 2 to cell 1. This value is needed for cell 1 to compute the value of the last position in the first portion of the second scan line. In this way every cell will eventually start its computation. Cell $i+1$ starts later than cell i by roughly the time needed to compute the values of positions in a portion of a scan line. The computed value for each position is stored at the cell that performs the computation.

After the red sweep is completed, the blue sweep starts similarly from cell k, the right-most cell. Red and blue sweeps are performed totally inside the linear array in an alternate manner, without host interaction. Note that bi-directional communications are needed between adjacent cells for each sweep.

5.2 Horitontal partitioning method

With the horizontal partitioning method each row of the map is assigned to a cell. Figure 7 illustrates this partitioning method for $k=n$. In this case cell i gets row i.

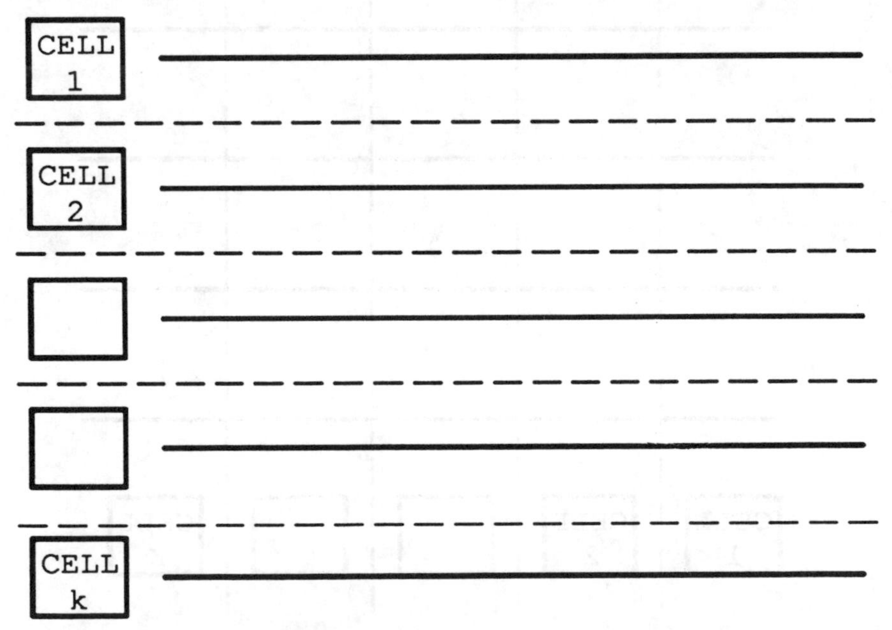

Figure 7 Horizontal partitioning, where the horizontal solid lines are scan lines.

For the red sweep, immediately after cell i has computed the values of two positions, it will pass the values to cell $i+1$ to get it started. Therefore the latency between the startup times of two adjacent cells is very small.

Suppose that k is smaller than n. Then a cell will have to compute values of positions in multiple rows. That is, cell i will be assigned to row $i+jk$ for $j=0,1,2,\ldots$. In this case cell 1 will need to receive computed values from cell k. For the Warp implementation this feed-back loop is accomplished through the host. Since cell k produces its first results before cell 1 finishes all its computations, the feedback time through the host is totally overlapped with the computation time of cell 1.

After the red sweep is completed, the blue sweep starts from cell k in a similar way, but the intermediate values will flow on the array in the opposite direction.

Note that the horizontal partitioning has a small startup latency between adjacent cells and only uni-directional communications are needed for each sweep. This partitioning method, however, has the complication of requiring the feed-back loop from cell k to cell 1, and vice versa.

5.3 Performance analysis

Suppose that a cell takes a unit time to perform the computation corresponding to Eq. (1) or (2). Then for both the mapping methods, the execution time for either the red or blue sweep is:

$$T=(n^2/k)+\text{overhead}$$

where the overhead is n or k for the vertical or horizontal partitioning method, respectively. The overhead is due to the latency of startup times of cells. We are interested in the ratio of the overhead to the total execution time. Using some algebra, we have:

$$\frac{\text{overhead}}{T}=\frac{\alpha}{1+\alpha} \quad \text{for the vertical partitioning method, and}$$

$$\frac{\text{overhead}}{T}=\frac{\alpha^2}{1+\alpha^2} \quad \text{for the horizontal partitioning method,}$$

where α is k/n. For both cases the ratio approaches 1/2 as α $(=k/n)$ approaches unity. However, when α is small, i.e., when k is a small fraction of n, we see that the horizontal partitioning method is more efficient than the vertical partitioning method.

5.4 Performance on Warp

We have implemented both partitioning methods on the Warp machine at Carnegie Mellon. On a 10-cell Warp, a single red or blue sweep for a 512×512 map takes 250 ms or 163 ms using the vertical or horizontal partitioning method, respectively. (We expect that these times can be reduced to 80 ms or less, by using more optimized code.)

6. CONCLUDING REMARKS

We have described a dynamic programming algorithm for solving the path planning problem. The algorithm accesses the map in a highly regular way, so it is suited for parallel implementation.

A linear array interconnection fits naturally for the parallel implementation of the algorithm. It does not seem that any other interconnection of higher dimensionality such as a 2-dimensional processor array can be more efficient. Both the vertical and horizontal partitioning methods have yielded efficient implementations of the algorithm on the linear systolic array in Warp. Basically, all the cells can do useful work most of the time during the algorithm execution. The only inefficiency is due to the latency in cell startup times. As pointed out in Section 5.3, this latency is at a minimum level for the horizontal partitioning method.

Several other architectural features of Warp have contributed to its effectiveness in the implementations.

- The bi-directional flow is useful for the vertical partitioning method, whereas the ring structure via the host supports the horizontal partitioning method.
- High-bandwidth inter-cell communication allows that intermediate values to be passed between adjacent cells efficiently.

- Floating-point operations in each cell provide the dynamic range needed to accumulate the costs of a large number of edges on a shortest path.
- Large local memory at each cell can store values that it computes in a sweep and the required traversability costs. Therefore cost and computed values need not be input to the cells for each new sweep, in order to reduce the I/O time.
- High degree of programmability of the cells allow rapid experiments of various partitioning methods, and the optimized Warp compiler produces efficient code for the Warp machine.

Results of this paper are not restricted to 2-dimensional, square maps. Generalizations to other computations that involve rectangular or 3-dimensional grids and use masks similar to those used in the paper are straightforward.

Acknowledgement

The authors wish to thank Chuck Thorpe at Carnegie Mellon for his comments on a draft of the paper.

References

[1] A. Aho, J. E. Hopcroft and J. D. Ullman, *The Design and Analysis of Computer Algorithms*, Addison-Wesley, Reading, Massachusetts, 1975.

[2] M. Annaratone, E. Arnould, T. Gross, H. T. Kung, M. Lam, O. Menzilcioglu and J. A. Webb, The Warp computer: architecture, implementation and performance, *IEEE Transactions on Computers C*-36 **12** (December 1987), 1523–1538.

[3] M. Annaratone, E. Arnould, R. Cohn, T. Gross, H. T. Kung, M. Lam, O. Menzilcioglu, K. Sarocky, J. Senko and J. Webb, Warp architecture: From prototype to production. *Proceedings of the 1987 National Computer Conference, AFIPS, 1987*, pp. 133–140.

[4] M. Annaratone, F. Bitz, J. Deutch, L. Hamey, H. T. Kung, P. Maulik, H. Ribas, P. Tseng and J. Webb, Applications experience on Warp. *Proceedings of the 1987 National Computer Conference, AFIPS, 1987*, pp. 149–158.

[5] K. E. Batcher, Bit-serial parallel processing systems, *IEEE Trans. Computer C-31* **5** (May 1982), 377–384.

[6] E. Clune, J. D. Crisman, G. J. Klinker and J. A. Webb, Implementation and performance of a complex vision system on a systolic array machine. Tech. Rept. CMU-RI-TR-87-16, Robotics Institute, Carnegie Mellon University, 1987, In: *Proc. of Conference on Frontiers in Computing*, Amsterdam, The Netherlands, December 1987.

[7] T. Gross and M. Lam, Compilation for a high-performance systolic array, *Proceedings of the SIGPLAN 86 Symposium* on Compiler Construction, ACM SIGPLAN, June 1986, pp. 27–38.

[8] T. Kanade and J. A. Webb, End of year report for parallel vision algorithm design and implementation, Robotics Institute, Carnegie Mellon University, 1987.

[9] C. Thorpe, M. Hebert, T. Kanade and S. Shafer, Vision and navigation for the CMU Navlab, Annual Reviews, Palo Alto, California, November 1987, pp. 521–556.

[10] D. L. Waltz, Applications of the connection machine, *IEEE Computer 20* **1** (January 1987), 85–97.

A SYSTOLIC SIMPLEX ALGORITHM

D. J. EVANS and G. M. MEGSON*

Parallel Algorithms Research Centre, Loughborough University of Technology, Loughborough, Leicestershire, U.K.

In this paper a systolic algorithm is presented for the Simplex algorithm as used in Linear Programming applications. In addition, to reduce computer storage requirements, a revised form of the algorithm is considered for systolic array implementation using the product form of the inverse.

KEY WORDS: Systolic array, parallel algorithm, Simplex method, linear programming.

C.R. CATEGORIES: B7.1, C1.2, G1.6.

1. INTRODUCTION

This use of systolic arrays to date has been largely confined to the more numerically compute intensive problems of Computational Linear Algebra. However there exist a class of table based algorithms such as interpolation and extrapolation techniques which also have wide uses in numerical computation and often produce results in tables of a triangular form (Evans and Megson [1]). This triangular structure and the manner in which the table elements are constructed indicate that systolic techniques for matrix problems may carry over to table based methods. Indeed, a table of elements is often represented as a matrix for easy and efficient manipulation on a computer system. Below certain similarities between matrix computations and extrapolation tables are developed to characterise table generation algorithms. The principles are then extended to table manipulation techniques for the more sophisticated Simplex and Linear Programming (LP) problems.

2. A SYSTOLIC SIMPLEX ALGORITHM

In this paper we consider table generating algorithms for Linear Programming and in particular the Simplex method. Linear Programming techniques are extremely useful in many diverse applications such as:

1) Agricultural applications—national and regional scale.
2) Procurement of contract awards.

---

*Computing Laboratory, University of Newcastle-Upon-Tyne.

3) Economic aids (Leontief inter-industry model).
4) Industrial applications (chemical, coal, airline, etc.).
5) Military applications (strategic and logistic).
6) Personnel assignment.
7) Production scheduling and inventory control.
8) Structural design.
9) Traffic analysis.
10) Transportation problems and network theory.
11) Travelling salesman problem.
12) Statistics, combinatorial analysis and graph theory.
13) Design of optical filters.

Thus a fast and efficient systolic design for LP problems is well justified.

A linear programming (LP) problem consists of a linear function,

$$H = c_1 x_1 + \cdots + c_n x_n, \tag{2.1}$$

which is to be minimised or maximised subject to certain constraints,

$$a_{i1} x_1 + \cdots + a_{in} x_n \leqq b_i, \quad 0 \leqq x_j, \quad i = 1(1)m, \quad j = 1(1)n. \tag{2.2}$$

The problem can be written in matrix vector notation as,

$$H(x) = C^T x = \text{minimum}, \quad Ax \leqq b, \quad 0 \leqq x, \tag{2.3}$$

and from linear programming theory it is known that the minimum (or maximum) occurs at an extreme feasible point. A point $(x_1, \ldots, x_n)$ is feasible if its coordinates satisfy all $(n+m)$ constraints, whereas an extreme feasible point forces at least n of the constraints to become equalities. By introducing slack variables $x_{n+1}, \ldots, x_{n+m}$ the constraints are converted to the form,

$$a_{i1} x_1 + a_{i2} x_2 + \cdots + a_{in} x_n + x_{n+i} = b_i, \quad i = 1(1)m \tag{2.4}$$

permitting extreme feasible points to be located by having n or more variables (including the slack variables) zero. A solution point is a minimum point of H; if there is more than one solution point, there is more than one extreme feasible point and any such point can be used as a solution.

To date systolic arrays for the LP problem have been limited to least squares approximation for linear systems, where,

$$Ax = b, \tag{2.5}$$

is the overdetermined $m \times n$ system (for $m > n$) in (2.3) and which satisfies the equations approximately in some "best" sense. Essentially we calculate the residual,

$$r = b - Ax, \tag{2.6}$$

and consider the function

$$\phi(x) = r^T r, \tag{2.7}$$

and choose x to minimise $\phi(x)$ (i.e. the sum of the squares). It follows that,

$$\phi(x) = (b - Ax)^T(b - Ax) = x^T A^T Ax - (bA^T x + x^T A^T b) + b^T b$$

$$= x^T AAx - 2x^T A^T b + b^T b \tag{2.8}$$

and (2.7) is minimised when the gradient vector $\text{grad}(\phi(x)) = 2A^T Ax - 2A^T b = 0$ hence,

$$A^T Ax = A^T b. \tag{2.9}$$

Thus, forming $A^T A$ and $A^T b$ produces an $n \times n$ matrix problem which can be solved by the arrays developed to manipulate matrices (Gentleman and Kung [2]). However the solution does not solve (2.5) exactly and we consider the more flexible Simplex algorithm which solves the original system by table manipulation.

The Simplex algorithm is a method which starts at some extreme feasible point and by a sequence of exchanges proceeds systematically by steadily reducing H to other extreme points until a solution point is found. The use of slack variables (which must be non-negative like the other x_i) allow the identification of extreme feasible points. Since the inequality in $Ax \leqq b$ implies a slack variable being zero, an extreme point is one where at least n of the variables $x_1, \ldots, x_{n+m}$ are zero. Alternatively, an extreme feasible point is one where at most m variables are non-zero. The matrix coefficients of (2.4) can be expressed as,

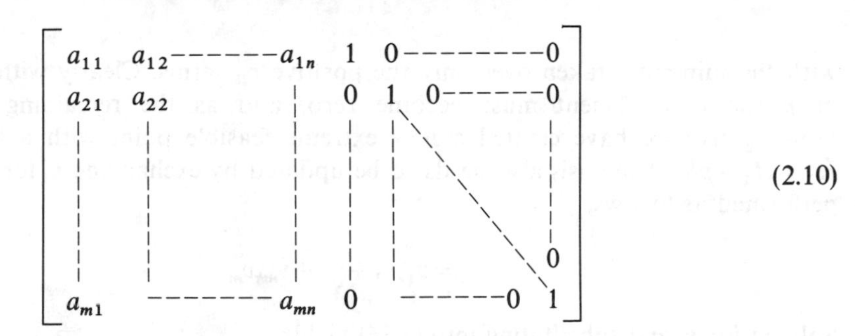

$$\begin{bmatrix} a_{11} & a_{12} & \cdots & a_{1n} & 1 & 0 & \cdots & 0 \\ a_{21} & a_{22} & & \vdots & 0 & 1 & 0 \cdots & 0 \\ \vdots & \vdots & & \vdots & \vdots & & \ddots & \vdots \\ & & & & & & & 0 \\ a_{m1} & \cdots & \cdots & a_{mn} & 0 & \cdots & 0 & 1 \end{bmatrix} \tag{2.10}$$

with the last m columns corresponding to slack variables. Thus, the $(n+m)$ columns of the matrix can be written as $v_1, v_2, \ldots, v_{n+m}$ and (2.4) written as,

$$x_1v_1+x_2v_2+\cdots+x_{n+m}v_{n+m}=b, \tag{2.11}$$

and if an extreme feasible point (say for simplicity) $x_{m+1}=\cdots=x_{m+n}=0$ is known there are at most m non-zero variables, hence,

$$x_1v_1+x_2v_2+\cdots+x_mv_m=b, \tag{2.12}$$

and

$$H=x_1c_1+x_2c_2+\cdots+x_mc_m. \tag{2.13}$$

If the vectors $v_1,\ldots,v_m$ are linearly independent, all $(n+m)$ vectors can be expressed in terms of this basis, viz.

$$v_j=v_{1j}v_1+\cdots+v_{mj}v_m, \quad j=1(1)n+m \tag{2.14}$$

also let,

$$h_j=v_{1j}c_1+\cdots+v_{mj}c_m-c_j, \quad j=1(1)n+m. \tag{2.15}$$

The Simplex method tries to reduce H by including some amount px_k for $k>m$ and p positive. Thus, in order to preserve the constraints we multiply (2.14) with $j=k$ by p and subtract (2.12) to get,

$$(x_1-pv_{1k})v_1+(x_2-pv_{2k})v_2+\cdots+(x_m-pv_{mk})v_m+pv_k=b, \tag{2.16}$$

and from (2.13) and (2.15) the new H is,

$$(x_1-pv_{1k}c_1+(x_2-pv_{2k})c_2+\cdots+(x_m-pv_{mk})c_m+pc_k=H_1-ph_k \tag{2.17}$$

Clearly to reduce $H_1-ph_k>0$, p must be as large as possible without making (x_i-pv_{ik}) negative hence,

$$x_l/v_{lk}=\min_i(x_i/v_{ik})=p, \tag{2.18}$$

with the minimum taken over only the positive v_{ik} terms. Clearly with this choice of p the c_l coefficient must become zero, and as the remaining points are non-negative we have created a new extreme feasible point with a better result $\bar{H}_1=H_1-ph_k$. The basis also needs to be updated by exchanging v_l for v_k, which is performed as follows,

$$v_k=v_{1k}v_1+\cdots+v_{mk}v_m. \tag{2.19}$$

Solving for v_l and substituting into (2.14) yields,

$$v_j=\bar{v}_{1j}v_1+\cdots+\bar{v}_{l-1,j}v_{l-1}+\bar{v}_{kj}v_k+v_{l+1,j}\bar{v}_{l+1}+\cdots+\bar{v}_{mj}v_m,$$

where,

$$\bar{v}_{ij}=\begin{cases} v_{ij}-(v_{lj}/v_{lk})v_{ik}, & i\neq l \\ v_{ij}/v_{lk}, & i=l \end{cases} \tag{2.20}$$

Substituting for v_l in (2.11) gives,

$$\bar{x}_1 v_1+\cdots+\bar{x}_{l-1}v_{l-1}+\bar{x}_k v_k+\bar{x}_{l+1}v_{l+1}+\cdots+\bar{x}_m v_m=b,$$

with

$$\bar{x}_i=\begin{cases} x_i-(x_l/v_{lk})v_{ik}, & i\neq l \\ x_i/v_{lk}, & i=1 \end{cases}. \tag{2.21}$$

Also,

$$\bar{h}_j=\bar{v}_{1j}c_1+\cdots+\bar{v}_{mj}c_m-c_j=h_j-(v_{lj}/v_{lk})h_k$$

with,

$$\bar{H}_1=H_1-(x_l/v_{lk})h_k.$$

The method is then iterated until either all the h_j are negative, or until for some $h_k>0$ no v_{ik} is positive. In the first case, the current point is as good as any adjacent extreme point. In the second case, p can be arbitrarily large and there is no minimum for H. For further reading on LP problems, their applications and the Simplex method (see Chavatal [3], Wu & Coppins [4], Llewellyn [5], Gass [6]).

The Simplex procedure can be represented compactly by the tabular form,

$$\begin{bmatrix} x_1 & v_{11} & v_{12} & \cdots & v_{1,n+m} \\ x_2 & v_{21} & v_{22} & \cdots & v_{2,n+m} \\ \vdots & \vdots & \vdots & & \vdots \\ x_m & v_{m1} & v_{m2} & \cdots & v_{m,n+m} \\ H_1 & h_1 & h_2 & \cdots & h_{n+m} \end{bmatrix}$$

and summarized by the following six steps:

i) Call v_{lk} the pivot (i.e. $p=x_l/v_{lk}$) the part to be added.

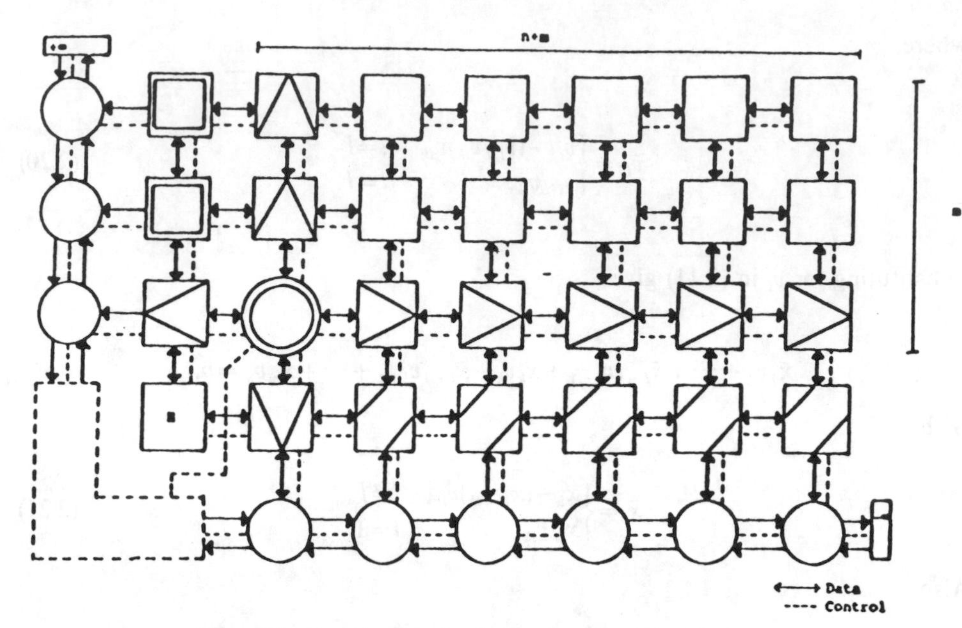

Figure 2.1 Systolic array for the Simplex algorithm ($m=3, n=3$).

ii) Divide the entries in the pivot row by the pivot.

iii) The pivot colum becomes zero except for 1 in the pivot position.

iv) All other entries are modified by the rectangle rule

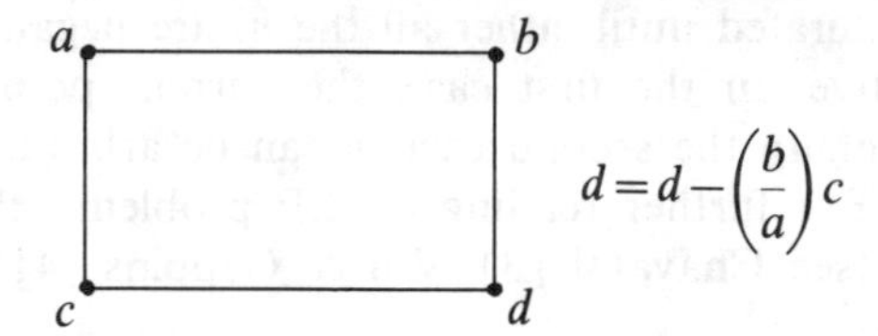

with $a=v_{lk}$ and $c=v_{ik}$.

v) Find the largest new h_j, $j=1(1)n+m$ which is positive (terminate if there are none).

vi) Find the new pivot according to (2.18) with the column indexed j. If none are positive then stop.

The global structure of a wavefront orientated architecture is shown in Figure 2.1, and basically consists of an $(m+1)*(n+m+1)$ orthogonally connected array representing Figure 2.2 and two boundary arrays which are used for sorting rows and columns of the table. All the connections are bi-directional except for the control lines which consist of two 2-bit one-way connections. These two control lines per link allow the specification of two superimposed and disjoint control networks on the array of processing elements. One network is used for row and column sorting, the other for the application of the rectangle rule and pivot

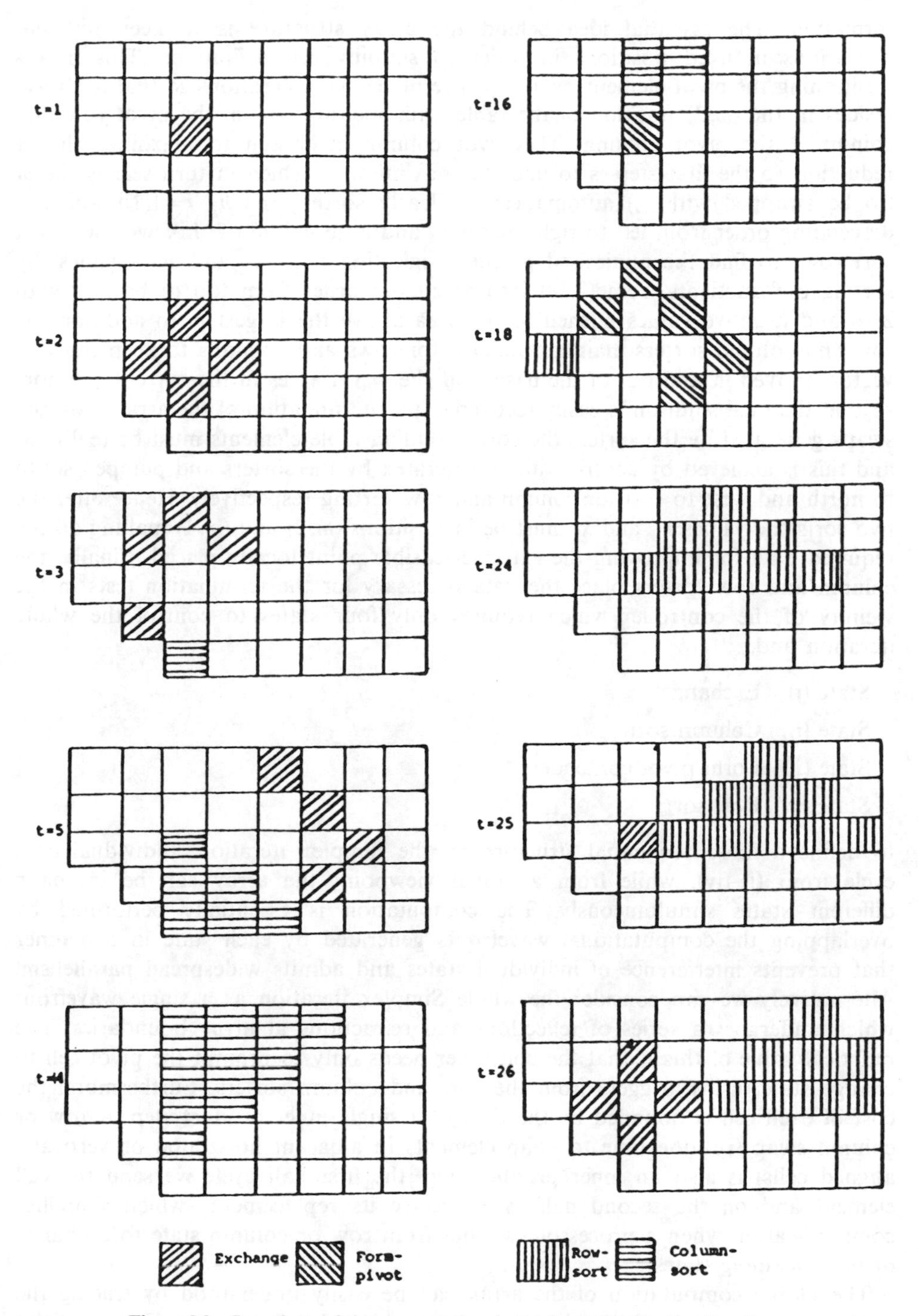

Figure 2.2 Snapshots of the computational wavefronts for Simplex iteration.

formation. The essential idea behind the array structure is to keep different operations in fixed positions in order to simplify cell definitions. This means positioning the pivot element by a sequence of systolic operations so that it always resides in the $(m, 2)$ position of the table, thus making row m the pivot row and column 2 the pivot column. The pivot column is chosen to maximize the H reduction so the first step is to find the maximum h_k which in turn selects the v_k (to be swapped with v_l) automatically. Clearly sorting the h_j, $j=1(1)n+m$ into descending order from left to right places h_k and v_k in column 2. Likewise we must sort rows to find the minimised p (after neglecting negative and zero values) by sorting the quotients from (2.18) into ascending order from top to bottom with zero and negative values pushed to the area above the largest p. In addition the row and column sorters maintain indexes for rows and columns to keep track of vectors moved in and out of the basis and the m variables in the current solution, so that the final solution is easily recovered after termination of the array. For any swaps generated by the sorters the corresponding table elements must be realigned and this is achieved by control values generated by the sorters and pumped south to north and west to east for column and row sorting respectively. Hence after the two sorts the values v_{lk} and x_l must be in positions $(m, 2)$ and $(m, 1)$ making all the required data for improving the extreme feasible point locally placed. Finally, the column and row sorters place the data necessary for the termination tests in the vicinity of the controller which requires only four states to control the whole iteration and,

State (i) Exchange

State (ii) Column sort

State (iii) Form pivot contenders

State (iv) Row sort,

to define a loose sequential structure on the Simplex iteration. Individual cells cycle from (i)–(iv), while from a global viewpoint the array can be in many different states simultaneously. The computation is essentially performed by overlapping the computational wavefronts generated by each state in a manner that prevents interference of individual states and admits widespread parallelism. Alternatively we can consider the whole Simplex iteration as a single wavefront which undergoes a series of reflections and refractions at array boundaries. The practical value of this is that the controller needs only to prompt the pivot cell to change state, using triggers from the row and column sorters. Furthermore, the cost of each cell is bounded by the time of a single inner product step. A row or column swap (i.e., the time to swap elements in adjacent horizontal or vertically aligned cells) is also an inner product step, the first half cycle we send the cell element and on the second half cycle receive its replacement—which simplifies communication when a processor switches from row or column state to exchange or pivot forming states.

The global computation of the array can be easily understood by tracing the wavefronts associated with each state as demonstrated in Figure 2.2. For simplicity $t=1$ (in Figure 2.2) represents the first cycle after the starting table has been loaded into the array, and $t=25$ depicts the start of the next cycle. It follows that wavefronts of different states never interfere and an estimate for the time of a Simplex iteration can be identified.

THEOREM 2.1 *A single change of an extreme feasible point in the standard Simplex algorithm with n unknowns and m constraints using an orthogonally connected array of* $0((m+2)(n+m))$ *cells requires*

$$T=(2n+4m+6) \text{ cycles.}$$

Proof (By observation of the dataflow in Figure 2.2).

(i) The exchange state requires $(n+m)$ cycles to reach the right hand array boundary, and an extra cycle before the last element can be loaded into the column sorter (i.e. $n+m+1$ cycles).

ii) If this last element is the largest h_j it will take $(n+m)$ cycles to reach the h_k position in the pivot column, by a sequence of interchanges.

iii) On the next cycle the last swap enters the $(m+1)$st row of the table going south to north and on the second cycle reaches the pivot cell.

iv) Thus, on the third cycle after the end of column sorting, the pivots drop into "form-pivot" state as the correct v_k from the pivot downwards have been formed.

v) After a further m cycles the last pivot contenders enter the row sorter, if this value is the smallest pivot (i.e. p) it requires a further m cycles to reach the pivot row.

vi) An additional 2 cycles sees the last row swap operations pass the pivot column. Thus all the columns to the left of and including the pivot are correct column and row sorted and the next exchange can start.
The H-cell also contains the improved minimum.

Summing these timings ensure the bound $T=(2n+4m+6)$ for a single Simplex update. The area bound is given by:

i) The table elements require $(m+1)(m+n)$ ips cells.

ii) Column sorting at most $(m+n)$ ips cell equivalents.

iii) Row sorting m ips cell equivalents.

iv) The unknown x_i values and H-cell

a) 2 ips for H and pivot row cell
b) $2(m-1)$ for remaining unknowns

yielding $(m+2)(m+n)+3m$ ips cell equivalents.

COROLLARY 2.1 *A search of z extreme points requires* $T=z(2n+4m+6)+2m$ *cycles using the Simplex method.*

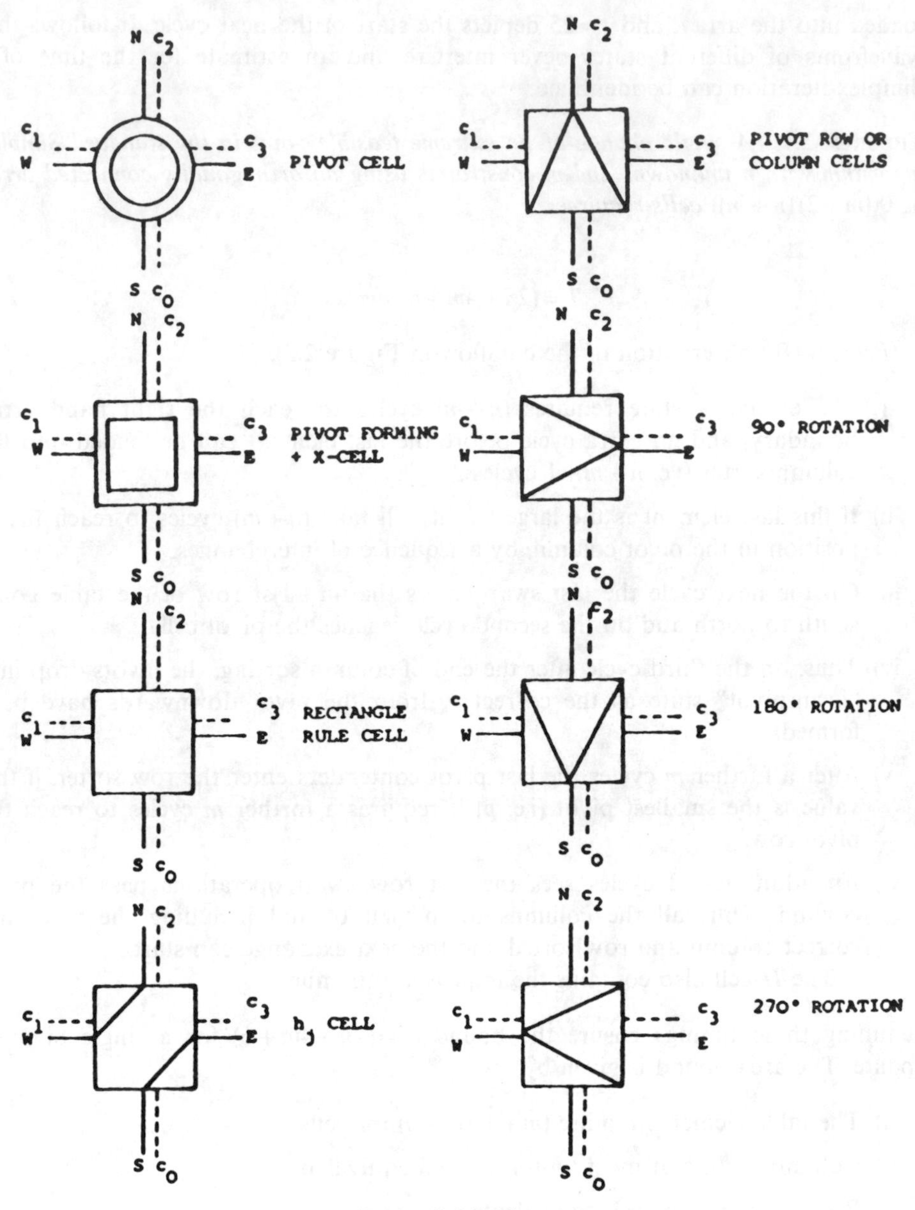

Array cell definitions.

Proof The loading and unloading of the starting and final tableaux requires at most an additional time of $2m$ cycles, then repeating the argument in Theorem (2.1) yields the timing immediately.

The Simplex array was simulated and tested using an OCCAM program. The array cell definitions can be found using the following key.

Snapshots of the array operation on a test example are given below, and the following pertinent remarks concerning array implementation may be of use.

The idea of column sorting is to move the column associated with the largest h_j to the pivot column position, and as stated above the problem reduces to sorting the h_j into descending order from left to right and performing the same swaps on the v_j column vectors. A linear systolic array for sorting is shown in Figure 2.3a and implements the well known odd–even transposition or parallel bubblesort and consists of $(n+m)$ cells. The sorting network is slightly different from the standard array because our starting values must be loaded sequentially and the sorting started systolically. The standard array assumed cells were already loaded and that array cells started simultaneously. Our sorting network also generates an $(n+m)$ control bit vector on every cycle (pumped south to north through the table) to control column swaps. Finally, the cells must keep track of the index j of v_j so that the row sorter can be informed which column is the new pivot column and hence determine the variable introduced to the solution. Figure 2.3b illustrates the sorter operation on the worst case list for an array of 5 cells. The key values to watch are the c_1 and $\bar{c}_1$ control values because these determine the array operation time. On the trip left to right c_1 loads the h_j values from the H-cells of the table portion of the array and starts the cells into odd and even operation mode. Where two cells decide to swap elements the control bit c_2 in each cell is set pumped up into the table to swap column elements. On reaching the rightmost sorting cell c_1 loads the last value and falls off the array, and a neutral element "$-\infty$" is used to prevent erroneous swaps at the array boundary. Next the $\bar{c}_1$ signal enters and moves right to left pushing the final maximum of h_j in front of it (by two cycles) and closing down the sorting cells. Thus after $2(m+n)=10$ (in this case) the maximum h_k resides in the leftmost cell along with its index k and the last column swap data is about to enter the tableau. After a further two cycles the $\bar{c}_1$ filter bit completes the startup–closedown control cycle and the last column swap has reached the pivot row (verifying the timing $2(n+m+1)$ for sorting above). It follows that the filter bit falling off the array can be used to prompt the controller into "form-pivot" state.

The row sorting mechanism works in an identical manner to the column sorter and requires $2m$ cycles to complete sorting and an additional 2 cycles to closedown all the cells (verifying the timing $2(m+1)$ in Theorem 2.1 for sorting). At the end of row sorting the pivot resides in cell $(m, 2)$ and the row label (i.e. the x_l index) has been placed in the bottom row sorting cell along with the minimum p. Hence the filter bit which has moved from top to bottom row closing down cells can be used to load the index k from the column sorter leftmost cell (via the controller) to overwrite the row label and set the controllers Exchange state simultaneously. Further complications arise when we consider sorting with negative p values, because under normal sorting conditions these will bubble to the bottom of the sorter. A forced swap for negative and zero p values implemented by a status flag set by the comparator determining the swap solves the problem simply, causing the undesirable values to bubble to the top of the array. If the bottom row sorting cell still contains a zero or negative value at the end of

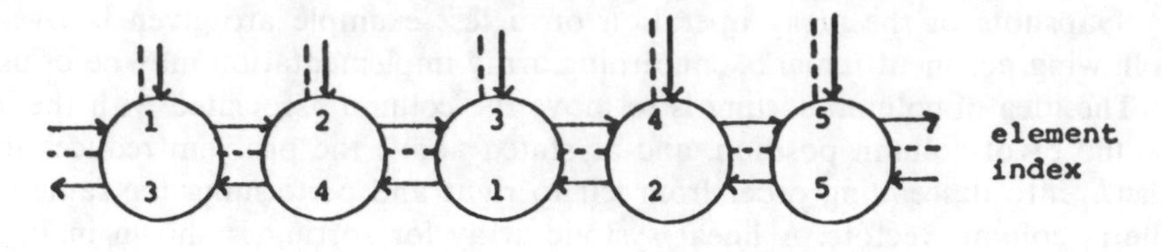

a) Odd-even sorter

ARRAY CELLS

t															
1	/	0	1	/	0		/	0		/	0		/	0	
		1			–			–			–			–	
		3			4			1			2			5	
2	0	0	0	/	0	1	/	0		/	0		/	0	
	1	1			2			–			–			–	
	3	3			4			1			2			5	
3	0	0	0	0	0	0	/	0	1	/	0		/	0	
		1	2	1	2			3			–			–	
		3	4	3	4			1			2			5	
4	0	1	0	0	2	0	0	0	0	/	0	1	/	0	
	2	2			1	3	1	3			4			–	
	4	4			3	1	3	1			2			5	
5	0	0	0	0	1	0	0	2	0	0	0	0	0	0	1 →
		2	3	2	3			1	4	1	4			5	–∞
		4	1	4	1			3	2	3	2			5	/
6	0	1	0	0	2	0	0	1	0	0	2	0	0	0	0
	3	3			2	4	2	4			1	5	1	5	
	1	1			4	2	4	2			3	5	3	5	
7	0	0	0	0	1	0	0	2	0	0	1	0	0	2	0 ←
		3	4	3	4			2	5	2	5			1	–∞
		1	2	1	2			4	5	4	5			3	/
8	0	1	0	0	2	0	0	1	0	0	2	0	3	0	0
	4	4			3	5	3	5			2	1	2	1	
	2	2			1	5	1	5			4	3	4	3	
9	0	0	0	0	1	0	0	2	0	3	0	0	0	0	0
		4	5	4	5			3	2	3	2			1	–∞
		2	5	2	5			1	4	1	4			3	/
10	0	1	0	0	2	0	3	0	0	0	0	0	0	0	0
		5			4	3	4	3			2	1	2	1	
		5			2	1	2	1			4	3	4	3	

C_1	C_2	C_1
W_1	a	E_1
W_2	j	E_2

b) Snapshots

Figure 2.3 Modified odd–even transposition sort.

sorting, all the values in the sorter must be non-positive. Hence the status bits of the bottom cell also flag a termination condition as no improvement to H is possible, and can inhibit the overwriting of the row label. Similarly a status flag in column sorter cells can be adopted to flag $h_k < 0$ and trap the second termination condition. Finally, some general remarks about table element swapping. Figure 2.4. illustrates the pipelining of both column and row swapping, it should be clear that the control vectors output by the sorters consist of the pairs $(1, 2)$ punctuated by pairs of zeroes where no swapping occurs. The special null vector therefore corresponds to no swaps, which can only be produced when a list is fully sorted or a sorter is switched off. It follows that even if some portion of the array drops into a sorting state before the sorter begins, row and column data will remain undamaged. This preservation of data is important because it allows flexible state transitions, and the use of column and row states as idling states between wavefronts. In particular it allows the overlapping of row and column table modifications.

Next, we consider the exchange process—which rewrites the table in terms of the new basis—essentially exchanging the vectors entering and leaving the basis. We can define three basic types of operation and by virtue of the static positioning of the pivot row and column three basic cell types. The basic operations are (a) find reciprocal in pivot cell, (b) divide pivot row by pivot element, (c) zero out the pivot column.

Using prompts from the controller the pivot cell in Figure 2.1 orchestrates the whole computation (including sorting) and is the source of the starting wavefronts. The pivot cell controls operations using the second control network by triggering cell states by pumping control signals through the network. Figure 2.5 illustrates the exchange control flow from which the following actions are defined:

i) Whenever a cell receives two true controls on the same cycle it performs the rectangle rule.
ii) If a single control value which is true arrives from the north or south, output the table value and zero the register.
iii) If a single value arrives from east or west perform a division by the pivot.
iv) If the cell is the pivot cell and the state is "row-sort", a control input sets state = "exchange" and:
    a) The reciprocal of the pivot is found, the result sent east and west.
    b) Overwrite the pivot with 1, and set all control outputs true,

which characterise the dataflow.

i) Control values falling off the southern boundary correspond to the startup procedure for the column sorter.
ii) A control value travelling horizontally or vertically continues to do so until it falls off the array.
iii) A control signal arriving in a pivot column cell is also refracted east and west.

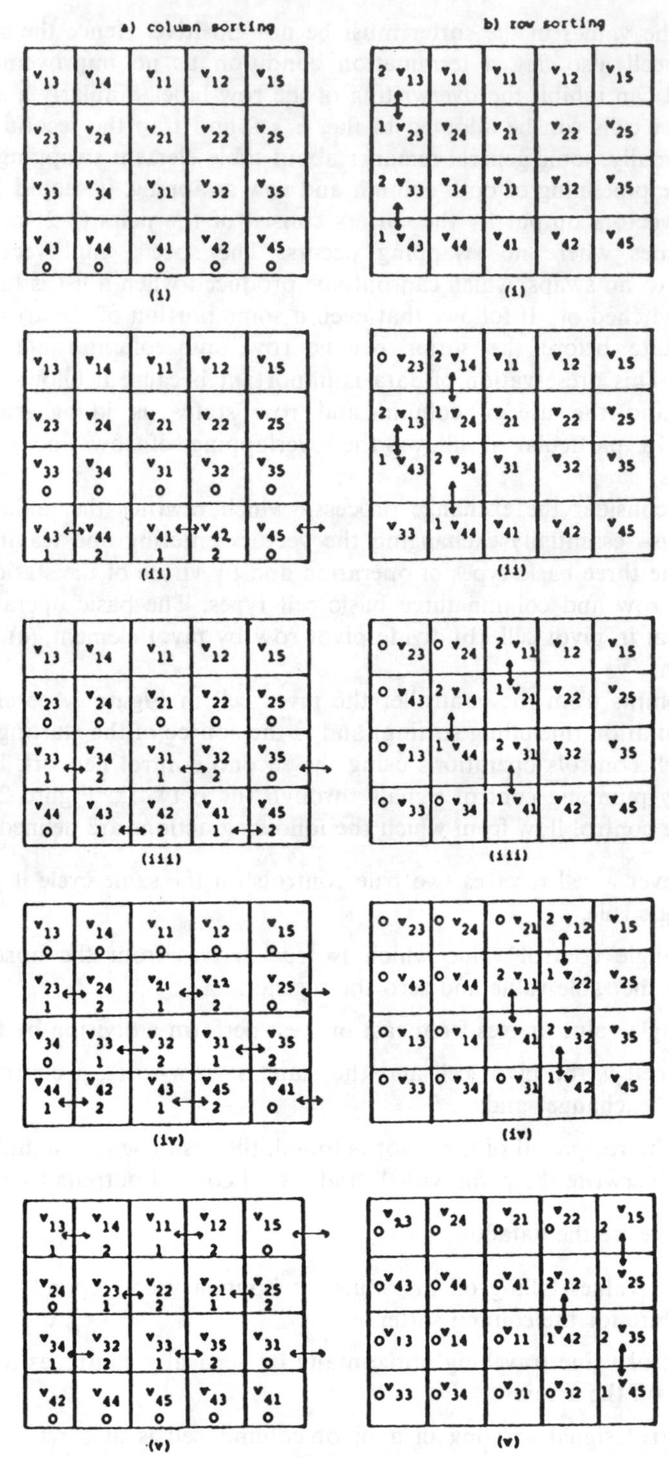

Figure 2.4 Control flow in sorting.

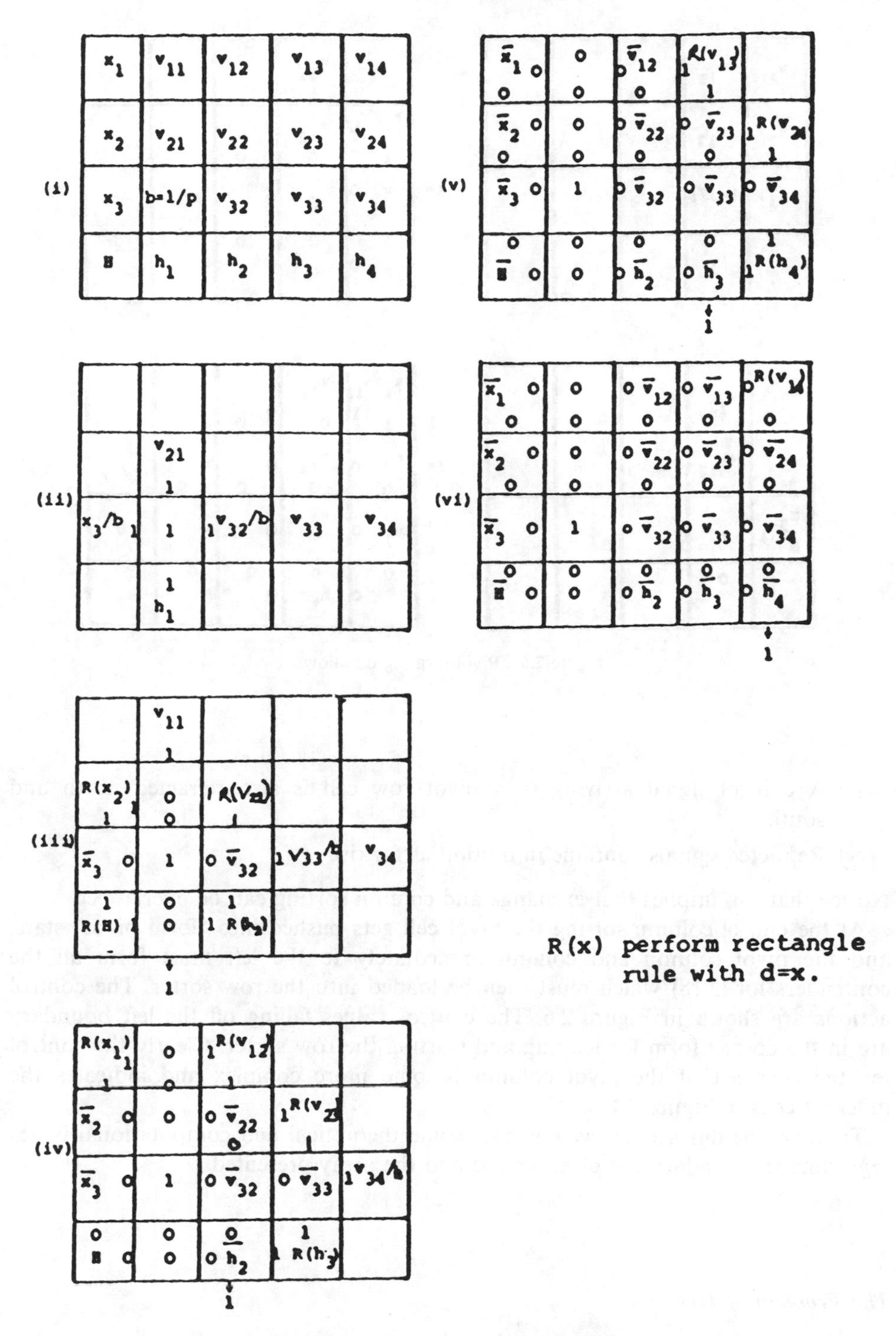

Figure 2.5 Control wavefront for exchange.

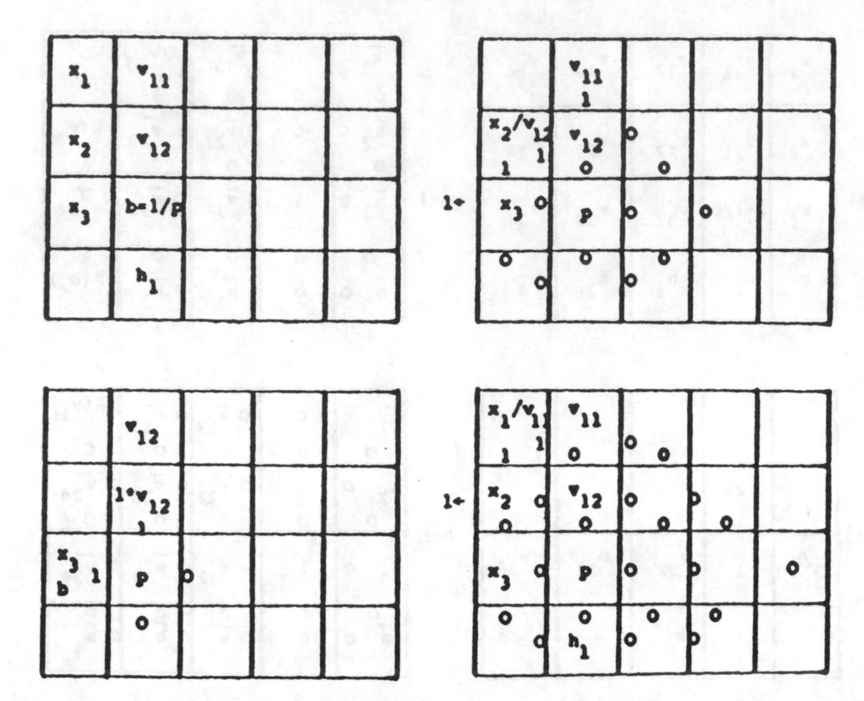

Figure 2.6 Pivot forming dataflow.

iv) A control signal arriving in a pivot row cell is also refracted north and south.

v) Refracted signals continue in motion as for (ii).

Notice that this implies that exchange and column sorting can be overlapped.

At the end of column sorting the pivot cell gets pushed into "form pivot" state, and the pivot column and column immediately to the left must form all the contenders for (2.18) which must then be loaded into the row sorter. The control actions are shown in Figure 2.6. The control values falling off the left boundary are in the correct form for loading and starting the row sorter. Clearly the control for the cells left of the pivot column become more complex and indicates the different cells of Figure 2.1.

To conclude this section we consider some theoretical and computational issues regarding the standard Simplex method and the array presented.

The Problem of Ties

During the column sort phase we select the vector v_j corresponding to $\max_j(h_j) = h_k$ to achieve the greatest immediate decrease in the objective function. A tie occurs when more than one j occurs with maximum h_j. The problem is resolved arbitrarily in the standard Simplex theory by choosing the lowest (or highest) index j—which proves to be a good choice. Although the current array description is adequate for breaking ties, we can incorporate this strategy by performing a swap according to

$$(h_j < h_i) \text{ or } ((h_j = h_i) \text{ and } (i > j))$$

in a column cell. This modification requires at most an additional comparator in each sorter cell, and is justified by the fact that tie breaking with this rule requires approximately m changes of basis to find the minimum. Thus with $z \approx m$ Corollary 2.1 gives a loose bound for the full Simplex calculation.

Degeneracy

A non-degenerate feasible solution is a feasible solution with exactly m positive x_i, if there are less than m positive x_i the solution is degenerate. If the above condition ocurs at least one x_i is zero and it would be possible to choose $p=0$ in (2.18), producing no reduction in H. If this lack of improvement continued for a number of Simplex iterations it is possible to repeat a basis and the solution process breaks down (the array would become stuck in an infinite loop). Degeneracy is indicated by less than m x_i values being positive, or (2.18) producing ties (and implying that more than one variable leaves the solution on a single iteration). Fortunately, to date degeneracy has only been exhibited by artificially constructed problems, and the normal course of action is to use $p=0$ when it occurs and break ties in a similar manner to the column sorter solution.

Artificial Basis Techniques

Throughout the array description we have assumed that a basis (hence extreme feasible point) was known. When this is not the case an artificial basis must be constructed which will produce a feasible basis for the original problem. The details of artificial basis can be found in standard texts on linear programming and are not discussed here, but the array of Figure 2.1 is easily upgraded to deal with them. Essentially, we add an additional row of cells in the $(m+2)$nd position which contain their own h_j type elements. The algorithm is then controlled by two phases. In the first phase the h_j values are sorted and the table updated until all the elements are non-positive. If all $h_j=0$ the resulting basis is feasible for the original problem, and if all $h_j<0$ the original problem was not feasible. In the former case we can continue by applying the sorting to the $(m+1)$th row or true h_j values to obtain a minimum. For the latter case the table is abandoned. Again, the extra hardware is justified by the fact that a full artificial basis of m columns requires approximately $z=2m$ iterations to find the minimum feasible solution which otherwise would not be solvable.

TEST EXAMPLE

Minimise

$$H = -2x_1 - x_2$$

subject to

$$0 \leqq x_1, \quad 0 \leqq x_2, \quad -x_1 + 2x_2 \leqq 2, \quad x_1 + x_2 \leqq 4.$$

Introducing slack variables we produce the following tables:

3	2	-1	2	1	0	0
4	4	1	1	0	1	0
5	3	1	0	0	0	1
	0	2	1	0	0	0
		1	2	3	4	5

3	3	0	0	1	-2	3
2	1	0	1	0	1	-1
1	3	1	0	0	0	1
	-7	0	0	0	-1	-1
		1	2	3	4	5

As the first action of the array is a modification, and the correct pivot is in the correct place variable x_1 is swapped with x_5 so we load the tableau with,

3	2	-1	2	1	0	0
4	4	1	1	0	1	0
1	3	1	0	0	0	1
	0	2	1	0	0	0
		1	2	3	4	5

After a few iterations (2) we get the following result from the OCCAM program.

a) With trace = on

1	3	1	0	0	0	1
3	3	0	0	1	-2	3
2	1	0	1	0	1	-1
	-7	0	0	0	-1	-1
		1	2	3	4	5

b) with trace = off

Results

[1] = 3.000000
[3] = 3.000000
[2] = 1.000000
[H] = −7.000000
[i] = variable i

which is a row and column permuted form of the correct final tableau.

3. A SYSTOLIC CYLINDER FOR THE REVISED SIMPLEX ALGORITHM

The above standard scheme is not the one usually chosen for computer implementation, instead a revised form of the Simplex algorithm is used. This new algorithm can be implemented in two ways:

a) The general form of the inverse.

b) The product form of the inverse.

The second technique is often used in practice because it minimises the amount of information to be recorded using the products of elemental matrices. Both techniques however reduce the amount of computation required to update the basis and Simplex tableau, and size of table recorded in the machine's main memory. This latter point of data compaction is important for large LP problems and we examine the possibilities of transferring these characteristics to systolic arrays.

At the start of the standard algorithm extra vectors (at most m) are added to the table to form a basis. The basis consists of m linearly independent vectors and it follows that,

$$B=(v_1, v_2, \ldots, v_m), \tag{3.1}$$

and that any other vector v_j is a linear combination of vectors from B, i.e.,

$$v_j=\alpha_{1j}v_1+\alpha_{2j}v_2+\cdots+\alpha_{mj}v_m$$

then,

$$\alpha_j=B^{-1}v_j, \tag{3.2}$$

where,

$$\alpha_j=(\alpha_{1j}, \alpha_{2j}, \ldots, \alpha_{mj}).$$

From (3.3) putting B as the first m vectors of A such that,

$$Bx_0=b, \quad x_0\geqq 0, \tag{3.3}$$

with $x_0=(x_{10}, x_{20}, \ldots, x_{m0})$ gives the first basic feasible solution

$$x_0=B^{-1}b, \tag{3.4}$$

and from (3.2) all the remaining vectors of A can be determined from B. The pieces used to determine the vector to be moved into the basis are given by, h_j, $j=1(1)n$ where,

$$\left.\begin{array}{l} \text{a) } h_j=z_j-c_j \text{ with} \\ \text{b) } z_j=c_1\alpha_{1j}+c_2\alpha_{2j}+\cdots+c_m\alpha_{mj} \end{array}\right\}. \tag{3.5}$$

Thus,

$$z_j=c_0\alpha_j=c_0B^{-1}v_j, \quad j=1(1)n \tag{3.6}$$

with $c_0=(c_1,\ldots,c_m)$ so with the feasible basis B we compute the corresponding z_j, and a pricing vector π_i can be defined as,

$$\pi=c_0B^{-1},\quad \pi=(\pi_1,\pi_2,\ldots,\pi_m) \tag{3.7}$$

hence for a vector not in the basis,

$$h_j=\pi v_j-c_j. \tag{3.8}$$

It follows that we have all the information to move from feasible solution to feasible solution, using only the original A and C values. The main idea is that rather than transferring all the elements of the Simplex tableau we need only to transform the elements of B^{-1}. The explicit form of B at each iteration can be constructed as follows. Let $B=(v_1,v_2,\ldots,v_l,\ldots,v_m)$ be the old basis differing from the new basis $\bar{B}$ by a single vector,

$$\bar{B}=(v_1,v_2,\ldots,v_k,\ldots,v_m),\quad v_l\neq v_k.$$

Then,

$$B^{-1}\bar{B}=B^{-1}(v_1,v_2,\ldots,v_l,\ldots,v_m)=I, \tag{3.9}$$

and,

$$B^{-1}\bar{B}=B^{-1}(v_1,v_2,\ldots,v_k,\ldots,v_m)=\begin{bmatrix} 1 & 0 & \cdots & \alpha_{1k} & \cdots & 0 \\ 0 & 1 & & \alpha_{2k} & & \vdots \\ \vdots & \vdots & & \vdots & & \vdots \\ & & & \alpha_{lk} & \cdots & 0 \\ \vdots & \vdots & & \vdots & & \vdots \\ 0 & 0 & \cdots & \alpha_{mk} & \cdots & 1 \end{bmatrix}. \tag{3.10}$$

Thus, an element b_{ij} of B^{-1} can be transformed to $\bar{b}_{ij}$ and element of $\bar{B}^{-1}$ in the corresponding position, by,

$$\begin{aligned} \bar{b}_{lj}&=\frac{b_{lj}}{\alpha_{lk}} \\ \bar{b}_{ij}&=b_{ij}-\bar{b}_{lj}\alpha_{ik},\quad i\neq l. \end{aligned} \tag{3.11}$$

The revised Simplex method is then constructed as follows,

i) introduce the additional variable $x_{n+m+1}=-H(x)$ from (5.3)

ii) allow for artificial vectors (using artificial basis techniques) by the redundant equation,

$$a_{m+2,1}x_1+a_{m+2,2}x_2+\cdots+a_{m+2,n}x_n+x_{m+n+2}=b_{m+2},$$

where,

$$a_{m+2,j}=-\sum_{i=1}^{m} a_{ij}, \quad j=1(1)n$$
$$b_{m+2}=-\sum_{i=1}^{m} b_i \tag{3.12}$$

and with $a_{m+1,j}=c_j$ we have the matrix problem,

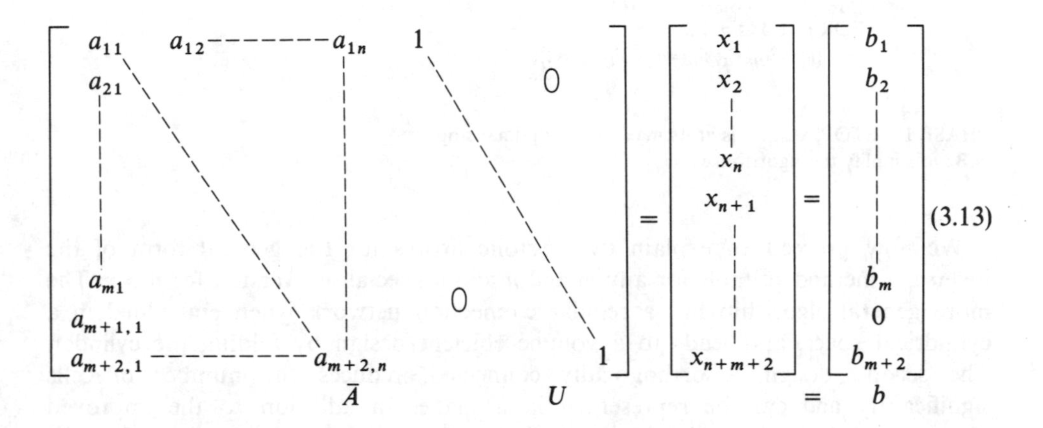

with $x_i \geq 0$, $i=1(1)n+m+2$.

The revised Simplex procedure can now be completely defined. In the algorithm definition we denote $\bar{A}_j$ as the columns of $\bar{A}$ and u_i the rows of the matrix U. As the procedure progresses the u_i represent the most recent update of the corresponding row. Row $m+2$ in $\bar{A}$ is used to evaluate h_j, while artificial variables are still in the solution, row $m+1$ when artificial variables have been removed.

/* revised simplex algorithm */
PHASE I: (Artificial variables in the solution, and all positive)
WHILE $x_{n+m+2}<0$ DO
{FOR $j=1$ TO n $\{\delta_j=u_{m+2}\bar{A}_j\}$;
IF ALL $\delta_j \geq 0$ THEN $\{x_{n+m+2}$ MAX NO FEASIBLE SOLUTION EXISTS}
ELSE $\{\delta_k=\min(\delta_j)\}$;
FOR $l=1$ TO $m+2$ $\{x_{ik}=u_i\bar{A}_k\}$;

$P=\min\limits_{1\leq i\leq m}\left(\dfrac{x_{i0}}{x_{ik}}\right)=\dfrac{x_{l0}}{x_{lk}};$

FOR $i=1$ TO $n+m+2$
{$\bar{x}_{ko}=x_{lo}/x_{lk}$; $\bar{x}_{io}=x_{io}-\bar{x}_{ko}x_{ik} i\neq k$
FOR $j=1$ TO $m+2$
{$\bar{u}_{lj}=u_{lj}/x_{lk}$; $\bar{u}_{ij}=u_{ij}-\bar{u}_{lj}x_{ik} i\neq l$}
};
};
PHASE II: (No positive artificial variables in solution)
{FOR $j=1$ TO n {$\gamma_j=u_{m+2}\bar{A}_j$};
WHILE $\gamma_j<0$ DO
{$\gamma_k=\min(\gamma_j)$};
FOR $i=1$ TO $m+2$ {$x_{ik}=u_i\bar{A}_k$};

$$P=\min_{1\leq i\leq m}\left(\frac{x_{io}}{x_{ik}}\right)=\left(\frac{x_{lo}}{x_{lk}}\right)$$

IF all $x_{ik}\leqq 0$ THEN {solution can be made arbitrarily large}
FOR $i=1$ TO $n+m+2$
{$\bar{x}_{ko}=x_{lo}/x_{lk}$; $\bar{x}_{io}=x_{io}-\bar{x}_{ko}x_{ik} i\neq k$;
FOR $j=1$ TO $m+2$
{$\bar{u}_{lj}=u_{lj}/x_{lk}$; $\bar{u}_{ij}=u_{ij}-\bar{u}_{lj}x_{ik} i\neq l$},
};
};
PHASE III: STOP; x_{n+m+1} is at its max value—optimal stop.
N.B.: to simplify the algorithm $x_i=x_{io}$.

We now proceed to explain two systolic arrays for the general form of the inverse, a method suitable for any m and n and a specialised version for $m>n$. The more general algorithm has a regular connection network when embedded in a cylindrical space, and leads to a volume efficient design by folding the cylinder. The second design is orthogonally connected, reduces the number of cells significantly and can be represented in a plane. In addition to the improved efficiency of the revised Simplex method these new algorithms recognize that the pivot row and column can be located and moved within the array without a full sort (or total ordering). A partial ordering to locate max or min elements is sufficient and can be implemented with simplified cells.

The global view of the systolic cylinder is shown in Figure 3.1 and can be considered as three individual sections, Part A, Part B and Part C, with dataflow around the cylinder interpreted as wavefronts across these sections.

Part A: is an $n*(m+2)$ matrix of cells, with an additional column of n boundary cells to the left. The array contains the elements of $\bar{A}$ stored in the order of $\bar{A}_j$ in the jth row $j=1(1)n$. The boundary cells are initially empty except for the column index.

Part B: This is an $(m+1)*(m+3)$ matrix of cells with a column of m boundary cells to the right. The array contains the $(m+2)*(m+2)$ basis matrix initially U, and can be hardwired to start up with $U=I$. The $(m+1)$st row contains two rows $(m+1)$ and $(m+2)$ for smooth dataflow, while the $(m+3)$rd column contains the starting solution vector. The column boundary cells on the right containing the indexes of the solution variables.

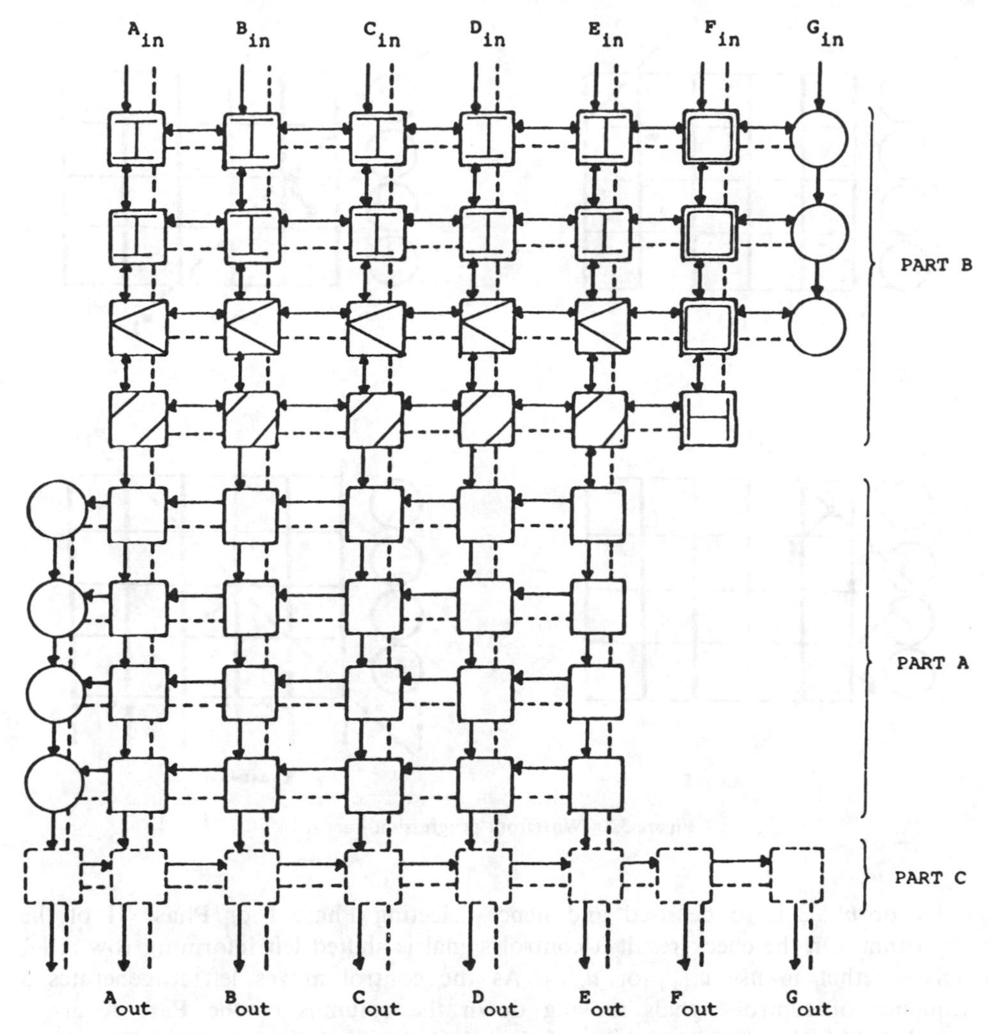

Figure 3.1 Cylindrical systolic array for revised Simplex method (general m and n).

Part C: This is a row of $(m+5)$ cells which wrap around the top row of Part B to the bottom row of Part A, forming the cylinder.

Notice that the two phases of the revised Simplex algorithm are almost identical, except that we use u_{m+1} instead of u_{m+2} and allow additional termination conditions. Thus placing both u_{m+1} and u_{m+2} in the same row of Part B reduces dataflow problems to only a single phase, with switching between phases controlled by the $(m+1)$st row of Part B. The boundary cells will be used to detect the remaining termination conditions.

The wavefronts for the cylinder computation are shown in Figures 3.2 and 3.3 and explained by the following commentary. At the start of computation the cell H in Figure 3.1 performs a check on x_{n+m+2} and x_{n+m+1} to determine which row

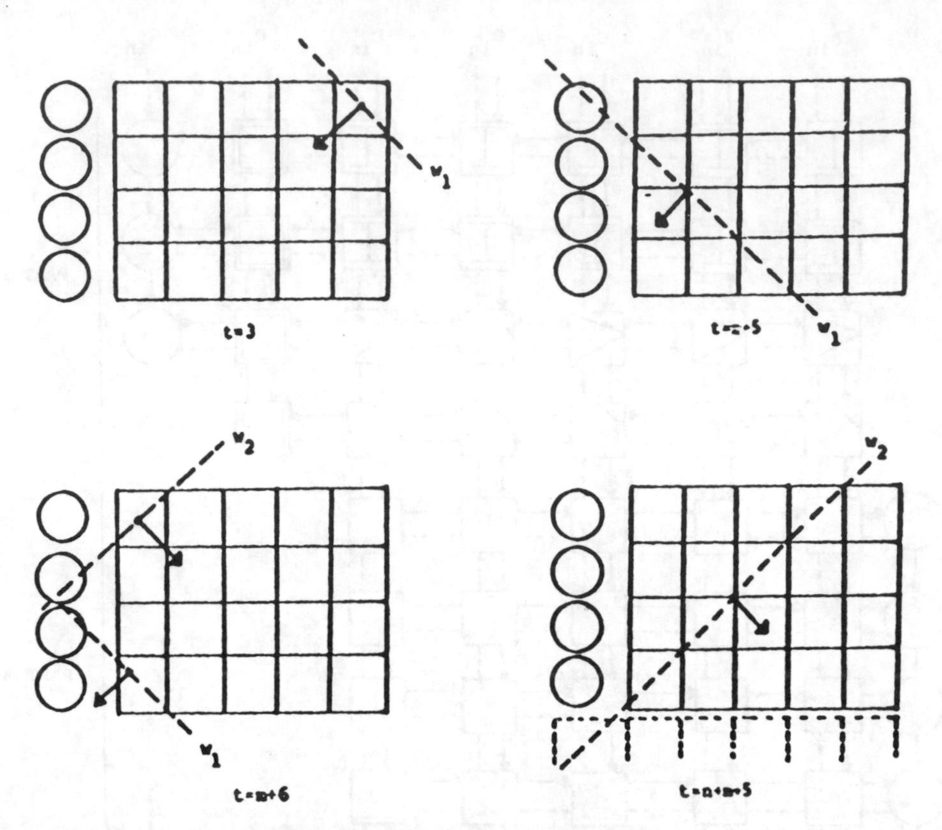

Figure 3.2 Wavefront progression Part A.

$m+1$ or $m+2$ is to be used and hence selecting Phase I or Phase II of the algorithm. On the check result a control signal is shifted left informing row $m+1$ cells whether to use u_{m+1} or u_{m+2}. As the control moves left it generates a sequence of control signals moving down the columns of the Part A array, together with the associated value of the selected row (u_{m+1} or u_{m+2}) elements. A wavefront (w_1 in Figure 3.2) spreads out from the top right of the Part A section generating δ_j (γ_j) depending on the phase, by accumulating partial products from right to left. On reaching the left boundary the δ_j (γ_j) values are loaded into the rows' boundary cell where it picks up its associated index j. w_1 is now reflected to form w_2 a wavefront moving from the top left to bottom right corner of A. w_2 computes the partial ordering of δ_j (γ_j) values pushing the minimum to the bottom of the boundary cell column, and issuing a sequence of controls left to right along each row of Part A transferring a copy of the corresponding A_j column towards the Part C array. It follows that on the $(n+m+5)$th cycle the best δ_j (γ_j) and its associated index j are in the leftmost cell of the Part C array. The next $(m+2)$ cycles see w_2 load the Part C cells with the column A_j. Thus, by the $(n+m+5)$th cycle we have identified k and have A_k moving systolically in the array. Next we must insert x_k into the solution vector and insert v_k into the basis ejecting x_l and v_l

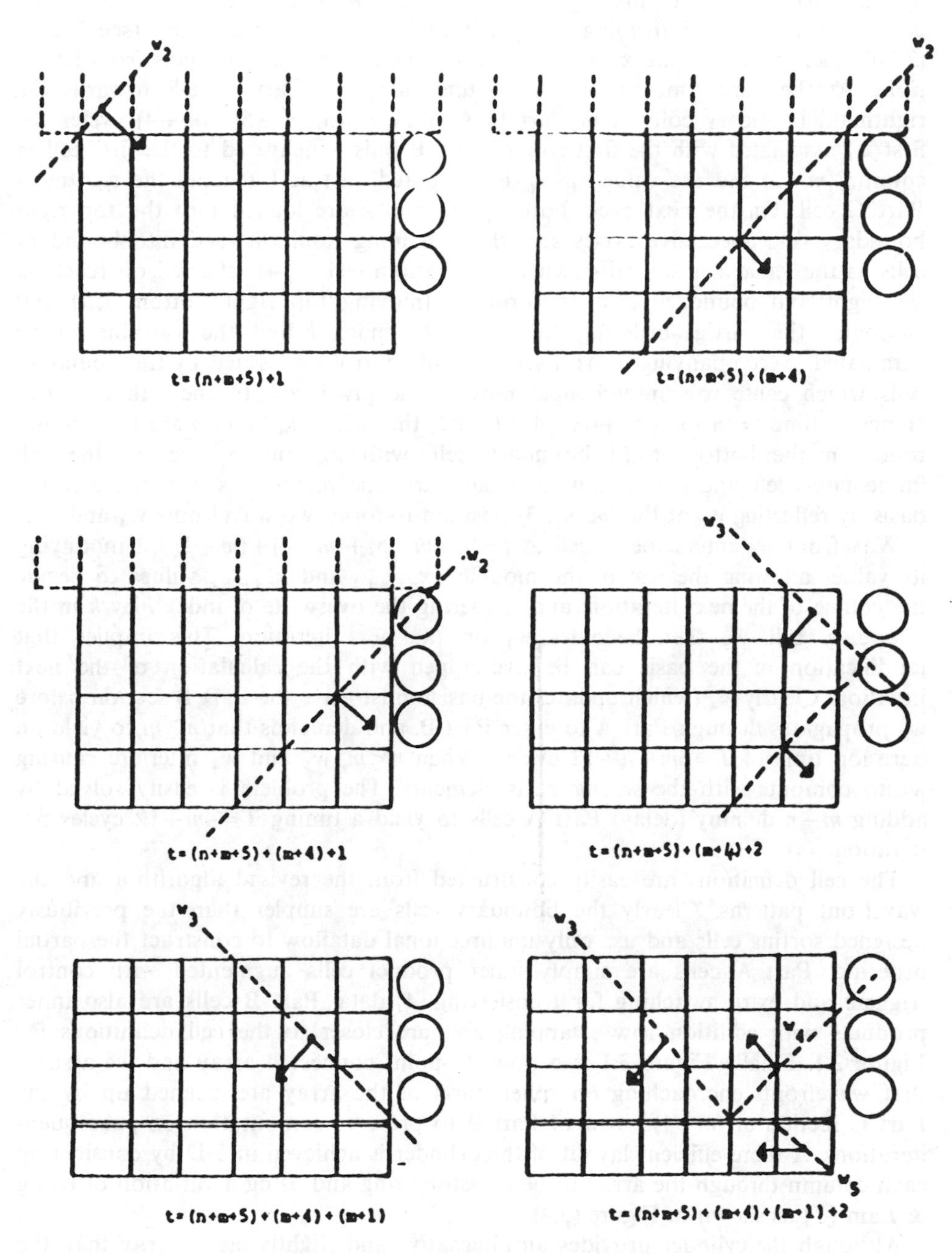

Figure 3.3 Wavefront progression Part B.

from the solution and basis respectively. This is achieved simply by using the cylinder arrangement to propagate w_2 from Part A to Part B using Part C. w_2 continues into Part B forming a top left to bottom right wavefront (see Figure (3.3)) which computes the x_{ik} values by accumulating partial products from left to right. At the same time the index k filters along the Part C cells towards the righthand boundary column of Part B. At time $t=(n+m+5)+(m+4)$ cycles the first x_{ik} associated with the first row of Part B cells is delivered to the first cell in column $(m+3)$ and the value x_{i0}/x_{ik} is computed, just as k reaches the rightmost Part C cell. On the next cycle both x_{i0}/x_{ik} and k are loaded into the top right boundary cell. Successive cycles sees the remaining results loaded into boundary cells as the value k is shifted down to the bottom cell. w_2 is reflected on reaching the righthand boundary cells to form w_3 (moving top right bottom left) and computes the partial ordering to locate the index l and the variable to be eliminated. Accompanying w_3 is a sequence of control bits issued by the boundary cells which cause row interchanges moving the pivot row to the mth cell row. Hence at time $t=(n+m+5)+(m+4)+(m+1)$ the values, k, l and p are known and reside in the bottom right boundary cell, with x_{lk} and x_{l0} set in the cell immediately left, and the basis update can start. The vector v_k is introduced to the basis by reflecting w_3 at the $(m, m+3)$ position to form two wavefronts w_4 and w_5.

Wavefront w_5 enters the H-cell at $t=(n+m+5)+(m+4)+(m+1)+2$ modifying its values allowing the test of the modified x_{n+m+2} and x_{n+m+1} values to decide the course of the next iteration, and triggering the overwrite of index l by k in the boundary cell. w_5 then becomes w_1 on the next iteration. This implies that modification of the basis can be overlapped with the calculation of the next iteration. Clearly w_4 (which updates the basis) must leave the Part B section before w_5 propagates through Part A to enter Part B, and demands that $n \geqq m$ to yield an iteration time of $T=3m+n+12$ cycles. When $n<m$, w_5 and w_4 interfere causing w_5 to compute with the wrong basis elements. The problem is easily solved by adding $m-n$ dummy (delay) Part A cells to yield a timing $T=4m+12$ cycles per iteration.

The cell definitions are easily constructed from the revised algorithm and the wavefront patterns. Clearly the boundary cells are simpler than the previously designed sorting cells and use only unidirectional dataflow to construct the partial ordering. Part A cells are simply inner product cells augmented with control triggers and extra switching for transferring A_j·data. Part B cells are also inner products with addition row swapping and are closer to the cell definitions for Figure 2.1. Finally Figure 3.1 is a point-to-point connected array and we assume that wavefronts encroaching on other parts of the array are cleaned up by the Part C section or $(m+1)$st row of Part B to preserve computation on subsequent iterations. A more efficient layout of the cylinder is achieved in 3-D by considering each column through the array to be a systolic ring and using a variation of Kung & Lam [7] as shown in Figure (3.4).

Although the cylinder provides an alternative and slightly faster array than the standard Simplex method of Theorem 2.1 we require $0(mn)$ inner product type cells to store the A matrix which the revised Simplex algorithm was designed to avoid. The compacted array in Figure 3.5a remedies this problem for $m>n$ by folding

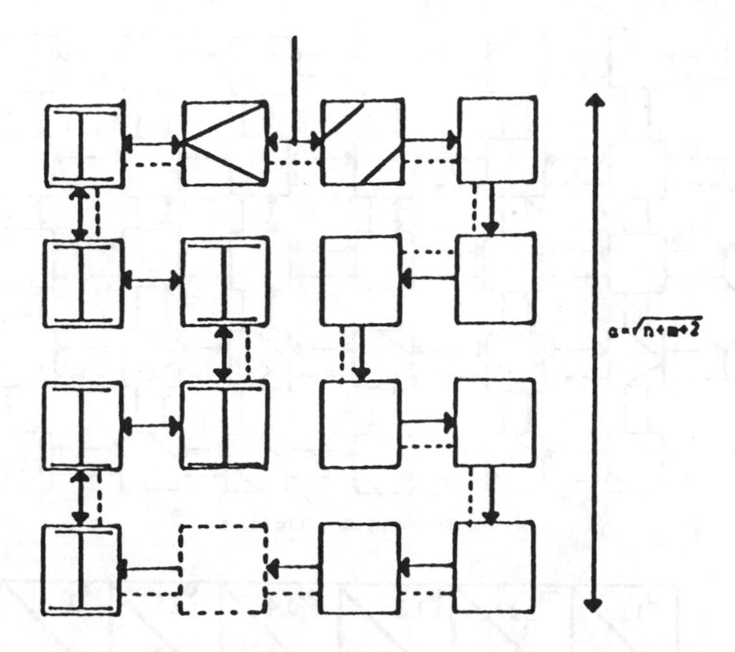

Figure 3.4a Ring segment of systolic cylinder ($n=7, m=7$).

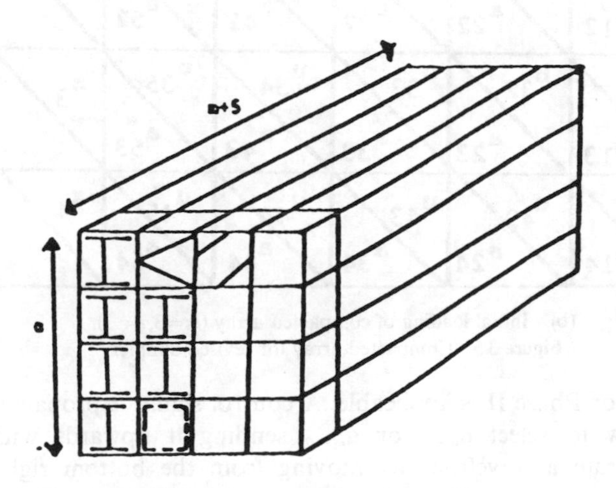

Figure 3.4b Cross-section of folded cylinder.

Figure 3.1 along the Part A, Part B partition and mapping cells containing A_{ij} elements into cells containing u_{ji} elements of the basis inverse (see Figure 3.5b). The basic idea is to save $0(mn)$ ips cells by adding extra control registers and switching to the $0(m^2)$ cells already required for recording the basis.

The start of an iteration, as before, begins in the H-cell, where we decide

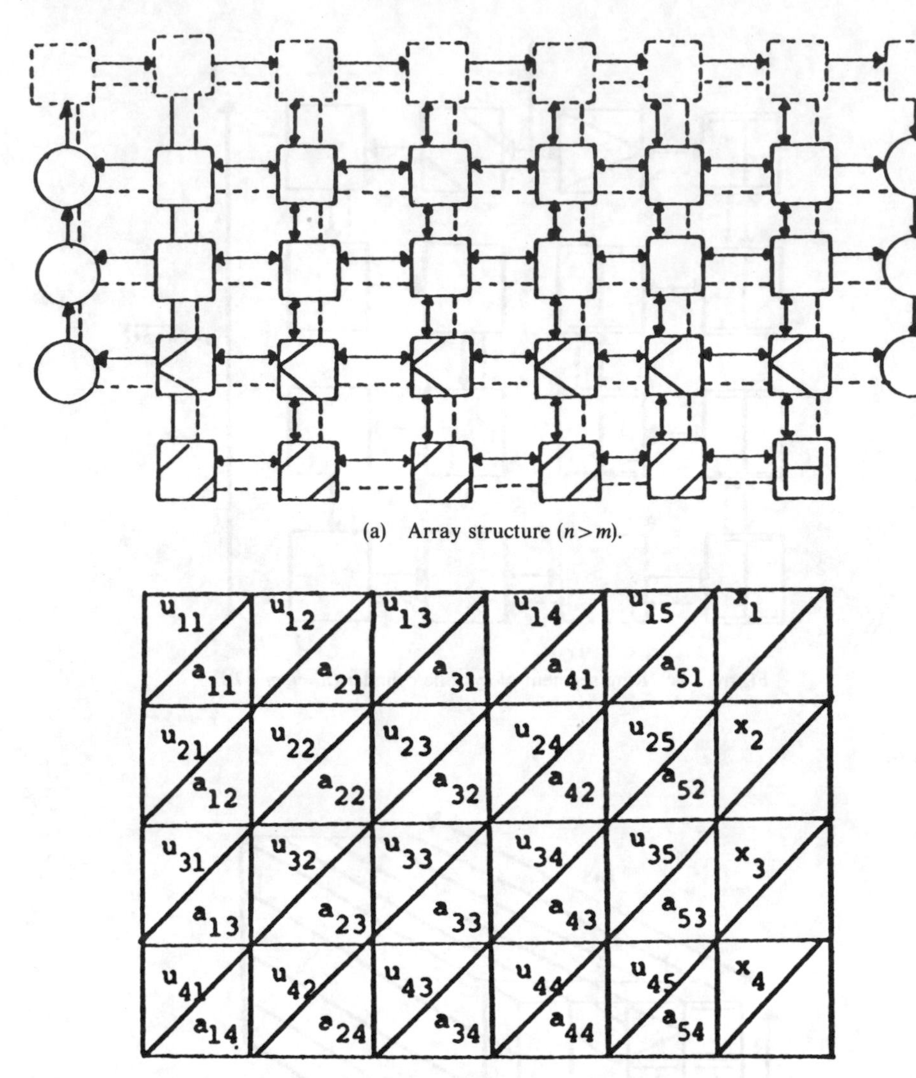

(a) Array structure ($n>m$).

(b) Initial loading of compacted array ($m=3, n=3$).

Figure 3.5 Compacted array for revised Simplex.

whether Phase I or Phase II is applicable. A control signal is propagated left along the $(m+1)$st row to select u_{m+1} or u_{m+2} sending it upwards with additional controls to generate a wavefront w_1 moving from the bottom right to top left corner of the array (see Figure 3.6). As w_1 partial products of δ_i (γ_i) are accumulated from right to left loading the values into their respective boundary cells on the left where a label j identifying column $\bar{A}_j$ also resides. w_1 is reflected by the boundary cells to become w_2 which propagates the value δ_k (γ_k) to the top of the boundary column as it moves to the right top corner of the array transferring column $\bar{A}_j$ to the top boundary (formerly Part C) cells. On reaching the top left corner of the array w_2 deposits the value δ_k (γ_k) and the index k into the top boundary cells, before being reflected to form w_3 (a wavefront headed for

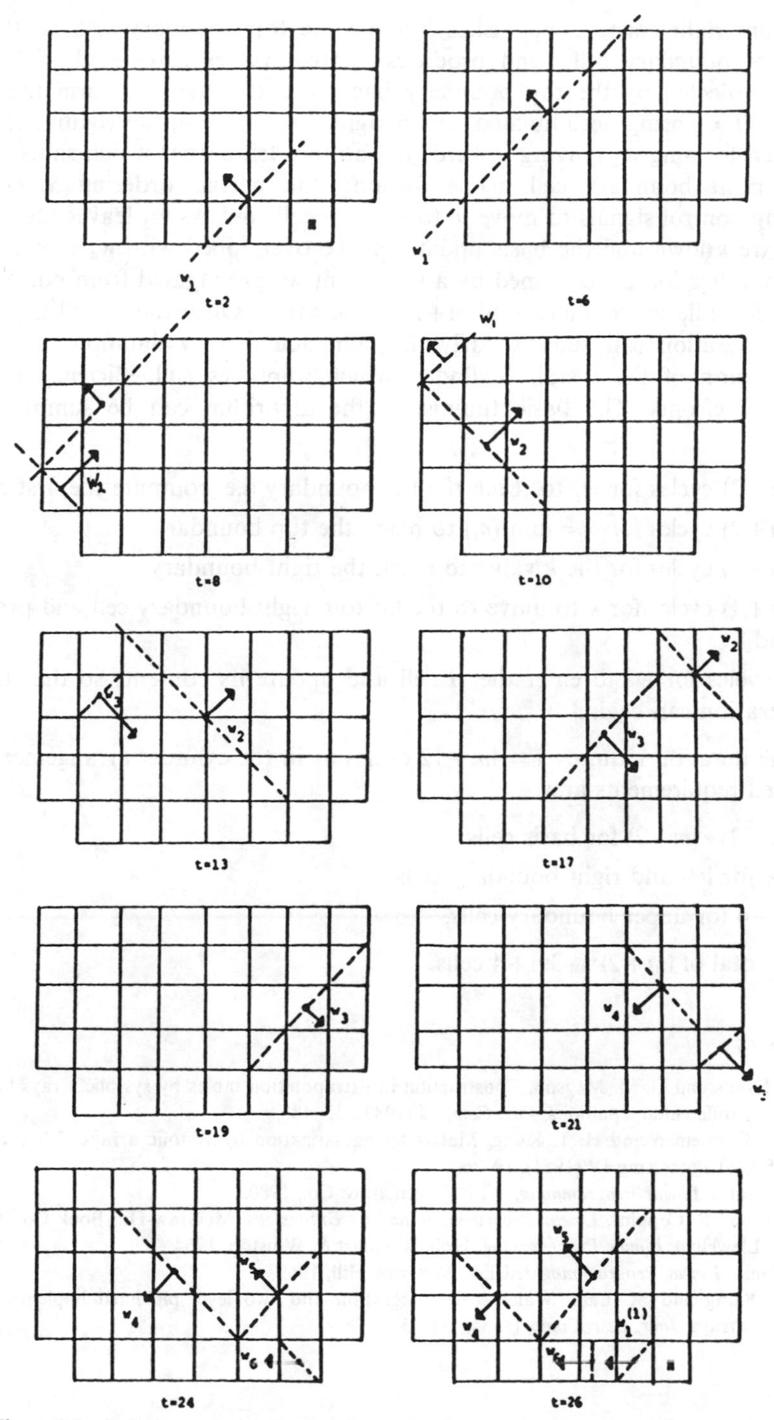

Figure 3.6 Snapshots of systolic wavefronts for the compacted revised Simplex array.

the bottom right corner). w_3 pushes k along the top boundary cells to the right column of boundary cells, and produces control values associated with the $\bar{A}_j$ elements reflected by the top boundary back into the array to form the partial products of x_{ik} being accumulated left to right. On reaching the rightmost cell w_3 is reflected forming w_4 moving towards the bottom left corner which shifts k to the bottom right boundary cell, while forming the partial ordering x_{l0}/x_{lk}, and producing control signals to move u_l to the mth cell row. As w_3 leaves the array, k, l and p are known and the basis update can be overlapped with w_4 replacing l by k. The modification is performed by a wavefront w_5 propagated from bottom right to top left while w_6 modifies row $m+2$ of the array. Once the H-cell is modified the next iteration can start. Clearly the compacted array dataflow is simply a folded version of the systolic cylinder, which improves cell efficiency by interleaving wavefronts. The basic timings of the algorithm can be summarised as follows:

i) $(m+2)$ cycles for w_1 to reach the left boundary (i.e. compute the first δ_i (γ_i).
ii) $(m+2)$ cycles for $\delta_k = \min_i(\delta_i)$ to reach the top boundary.
iii) $(m+3)$ cycles for the k value to reach the right boundary.
iv) $(m+1)$ cycles for k to move to the bottom right boundary cell and produce p and l.
v) 4 cycles for w_5 to enter the H cell and update its contents so that the next iteration can begin.

Thus one iteration requires $T = 4m + 12$ cycles as in the cylinder arrangement.
The cell requirements are:

i) $(m+2)*(m+2)$ for basis cells
ii) $2m$ for left and right boundary cells
iii) $m+4$ for upper boundary cells,

giving a total of $(m+2)^2 + 3m + 4$ cells.

References

[1] D. J. Evans and G. M. Megson, Construction of extrapolation tables by systolic arrays for solving ordinary differential equations, *Par. Comp.* **4** (1987), 33–48.
[2] W. M. Gentleman and H. T. Kung, Matrix triangularization by systolic arrays, *SPIE* **298**, Real-time Signal Processing IV (1981), 19–26.
[3] V. Chavatal, *Linear Programming*, W. H. Freeman & Co., 1980.
[4] N. Wu and R. Coppins, *Linear Programming and its Extensions*. McGraw-Hill Book Co., 1981.
[5] R. W. Llewellyn, *Linear Programming*, Holt, Rinehart & Winston, 1964.
[6] S. I. Gass, *Linear Programming* 3rd Ed., McGraw-Hill, 1969.
[7] H. T. Kung and M. Lam, Wafer scale integration and two level pipelined implementation of systolic arrays, *Jour. Dist. Comp.* (1984), 32–63.

INDEX